VOLUME ONE HUNDRED AND TWENTY EIGHT

CURRENT TOPICS IN DEVELOPMENTAL BIOLOGY

Cell Fate in Mammalian Development

CURRENT TOPICS IN DEVELOPMENTAL BIOLOGY

"A meeting-ground for critical review and discussion of developmental processes"
A.A. Moscona and Alberto Monroy (Volume 1, 1966)

VOLUME ONE HUNDRED AND TWENTY EIGHT

CURRENT TOPICS IN DEVELOPMENTAL BIOLOGY

Cell Fate in Mammalian Development

Edited by

BERENIKA PLUSA
Faculty of Biology, Medicine and Health,
The University of Manchester,
Manchester, United Kingdom;
Institute of Genetics and Animal Breeding,
Polish Academy of Sciences,
Jastrzebiec, Poland

ANNA-KATERINA HADJANTONAKIS
Sloan Kettering Institute,
Memorial Sloan Kettering Cancer Center,
New York, NY, United States

ELSEVIER

ACADEMIC PRESS
An imprint of Elsevier

Academic Press is an imprint of Elsevier
50 Hampshire Street, 5th Floor, Cambridge, MA 02139, United States
525 B Street, Suite 1800, San Diego, CA 92101-4495, United States
The Boulevard, Langford Lane, Kidlington, Oxford OX5 1GB, United Kingdom
125 London Wall, London, EC2Y 5AS, United Kingdom

First edition 2018

ISBN: 978-0-12-804252-6
ISSN: 0070-2153

Publisher: Zoe Kruze
Acquisition Editor: Zoe Kruze
Editorial Project Manager: Shellie Bryant
Production Project Manager: Denny Mansingh
Cover Designer: Greg Harris

Typeset by SPi Global, India

CONTENTS

CONTRIBUTORS

Ramiro Alberio
School of Biosciences, University of Nottingham, Nottingham, United Kingdom

Yanina D. Alvarez
Institute of Molecular and Cell Biology, A*STAR, Singapore, Singapore; Facultad de Ciencias Exactas y Naturales, Universidad de Buenos Aires, Conicet, Buenos Aires, Argentina

Cécilia Bassalert
GReD, Université Clermont Auvergne, CNRS, INSERM, Clermont-Ferrand, France

Stephanie Bissiere
Institute of Molecular and Cell Biology, A*STAR, Singapore, Singapore

Jörg Burgstaller
British Heart Foundation Centre for Research Excellence, National Heart and Lung Institute, Imperial Centre for Translational and Experimental Medicine, Imperial College London, London, United Kingdom; Biotechnology in Animal Production, Department for Agrobiotechnology, IFA Tulln, Tulln, Austria

Claire Chazaud
GReD, Université Clermont Auvergne, CNRS, INSERM, Clermont-Ferrand, France

Stephen Frankenberg
School of BioSciences, University of Melbourne, Parkville, VIC, Australia

Maxime Gasnier
Institute of Molecular and Cell Biology, A*STAR, Singapore, Singapore

Anna-Katerina Hadjantonakis
Developmental Biology Program, Sloan Kettering Institute, Memorial Sloan Kettering Cancer Center, New York, NY, United States

Anna Kasperczuk
Faculty of Biology, University of Warsaw, Warsaw, Poland

Katarzyna Klimczewska
Faculty of Biology, University of Warsaw, Warsaw, Poland

Ana Lima
British Heart Foundation Centre for Research Excellence, National Heart and Lung Institute, Imperial Centre for Translational and Experimental Medicine, Imperial College London; Cell Stress Group, MRC London Institute of Medical Sciences (LMS), London, United Kingdom

Alyson Lokken
Michigan State University, East Lansing, MI, United States

Zofia E. Madeja
Faculty of Veterinary Medicine and Animal Sciences, Poznan University of Life Sciences, Poznan, Poland

Miguel Manzanares
Centro Nacional de Investigaciones Cardiovasculares (CNIC), Madrid, Spain

Sergio Menchero
Centro Nacional de Investigaciones Cardiovasculares (CNIC), Madrid, Spain

Daniel M. Messerschmidt
Developmental Epigenetics and Disease Group, Institute of Molecular and Cell Biology (IMCB), Agency for Science, Technology and Research (A*STAR), Singapore, Singapore

Alexandra Moauro
Graduate Program in Physiology, Michigan State University, East Lansing, MI, United States

Kathy K. Niakan
Human Embryo and Stem Cell Laboratory, The Francis Crick Institute, London, United Kingdom

Jennifer Nichols
Wellcome Trust - Medical Research Council Stem Cell Institute, University of Cambridge, Cambridge, United Kingdom

Anna Piliszek
Institute of Genetics and Animal Breeding, Polish Academy of Sciences, Jastrzebiec, Poland

Nicolas Plachta
Institute of Molecular and Cell Biology, A*STAR; National University of Singapore, Singapore, Singapore

Berenika Plusa
Faculty of Biology, Medicine and Health, The University of Manchester, Manchester, United Kingdom; Institute of Genetics and Animal Breeding, Polish Academy of Sciences, Jastrzebiec, Poland

Amy Ralston
Program in Reproductive and Developmental Sciences, Michigan State University, East Lansing, MI, United States

Priscila Ramos-Ibeas
School of Biosciences, University of Nottingham, Nottingham, United Kingdom

Tristan A. Rodríguez
British Heart Foundation Centre for Research Excellence, National Heart and Lung Institute, Imperial Centre for Translational and Experimental Medicine, Imperial College London, London, United Kingdom

Deepak Saini
Goodman Cancer Research Centre, McGill University, Montreal, QC, Canada

Julio Sainz de Aja
Centro Nacional de Investigaciones Cardiovasculares (CNIC), Madrid, Spain

Juan M. Sanchez-Nieto
British Heart Foundation Centre for Research Excellence, National Heart and Lung Institute, Imperial Centre for Translational and Experimental Medicine, Imperial College London, London, United Kingdom

Michelle K.Y. Seah
Developmental Epigenetics and Disease Group, Institute of Molecular and Cell Biology (IMCB), Agency for Science, Technology and Research (A*STAR), Singapore, Singapore

Shankar Srinivas
University of Oxford, Oxford, United Kingdom

Matthew J. Stower
University of Oxford, Oxford, United Kingdom

Aneta Suwińska
Faculty of Biology, University of Warsaw, Warsaw, Poland

Lorena Valverde-Estrella
GReD, Université Clermont Auvergne, CNRS, INSERM, Clermont-Ferrand, France

Sissy E. Wamaitha
Human Embryo and Stem Cell Laboratory, The Francis Crick Institute, London, United Kingdom

Jennifer Watts
Program in Reproductive and Developmental Sciences; Graduate Program in Physiology, Michigan State University, East Lansing, MI, United States

Yojiro Yamanaka
Goodman Cancer Research Centre, McGill University, Montreal, QC, Canada

CHAPTER ONE

Introduction

Berenika Plusa*,†,1, Anna-Katerina Hadjantonakis‡,1
*Faculty of Biology, Medicine and Health, The University of Manchester, Manchester, United Kingdom
†Institute of Genetics and Animal Breeding, Polish Academy of Sciences, Jastrzebiec, Poland
‡Developmental Biology Program, Sloan Kettering Institute, Memorial Sloan Kettering Cancer Center, New York, NY, United States
[1]Corresponding authors: e-mail address: Berenika.Plusa@manchester.ac.uk; hadj@mskcc.org

Mammalian embryos spend the first few days of their development floating freely in their mother's oviduct. They will subsequently establish a connection with their maternal tissues through a process called implantation. Thus, this earliest phase of mammalian development is often referred to as the preimplantation period. To guarantee successful embryo implantation into the maternal uterus two different types of lineages—embryonic and extraembryonic—must be specified. Two extraembryonic lineages—the trophectoderm (TE) and the primitive endoderm (PrE)—together encapsulate the embryonic lineage—the pluripotent epiblast. The embryonic epiblast is the tissue which will give rise to somatic cells and germ cells. By contrast, the TE and PrE will not contribute majorly to the embryo per se, however they will eventually give rise to the fetal portion of the placenta and extraembryonic membranes. These extraembryonic tissues are essential for embryonic development, but dispensable postnatally. Correct specification and differentiation of both embryonic and extraembryonic lineages ensures the formation of a functional connection with the mother, allowing the exchange of nutrients and waste metabolites between a mother and her embryo.

As a fertilized egg transforms into an embryo and the process of the lineage specification commences. A central overarching question is that of how cells acquire distinct identities over time and in space, and how this is done in a robust and invariable fashion between individuals. Totipotency—which can be considered as the ability of a cell to give rise to both embryonic and extraembryonic tissues—is gradually lost as the fertilized egg divides, and its daughter cells, and subsequently their daughter cells, acquire distinct fates. The subject of how the cells of early mammalian embryos make decisions, and how they make choices between alternative fates is the subject of this collection of reviews.

Current Topics in Developmental Biology, Volume 128
ISSN 0070-2153
https://doi.org/10.1016/bs.ctdb.2017.12.002

Studies on early mammalian embryos are exemplified by the mouse model, but also include the human and other nonclassical models. Though the overarching organization of embryos of different mammalian species may at first glance appear comparable, there is a remarkable degree of diversity and species specificity, perhaps resulting from temporal and/or environmental adaptations. Armed with a knowledge of the timing of successive, and oftentimes conserved, morphological landmarks, investigators have now begun to piece together the molecular frameworks driving the critical events surrounding the early development of viviparous mammalian embryos. It is now recognized that there are many common principles, and also several divergent mechanisms. Indeed, the earliest developmental events are morphogenetic, and are common across different mammals. They involve processes of cellular polarization and the emergence of the first epithelial lineages, as well as formation of a fluid-filled cavity (the blastocoel) in a process referred to as cavitation. These early events in mammalian development appear to be controlled by a similar set of transcription factors and signaling molecules. However, the exact timing and the involvement of specific factors in any given event may vary between different species.

This collection of chapters reviews our current understanding of the earliest stages of mammalian development, as well as some of the key features of stem cells that can be propagated in culture but which are derived directly from early mammalian embryos. The focus of these reviews is broad covering both classical and emergent mammalian model systems, and ranges from discussions of the details of the cellular mechanics, to the intersection of transcriptional programs with signaling pathways, as well as metabolic and epigenetic mechanisms.

The first set of articles address the earliest decisions made by cells to particular lineages, the changes in cellular properties that facilitate these fate decisions, and the integration of such information with key signaling pathways, as well as how metabolism and mechanics influence cell states and ultimately cell fates. The ability to derive lineage-specific self-renewing stem cells from pregastrulation mammalian embryos opens up the opportunity of mimicking early developmental events in a dish, and there is a discussion of the various stem cell types that have been isolated from different mammalian species. These early events are compared and contrasted in various nonclassical model organisms (domestic animals and noneutherian mammals), as well as in the human. Embryo implantation is followed by a break in bilateral symmetry and the establishment of an anterior–posterior axis. In the final chapter of this collection, the emergence of anterior–posterior pattern is reviewed in the mouse, this being mammalian species for which the most is currently known.

Yojiro Yamanaka's work explores the role of physical and molecular cues in cell allocation during the first lineage specification and symmetry-breaking event taking place in mammalian development: the formation of outside and inside cells that in turn give rise to the TE and the inner cell mass (ICM), the latter differentiating into the epiblast and PrE after blastocyst formation. In their chapter entitled "Cell polarity-dependent regulation of cell allocation and the first lineage specification in the preimplantation mouse embryo" Saini and Yamanaka explain how the interplay between compaction, polarization, and asymmetric cell division affects differential regulation of the Hippo signaling pathway between inside and outside cells. The authors explain how initiation of cell polarization along the apical–basal axis of cells (also referred to as blastomeres) at the 8-cell stage leads to restriction of certain proteins to the apical domain of cells exposed to the outside environment. Subsequent cell divisions (from 8- to 16- and 16- to 32-cell stages) result in formation of either two polar cells (if both daughter cells inherit the apical domain after division), or one polar and one apolar cell (if only one daughter cell inherits the apical domain). Individual cells use the presence or absence of the apical domain as a mechanism to sense their position within the embryo in order to activate or repress the Hippo signaling pathway. Importantly the presence of the apical domain creates two distinct cortical subdomains of surface contractility within any one cell, which in turn influence that cell's position within the embryo. The inheritance of the apical domain results in a less contractile cell that retains an outside position and has a tendency to spread over highly contractile apolar cells, leading to the complete internalization of the latter.

Studying molecular dynamic events in the intact embryo remains a challenge. The work of Nicolas Plachta confronts this issue by exploiting high-resolution quantitative microscopy combined with biophysical methods to investigate mechanisms directing cells to make fate choices in intact embryos. These approaches are complemented with mechanical perturbations of cells (for example, laser ablations), as well as the use of sophisticated image analysis (including segmentation) software tools. Recent studies from Plachta and colleagues have used live imaging to reveal novel (filopodial-based) morphogenetic mechanisms driving compaction in the mouse morula, anisotropies in cortical tension drive formation of the ICM, and demonstrated how two key transcription factors (Sox2 and Oct4) search and dynamically bind to DNA to control cell fates in vivo in embryos.

In their chapter entitled "Cell fate decisions during preimplantation mammalian development" Plachta and colleagues review progress linking molecular dynamic events occurring at the level of the single cell in vivo, to several key morphogenetic changes occurring during early mouse development. They start by overviewing some of the insights gleaned from studies on mouse mutants exhibiting defects in cell fate assignments. It is widely established that changes in gene expression control cell fate decisions, yet gene expression is regulated at multiple levels. They discuss how the levels of expression of key transcription factors, and indeed their differential levels in any given cell impact that cell's fate choice. A further level of regulation is introduced—namely, the subcellular localization of these transcription factors—and there is a discussion of a key feature of mammalian early development which likely underlies the regulative nature of early mammalian embryos—namely, cell-to-cell variability. Furthermore, the authors discuss an additional level of control, this being at the epigenetic level, and finally they speculate how these different levels of genetic control might impact cell shape and position.

The TE, a lineage which will give rise to the fetal portion of the placenta and extraembryonic membranes, is the first cell population to appear in mammalian embryos. TE cells are specified when totipotent cells (known as blastomeres) located on the embryo's outer surface differentiate and concomitantly acquire apical–basal polarity as they epithelialize. Indeed, the specification of TE represents a convergence of differentiation and morphogenetic programs. How cells acquire a TE identity is the subject of the research of Miguel Manzanares and his group. In their review entitled "Our first choice: Cellular and genetic underpinnings of trophectoderm identity and differentiation in the mammalian embryo" Manzanares and colleagues discuss the gene regulatory networks that converge to drive lineage establishment and maintenance, and leading to the maturation of the TE program. These integrate inputs from key signaling pathways (such as Hippo and Notch) on essential transcriptional regulators (such as CDX2). They discuss the integration of these gene regulatory networks with the key morphogenetic behaviors, involving cellular compaction and epithelialization, which are stereotypical for TE development.

Claire Chazaud has a long-standing interest in the processes that ensure specification of the two ICM lineages: the pluripotent epiblast and the PrE. Her pioneering work challenged the existing model of PrE formation, which posited that inductive signals from the blastocyst cavity were responsible for the specification of PrE cells on the surface of the ICM. Instead,

Chazaud and colleagues demonstrated that PrE and epiblast cells are specified within the ICM in a seeming random, so-called salt-and-pepper, manner, and only later become segregated into two distinct layers, underlying the cavity PrE and the epiblast encapsulated between PrE and TE. In their chapter entitled "Primitive endoderm differentiation: From specification to epithelialization" Chazaud and colleagues describe how our understanding of ICM lineage segregation evolved from the simple, position-based model to the model integrating stochastic events, FGF signaling activity, and the mutual inhibitory action between PrE-specific transcription factor GATA6 and epiblast-specific NANOG to create salt-and-pepper distribution of PrE and epiblast precursors. In the most recent model of PrE and epiblast specification, asynchronous specification of PrE and epiblast progenitors is achieved via FGF4 action on FGFR1 and FGFR2. Interestingly, although Fgfr2 expression is associated with PrE cells, it is the ubiquitously expressed Fgfr1 that is critical for establishment of PrE. Activation of the FGF pathway leads to elevation of GATA6/NANOG ratio in PrE precursors, whereas in epiblast precursors GATA6 is donwnregulated while NANOG expression is maintained. Finally, the authors discuss the role of cell adhesion and polarity in the maturation of the PrE lineage and the sorting of the epiblast and PrE to their appropriate positions whereby they form two adjacent tissue layers.

The regulative nature of mammalian development and associated embryonic cell plasticity is the main theme of Aneta Suwińska's work. In her experiments, she uses embryonic chimeras and lineage tracing by cell labeling and time-lapse movies to understand the mechanisms governing early embryonic plasticity. She has followed in the footsteps of her mentor, the late Professor Andrzej Krzysztof Tarkowski, whose elegant experiments and pioneering studies established the foundation for our understanding of the regulative nature of mammalian development. In their chapter entitled "Regulative nature of mammalian embryos" Suwińska and colleagues recall several of these ground-breaking experiments that demonstrated that early embryonic cells, called blastomeres, remain totipotent (defined in this case as being able to contribute to both embryonic and extraembryonic lineages) for a substantial period during preimplantation development. Moreover, single blastomeres from the 2-cell stage embryo in mouse, and 4- or even 8-cell stage from other mammalian species, retain the ability to develop into a normally sized adult organism, underlining the high degree of plasticity associated with early mammalian development. Despite the evidence that all early blastomeres have full developmental potency, Suwińska and colleagues point out several reports describing the molecular differences that exist between blastomeres starting from the 2-cell stage. Blastomeres differ in gene expression, possibly developmental potency and potency to generate ESCs, and this molecular heterogeneity between blastomeres may bias their future fate when they are in the competitive environment of an embryo. Finally, the authors describe how the extraordinary plasticity of mammalian embryos can be harnessed in animal breeding to maintain high quality and healthy livestock with desirable genetic characteristics as well as for preimplantation genetic diagnostics in humans to identify healthy embryos for transfer.

Jennifer Nichols has a long-standing interest in cell potency during the earliest stages of mammalian development. Her laboratory's recent work exploiting single cell stem cell derivations and single cell transcriptomics has been instrumental in definitively demonstrating that mouse embryonic stem cells are the in vitro counterpart of the epiblast cells of blastocyst-stage embryo. Studies from Nichols and colleagues have championed a primate (the marmoset) as an informative mammalian model for forging similarities and distinctions in the transcription factors identifying pluripotent cells in different species, and their species-divergent signaling requirement. In a chapter entitled "States and origins of mammalian embryonic pluripotency in vivo and in a dish" Nichols and her colleagues present a historical context and discuss both recent developments in the understanding early embryogenesis, and specifically relating to the emergence of pluripotency, in different mammalian species. Considering pluripotency as a continuum, which spans from the blastocyst until gastrulation, they discuss the distinct successive states of pluripotency. They integrate this knowledge with some of the notable differences in embryo development—ranging from embryo shape, timing of implantation, to the expression of key pluripotency-associated genes—in three cardinal mammals, the mouse, the primate, and the pig.

Amy Ralston's interests center around the identification of key molecules that instruct cell behavior in early mammalian embryos. Ralston and colleagues have investigated some of the key transcription factors (including Oct4 and Sox2) driving key cell fate decisions taking place in the mouse preimplantation embryo, but also how the different cell fates that arise can be captured, manipulated, and interconverted in vitro, namely, how stem cells representing each of the three lineages of the mouse blastocyst can be represented in cell culture paradigms. In their chapter entitled, "Capturing and interconverting embryonic cell fates in a dish" Ralston and colleagues introduce the reader to the three embryo-derived stem distinct cell types—ES cells, TS cells, and XEN cells—representing bona fide in vitro counterparts of the epiblast, trophoblast, and PrE (hypoblast), respectively, namely, the three lineages of the mammalian blastocyst. As stem cells, ES, TS, and XEN cells are capable of self-renewing or differentiating in a lineage-appropriate manner. These stem cell types have been isolated from the mouse where they serve as a paradigm for defining the genes and pathways that specify and maintain lineage and stem cell identity. They discuss key studies elucidating mechanisms driving the activation of gene expression programs that promote lineage specificity, with concomitant repression of programs associated with alternative lineage identities, and discuss these mechanisms in the context of in vitro lineage (and stem cell) interconversion.

Cell fate commitment is achieved by the establishment of cell type-specific transcriptional programs, which in turn are guided, reinforced, and ultimately locked-in by epigenetic mechanisms. While the fertilized egg and very earliest blastomeres are totipotent, progressive stages of embryonic development lead to restrictions in cell fates and developmental potential. Interestingly, there is one outlier population—the germ cells—which, by contrast to other animals, are not set aside early in mammals, but instead are specified from derivatives of the epiblast. As development proceeds germ cell precursors become reprogrammed, and thereafter become specialized as they differentiate into sex-specific gametes exhibiting specialized epigenomes. Erasure of these gamete-specific features is then necessary to enable acquisition of a totipotent state in the zygote. Daniel Messerschmidt is interested in how epigenetic mechanisms impact cell fates within early mammalian embryos. In their chapter entitled "From germline to soma: Epigenetic dynamics in the mouse preimplantation embryo" Messerschmidt and Seah discuss recent insights into dynamic epigenetic mechanisms of DNA methylation and histone modifications with respect to embryonic epigenetic reprogramming, an event which takes place in the precursors of germ cells.

Stephen Frankenberg focuses his research on understanding how early lineage specification is executed in noneutherian mammals, and its connection with the evolution of placentation and viviparity. His chapter entitled "Pregastrula development of noneutherian mammals" starts by reviewing what is known of early lineage formation in monotremes and marsupials, and explains how this information can provide important insights into how extraembryonic lineages evolved, and which mechanisms might be conserved between different mammalian and vertebrate groups. He proposes a model of how conceptuses of the oviparous mammalian ancestors transitioned into those of humans and mice. Frankenberg also hypothesizes how the relative timing of extraembryonic lineage specifications could influence the molecular mechanisms that regulate them. These timing differences are themselves driven by selection pressures related to the contribution of yolk vs a placenta for supporting development. The molecular pathways discussed include: the Hippo signaling and its role in trophoblast specification, as well as class V POU domain transcription factors (POU5F1 and POU5F3, also referred to as OCT4), which are key regulators of pluripotency and germ cell function. Frankenberg's review highlights the deep insights that can be gained from studying comparative early development in nonstandard model species.

Anna Piliszek's research seeks to understand how the first three lineages differentiate in different model organisms, with a particular emphasis on mouse and rabbit development. In their chapter entitled "Preimplantation development of domestic animals" Piliszek and Madeja review lineage specification in domestic animals and highlight some of the major similarities and differences with the mouse. The authors speculate that differences in the timing of zygotic genome activation (ZGA, also referred to as embryonic genome activation, EGA) might be linked to the differences in timing of other developmental events such as the specification of lineages, and might, at least partially, account for species-specific differences in the dynamic expression of lineage-specific transcription factors. One of the major differences highlighted by the authors is that OCT4/CDX2 mutual restriction described in the mouse does not seem to be preserved in nonrodent mammals, and downregulation of OCT4 in outside cells does not seem to be prerequisite for TE formation other than in the mouse. It has been postulated that this divergence in CDX2/OCT4 dynamics might result from differences in OCT4 regulatory regions between mice and other mammalian species. As with TE vs ICM specification, Piliszek and Madeja point out the existing differences in specification of PrE (called in other mammals the hypoblast) and epiblast between different mammalian species. One of the sticking differences is that mutual inhibition of GATA6 and NANOG might not be a strict requirement for epiblast vs PrE specification in all mammals as exemplified in rabbit and porcine development. Moreover, up to date none of the studied nonrodent species seems to share full dependence of the FGF/Erk signaling in PrE formation, suggesting an alternative mechanism may operate in other mammals.

Kathy Niakan seeks to understand how the first lineages are specified during preimplantation human development. Her lab was the first to harness

CRISP/Cas9 technology to create human embryos with a mutated *Oct4* gene to understand its role in human embryonic development. In their chapter entitled "Human pre-gastrulation development" Niakan and Wamaitha present a comprehensive overview of the early events in human pregastrulation development. The authors focus on the role of extracellular signaling in specification of the TE, PrE, and epiblast in humans, and discuss how the regulation of lineage specification differs from the mouse model. Current data on the early pattern of expression of TE and ICM makers suggest that unlike in mouse, the Hippo pathway may not be necessary for TE vs ICM specification in human. Moreover, inhibiting FGF/Erk signaling in human embryos has no effect on PrE or EPI formation in striking contrast with observations made in the mouse. The differences in the signaling pathways that direct early lineage decision in mouse and human embryos may account for the fact that human ES cells are distinct from mouse ES cells both in their gene expression and signaling requirements. Potentially, these differences may also account for the fact that derivation of human equivalent XEN and TS cells from blastocyst-stage embryos has not been successful to date.

Though much of our understanding of cell fate decisions comes through studies investigating the intersection of transcription factors with signaling pathways, how epigenetic mechanisms lock-in cell fates and increasingly how mechanical cues feed into these decisions, little attention has been paid to the cell fitness and the intrinsic mechanisms that ensure cells are capable of contributing to the developing mammalian embryo. Studies in invertebrates (namely Drosophila) have identified cell competition as a noncell autonomous mechanism for sensing differences between neighboring cells, resulting in the selective elimination of one (less fit) population. Work from Tristan Rodríguez and colleagues has demonstrated that the phenomenon of cell competition is also intrinsic to early mammalian embryos. In their chapter entitled "The mitochondria and the regulation of cell fitness during early mammalian development" Rodríguez and colleagues provide an overview of how cell fitness is selected for during early mammalian development. They discuss quality control mechanisms operating to ensure optimal mitochondrial activity, how changes affect mitochondria, and how they impact on the energy metabolism and apoptotic response of the early mammalian embryo.

The early mammalian embryo exhibits a bilateral symmetry. Breaking this symmetry results in the establishment of the anterior–posterior axis shortly after implantation of the mammalian blastocyst into the maternal uterus. The pluripotent epiblast is patterned through signaling interactions with its adjacent predominantly extraembryonic tissues, including the PrE-derived visceral endoderm and the trophectoderm-derived extraembryonic ectoderm. The establishment of the anterior–posterior axis dictates the future site of gastrulation and the emergence of the three definitive germ layers. The work of Shankar Srinivas and colleagues has used time-lapse imaging and genetic methods to investigate the mechanisms of axis establishment in mice. A specialized population of cells, the anterior visceral endoderm (AVE), undergoes a transepithelial collective cell migration that converts the proximal–distal axis of the embryo to an anterior–posterior pattern. In the final chapter entitled "The head's tale: anterior–posterior axis formation in the mouse embryo" Stower and Srinivas focus on the role of the AVE, as they summarize the current understanding of the generation of anterior pattern in mammals. They also go on to outline some of the many open questions regarding the mechanism by which the AVE is induced, and how it collectively migrates within the visceral endoderm epithelium.

We hope that this collection of chapters conveys the fact that the very earliest stages of mammalian development continue to be a subject of intense investigation. Deepening our understanding of the fundamental processes involved in cell fate plasticity and lineage commitment, embryo self-organization and implantation continues to yield important insights into underlying causes of infertility, early pregnancy loss, and developmental disorders in humans, as well as assisting the development of improved reproductive treatments (including in vitro fertilization, embryo cryopreservation, and preimplantation genetic diagnosis). Moreover, it was the pioneering studies on the fundamental aspects of the earliest stages of mammalian development which ignited the field of pluripotent stem cell research, and helped develop efficient embryo storage and culture procedures which have been adapted for use in domestic animal husbandry and endangered species preservation.

We are indebted to the following individuals for assisting with the review of the chapters comprising this book: Effie Apostolou, Alex Bruce, Laina Freyer, Lydia Finley, Vidur Garg, Mariana Kruithof de Julio, Jean Leon Maitre, Sophie Morgani, Silvia Munoz-Descalzo, Jie Na, Theresa Rayon, Nestor Saiz, Jose Silva, Claire Simon, Katarzyna Szczepanska, Katarzyna Filimonow, Minjung Kang, Thorsten Boroviak, Carla Mulas, and Folrence Wianny. Their unique and collective insights proved invaluable in the preparation of this content. Finally, we would like to thank our editor Shellie Bryant for assisting with preparation of the content of this book and for making sure deadlines were met.

CHAPTER TWO

Cell Polarity-Dependent Regulation of Cell Allocation and the First Lineage Specification in the Preimplantation Mouse Embryo

Deepak Saini, Yojiro Yamanaka[1]
Goodman Cancer Research Centre, McGill University, Montreal, QC, Canada
[1]Corresponding author: e-mail address: yojiro.yamanaka@mcgill.ca

Contents

Abstract

During the first few days in the mouse preimplantation embryo, two types of cells, polar and apolar cells are generated from spherical totipotent blastomeres. Sequential morphogenetic events, polarization, compaction, and asymmetric division, are essential for the generation of the first distinct cell populations, polar and apolar cells, which establish the outer/inner configuration within the embryo. This leads to position-dependent activation of the Hippo signaling pathway and lineage-specific gene expression to form the trophectoderm and inner cell mass in a blastocyst. It is still unknown how each morphogenetic event is initiated and how it influences subsequent events. In this chapter, we briefly review the two classic models of mouse preimplantation development and

ISSN 0070-2153
https://doi.org/10.1016/bs.ctdb.2017.10.008

discuss recent studies providing novel insights into the self-organizing ability of the preimplantation mouse embryo. Advances in live cell imaging and mathematical modeling contribute a greater understanding to lineage specification and cell fate commitment at the single cell level. Differential molecular and mechanistic characteristics created by the presence and absence of the apical domain in polar and apolar cells, respectively, dictate cell allocation, divisional orientation, and differential activation of the Hippo signaling pathway.

1. INTRODUCTION

Early mammalian development is a series of processes generating various cell types and allocating them in proper 3D tissue and organ space. After fertilization, the zygote undergoes rapid cell divisions without cell growth, called cleavage, which increases the total cell number but the overall volume of the embryo remains the same (Aiken, Swoboda, Skepper, & Johnson, 2004) (Fig. 1). Until the 8-cell stage, early blastomeres within the mouse preimplantation embryo are morphologically indistinguishable in cell size and shape (Chazaud & Yamanaka, 2016). During the 8- to 32-cell morula stage, the first two distinct cell populations emerge through asymmetric division: polar cells, the cells with apico-basal cell polarity, and apolar cells, the cells without it. The polar cells, which take an outer position exposed to the environment, become the trophectoderm (TE) in the blastocyst and gives rise to the placenta after implantation. On the other hand, the apolar cells, which take an inner position within the embryo, become the inner cell mass (ICM),

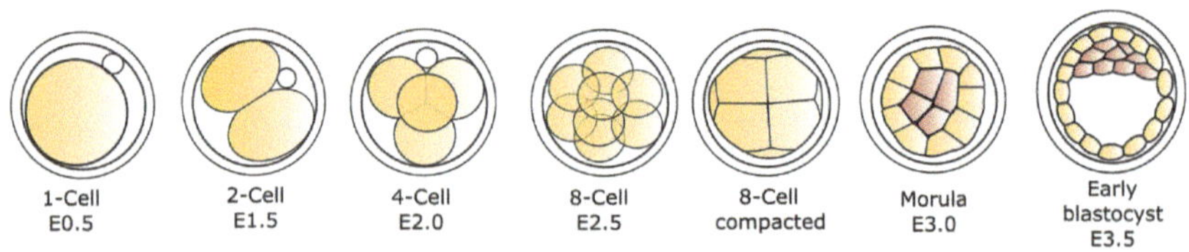

Fig. 1 Preimplantation mouse development. The zygote undergoes cleavage, which is rapid relatively synchronous cell division without cell growth to increase the number of cells in the embryo. Up until the 8-cell stage, all blastomeres are morphologically indistinguishable. The first two morphologically distinct cell populations, outer (*yellow*) and inner (*red*) cells, are generated around the 16-cell stage. The outer cells become the TE and the inner cells become the ICM in a blastocyst. The ICM will be further specified into the PE and EPI, giving rise to the yolk sac and fetal tissues, respectively, but not shown in this diagram.

which further develops into the epiblast (EPI) and the primitive endoderm (PE) to give rise to the fetus and the yolk sac, respectively (Fig. 1).

In this chapter, we will be discussing three key morphogenetic events taking place during the 8- to 32-cell stage leading to the formation of the TE/ICM lineages: compaction, polarization, and asymmetric division and how they affect differential regulations of the Hippo signaling pathway.

1.1 Models for Early Lineage Specification in the Preimplantation Mouse Embryo

1.1.1 The Inside-Outside Model

The inside-outside model proposed by Tarkowski and colleagues in 1967 is still valid to explain the events of the first lineage specification in the mouse embryo (Tarkowski & Wróblewska, 1967). They performed a series of disaggregation experiments of early preimplantation embryos and found that isolated 2-cell and 4-cell blastomeres frequently develop into miniblastocysts, carrying both the TE and ICM. In contrast, isolated 8-cell blastomeres often develop into trophoblast vesicles without the ICM (Tarkowski, 1959; Tarkowski & Wróblewska, 1967). Their interpretation of this observation was that isolated 8-cell blastomeres could not produce enough cells to form the inner population. Smaller-sized cell aggregates increase the likelihood for cells to remain in an outer position and the outer cells exposed to external environment become the TE (Tarkowski & Wróblewska, 1967). This interpretation leads to the inside-outside model, in which the fates of individual cells in the morula embryo are determined based upon their relative position to the external environment; outer cells become the TE, while inner cells become the ICM. This model raises interesting questions: how many cells are enough to form the inner cells and how do individual cells within the embryo sense their position?

1.1.2 The Cell Polarity Model

A series of studies by Johnson and colleagues further explored the inside-outside model, by examining whether there are any cellular modifications that may be influencing the position of blastomeres within the embryo. Using scanning electron microscopy, they carefully analyzed 8-cell stage embryos and isolated 8-cell blastomeres (Johnson & Ziomek, 1981b). They observed that each 8-cell blastomere undergoes polarization from an initial nonpolarized state, which is recognized as the formation of microvilli on a portion of the cell surface indicating the apical domain (Johnson & Ziomek, 1981a). Interestingly, when an isolated 8-cell blastomere divides to form a couplet of 16-cell blastomeres, two types of couplets are observed; a pair of polar/polar cells both carrying microvilli or a polar/apolar pair consisting of one polar cell and one apolar cell without microvilli (Johnson & Ziomek, 1981b). This indicates that there are two types of cell divisions, symmetric and asymmetric, defined by the inheritance of the apical domain from the parental 8-cell blastomere. In the former couplets, the two polar cells keep their apposed position (i.e., both are exposed to environment) while in the latter couplets, the apolar cell internalizes to take an inner position, resembling the outer/inner cell configuration in intact embryos (Anani, Bhat, Honma-Yamanaka, Krawchuk, & Yamanaka, 2014; Dietrich & Hiiragi, 2007; Korotkevich et al., 2017), suggesting that the cell number within the aggregates is not the determinant for the formation of outer/inner cells. These results indicate that the first lineage specification in the mouse embryo is not simply regulated by positional differences within the embryo, but rather by more dynamic cellular activities such as cell polarity, asymmetric division, and active cell sorting.

These two models were proposed solely based on manipulative experiments on isolated blastomeres. It was not possible to test the models in intact embryos, because there was no way to track individual cells and their behavior within the developing embryo at the time. Recent advances in live imaging technology including various fluorescent reporters (Day & Davidson, 2009), state of art microscopic techniques (Ntziachristos, 2010), and 3D imaging analysis (Fischer, Wu, Kanchanawong, Shroff, & Waterman, 2011) allow for visualization and quantification of dynamic morphogenetic processes in intact developing embryos. The quantification of spatiotemporal changes in cellular and embryonic morphology, cell position, and gene expression, is serving as essential experimental data for mathematical modeling. These new observations and analyses have provided new insights into the two classic models, further increasing our understanding of molecular and cellular mechanisms involved in the first lineage specification of mammalian development.

2. COMPACTION AND POLARIZATION: TWO KEY MORPHOGENETIC EVENTS INFLUENCING THE FORMATION OF OUTER/INNER CELLS

At the 8-cell stage, the mouse embryo undergoes compaction, a cellular event in which each blastomere flattens to increase cell–cell contacts between neighboring cells and minimize the whole surface area of the embryo (Ducibella & Anderson, 1975). Before compaction, individual spherical 8-cell blastomeres are easily distinguished from each other; while after compaction, they become indistinguishable. The compacted embryo appears as a single spherical ball.

What mediates this dynamic morphological change? It was initially found that extracellular calcium (Ca^{2+}) plays a predominant role in compaction (Ducibella & Anderson, 1975). Culturing embryos in Ca^{2+}-free media not only prevents compaction but also reverts compacted embryos back into an uncompacted state. E-cadherin (encoded by *Cdh1*) is a Ca^{2+}-dependent cell–cell adhesion molecule and blocking its function with a neutralizing antibody prevents compaction (Shirayoshi, Okada, & Takeichi, 1983; Winkel, Ferguson, Takeichi, & Nuccitelli, 1990). This suggests that E-cadherin is a major component regulating compaction. However, zygotic $Cdh1^{-/-}$ mutant embryos do not show a compaction defect due to maternally deposited *Cdh1* mRNA and protein, although they fail to form the TE epithelial layer (Larue, Ohsugi, Hirchenhain, & Kemler, 1994). The importance of E-cadherin function in compaction was fully confirmed in maternal zygotic (MZ) $Cdh1^{-/-}$ mutant embryos, where E-cadherin is completely absent (de Vries et al., 2004; Stephenson, Yamanaka, Rossant, 2010). Consistent with this, the complete removal of β-catenin (MZ $Ctnnb1^{-/-}$), a cytoplasmic binding partner of E-cadherin essential for E-cadherin-mediated cell–cell adhesion (Gumbiner, 2005), also leads to failure in compaction. Interestingly, some cells display minimal cell–cell contacts forming small aggregated clusters in MZ $Cdh1^{-/-}$ mutant embryos, suggesting the involvement of other cell adhesion molecule(s) in early embryos (Stephenson et al., 2010).

The treatment of conventional protein kinase C (PKC) inhibitors or activators can modulate the timing of compaction (Bloom, 1989; Winkel et al., 1990). Conventional PKCs are serine/threonine kinases that require both Ca^{2+} and diacylglycerol (DAG) for their activation. Phorbol myristate acetate (PMA), a potent activator but subsequently causing downregulation of PKC (Nishizuka, 1984), can induce premature compaction within 30 min and then decompaction after 1 h when treated at the 4-cell stage (Winkel et al., 1990). This decompaction is likely due to E-cadherin internalization caused by the prolonged treatment of PMA. In contrast, treatment with DAG analogs, diC_8 or OAG, which only activates PKC, can induce sustained premature compaction (Winkel et al., 1990). Treatment with sphingosine, an inhibitor of PKC, can prevent the DAG-induced premature compaction as well as normal compaction at the 8-cell stage (Winkel et al., 1990). These data indicate that the activation of PKC can facilitate compaction while its inhibition can prevent it. Although these inhibitors and activators are quite useful tools to dissect molecular events, their specificities would be of concern. Further genetic studies will be required to identify which PKC subfamily is responsible for regulating compaction and how their activity is regulated in the early embryo.

Recent advances in live imaging technology allow direct visualization of this dynamic morphogenetic process in intact embryos. Two recent studies proposed novel mechanisms of compaction (Fierro-González, White, Silva, & Plachta, 2013; Maître, Niwayama, Turlier, Nédélec, & Hiiragi, 2015). The first study revisited the formation of filopodia-like protrusions from individual 8-cell blastomeres, originally identified in scanning electron microscopy (Calarco & Epstein, 1973), which stretch across the cell surface of their neighboring cells (Fierro-González et al., 2013). These filopodia create a pulling force to support the flattening of neighboring cells facilitating compaction by anchoring to maintain tension across the embryo surface (Fig. 2A). Approximately 60% of 8-cell blastomeres exhibit the filopodia protrusions. The filopodia in the embryos consists of E-cadherin, F-actin, and myosin-10 (Myo10), similar to those in other polarized and nonpolarized epithelial cells (Liu, Jacobs, Dunn, Fanning, & Cheney, 2012; Mattila & Lappalainen, 2008). Laser ablations of the filopodia induce a cell shape change from flattening to rounding, indicating that they are required for maintaining a compacted morphology. Downregulation of E-cadherin in only half of the embryos exhibited noncompacted morphology without the filopodia formation, whereas nonaffected cells still produced filopodia and underwent compaction, suggesting a cell-autonomous nature of this process. Removal of E-cadherin, α/β-catenins or Myo10 by siRNAs prevents or delays compaction. Myo10 localization within filopodia is required for filopodia formation and its overexpression induces premature filopodia formation and compaction in the 5- to 7-cell stage (Fierro-González et al., 2013). However, a recent study (Maître et al., 2016) showed that Myo10 null embryos (MZ *Myh10*) exhibit normal blastocyst formation, suggesting that Myo10 would be dispensable for compaction. Further studies will be required to reconcile this discrepancy.

The second study focused on the surface contractility at the cell-medium interphase in individual cells (Maître et al., 2015). Using micropipette aspiration techniques, Maître et al. showed that surface contractility at the noncontact surface (i.e., an exposed surface) of cells increases during compaction. Blocking actomyosin contractility by blebbistatin, an inhibitor of myosin II, prevents compaction but does not disrupt previously formed cell–cell contacts (Maître et al., 2015), suggesting that this actomyosin-mediated surface

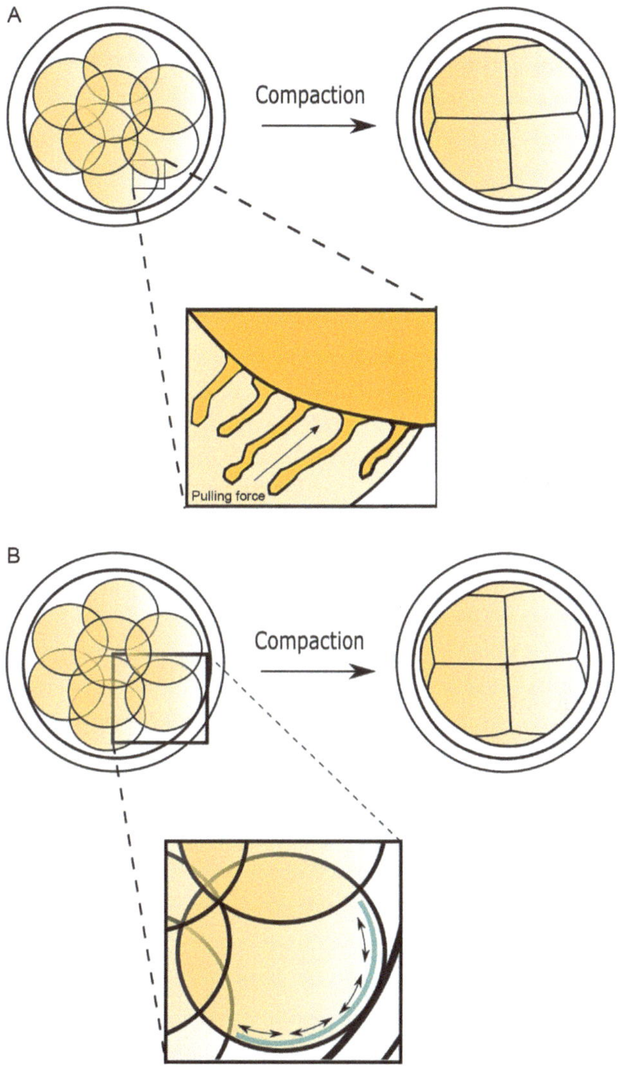

Fig. 2 Two new models of compaction in the preimplantation mouse embryo at the 8-cell stage. (A) E-cadherin-dependent filopodia formation creates a pulling force to facilitate cell flattening for compaction. These filopodia express E-cadherin, F-actin, and myosin-10. (B) The increase of surface contractility (*arrows*) of the noncontact surface of a cell initiates cell flattening for compaction. The actomyosin network (*blue*) of the noncontact surface is increased while no contractility increases at cell–cell contacts.

contractility provides a driving force for cell flattening. Notably, this new view does not discount the influence of cell adhesion during compaction. Formation of cell–cell contacts, which are devoid of contractility, is also essential for compaction (Fig. 2B).

Compaction is easily observed under a dissecting microscope as a morphological change of the embryo. However, its molecular mechanism is still largely unknown. Although the two recent studies discussed earlier brought new mechanistic views of compaction, what initiates those mechanistic changes at the 8-cell stage is elusive. It will be very important to identify molecular differences before and after compaction that regulate the morphological change within individual 8-cell blastomeres.

In parallel to compaction, polarization also occurs at the 8-cell stage. Nonpolarized 8-cell blastomeres undergo polarization recognized by the formation of the apical domain (Anani et al., 2014; Johnson & Ziomek, 1981a; Korotkevich et al., 2017). The formation of apico-basal cell polarity permits asymmetric distributions of proteins and their activities within a cell (Yamanaka, Ralston, Stephenson, & Rossant, 2006). Interestingly, polarization appears to be independent of compaction. Both the MZ *Cdh1*$^{-/-}$ mutant embryos and the embryos cultured in Ca^{2+}-free media, which cannot undergo compaction, are still able to polarize (Stephenson et al., 2010; Vinot et al., 2005). In addition, induction of premature compaction in the 4-cell embryo through modulation of PKC activity does not concurrently induce premature polarization (Ohsugi, Ohsawa, & Yamamura, 1993; Winkel et al., 1990).

What molecules are responsible for the initiation of polarization? The small Rho GTPase family proteins, such as Cdc42 and Rho, are known to play important roles in cell polarity in many other model systems such as yeast, *Caenorhabditis elegans* and tissue culture cells, through dynamic regulation of cytoskeletal organization (Rodriguez-Boulan & Macara, 2014). Microinjection of activated forms of Rho and Cdc42 into 4-cell blastomeres leads to premature polarization indicated by actin enrichment at the apical pole, suggesting their activation is sufficient to initiate polarization (Clayton, Hall, & Johnson, 1999). On the other hand, inhibition of Rho with the botulinum C3-transferase prevents polarization at the 8-cell stage (Clayton et al., 1999). In MZ *Cdc42* mutant embryos, the apical domain is severely compromised (Korotkevich et al., 2017). Although it is still unknown where these small GTPase proteins are activated during polarization, these results suggest their involvement in polarization at the 8-cell stage.

Ezrin, an ERM (ezrin/radixin/moesin) family protein, is a linker between a membrane protein and the actin cytoskeleton and is involved in microvilli formation in mature epithelial cells (Namgoong & Kim, 2016; Viswanatha, Bretscher, & Garbett, 2014). Phosphorylation of the threonine 567 residue of Ezrin by Rho kinase (ROCK) activates it from a cytoplasmic dormant form to an active linker form (Fehon, McClatchey, & Bretscher, 2010; Matsui et al., 1998; Ponuwei, 2016). The localization of Ezrin phoshorylation is restricted at the apical domain of embryos during polarization (Anani et al., 2014; Dard, Louvet-Vallée, Santa-Maria, & Maro, 2004). The PAR3, PAR6, and atypical PKC (aPKC) complex, an evolutionarily conserved polarity protein complex, is also localized at the apical domain (Goldstein & Macara, 2007). Knockdown of PAR3 and PAR6 by shRNA microinjection or overexpression of a dominant negative form of aPKC is sufficient to prevent proper polarization and enhance the contribution to the ICM lineage (Alarcon, 2010; Plusa et al., 2005). Interestingly, the treatment of Y27632, a ROCK inhibitor, also prevents the apical localization of aPKC and PARD6B, and TE-specific CDX2 expression (Kono, Tamashiro, & Alarcon, 2014). This suggests that ROCKs might be involved as a link between Rho activation and the apical recruitment of the PAR complex during polarization. However, because Y27632 can also efficiently inhibit aPKC activity in *Drosophila* (Atwood & Prehoda, 2009), and zygotic *Rock1−/−; Rock2−/−* double mutant embryos can form normal mouse blastocysts (Kamijo et al., 2011), further genetic analyses will be required to confirm the involvement of ROCKs in polarization at the 8-cell stage.

3. ASYMMETRIC DIVISION AND THE ESTABLISHMENT OF THE OUTER/INNER CONFIGURATION OF POLAR/APOLAR CELLS

During the morula stage, there are two rounds of asymmetric division at the fourth (from 8- to 16-cell) and fifth (from 16- to 32-cell) cell division. After asymmetric division, the first two distinct cell populations emerge as cells that inherit the apical domain from their parental 8-cell blastomeres and ones that do not, resulting in polar and apolar cells, respectively. The polar cells take an outer position to become the TE and the apolar cells take an inner position to become the ICM, leading to formation of the outer/inner cells within the embryo.

Although asymmetric division is essential for the first lineage specification, the frequency of asymmetric division of 8-cell blastomeres appears to be context dependent. Isolated single 8-cell blastomeres exhibit a higher frequency of asymmetric division, approximately 82%, while couplets of

8-cell blastomeres display a frequency of only 50% (Johnson & Ziomek, 1981b). In intact embryos, the average asymmetric division frequency during the 8–16 cell division is approximately 60%–70% in total (Anani et al., 2014; Fleming, McConnell, Johnson, & Stevenson, 1989). This suggests that on average five or six blastomeres of the 8-cell embryo divide asymmetrically. However, in individual embryos, the frequency of asymmetric division is quite variable, ranging from only one blastomere to all eight dividing asymmetrically. What controls the decision between symmetric or asymmetric division in individual blastomeres is not fully known yet. In an early study by Pickering et al., they suggested that the size of the apical domain and the position of cell–cell contact influence the decision for asymmetric division (Johnson, 2009; Pickering, Maro, Johnson, & Skepper, 1988). In their model, the larger apical domain has a higher chance to divide symmetrically while the smaller one has a higher chance to divide asymmetrically. On the other hand, more recent studies suggest that cellular characteristics such as the cell shape during division (Dard, Le, Maro, & Louvet-Vallée, 2009) and the nuclear position before division (Ajduk, Biswas Shivhare, & Zernicka-Goetz, 2014) show some correlation with divisional orientation in intact embryos.

A recent study analyzing the orientation of mitotic spindles during the 8–16 cell division in isolated blastomeres demonstrated that the majority of cells aligned their mitotic spindle to the apico-basal axis regardless of the apical area encompassing the cell surface (Korotkevich et al., 2017). Destabilization of the apical domain in *Cdc42*$^{-/-}$ mutant embryos leads to randomization of their orientation. In conjunction with this, the live imaging analysis of SAS4-GFP transgenic mouse embryos, which visualize microtubule-organizing centers (MTOC) localization, shows that acentrosomal spindle assembly occurs subapically during de novo apical domain formation. This suggests that the apical domain somehow recruits MTOC to regulate spindle orientation to facilitate asymmetric division, ensuring the production of the two cell populations at the 16-cell stage (Korotkevich et al., 2017). This notion is interesting because there are no astral microtubules in mitotic spindles of the early mouse embryo (Hiraoka, Golden, & Magnuson, 1989; Szollosi, Calarco, & Donahue, 1972). Astral microtubules are an essential component to reorient the mitotic spindle and align its orientation with cell polarity during asymmetric division in other systems (Bergstralh, Dawney, & St Johnston, 2017). Reorientation of the mitotic spindles has not yet been observed in early mouse embryos. Are mitotic spindles formed in an oriented manner without reorientation? High-resolution time-lapse analysis of spindle formation in early embryos will answer this question in the near future.

It was originally considered that the outer/inner configuration of the polar/apolar cells is regulated by divisional orientation (Johnson, 2009). When cell division is planar, both daughter cells are polar and placed at an outer position, while when cell division is orthogonal, one daughter cell inherits the whole surface apical domain to become an outer polar cell and the other daughter cell is deposited directly inside of the embryo, becoming an apolar inner cell. This simple idea was proposed based on the results of isolated blastomeres without direct observation of intact embryos. Apart from orthogonal and planar divisions, the existence of frequent oblique divisions was noticed in an early time-lapse study (Sutherland, Speed, & Calarco, 1990). This early observation was fully confirmed in recent studies using more sophisticated live-imaging techniques (Anani et al., 2014; McDole, Xiong, Iglesias, & Zheng, 2011; Samarage et al., 2015; Watanabe, Biggins, Tannan, & Srinivas, 2014; Yamanaka, Lanner, & Rossant, 2010). In intact embryos, many 8-cell blastomeres divide in an oblique orientation with respect to the embryo surface. Careful positional analysis revealed that in 16-cell stage embryos, only one or two cells fully take an inner position (Anani et al., 2014; Dietrich & Hiiragi, 2007). Additionally, there is a unique population that appears to occupy the intermediate nuclear position between the outer and inner positions (Anani et al., 2014; McDole et al., 2011; Watanabe et al., 2014). Many of these cells internalize into the inner position before the next division cycle (16–32 cell division). This suggests that divisional orientation is not a good predictor of their final cell allocation (Anani et al., 2014; Dard et al., 2009; Samarage et al., 2015; Watanabe et al., 2014). On the other hand, many divisions in an oblique orientation are asymmetric in terms of the inheritance of the apical domain (Anani et al., 2014; Yamanaka et al., 2010). Both daughter cells take an outer position soon after division, despite one of them being an apolar cell. This outer apolar cell later internalizes to take an inner position (Anani et al., 2014; Korotkevich et al., 2017; Maître et al., 2016). Interestingly, this is not fully deterministic; as some cells can occasionally repolarize and stay at an outer position to adopt the TE fate.

This raises an interesting question, why do only outer apolar cells internalize and not polar cells? This internalization process is mediated by the increase in surface contractility at the noncontact surface of outer apolar cells. Phosphorylation of myosin II is locally enriched at the noncontact surface of outer apolar cells (Anani et al., 2014; Maître et al., 2016) and blocking actomyosin contractility with blebbistatin, a myosin inhibitor, prevents internalization of the outer apolar cells (Maître et al., 2016; Samarage et al., 2015). The maternal MYH9 is one of the key components for this actomyosin contractility. Aggregating a wild-type blastomere with a maternal *Myh9* null embryo causes the donor wild-type blastomere to internalize without division. Conversely, aggregating a *Myh9* null blastomere with a wild-type embryo causes the donor *Myh9* null blastomere to spread and remains on the surface of the wild-type embryo (Maître et al., 2016). A mathematical simulation revealed that a 1.5:1 fold asymmetry in surface contractility between apolar and polar cells is sufficient to drive the apolar cell to internalize (Fig. 3, Maître et al., 2016). These studies suggest that difference in surface contractility of polar/apolar cells dictates the establishment of the outer/inner cell configuration in the embryo. Therefore, once polar and apolar cells are generated through asymmetric division, they can self-organize to form the outer/inner populations within the embryo (Anani et al., 2014; Maître et al., 2016).

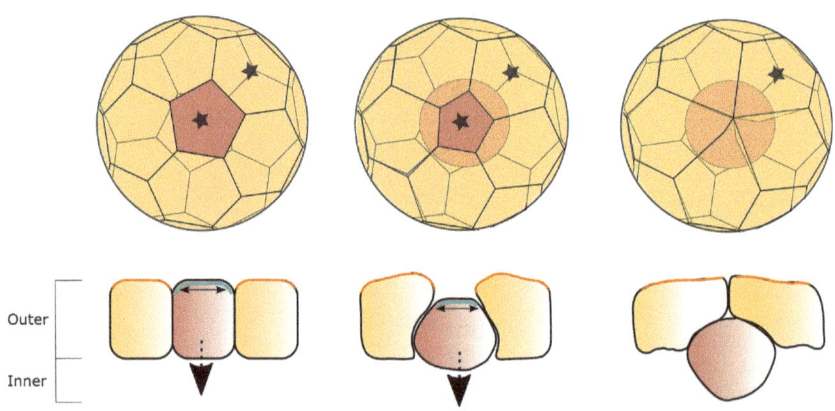

Fig. 3 Establishing the outer/inner configuration of polar/apolar cells. Soon after asymmetric division, some apolar cells are placed at the embryo surface (*red* cells), similar to polar cells (*yellow*). Sister cells are marked with stars. However, because the outer apolar cells have an enriched actomyosin network (*blue*) at the noncontact surface, leading to higher surface contractility (*solid arrow*) compared to polar cells. The outer apolar cells initiate internalization (*dotted arrow*) to take an inner position within the embryo. The two adjacent polar cells will be brought closer and eventually fully encompass the surface of the internalizing apolar cell.

4. ACTIVATION OF LINEAGE-SPECIFIC GENETIC PROGRAMS IN THE OUTER/INNER CELLS TO FORM THE TE AND ICM LINEAGES

CDX2 is a *caudal*-related homeobox transcriptional factor (TF) essential for TE formation during preimplantation development (Blij, Frum, Akyol, Fearon, & Ralston, 2012; Strumpf et al., 2005) and its expression is restricted to the TE in the blastocyst. Overexpression of CDX2 in embryonic stem cells (ESCs) is sufficient to differentiate ESCs into trophoblast stem cells and the TE derivatives like giant cells (Niwa et al., 2005). On the other hand, *Cdx2* null embryos fail to implant because no functional TE is formed (Strumpf et al., 2005). Interestingly, *Oct4* and *Nanog*, the genes encoding ICM-specific TFs, are ectopically expressed in the outer cells of *Cdx2* null embryos. This suggests that CDX2 is required for suppression of *Oct4* and *Nanog* expression in the TE (Strumpf et al., 2005). Conversely, OCT4 is a POU-family TF essential for pluripotency in vivo and in in vitro pluripotent stem cells (Niwa et al., 2005; Young, 2011). The expression of OCT4 in the late blastocyst is restricted to the ICM and *Oct4* null embryos fail to form a functional ICM. Although the *Oct4* null blastocyst can form the ICM morphologically, their fate is not properly specified and thus is transformed into the TE in blastocyst outgrowth assays (Nichols et al., 1998). These studies suggest that OCT4 and CDX2 negatively regulate each other's expression to establish a mutually exclusive pattern in the embryo. However, unlike the late blastocyst, the initial expression of both TFs in the early morula stage are not restricted to specific cell positions but express relatively uniformly throughout the embryo (Niwa et al., 2005). During the formation of outer/inner cells and subsequent blastocyst formation, CDX2 expression becomes restricted first in the outer polar cells and OCT4 expression becomes restricted in the ICM later by the late blastocyst stage (Szczepanska, Stanczuk, & Maleszewski, 2011). OCT4 and CDX2 can form a complex for the reciprocal repression of their target genes in ESCs, suggesting that the mutual inhibition of the two TFs could be a part of the mechanism to establish the TE/ICM lineage-specific gene expression (Niwa et al., 2005).

Besides their mutual inhibition, how is CDX2 expression restricted in outer cells? The key signaling pathway connecting CDX2 expression to cell position is the Hippo signaling pathway. The Hippo signaling pathway was originally identified in *Drosophila* as the signaling pathway controlling organ size (Halder & Johnson, 2011; Irvine & Harvey, 2015). This pathway is well conserved in mammals and regulates cell proliferation, apoptosis, and stem cell maintenance. The Hippo signaling pathway is regulated by a series of serine/threonine phosphorylations, which are mediated by several kinases. The transcriptional effectors of the Hippo signaling pathway are cotranscriptional activators, YAP and TAZ, which can shuttle between the cytoplasm and the nucleus. Upon activation of the Hippo signaling pathway, LATS1/2 kinases phosphorylate YAP and TAZ to sequestrate them in the cytoplasm, preventing their translocation into the nucleus. On the other hand, when the Hippo signaling pathway is suppressed, YAP and TAZ translocate into the nucleus to bind to the TEAD family DNA-binding proteins to activate YAP target genes (Halder & Johnson, 2011; Irvine & Harvey, 2015).

Involvement of the Hippo signaling pathway in the TE/ICM specification was first reported in the *Tead4* knockout studies (Nishioka et al., 2008; Yagi et al., 2007) where the *Tead4* null embryos failed to form a blastocyst but an abnormal cell aggregate. Although the initial CDX2 expression at the 8-cell stage is observed, it is soon downregulated and absent throughout the rest of development. The phenotype of the *Tead4* null embryos is more severe than that of the *Cdx2* null embryos, which still can form a blastocoel (Blij et al., 2012; Strumpf et al., 2005). Similar to the CDX2 null embryo, the outer cells of the *Tead4* null embryo maintain OCT4 expression (Nishioka et al., 2008; Yagi et al., 2007). On the other hand, *Tead4* null ESCs can be isolated from the *Tead4* null embryos indicating that TEAD4 is not required for ICM formation and establishing pluripotency. Modulation of TEAD4-dependent transcriptional activity using transcriptionally active and repressive forms of TEAD4 is sufficient to induce ectopic CDX2 expression in inner cells and to suppress endogenous CDX2 expression in outer cells, respectively, without changing their cell positions (Nishioka et al., 2009). Since TEAD4 is expressed uniformly in all nuclei within the preimplantation embryo (Hirate, Cockburn, Rossant, & Sasaki, 2012; Nishioka et al., 2008), additional factor(s) is/are required to explain the TE-specific requirement of TEAD4. As mentioned earlier, YAP is a transcriptional coactivator for the TEAD family DNA-binding proteins and can shuttle between the cytoplasm and nucleus in a phosphorylation-dependent manner (Halder & Johnson, 2011; Irvine & Harvey, 2015). During the morula and blastocyst stages, YAP is localized in the nucleus of outer TE cells, while in inner cells YAP is localized in the cytoplasm and excluded from the nucleus (Nishioka et al., 2009) (Fig. 4). Cytoplasmic YAP is phosphorylated at the serine 127 residue, suggesting activation of the Hippo signaling pathway in the inner

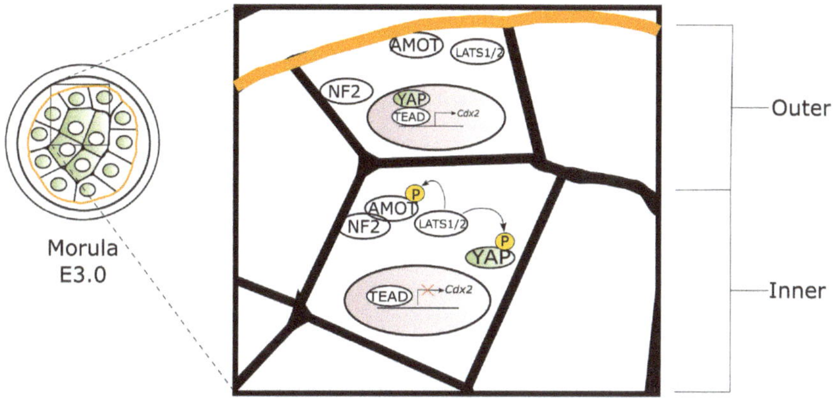

Fig. 4 Differential regulation of the Hippo signaling pathway in the outer polar and inner apolar cells to specify the TE/ICM lineages. Differential regulation of the Hippo signaling pathway is essential for TE/ICM formation in the blastocyst. In outer polar cells, AMOT is sequestered at the apical domain (*yellow*) to suppress the Hippo signaling pathway. Thus, YAP translocates into the nucleus to induce TE-specific genes like *Cdx2*. In contrast, in inner apolar cells, AMOT forms an active complex with NF2 and LATS1/2 at the basolateral domain (*black*) to facilitate YAP phosphorylation by LATS1/2. Phosphorylated YAP is excluded from the nucleus and sequestered in the cytoplasm.

cells. Similar to *Drosophila*, LATS1/2 kinases are responsible for this phosphorylation. Expression of the dominant negative form of LATS2 is sufficient to prevent YAP phosphorylation and permit YAP nuclear translocation in the inner cells to induce ectopic CDX2 expression (Nishioka et al., 2009). Conversely, overexpression of wt-LATS2 kinase is sufficient to suppress the endogenous CDX2 expression in the outer cells. SOX2 is another ICM-specific TF and shows restricted expression in the inner cells at the morula stage (Wicklow et al., 2014). Interestingly, the Hippo signaling pathway also regulates this inner cell-specific expression of SOX2. When the pathway is ectopically activated by overexpression of LATS2 kinase, SOX2 is expressed ectopically in outer cells. Unlike OCT4 and NANOG, SOX2 suppression in the outer cells is not mediated by CDX2 because SOX2 ectopic expression is not observed in *Cdx2* null embryos. This suggests another mechanism to suppress inner cell-specific genes in the outer cells downstream of the YAP/TEAD4 target genes.

Taken together, modulation of the Hippo signaling pathway cannot alter cell position in a blastocyst but can disengage the link between cell position and TE/ICM-specific TF expression. Thus, the Hippo signaling pathway acts downstream of cell position to control lineage-specific gene expression.

5. HOW DO THE CELLS WITHIN THE EMBRYO SENSE THEIR POSITION TO PROPERLY REGULATE HIPPO SIGNALING?

What controls the Hippo signaling pathway downstream of cell position within the embryo? Angiomotin (AMOT), a scaffolding protein and its family protein, AMOTL2, play important roles upstream of LATS kinases in TE/ICM specification (Hirate et al., 2013; Leung & Zernicka-Goetz, 2013). In outer/TE cells, AMOT exclusively localizes to the apical domain, while in inner/ICM cells, AMOT localizes to the whole cell cortex/membrane (Hirate et al., 2013). When both AMOT and AMOTL2 are removed from the embryo (AMOT-free), YAP nuclear localization and CDX2 expression are uniformly observed in all cells, suggesting that AMOT and AMOTL2 are required for activation of the Hippo signaling pathway. Interestingly, in the inner cells, the serine 176 residue (S176) of AMOT is phosphorylated by LATS kinases. S176 phosphorylation suppresses the actin-binding activity of AMOT and enhances its interaction with NF2 and LATS kinases to facilitate LATS-dependent YAP phosphorylation (Fig. 4). The S176A mutant of AMOT, which cannot be phosphorylated by LATS, can still bind to NF2 but shows a very weak rescue activity in the AMOT-free embryos. Conversely, the phosphomimetic form of AMOT (S176E) can induce YAP nuclear exclusion and phosphorylation not only in the inner cells but also in the outer cells, suggesting that the S176E AMOT acts as a constitutively active form of the Hippo signaling pathway. These suggest that LATS-dependent phosphorylation at the S176 of AMOT is the key event for the differential YAP transcriptional activity in outer and inner cells. In the outer cells, the apical domain recruits AMOT to sequester from LATS-dependent phosphorylation. This, in turn, leads to the inactivation of the Hippo signaling pathway to permit YAP nuclear localization to induce TE-specific genes.

In addition to AMOT, NF2, a membrane-associated FERM domain protein (Zhang et al., 2010), is also shown to work upstream of LATS kinases to activate the Hippo signaling pathway. Due to maternal contribution, zygotic *Nf2* mutant embryos show only a partial phenotype of ectopic CDX2 expression in a few ICM cells at the blastocyst stage (Cockburn, Biechele, Garner, & Rossant, 2013). In the MZ *Nf2* mutant embryos, all ICM cells ectopically express CDX2 but still form a morphologically normal blastocyst. Consistent with this, no *Nf2* null ESC has been established from the MZ *Nf2* mutant embryos while the paternal allele is sufficient to rescue all these deficiencies. In contrast to AMOT, which has distinct localization and phosphorylation patterns in outer and inner cells, NF2 appears to be evenly localized throughout the whole cortex/membrane in both outer and inner cells, suggested by the localization of the NF2–GFP fusion protein. Thus, it is currently unknown whether NF2 activity is also differentially regulated in outer and inner cells, like AMOT.

Soon after asymmetric division, as described in the previous section, both polar and apolar cells are often placed at the outer position and then the outer apolar cells internalize (Anani et al., 2014; Samarage et al., 2015). Interestingly, before fully internalizing within the embryo, the outer apolar cells have higher Hippo signaling activity indicated by cytoplasmic YAP phosphorylation, at levels comparable to inner apolar cells (Anani et al., 2014; Hirate et al., 2015; Korotkevich et al., 2017). This suggests that the absence or presence of the apical domain is sufficient for the activation or suppression of the Hippo signaling pathway rather than their cell position.

What are the key molecules in the apical domain to suppress the Hippo signaling pathway? Knockdown of *Pard6b* showed that the apical localization of PARD6B is required for apical enrichment of aPKC, YAP nuclear localization, and the TE/ICM lineage specification (Alarcon, 2010; Hirate et al., 2013). PKCzeta and PKC lambda/iota, are the two isoforms of aPKC, encoded in *Prkcz* and *Prkci*, respectively. Studies on the role of aPKC in polarization show conflicting data (Hirate et al., 2013; Korotkevich et al., 2017), but both support the importance of the apical domain for Hippo signaling regulation. In Hirate et al., *Prkcz*$-/-$; *Prkci*$-/-$ double mutant embryos were generated by double heterozygous intercross (Hirate et al., 2013). These embryos fail to form a blastocyst and there is no YAP nuclear localization in the outer cells. On the other hand, Korotkevich et al. generated MZ *Prkci*$^{-/-}$;*Prkcz*$^{-/-}$ mutant embryos by crossing *Prkci*$^{flox/flox}$; *Zp3-Cre; Prkcz*$^{-/-}$ females with *Prkci*$^{+/-}$; *Prkcz*$^{-/-}$ males (Korotkevich et al., 2017). Although their mutant embryos also fail to form a blastocyst, the apical localization of PARD6B and Radixin, as well as CDX2 expression in the outer cells are still observed. Additionally, mutually exclusive expression of CDX2 and SOX2 are established, suggesting that the functional apical domains are still formed without aPKC. MZ *Cdc42*$^{-/-}$ mutant embryos display a severely compromised outer surface localization of PARD6B and Radixin, indicating the importance of CDC42 in the recruitment of PARD6B in the apical domain and microvilli formation. The MZ *Cdc42*$^{-/-}$ mutant embryos also do not form a blastocyst and have very weak CDX2 expression in a few outer cells. It is not clear what causes the apparent discrepancy between the two aPKC double mutant studies regarding formation of the apical domain, CDX2 expression, and cell polarity. However, these two studies further confirmed a tight correlation between the presence of the apical domain and YAP nuclear localization, or CDX2 expression in outer cells. Interestingly, in the double shRNA knockdown embryos of basolateral regulators, *Par1a/b* (Suzuki & Ohno, 2006), AMOT localization is not restricted to the apical domain but expands to the lateral membrane of the outer cells, although no change is observed in aPKC and PARD6B apical localization (Hirate et al., 2013). Consistent with this AMOT localization pattern, YAP is phosphorylated and excluded from the nucleus, suggesting that PAR1 is required for AMOT apical sequestration to suppress the Hippo signaling pathway in the outer cells. Another mammalian PAR1 ortholog found in breast cancer cells, MARK4, inhibits the Hippo signaling pathway through phosphorylation of the Hippo pathway kinases, MST and SAV (Heidary Arash, Shiban, Song, & Attisano, 2017). MARK4 can bind to MST/SAV, leading to their phosphorylation, and attenuates the formation of a complex between MST/SAV and LATS. Similar molecular mechanisms might be used at the lateral domain of the outer cells to regulate AMOT localization and the suppression of the Hippo singling pathway.

When the apical domain is transplanted onto apolar cells, AMOT is relocated from the whole cortex/membrane to the ectopically transplanted apical domain (Korotkevich et al., 2017). Although the mechanism in which the apical domain recruits AMOT is still elusive, the presence of the apical domain is sufficient for suppression of the Hippo signaling pathway to permit YAP nuclear localization for outer TE specification. This suggests that the individual cells use the presence or absence of the apical domain as the mechanism to sense their position within the embryo to activate or suppress the Hippo signaling pathway.

Although cell polarity dictates cell allocation and differential Hippo signaling activity. Cells in the morula or the early blastocyst retain their plasticity and can still change into the alternative cell fate when experimentally manipulated. When the outer cells or the TE cells are immunosurgically removed in the morula or the early blastocyst, the TE layer regenerates through repolarization of the surface cells, leading to YAP nuclear localization for CDX2 expression (Handyside, 1978; Rossant & Lis, 1979; Stephenson et al., 2010; Suwińska, Czołowska, Ozdzeński, & Tarkowski, 2008). On the other hand, the outer cells retain some plasticity to become the ICM cells up to the 32-cell stage (Rossant & Vijh, 1980; Tarkowski, Suwińska, Czolowska, & Ozdzezski, 2010). A recent study analyzing the timing of TE/ICM fate commitment showed that the individual cells progressively lose their plasticity during the 16- to 64-cell stage, which correlates well with the level of CDX2 expression (Posfai et al., 2017). By the early 32-cell stage in the single cell transplantation experiments, the transplanted cells can still change their fate without cell division to adopt the new environment, despite the initial levels of CDX2 expression. However, this ability disappears by the late 32-cell stage. It is an interesting question to address how the cells lose the ability to polarize/depolarize in the future. In addition, it is very interesting to know what permits the cells to divide asymmetrically between the 8- and 32-cell stages. These would be important for further understanding of the unique lineage plasticity of early blastomeres.

6. FUTURE PERSPECTIVES

Formation of the apical domain creates two distinct cortical subdomains of surface contractility within a cell as well as provides the domain for sequestration of AMOT to prevent the activation of the Hippo signaling pathway. The subsequent asymmetric divisions separate the two subdomains into the polar and apolar cells. The polar and apolar cells self-organize through an active cell sorting process and differentially regulate the Hippo signaling pathway to induce lineage-specific gene expression as the outer polar cells become the TE and the inner apolar cells become the ICM.

While significant progress has been made in the last decade in understanding molecular and cellular mechanisms of preimplantation mouse development, there are still questions that remain unanswered. We still have not fully grasped the molecular and cellular mechanisms of compaction and polarization. We know what molecules are involved in maintaining the apical domain but little is known about how compaction and polarization are initiated and why it occurs at the 8-cell stage. It is unknown what controls the frequency of asymmetric division in individual embryos and whether the embryo has compensatory mechanisms to adopt the low or high numbers of asymmetric division in the first round. In addition, how the Hippo signaling pathway is activated in inner cells and whether there is any instructive signaling to activate the ICM-specific genes are still unclear.

Mammalian preimplantation development has been considered as unique because of its regulative and plastic nature. Revisiting classic experiments that originally identified the uniqueness of mammalian development with cutting edge technologies, like direct visualization of cell heterogeneity with fluorescent reporters as well as high-resolution single cell transcriptome and epigenome analyses would be important for further understanding of molecular mechanisms involved in developmental plasticity and self-organizing ability in the mammalian embryo.

ACKNOWLEDGMENTS

We would like to thank Katie Teng for careful reading of this manuscript. D.S. is supported by McGill CRRD and Faculty of Medicine studentships. Y.Y. is supported by the Canadian Institutes of Health Research (CIHR) [MOP111197] and the Natural Sciences and Engineering Research Council of Canada (NSERC) [RGPIN41870].

REFERENCES

Aiken, C. E. M., Swoboda, P. P. L., Skepper, J. N., & Johnson, M. H. (2004). The direct measurement of embryogenic volume and nucleo-cytoplasmic ratio during mouse pre-implantation development. *Reproduction, 128*, 527–535. https://doi.org/10.1530/rep.1.00281.

Ajduk, A., Biswas Shivhare, S., & Zernicka-Goetz, M. (2014). The basal position of nuclei is one pre-requisite for asymmetric cell divisions in the early mouse embryo. *Developmental Biology, 392*, 133–140. https://doi.org/10.1016/j.ydbio.2014.05.009.

Alarcon, V. B. (2010). Cell polarity regulator PARD6B is essential for trophectoderm formation in the preimplantation mouse Embryo. *Biology of Reproduction, 83*, 347–358. https://doi.org/10.1095/biolreprod.110.084400.

Anani, S., Bhat, S., Honma-Yamanaka, N., Krawchuk, D., & Yamanaka, Y. (2014). Initiation of Hippo signaling is linked to polarity rather than to cell position in the pre-implantation mouse embryo. *Development, 141*, 2813–2824. https://doi.org/10.1242/dev.107276.

Atwood, S. X., & Prehoda, K. E. (2009). aPKC phosphorylates Miranda to polarize fate determinants during neuroblast asymmetric cell division. *Current Biology, 19*, 723–729. https://doi.org/10.1016/j.cub.2009.03.056.

Bergstralh, D. T., Dawney, N. S., & St Johnston, D. (2017). Spindle orientation: A question of complex positioning. *Development, 144*, 1137–1145. https://doi.org/10.1242/dev.140764.

Blij, S., Frum, T., Akyol, A., Fearon, E., & Ralston, A. (2012). Maternal Cdx2 is dispensable for mouse development. *Development, 139*, 3969–3972. https://doi.org/10.1242/dev.086025.

Bloom, T. L. (1989). The effects of phorbol ester on mouse blastomeres: A role for protein kinase C in compaction? *Development, 106*, 159–171.

Calarco, P. G., & Epstein, C. J. (1973). Cell surface changes during preimplantation development in the mouse. *Developmental Biology, 32*, 208–213. https://doi.org/10.1016/0012-1606(73)90233-9.

Chazaud, C., & Yamanaka, Y. (2016). Lineage specification in the mouse preimplantation embryo. *Development, 143*, 1063–1074. https://doi.org/10.1242/dev.128314.

Clayton, L., Hall, A., & Johnson, M. H. (1999). A role for Rho-like GTPases in the polarisation of mouse eight-cell blastomeres. *Developmental Biology, 205*, 322–331. https://doi.org/10.1006/dbio.1998.9117.

Cockburn, K., Biechele, S., Garner, J., & Rossant, J. (2013). The Hippo pathway member Nf2 is required for inner cell mass specification. *Current Biology*, *23*, 1195–1201. https://doi.org/10.1016/j.cub.2013.05.044.

Dard, N., Le, T., Maro, B., & Louvet-Vallée, S. (2009). Inactivation of aPKClambda reveals a context dependent allocation of cell lineages in preimplantation mouse embryos. *PLoS One*, *4*, e7117. https://doi.org/10.1371/journal.pone.0007117.

Dard, N., Louvet-Vallée, S., Santa-Maria, A., & Maro, B. (2004). Phosphorylation of ezrin on threonine T567 plays a crucial role during compaction in the mouse early embryo. *Developmental Biology*, *271*, 87–97. https://doi.org/10.1016/j.ydbio.2004.03.024.

Day, R. N., & Davidson, M. W. (2009). The fluorescent protein palette: Tools for cellular imaging. *Chemical Society Reviews*, *38*, 2887. https://doi.org/10.1039/b901966a.

de Vries, W. N., Eviskov, A. V., Haac, B. E., Fancher, K. S., Holbrook, A. E., Kemler, R., et al. (2004). Maternal beta-catenin and E-cadherin in mouse development. *Development*, *131*, 4435–4445. https://doi.org/10.1242/dev.01316.

Dietrich, J., & Hiiragi, T. (2007). Stochastic patterning in the mouse pre-implantation embryo. *Development*, *134*, 4219–4231. https://doi.org/10.1242/dev.003798.

Ducibella, T., & Anderson, E. (1975). Cell shape and membrane changes in the eight-cell mouse embryo: Prerequisites for morphogenesis of the blastocyst. *Developmental Biology*, *47*, 45–58. https://doi.org/10.1016/0012-1606(75)90262-6.

Fehon, R. G., McClatchey, A. I., & Bretscher, A. (2010). Organizing the cell cortex: The role of ERM proteins. *Nature Reviews. Molecular Cell Biology*, *11*, 674. https://doi.org/10.1038/nrm2955.

Fierro-González, J. C., White, M. D., Silva, J. C., & Plachta, N. (2013). Cadherin-dependent filopodia control preimplantation embryo compaction. *Nature Cell Biology*, *15*, 1424–1433. https://doi.org/10.1038/ncb2875.

Fischer, R. S., Wu, Y., Kanchanawong, P., Shroff, H., & Waterman, C. M. (2011). Microscopy in 3D: A biologist's toolbox. *Trends in Cell Biology*, *21*, 682–691. https://doi.org/10.1016/j.tcb.2011.09.008.

Fleming, T. P., McConnell, J., Johnson, M. H., & Stevenson, B. R. (1989). Development of tight junctions de novo in the mouse early embryo: Control of assembly of the tight junction-specific protein, ZO-1. *The Journal of Cell Biology*, *108*, 1407–1418.

Goldstein, B., & Macara, I. G. (2007). The PAR proteins: Fundamental players in animal cell polarization. *Developmental Cell*, *13*, 609–622. https://doi.org/10.1016/j.devcel.2007.10.007.

Gumbiner, B. M. (2005). Regulation of cadherin-mediated adhesion in morphogenesis. *Nature Reviews. Molecular Cell Biology*, *6*, 622–634. https://doi.org/10.1038/nrm1699.

Halder, G., & Johnson, R. L. (2011). Hippo signaling: Growth control and beyond. *Development*, *138*, 9–22. https://doi.org/10.1242/dev.045500.

Handyside, A. H. (1978). Time of commitment of inside cells isolated from preimplantation mouse embryos. *Journal of Embryology and Experimental Morphology*, *45*, 37–53.

Heidary Arash, E., Shiban, A., Song, S., & Attisano, L. (2017). MARK4 inhibits Hippo signaling to promote proliferation and migration of breast cancer cells. *EMBO Reports, 18*, e201642455. https://doi.org/10.15252/embr.201642455.

Hiraoka, L., Golden, W., & Magnuson, T. (1989). Spindle-pole during early mouse development. *Developmental Biology*, *133*, 24–36.

Hirate, Y., Cockburn, K., Rossant, J., & Sasaki, H. (2012). Tead4 is constitutively nuclear, while nuclear vs. cytoplasmic Yap distribution is regulated in preimplantation mouse embryos. *Proceedings of the National Academy of Sciences of the United States of America*, *109*, E3389–E3390. https://doi.org/10.1073/pnas.1211810109.

Hirate, Y., Hirahara, S., Inoue, K.-I., Kiyonari, H., Niwa, H., & Sasaki, H. (2015). Par-aPKC-dependent and -independent mechanisms cooperatively control cell polarity,

Hippo signaling, and cell positioning in 16-cell stage mouse embryos. *Development, Growth & Differentiation, 57*, 544–556. https://doi.org/10.1111/dgd.12235.

Hirate, Y., Hirahara, S., Inoue, K.-I., Suzuki, A., Alarcon, V. B., Akimoto, K., et al. (2013). Polarity-dependent distribution of angiomotin localizes Hippo signaling in preimplantation embryos. *Current Biology, 23*, 1181–1194. https://doi.org/10.1016/j.cub.2013.05.014.

Irvine, K. D., & Harvey, K. F. (2015). Control of organ growth by patterning and Hippo signaling in drosophila. *Cold Spring Harbor Perspectives in Biology*, 7, 1–16. https://doi.org/10.1101/cshperspect.a019224.

Johnson, M. H. (2009). From mouse egg to mouse embryo: Polarities, axes, and tissues. *Annual Review of Cell and Developmental Biology, 25*, 483–512. https://doi.org/10.1146/annurev.cellbio.042308.113348.

Johnson, M. H., & Ziomek, C. A. (1981a). Induction of polarity in mouse 8-cell blastomeres: Specificity, geometry, and stability. *The Journal of Cell Biology, 91*, 303–308. https://doi.org/10.1083/jcb.91.1.303.

Johnson, M. H., & Ziomek, C. A. (1981b). The foundation of two distinct cell lineages within the mouse morula. *Cell, 24*, 71–80. https://doi.org/10.1016/0092-8674(81)90502-X.

Kamijo, H., Matsumura, Y., Thumkeo, D., Koike, S., Masu, M., Shimizu, Y., et al. (2011). Impaired vascular remodeling in the yolk sac of embryos deficient in ROCK-I and ROCK-II. *Genes to Cells, 16*, 1012–1021. https://doi.org/10.1111/j.1365-2443.2011.01546.x.

Kono, K., Tamashiro, D. A. A., & Alarcon, V. B. (2014). Inhibition of RHO-ROCK signaling enhances ICM and suppresses TE characteristics through activation of Hippo signaling in the mouse blastocyst. *Developmental Biology, 394*, 142–155. https://doi.org/10.1016/j.ydbio.2014.06.023.

Korotkevich, E., Niwayama, R., Courtois, A., Friese, S., Berger, N., Buchholz, F., et al. (2017). The apical domain is required and sufficient for the first lineage segregation in the mouse embryo. *Developmental Cell, 40*, 235–247.e7. https://doi.org/10.1016/j.devcel.2017.01.006.

Larue, L., Ohsugi, M., Hirchenhain, J., & Kemler, R. (1994). E-cadherin null mutant embryos fail to form a trophectoderm epithelium. *Proceedings of the National Academy of Sciences of the United States of America, 91*, 8263–8267. https://doi.org/10.1073/pnas.91.17.8263.

Leung, C. Y., & Zernicka-Goetz, M. (2013). Angiomotin prevents pluripotent lineage differentiation in mouse embryos via Hippo pathway-dependent and -independent mechanisms. *Nature Communications, 4*, 1–11. https://doi.org/10.1038/ncomms3251.

Liu, K. C., Jacobs, D. T., Dunn, B. D., Fanning, A. S., & Cheney, R. E. (2012). Myosin-X functions in polarized epithelial cells. *Molecular Biology of the Cell, 23*, 1675–1687. https://doi.org/10.1091/mbc.E11-04-0358.

Maître, J.-L., Niwayama, R., Turlier, H., Nédélec, F., & Hiiragi, T. (2015). Pulsatile cell-autonomous contractility drives compaction in the mouse embryo. *Nature Cell Biology, 17*, 849–855. https://doi.org/10.1038/ncb3185.

Maître, J., Turlier, H., Illukkumbura, R., Eismann, B., Niwayama, R., Nédélec, F., et al. (2016). Asymmetric division of contractile domains couples cell positioning and fate specification. *Nature, 536*, 344–348. https://doi.org/10.1038/nature18958.

Matsui, T., Maeda, M., Doi, Y., Yonemura, S., Amano, M., Kaibuchi, K., et al. (1998). Rho-kinase phosphorylates COOH-terminal threonines of ezrin/radixin/moesin (ERM) proteins and regulates their head-to-tail association. *The Journal of Cell Biology, 140*, 647–657. https://doi.org/10.1083/jcb.140.3.647.

Mattila, P. K., & Lappalainen, P. (2008). Filopodia: Molecular architecture and cellular functions. *Nature Reviews. Molecular Cell Biology, 9*, 446–454. https://doi.org/10.1038/nrm2406.

McDole, K., Xiong, Y., Iglesias, P. A., & Zheng, Y. (2011). Lineage mapping the pre-implantation mouse embryo by two- photon microscopy, new insights into the segregation of cell fates. *Developmental Biology*, *355*, 239–249.
Namgoong, S., & Kim, N. (2016). Roles of actin-binding proteins in mammalian oocyte maturation and beyond. *Cell Cycle*, *4101*, 1830–1843. https://doi.org/10.1080/15384101.2016.1181239.
Nichols, J., Zevnik, B., Anastassiadis, K., Niwa, H., Klewe-Nebenius, D., Chambers, I., et al. (1998). Formation of pluripotent stem cells in the mammalian embryo depends on the POU transcription factor Oct4. *Cell*, *95*, 379–391.
Nishioka, N., Inoue, K., Adachi, K., Kiyonari, H., Ota, M., Ralston, A., et al. (2009). The Hippo signaling pathway components Lats and Yap pattern Tead4 activity to distinguish mouse trophectoderm from inner cell mass. *Developmental Cell*, *16*, 398–410. https://doi.org/10.1016/j.devcel.2009.02.003.
Nishioka, N., Yamamoto, S., Kiyonari, H., Sato, H., Sawada, A., Ota, M., et al. (2008). Tead4 is required for specification of trophectoderm in pre-implantation mouse embryos. *Mechanisms of Development*, *125*, 270–283. https://doi.org/10.1016/j.mod.2007.11.002.
Nishizuka, Y. (1984). The role of protein kinase C in cell surface signal transduction and tumour promotion. *Nature*, *308*, 693–698. https://doi.org/10.1038/308693a0.
Niwa, H., Toyooka, Y., Shimosato, D., Strumpf, D., Takahashi, K., Yagi, R., et al. (2005). Interaction between Oct3/4 and Cdx2 determines trophectoderm differentiation. *Cell*, *123*, 917–929. https://doi.org/10.1016/j.cell.2005.08.040.
Ntziachristos, V. (2010). Going deeper than microscopy: The optical imaging frontier in biology. *Nature Methods*, *7*, 603–614. https://doi.org/10.1038/nmeth.1483.
Ohsugi, M., Ohsawa, T., & Yamamura, H. (1993). Involvement of protein kinase C in nuclear migration during compaction and the mechanism of the migration: Analysis in two-cell mouse embryos. *Developmental Biology*, *156*, 146–154.
Pickering, S. J., Maro, B., Johnson, M. H., & Skepper, J. N. (1988). The influence of cell contact on the division of mouse 8-cell blastomeres. *Development*, *103*, 353–363.
Plusa, B., Frankenberg, S., Chalmers, A., Hadjantonakis, A.-K., Moore, C. A., Papalopulu, N., et al. (2005). Downregulation of Par3 and aPKC function directs cells towards the ICM in the preimplantation mouse embryo. *Journal of Cell Science*, *118*, 505–515. https://doi.org/10.1242/jcs.01666.
Ponuwei, G. A. (2016). A glimpse of the ERM proteins. *Journal of Biomedical Science*, *23*, 35. https://doi.org/10.1186/s12929-016-0246-3.
Posfai, E., Petropoulos, S., de Barros, F. R. O., Schell, J. P., Jurisica, I., Sandberg, R., et al. (2017). Position- and Hippo signaling-dependent plasticity during lineage segregation in the early mouse embryo. *eLife*, *6*, 1–24. https://doi.org/10.7554/eLife.22906.
Rodriguez-Boulan, E., & Macara, I. G. (2014). Organization and execution of the epithelial polarity programme. *Nature Reviews. Molecular Cell Biology*, *15*, 225–242. https://doi.org/10.1038/nrm3775.
Rossant, J., & Lis, W. T. (1979). Potential of isolated mouse inner cell masses to form trophectoderm derivatives in vivo. *Developmental Biology*, *70*, 255–261. https://doi.org/10.1016/0012-1606(79)90022-8.
Rossant, J., & Vijh, K. M. (1980). Ability of outside cells from preimplantation mouse embryos to form inner cell mass derivatives. Aggre- outside cells from late morulae outside cell aggregates in Zonae empty zonae were obtained by sucking. *Developmental Biology*, *482*, 475–482.
Samarage, C. R., White, M. D., Álvarez, Y. D., Fierro-González, J. C., Henon, Y., Jesudason, E. C., et al. (2015). Cortical tension allocates the first inner cells of the mammalian embryo. *Developmental Cell*, *34*, 435–447. https://doi.org/10.1016/j.devcel.2015.07.004.

Shirayoshi, Y., Okada, T. S., & Takeichi, M. (1983). The calcium-dependent cell-cell adhesion system regulates inner cell mass formation and cell surface polarization in early mouse development. *Cell, 35*, 631–638. https://doi.org/10.1016/0092-8674(83)90095-8.

Stephenson, R. O., Yamanaka, Y., & Rossant, J. (2010). Disorganized epithelial polarity and excess trophectoderm cell fate in preimplantation embryos lacking E-cadherin. *Development, 137*, 3383–3391. https://doi.org/10.1242/dev.050195.

Strumpf, D., Mao, C., Yamanaka, Y., Ralston, A., Chawengsaksophak, K., Beck, F., et al. (2005). Cdx2 is required for correct cell fate specification and differentiation of trophectoderm in the mouse blastocyst. *Development, 132*, 2093–2102. https://doi.org/10.1242/dev.01801.

Sutherland, A. E., Speed, T. P., & Calarco, P. G. (1990). Inner cell allocation in the mouse morula: The role of oriented division during fourth cleavage. *Developmental Biology, 137*, 13–25. https://doi.org/10.1016/0012-1606(90)90003-2.

Suwińska, A., Czołowska, R., Ozdzeński, W., & Tarkowski, A. K. (2008). Blastomeres of the mouse embryo lose totipotency after the fifth cleavage division: Expression of Cdx2 and Oct4 and developmental potential of inner and outer blastomeres of 16- and 32-cell embryos. *Developmental Biology, 322*, 133–144. https://doi.org/10.1016/j.ydbio.2008.07.019.

Suzuki, A., & Ohno, S. (2006). The PAR-aPKC system: Lessons in polarity. *Journal of Cell Science, 119*, 979–987. https://doi.org/10.1242/jcs.02898.

Szczepanska, K., Stanczuk, L., & Maleszewski, M. (2011). Isolated mouse inner cell mass is unable to reconstruct trophectoderm. *Differentiation, 82*, 1–8. https://doi.org/10.1016/j.diff.2011.04.001.

Szollosi, D., Calarco, P., & Donahue, R. P. (1972). Absence of centrioles in the first and second meiotic spindles of mouse oocytes. *Journal of Cell Science, 11*, 521–541.

Tarkowski, A. K. (1959). Experiments on the development of isolated blastomers of mouse eggs. *Nature, 184*, 1286–1287. https://doi.org/10.1038/1841286a0.

Tarkowski, A. K., Suwińska, A., Czolowska, R., & Ozdzezski, W. (2010). Individual blastomeres of 16- and 32-cell mouse embryos are able to develop into foetuses and mice. *Developmental Biology, 348*, 190–198. https://doi.org/10.1016/j.ydbio.2010.09.022.

Tarkowski, A. K., & Wróblewska, J. (1967). Development of blastomeres of mouse eggs isolated at the 4- and 8-cell stage. *Journal of Embryology and Experimental Morphology, 18*, 155–180.

Vinot, S., Le, T., Ohno, S., Pawson, T., Maro, B., & Louvet-Vallée, S. (2005). Asymmetric distribution of PAR proteins in the mouse embryo begins at the 8-cell stage during compaction. *Developmental Biology, 282*, 307–319. https://doi.org/10.1016/j.ydbio.2005.03.001.

Viswanatha, R., Bretscher, A., & Garbett, D. (2014). Dynamics of ezrin and EBP50 in regulating microvilli on the apical aspect of epithelial cells. *Biochemical Society Transactions, 42*, 189–194. https://doi.org/10.1042/BST20130263.

Watanabe, T., Biggins, J. S., Tannan, N. B., & Srinivas, S. (2014). Limited predictive value of blastomere angle of division in trophectoderm and inner cell mass specification. *Development, 141*, 2279–2288. https://doi.org/10.1242/dev.103267.

Wicklow, E., Blij, S., Frum, T., Hirate, Y., Lang, R. A., Sasaki, H., et al. (2014). HIPPO pathway members restrict SOX2 to the inner cell mass where it promotes ICM fates in the mouse blastocyst. *PLoS Genetics, 10*. https://doi.org/10.1371/journal.pgen.1004618.

Winkel, G. K., Ferguson, J. E., Takeichi, M., & Nuccitelli, R. (1990). Activation of protein kinase-C triggers premature compaction in the 4-cell stage mouse embryo. *Dev. Biol., 138*, 1–15. https://doi.org/10.1016/0012-1606(90)90171-e.

Yagi, R., Kohn, M. J., Karavanova, I., Kaneko, K. J., Vullhorst, D., DePamphilis, M. L., et al. (2007). Transcription factor TEAD4 specifies the trophectoderm lineage at the beginning of mammalian development. *Development*, *134*, 3827–3836. https://doi.org/10.1242/dev.010223.

Yamanaka, Y., Lanner, F., & Rossant, J. (2010). FGF signal-dependent segregation of primitive endoderm and epiblast in the mouse blastocyst. *Development*, *137*, 715–724. https://doi.org/10.1242/dev.043471.

Yamanaka, Y., Ralston, A., Stephenson, R. O., & Rossant, J. (2006). Cell and molecular regulation of the mouse blastocyst. *Developmental Dynamics*, *235*, 2301–2314. https://doi.org/10.1002/dvdy.20844.

Young, R. A. (2011). Control of the embryonic stem cell state. *Cell*, *144*, 940–954. https://doi.org/10.1016/j.cell.2011.01.032.

Zhang, N., Bai, H., David, K. K., Dong, J., Zheng, Y., Cai, J., et al. (2010). The Merlin/NF2 tumor suppressor functions through the YAP oncoprotein to regulate tissue homeostasis in mammals. *Developmental Cell*, *19*, 27–38. https://doi.org/10.1016/j.devcel.2010.06.015.

FURTHER READING

Hogan, B., & Tilly, R. (1978). In vitro development of inner cell masses isolated immunosurgically from mouse blastocysts. II. Inner cell masses from 3.5- to 4.0-day p.c. blastocysts. *Journal of Embryology and Experimental Morphology*, *45*, 107–121.

Kerber, M. L., & Cheney, R. E. (2011). Myosin-X: A MyTH-FERM myosin at the tips of filopodia. *Journal of Cell Science*, *124*, 3733–3741. https://doi.org/10.1242/jcs.023549.

Lorthongpanich, C., Messerschmidt, D. M., Chan, S. W., Hong, W., Knowles, B. B., & Solter, D. (2013). Temporal reduction of LATS kinases in the early preimplantation embryo prevents ICM lineage differentiation. *Genes & Development*, *27*, 1441–1446. https://doi.org/10.1101/gad.219618.113.

Morey, L., Santanach, A., & Di Croce, L. (2015). Pluripotency and epigenetic factors in mouse embryonic stem cell. *Molecular and Cellular Biology*, *35*, 2716–2728. https://doi.org/10.1128/MCB.00266-15.

Niwa, H., Miyazaki, J., & Smith, A. G. (2000). Quantitative expression of Oct-3/4 defines differentiation, dedifferentiation or self-renewal of ES cells. *Nature Genetics*, *24*, 2–6.

Ralston, A., Cox, B. J., Nishioka, N., Sasaki, H., Chea, E., Rugg-gunn, P., et al. (2010). Gata3 regulates trophoblast development downstream of Tead4 and in parallel to Cdx2. *Development*, *137*, 395–403. https://doi.org/10.1242/dev.038828.

Ralston, A., & Rossant, J. (2008). Cdx2 acts downstream of cell polarization to cell-autonomously promote trophectoderm fate in the early mouse embryo. *Development, Growth & Differentiation*, *313*, 614–629. https://doi.org/10.1016/j.ydbio.2007.10.054.

Spindle, A. I. (1978). Trophoblast regeneration by inner cell masses isolated from cultured mouse embryos. *The Journal of Experimental Zoology*, *203*, 483–489. https://doi.org/10.1002/jez.1402030315.

Sterneckert, J., Höing, S., & Schöler, H. R. (2012). Concise review: Oct4 and more: The reprogramming expressway. *Stem Cells*, *30*, 15–21. https://doi.org/10.1002/stem.765.

CHAPTER THREE

Cell Fate Decisions During Preimplantation Mammalian Development

Stephanie Bissiere*, Maxime Gasnier*, Yanina D. Alvarez*,†, Nicolas Plachta*,‡,1

*Institute of Molecular and Cell Biology, A*STAR, Singapore, Singapore
†Facultad de Ciencias Exactas y Naturales, Universidad de Buenos Aires, Conicet, Buenos Aires, Argentina
‡National University of Singapore, Singapore, Singapore
[1]Corresponding author: e-mail address: plachtan@imcb.a-star.edu.sg

Contents

Abstract

The early mouse embryo offers a phenomenal system to dissect how changes in the mechanisms controlling cell fate are integrated with morphogenetic events at the single-cell level. New technologies based on live imaging have enabled the discovery of dynamic changes in the regulation of single genes, transcription factors, and epigenetic mechanisms directing early cell fate decision in the early embryo. Here, we review recent progress in linking molecular dynamic events occurring at the level of the single cell in vivo, to some of the key morphogenetic changes regulating early mouse development.

Current Topics in Developmental Biology, Volume 128
ISSN 0070-2153
https://doi.org/10.1016/bs.ctdb.2017.11.001

1. INTRODUCTION

Understanding how a single cell produces daughter cells that will go on to adopt distinct fates and morphologies after a few divisions remains one of the most fascinating open challenges in developmental biology. The mouse embryo offers an excellent system to study how these early fate decisions are made, as it is highly accessible for ex utero culture, genetic manipulation, and live imaging. Between fertilization and implantation, the mammalian embryo undergoes compaction, generates three separate lineages, and grows about three times its size.

It is widely established that changes in gene expression control cell fate decisions. Yet, gene expression is regulated at multiple levels, including, for example, various epigenetic modifications as well as dynamic binding of multiple gene-regulatory proteins to DNA (Hager, McNally, & Misteli, 2009; Mueller, Stasevich, Mazza, & McNally, 2013). Studying these molecular dynamic events in the intact embryo remains challenging, and therefore in past years there has been considerably more research in cell culture systems such as embryonic stem (ES) cells. With the advancement of new imaging technologies and the rise of system biology approaches in vivo, we have recently obtained new insights into how dynamic changes in the organization and function of the cell nucleus as well as in gene regulation control cell fate decision in the embryo.

2. GENE KNOCKOUTS PAVE THE WAY TO UNDERSTAND CELL FATE DECISIONS

The mouse embryo offers an excellent model system for genetics. Pioneering work using transgenic approaches and the subsequent development of ES cell technologies enabled the production of thousands of mouse models carrying a vast range of genetic modifications (Evans & Kaufman, 1981; Martin, 1981). The finding of mutations causing early embryonic arrest or lethality has led to the identification of multiple gene-regulatory proteins essential for cell fate during preimplantation development (Rossant, 2004). Many of these genes have turned out to encode transcription factors (TFs).

Some of the most widely known examples include the TFs required for the generation or maintenance of the pluripotent cell lineage, such as Oct4, Sox2, and Nanog (Avilion et al., 2003; Nichols et al., 1998;

Scholer, Dressler, Balling, Rohdewohld, & Gruss, 1990). Mutation of these genes produces embryos with disrupted inner cells number or specification. Other TFs required for the development of the extraembryonic lineage, which later contributes to the placenta include Cdx2, TEAD4, and EOMES (Strumpf et al., 2005; Yamanaka, Ralston, Stephenson, & Rossant, 2006). Many of these TFs have been more extensively studied at the functional level in stem cell assays using ES cells as well as induced pluripotent stem cells (iPS). These cell types offer a robust in vitro system to test the role of these proteins in controlling pluripotency and have enable for instance the characterization of interactions between Oct4 and Cdx2 for the segregation of the trophectoderm and pluripotent lineages (Niwa et al., 2005). The subsequent establishment of trophectoderm stem cell lines has then enabled studies of TFs, which were important for extra-embryonic lineage development at the biochemical level (Tanaka, Kunath, Hadjantonakis, Nagy, & Rossant, 1998).

3. DIFFERENT PATTERNS OF TF EXPRESSION LEVELS

Histological analyses using in situ hybridization or immunofluorescence techniques have represented one of the most straightforward ways to infer in which cell types a particular TF has important functions for development. For example, Cdx2 starts to be expressed at high levels in outer cells of the mouse embryo just prior to blastocyst formation and is detected at significantly lower levels in the inner cells of the embryo which will form the pluripotent inner mass (Strumpf et al., 2005) (Fig. 1). Other TFs like Nanog display a more salt-and-pepper expression pattern (Chambers et al., 2003; Mitsui et al., 2003) (Fig. 1).

Oct4 and Sox2 are particularly important for the pluripotent inner cells of the embryo (Boyer et al., 2005) and have well-established roles in maintaining pluripotency in ES cell and iPS cell systems (Nichols & Smith, 2012; Takahashi & Yamanaka, 2016). However, these proteins appear to be expressed at similarly high levels by all cells in the embryo, and they only start to become restricted to the inner cells by blastocyst stages (i.e., after the 32-cell stage) (Fig. 1) (Avilion et al., 2003; White, Angiolini, et al., 2016; White, Bissiere, Alvarez, & Plachta, 2016; Wicklow et al., 2014). Furthermore, their expressions can be detectable as early as the two- to four-cell stage of preimplantation development (Avilion et al., 2003; White, Angiolini, et al., 2016; White, Bissiere, et al., 2016.

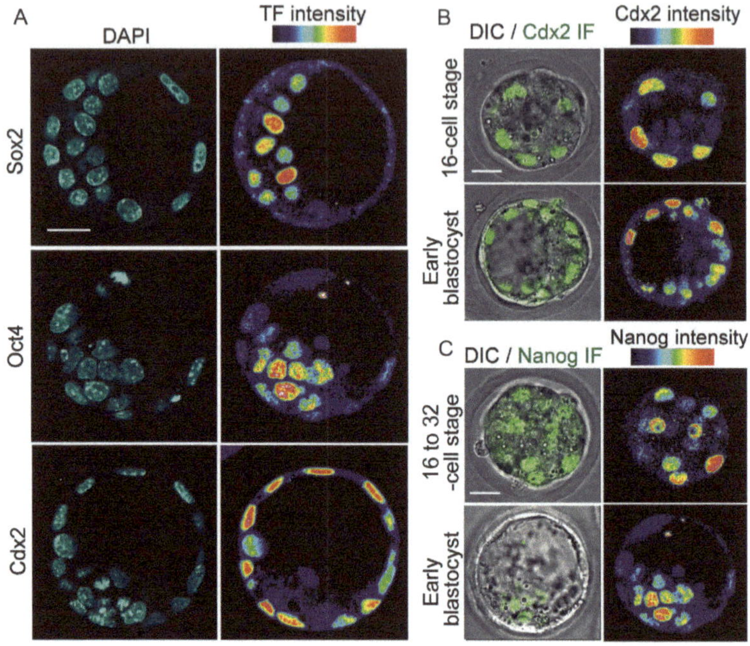

Fig. 1 TFs display different patterns of expression in the embryo. (A) Images show immunolabeling for the TFs Sox2, Oct4, and Cdx2 at the blastocysts stage. These TFs display a lineage-specific localization. Sox2 and Oct4 localize to the inner cell mass, whereas Cdx2 labels the cells of the outer trophectoderm layer. (B) The onset of Cdx2 expression occurs at the ~16-cell stage. Its expression is restricted to the outer cells of the embryo from this stage onward. (C) Unlike Cdx2, Nanog displays a more salt-and-pepper expression pattern at the 16-cell stage, before becoming restricted to the inner cell mass of the blastocyst. Images show confocal sections. DIC, differential interference contrast; DAPI (4′,6-diamidino-2-phenylindole) stains the DNA within cell nuclei. Scale bar represents 20 μm.

It is therefore more difficult to dissect the different functions of these TFs in the early embryo. One possibility is that the TFs are regulated by multiple cofactors and translational modification in different cell lineages. Some of these mechanisms have been explored using biochemical assays in culture stem cells (Yeap, Hayashi, & Surani, 2009; Yuan et al., 2009). Furthermore, some TFs such as Oct4 and Cdx2 have mutually antagonistic effects on trophectoderm development (Niwa et al., 2005). Another important level of regulation, which has remained more challenging to study, is the dynamic changes in the biophysical properties of these TFs within the cell nucleus (Hager et al., 2009).

4. GENE EXPRESSION DYNAMICS

Pioneering work with the lac operon system in bacteria showed that small differences in the concentration of DNA-binding proteins rapidly change gene expression (Jacob & Monod, 1961). However, subsequent work in eukaryotic cells from the field of systems biology has revealed an additional important level of gene regulation, which is dictated by relative changes in the concentration or quantity of gene-regulatory proteins, as well as in the spatiotemporal changes in the distribution of these proteins within the cell (Zhao, White, Bissiere, Levi, & Plachta, 2016).

For instance, it has been reported that gene transcription is not linear, but can occur in bursts. Transcriptional bursts result from stochastic fluctuations in the on-and-off state of gene promoters and can lead to the synthesis of mRNAs in bursts (Bahar Halpern et al., 2015; Bartman, Hsu, Hsiung, Raj, & Blobel, 2016; Dar et al., 2012; Larson et al., 2013; Molina et al., 2013; Raj, Peskin, Tranchina, Vargas, & Tyagi, 2006; Raj & van Oudenaarden, 2008; Senecal et al., 2014; Singer et al., 2014; Singh, Razooky, Cox, Simpson, & Weinberger, 2010; Suter et al., 2011). Bursty transcription itself is a complex mode of gene regulation that has been recently shown to be controlled at multiple levels. For example, cells can differentially control (1) the quantity of mRNAs produced in each burst, (2) the on/off rate of mRNA production itself, or (3) the refractory period between bursts and the external stimuli for transcription (Dar et al., 2012; Larson et al., 2013; Molina et al., 2013; Suter et al., 2011).

Yet despite these exciting findings, little is known about bursty transcription mechanisms operating during early mammalian development. The main consequence of bursty transcription is the establishment of noise-based mechanisms that could drive cell fate decisions (Eldar & Elowitz, 2010). One of the most interesting examples that such mechanisms may operate during normal mammalian development comes from work on ES cells, demonstrating the bursty transcription of the TF Nanog during differentiation (Abranches et al., 2014).

5. TF INTRACELLULAR TRANSPORT DYNAMICS

An additional level of gene regulation involves the more dynamics changes in the distribution of a TF within the cell. Like most

DNA-regulatory proteins, TFs are highly mobile molecules. Although traditionally thought to reside mostly inside the cell nucleus and to bind DNA for relatively long-time periods (in the order of minutes to hours), an increasing number of studies in various cellular systems have revealed a more dynamic picture (Mueller et al., 2013). Importantly, these studies have demonstrated that the biological functions of various TFs can be controlled by rapid changes in nuclear export and nuclear import rates, as well as the rates and frequency of DNA binding.

Work in immortalized cell lines has demonstrated that the dynamics of nuclear export–import of the TF nuclear factor kappa-light-chain-enhancer of activated B cells (NF-KB) control its target genes (Ashall et al., 2009; Nelson et al., 2004). Furthermore, it has been shown that its transport behavior can be regulated at different levels, including the number of nucleocytoplasmic translocation events and their amplitude (Ashall et al., 2009; Tay et al., 2010; Turner et al., 2010). Similar dynamic behaviors for other TFs have been found in other systems, such as, for example, the tumor suppressor p53 in cancer cell lines (Lahav et al., 2004).

6. TF NUCLEAR–CYTOPLASMIC SHUTTLING IN EMBRYOS

The transport dynamics of the TF Oct4 have been demonstrated in live mouse embryos (Plachta, Bollenbach, Pease, Fraser, & Pantazis, 2011). For this study, Oct4 was fused to the photoactivatable form of GFP, termed paGFP (Patterson & Lippincott-Schwartz, 2002), and a femtosecond laser was used to selectively photoactivate a defined fraction of Oct4-paGFP proteins within single nuclei deep inside live mouse embryos. The advantage of using paGFP is that enables photoactivation within selective volumes (Pantazis & Gonzalez-Gaitan, 2007). The photoactivated TFs can then be followed using time-lapse microscopy and by monitoring changes in overall fluorescence within the cell nucleus and cytoplasm the rates of nuclear transport, and the overall nuclear immobile fraction can be determined (Fig. 2).

This study identified differences in the transport and overall immobile fractions of Oct4 in the cells of the embryo at the eight-cell stage. Because these experiments can be performed in live embryos without perturbing normal development, the fate of the cells could be tracked down, and the cells displaying more stable Oct4-paGFP dynamics were found to produce more progeny to the inner mass of the embryo by the 16-cell stage.

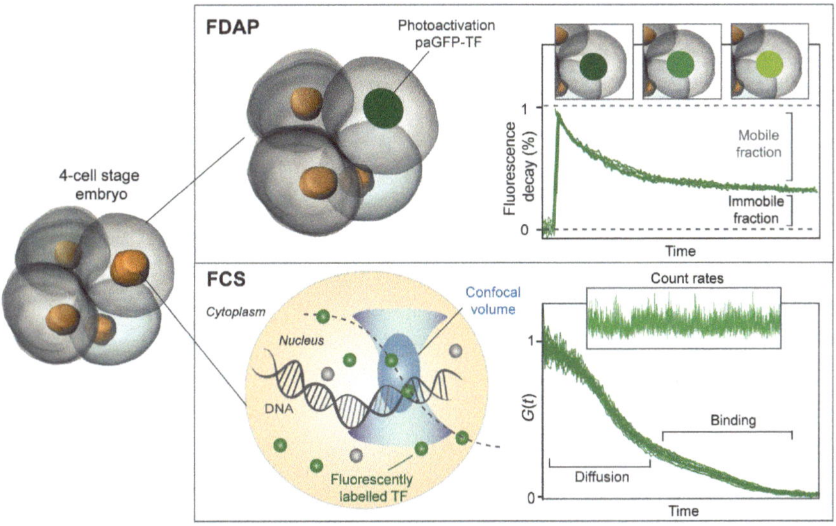

Fig. 2 Imaging technologies enable tracking of TF dynamics in the living mouse embryo. Examples of recently developed technologies to study TF dynamics in vivo at the single-cell level. (*Upper panel*) In FDAP, the cell nucleus of an embryo expressing a TF fused to paGFP is selectively photoactivated. The intensity of fluorescence within the cell nucleus is then monitored using time lapse imaging. By measuring the mean fluorescence decay over time, the rates of nuclear export–import and the nuclear immobile fraction can be obtained. (*Lower panel*) paFCS enables measurements of TF–DNA-binding dynamics. In this method, only a selective pool of TFs is photoactivated using a laser light different from that used for imaging the TFs. By controlling the level of fluorescently labeled species, the signal-to-noise ratio can be optimized. Fluorescent fluctuations are then measured when the TFs diffuse and bind to DNA within a small confocal volume inside the cell nucleus. These fluctuations are then autocorrelated to obtain key parameters describing the behavior of the TFs when interact with the DNA, including, for example, the mean binding times and fraction of molecules involved in the binding events with the DNA.

These experiments suggested that the cells of the embryo might have differences in the accessibility of target genes before the physical segregation of inner and outer cells.

7. CELL-TO-CELL VARIABILITY IN THE EARLY EMBRYO

Unlike other nonmammalian species, the early mouse embryo lacks clear morphological asymmetric indicative of future cell fates. Therefore, it remains difficult to predict which cell will contribute more progeny to the inner and outer cell lineages of the embryo, as these lineages become physically segregated starting at the 8- to 16-cell stage (Rossant & Tam, 2009; Yamanaka et al., 2006; Zernicka-Goetz, Morris, & Bruce, 2009). Some studies proposed that cell fates are determined after the eight-cell stage of development. These were mostly based on analyses of TF expression (Dietrich & Hiiragi, 2007) or gene expression patterns in fixed embryos (Guo et al., 2010), or cell tracking in live embryos (Kurotaki, Hatta, Nakao, Nabeshima, & Fujimori, 2007; Louvet-Vallee, Vinot, & Maro, 2005; Strnad et al., 2015). By contrast, others found that cellular heterogeneities appearing before the eight-cell stage can predict cell fates. For example, early studies identified differences in the orientation of cleavage patterns that predict cell fates in some blastomeres, as early as the four-cell stage of development (Bischoff, Parfitt, & Zernicka-Goetz, 2008; Gardner, 2002; Morris, Guo, & Zernicka-Goetz, 2012; Piotrowska-Nitsche, Perea-Gomez, Haraguchi, & Zernicka-Goetz, 2005; Piotrowska-Nitsche & Zernicka-Goetz, 2005). Cell-to-cell variability has also been found using the Rainbow mouse system (Tabansky et al., 2013). For this study, the cells of the embryo were fluorescently labeled using a combinatorial, recombination-mediated technique, which revealed lineage-dependent bias in early development.

Using RNA-seq, studies have also revealed heterogeneities in gene expression at the two- and four-cell stage of development (Biase, Cao, & Zhong, 2014; Shi et al., 2015). The use of single-cell transcriptomics offers powerful new ways to interrogate cell-to-cell variability and make predictions about the functional relevance of some differentially regulated genes (Tang et al., 2010; Wang, Gerstein, & Snyder, 2009). A recent study extensively analyzed single-cell gene expression profiles using this approach and found the TF Sox21 to be differentially regulated in four-cell blastomeres (Goolam et al., 2016). Interfering with Sox21 expression promoted the acquisition of extraembryonic fates, at the expense of pluripotent inner fates, likely as a consequence of changes in the balance of asymmetric cell division and internalization events, or upregulation of trophectoderm TFs like Cdx2. Importantly, the differences in Sox21 expression were associated with epigenetic heterogeneities at the level of histone methylation (Goolam et al., 2016).

Previous studies examining differences in methylation profiles have also shown differences in the levels of histone H3 arginine 26 methylation (H3R26me2) in the four-cell embryo (Torres-Padilla, Parfitt, Kouzarides, & Zernicka-Goetz, 2007). This epigenetic modification is mediated by the enzyme coactivator-associated arginine methyltransferase 1 (CARM1)

(White, Angiolini, et al., 2016; White, Bissiere, et al., 2016). Furthermore, disruption of CARM1 function and reduction in H3R26me methylation affected both, Sox21 expression and cell-fate allocation (Goolam et al., 2016).

Overall, these studies suggest that during early developmental stages, the fate of single blastomeres starts to be biased toward an inner or outer lineage of the embryo early on. It would be important to continue to examine to which extent early developmental heterogeneities, like those described here, relate to cell-fate lineage trees using live imaging experiments. Nevertheless, a general picture emerges in which cells begin to develop epigenetic differences and distinct gene expression patterns. The different epigenetic modifications could act by changing the accessibility of TF binding to their distinct specific binding sites on the genome, yet it had remained a challenge to probe how TFs search and bind to DNA in real time and in vivo.

8. TF–DNA BINDING DYNAMICS IN THE EMBRYO

TF–DNA binding has traditionally been probed with biochemical methods, such as immunoprecipitation and chromatin immunoprecipitation (Chip) analysis in cell homogenates. These techniques, however, cannot reveal cell-to-cell variability, as they only provide results from large cell populations. Therefore, it was traditionally inferred that TFs undergo stable interactions with DNA in the order of minutes to hours (Perlmann, Eriksson, & Wrange, 1990).

Using live-cell imaging, a number of studies have revealed a much more dynamic picture (Gorman & Greene, 2008; Hager et al., 2009; Halford & Marko, 2004; Mueller et al., 2013). Using single-molecule tracking methods in cultured cells, studies demonstrated that TF–DNA interactions cannot be interpreted as a homogeneous population of molecules binding with the same dynamics to DNA. Instead, TF–DNA binding has been estimated to range from milliseconds to seconds (Chen et al., 2014; Gebhardt et al., 2013; Normanno et al., 2015).

Importantly, current models of TF–DNA binding propose that the shorter-lived DNA-binding events occur at nonconsensus "off-target" sites. Such rapid events likely result from a combination of one-dimensional (1D)-sliding of the TF along the DNA, and small TF hops and intersegmental transfers between DNA segments (Chen et al., 2014; Gorman & Greene, 2008; Mueller et al., 2013; von Hippel & Berg, 1989). The key advantage

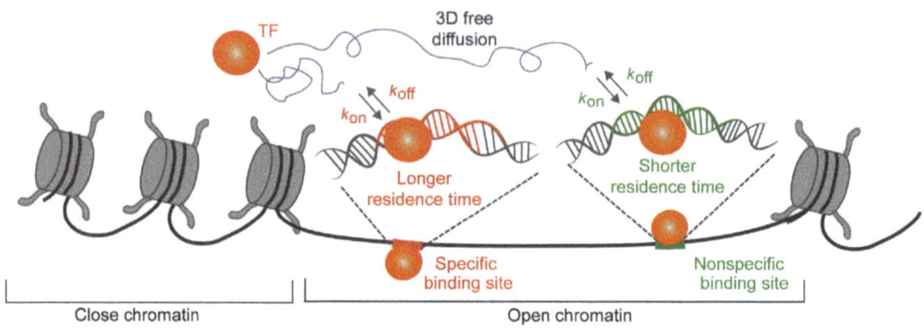

Fig. 3 Current model of TF–DNA interactions. TFs employ three-dimensional (3D) diffusion to search for DNA within the cell nucleus. Upon binding to the DNA, the TF can engage in two different types of binding interactions. Binding to sites for which the TF has low affinity (nonspecific site) results in more transiently lived interaction. Typically these sites enable the rapid scanning of the DNA via 1D diffusion or sliding. When the TF encounters a site for which it has high affinity (specific site), the binding event is more long-lived or stable. The accessibility of the TF to the DNA is regulated by the local organization of the chromatin, which is in turn regulated via multiple epigenetic mechanisms. The terms k_{on} and k_{off} indicate the rates of DNA binding and unbinding.

of such TF dynamics is that they enable the protein to scan the DNA more rapidly than using only Brownian 3D diffusion. In contrast to the short-lived binding events, longer-lived TF–DNA interactions are said to occur at DNA sites for which the protein has higher affinity. These interactions are therefore likely to represent the specific binding events, which ultimately promote gene transcription (Hager et al., 2009; Mueller et al., 2013) (Fig. 3).

Single-molecule tracking methods have recently started to work very efficiently to analyze TF–DNA binding interactions in cultured cells, yet the application of these methods in vivo has remained a challenge, as the signal-to-noise resolution becomes unsatisfactory when applying these methods in embryos or tissues. Recent studies have instead used fluorescence correlation spectroscopy (FCS) techniques to test this model of TF–DNA binding in living mouse embryos. FCS probes molecular motions by measuring intensity in fluorescence fluctuations produced when fluorescent molecules move in and out of a small observation volume (Digman & Gratton, 2011; Machan & Wohland, 2014). An autocorrelation method is then used to analyze these fluctuation traces and to provide a quantitative description of the molecular dynamics (Fig. 2). Although FCS has been extensively used to obtain precise measurements of molecular dynamics in solution, an important problem is that the technique fails to provide satisfactory signal to noise when the concentration of fluorescent molecules studied is high. To circumvent this problem, FCS was recently combined with photoactivation. This approach enables control over the concentration of fluorescent species by photoactivation of a small selective fraction of TFs fuse to paGFP (Kaur et al., 2013). In combination with mathematical modeling technique (Angiolini, Plachta, Mocskos, & Levi, 2015), the binding dynamics of Oct4, Sox2, and Cdx2 were probed in the preimplantation embryo (Zhao et al., 2017).

In blastocysts, the TFs Oct4 and Sox2 bind DNA more stably in pluripotent than in extraembryonic cells (White, Angiolini, et al., 2016; White, Bissiere, et al., 2016). These results demonstrate the dynamic repartitioning of TFs between DNA sites driven by physiological epigenetic changes. However, in the four-cell embryo, Sox2 was found to engage in more long-lived interactions with the DNA than Oct4. Importantly, Sox2 long-lived binding varies between cells at the four-cell stage. Moreover, the binding dynamics are regulated by H3R26 methylation. Because FCS is a noninvasive technique, the cells displaying more Sox2 binding to DNA could be tracked using time-lapse imaging and were found to contribute more progeny to the inner mass of the 16-cell stage embryo (White, Angiolini, et al., 2016; White, Bissiere, et al., 2016).

Together with the studies showing differences in gene expression, these findings suggest that during early development, the cells of the embryo establish differences in the accessibility of some TFs to their DNA targets, which are in turn controlled by differential epigenetic regulation at the level of histone modifications. A key open question for the future is to understand how the first epigenetic differences arise in the embryo, and to which extent they may rely on yet unidentified deterministic mechanisms, or the action of noise-based processes, as those found in cell fate-decision making in simpler experimental systems (Eldar & Elowitz, 2010).

9. EPIGENETIC REGULATION

The extent to which the entire epigenetic state is regulated remains a fascinating open question due to the large-scale reorganization of the epigenome during early embryonic life. Following fertilization, the mouse and human embryo undergoes very rapid changes in epigenetic regulation. One of the most prominent processes includes the extensive DNA demethylation occurring immediately after fertilization, which is differentially regulated between the maternal and paternal genomes (Burton & Torres-Padilla, 2014; Smallwood et al., 2011). This is followed by a complex pattern of genome methylation that has been studied until the blastocysts stage

(Beaujean et al., 2004; Mayer, Niveleau, Walter, Fundele, & Haaf, 2000). Furthermore, the maternal and paternal genomes also differ in the pattern of associated histones (Brykczynska et al., 2010; Daujat et al., 2009; Lepikhov & Walter, 2004; Santenard et al., 2010), as well as on the presence of repressive markers such as the Polycomb repressive complex 2 (PRC2) (Puschendorf et al., 2008; Santos, Peters, Otte, Reik, & Dean, 2005). Most of these complex changes in histone modifications enable the substitution of the parental epigenetic state and the removal of protamines from the paternal genome, for a new set of epigenetic modifications (Burton & Torres-Padilla, 2014). A key change in epigenetic regulation of the early embryo is the loss of maternal heterochromatin features, including, for example, the removal of histone H4 lysine 20 methylation H4K20me3 and histone H3 lysine 64 methylation H3K64me3 (Daujat et al., 2009; Oda et al., 2009). These changes are thought to promote an open chromatic state, which is more permissive for gene transcription.

Furthermore, early cell-to-cell variability has also been associated with the differential expression of protein enzymes responsible for epigenetic regulation. For example, four-cell blastomeres display heterogeneous expression of the protein CARM1, which mediates H3R26me methylation (Torres-Padilla et al., 2007) and controls the binding of Sox2 to the DNA (White, Angiolini, et al., 2016; White, Bissiere, et al., 2016). Other proteins such as PR Domain Zinc Finger Protein 14 (PRDM14) have also been added to the list of differentially expressed epigenetic regulators (Burton et al., 2013). Finally, as the first cells of the embryo become segregated into their inner and outer position, they also display important differences in DNA and histone methylation.

10. ADDITIONAL LEVELS OF GENE REGULATION

Some of the mechanisms outlined earlier likely represent only a fraction of the multiple processes controlling gene expression and cell fate in vivo. Although we have recently learned new insights about DNA and histone modifications, and the role of protein dynamics, far less is known about the regulation of RNA molecules and many of their diverse functions in vivo. Other levels of regulation will be based on the oligomerization dynamics of TFs. It would be important to push the applications of new FCS-based techniques in vivo, similar to those used recently to reveal how the dynamics of p53 oligomerization control cell fates in cell culture systems (Gaglia, Guan, Shah, & Lahav, 2013), or how other gene-regulatory proteins like signal transducer and activator of transcription 3 (STAT3) regulate their nucleocytoplasmic transport (Hinde et al., 2016).

Recent advances in techniques enabling highly multiplexed imaging of proteins, DNA, and RNA will also provide exciting new ways to analyze early mammalian development (Chen, Boettiger, Moffitt, Wang, & Zhuang, 2015; Coskun & Cai, 2016; Jungmann et al., 2014). It would also be exciting to determine if TFs, or other nuclear proteins, are regulated at the level of nucleocytoplasmic oscillations. Finally, it would be interesting to examine the function of the dynamic changes in the distribution and movement of RNA polymerases in vivo (Cho et al., 2016; Cisse et al., 2013; Ghamari et al., 2013; Zhao et al., 2014).

Lastly, it would also be important to further explore the role of differential allelic expression of some genes. While Nanog has been shown to be expressed from a single allele, Oct4 is more evenly expressed in the early embryo during the two- to eight-cell stage (Deng, Ramskold, Reinius, & Sandberg, 2014; Miyanari & Torres-Padilla, 2012). Future work should elucidate the functional implications of monoallelic vs biallelic regulation of gene expression as the embryo develops.

11. HOW GENE REGULATION IMPACTS ON CELL SHAPE AND POSITION

One of the most important open questions is how early changes in gene expression actually control the main morphogenetic events occurring during early mammalian development (Fig. 4). The early mouse embryo offers an excellent system to address this question, as it enables the monitoring of very precise changes in shape and position in real time. The first key change in cell morphology occurs during embryo compaction. This process starts at the eight-cell stage, as the cells flatten their external surfaces against each other to adopt a more compacted morphology (White, Angiolini, et al., 2016; White, Bissiere, et al., 2016). Some of the main changes required for compaction are the increase in cell–cell adhesion, typically mediated by the action of E-cadherin (Fierro-Gonzalez, White, Silva, & Plachta, 2013; Larue, Ohsugi, Hirchenhain, & Kemler, 1994; Stephenson, Yamanaka, & Rossant, 2010), as well as changes in cortex tension (Maitre, Niwayama, Turlier, Nedelec, & Hiiragi, 2015). In addition, it was also discovered that specifically during the compaction process, some cells in the embryo

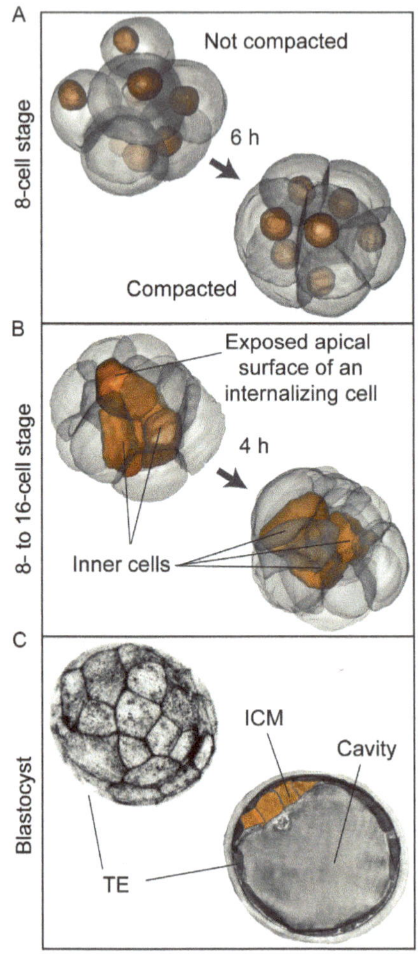

Fig. 4 Main morphogenetic events occurring during preimplantation development. (A) Embryo compaction is the key event starting at the eight-cell stage. During compaction, the cells flatten their membranes against each other and the embryo transforms into a more tightly packed structure. (B) During the 8- to 16-cell stage, the inner mass starts to form. Cells are internalized, as a direct result of the orientation of cell division cleavage plane, or cells adopt the internal position after dividing along the surface of the embryo and undergoing a subsequent apical constriction event. At the 16-cell stage, embryos have three cells inside that will form the ICM. The images show an embryo at earlier and later stages of development, highlighting the completion of internalization of three cells (*brown colored*). (C) Finally, at blastocysts stage, the first internal cavity of the embryo forms. The cell cortex is mechanically coupled to the cell cortex of neighboring cells for the establishment of an epithelial barrier to seal the embryo. The sealing of the outer layer of cells is required for the hydrostatic pressure to expand the internal cavity of the embryo.

extend long filopodia protrusions which are used to pull their neighboring cells closer and stretch part of their apical membrane (Fierro-Gonzalez et al., 2013). Yet, the mechanisms linking such changes in cell shape to the cell nucleus and gene regulation still remain largely unknown.

Following compaction, some of the cells in the embryo start to become internalized to form the pluripotent inner mass (Fig. 4). The internalization events start after the division of eight-cell stage blastomeres. Although some of these divisions can be spatially oriented in a way that favor the internalization of one of the daughter cells, as a direct result of the scission, recent studies showed that many cells reach the internal position after dividing along the surface of the embryo and undergoing a subsequent apical constriction event (Anani, Bhat, Honma-Yamanaka, Krawchuk, & Yamanaka, 2014; McDole, Xiong, Iglesias, & Zheng, 2011; Plusa et al., 2005; Samarage et al., 2015).

These internalization events are likely to interact with the Hippo signaling pathway, which has been associated with cellular proliferation, organ size, and cell fate decision-making in various species (Cockburn, Biechele, Garner, & Rossant, 2013; Hirate et al., 2013; Leung & Zernicka-Goetz, 2013; Nishioka et al., 2009). When the Hippo signaling pathway is inactive, the Yes-associated protein 1 (Yap1) transcriptional coactivator promotes trophectoderm fates in the embryo. But when Hippo signaling is activated, Yap1 moves from the cell nucleus to cytoplasm, promoting pluripotent cell fates (Cockburn et al., 2013; Leung & Zernicka-Goetz, 2013; Nishioka et al., 2009). Importantly, it was also reported that the Hippo signaling pathway can be regulated by cell–cell interactions and can actively respond to changes in the rigidity of the extracellular matrix as cells change their geometry and polarity (Cockburn et al., 2013; Leung & Zernicka-Goetz, 2013; Nishioka et al., 2009). However, the precise mechanism by which Yap1 changes in response to cell–cell adhesion or cortex tension as cells gain their fate in the early preimplantation embryo remains unclear. It will therefore be important to determine how mechanical forces affect the subcellular dynamics of Yap1 in vivo. It is also plausible that Hippo signaling may be triggered by the establishment of cell polarity prior to sensing changes in cell position (Anani et al., 2014; Hirate et al., 2015, 2013). Furthermore, it would also be important to reveal how early changes at the level of the cell nucleus affect the establishment of the first forms of cell polarity, as studies have shown that deregulation of Par3 and aPKC disrupts the number of inner–outer cells (Plusa et al., 2005). It would also be interesting to dissect the role of asymmetrically localized mRNAs, such as that of Cdx2, shown to influence the first lineage segregation events (Jedrusik et al., 2008).

Finally, at the 32-cell stage, the embryo starts to form the blastocyst. The transition from morula to blastocyst is another complex morphogenetic event that requires the sealing of intercellular spaces in order to create the first internal cavity of the embryo, called the blastocoel (Marikawa & Alarcon, 2012). As it remains challenging to image embryos at these advanced stages, little is known about changes in gene expression dynamics or the gene-regulatory basis for blastocyst formation. It will be important to study how genes implicated in inner vs outer cell fate affect the regulation of genes required for adherens and tight junction formation, cortex tension, and the formation of the molecular pump machinery used to expand the internal cavity, via changes in osmotic pressure.

ACKNOWLEDGMENTS

This work was supported by grants from the Agency for Science, Technology and Research (A*STAR), European Molecular Biology Organization (EMBO), and Howard Hughes Medical Institute (HHMI)-Wellcome Trust to N.P.

REFERENCES

Abranches, E., Guedes, A. M. V., Moravec, M., Maamar, H., Svoboda, P., Raj, A., et al. (2014). Stochastic NANOG fluctuations allow mouse embryonic stem cells to explore pluripotency. *Development*, *141*(14), 2770–2779.

Anani, S., Bhat, S., Honma-Yamanaka, N., Krawchuk, D., & Yamanaka, Y. (2014). Initiation of Hippo signaling is linked to polarity rather than to cell position in the pre-implantation mouse embryo. *Development*, *141*(14), 2813–2824.

Angiolini, J., Plachta, N., Mocskos, E., & Levi, V. (2015). Exploring the dynamics of cell processes through simulations of fluorescence microscopy experiments. *Biophysical Journal*, *108*(11), 2613–2618.

Ashall, L., Horton, C. A., Nelson, D. E., Paszek, P., Harper, C. V., Sillitoe, K., et al. (2009). Pulsatile stimulation determines timing and specificity of NF-kappaB-dependent transcription. *Science*, *324*(5924), 242–246.

Avilion, A. A., Nicolis, S. K., Pevny, L. H., Perez, L., Vivian, N., & Lovell-Badge, R. (2003). Multipotent cell lineages in early mouse development depend on SOX2 function. *Genes & Development*, *17*(1), 126–140.

Bahar Halpern, K., Tanami, S., Landen, S., Chapal, M., Szlak, L., Hutzler, A., et al. (2015). Bursty gene expression in the intact mammalian liver. *Molecular Cell*, *58*(1), 147–156.

Bartman, C. R., Hsu, S. C., Hsiung, C. C., Raj, A., & Blobel, G. A. (2016). Enhancer regulation of transcriptional bursting parameters revealed by forced chromatin looping. *Molecular Cell*, *62*(2), 237–247.

Beaujean, N., Hartshorne, G., Cavilla, J., Taylor, J., Gardner, J., Wilmut, I., et al. (2004). Non-conservation of mammalian preimplantation methylation dynamics. *Current Biology*, *14*(7), 266–267.

Biase, F. H., Cao, X., & Zhong, S. (2014). Cell fate inclination within 2-cell and 4-cell mouse embryos revealed by single-cell RNA sequencing. *Genome Research*, *24*(11), 1787–1796.

Bischoff, M., Parfitt, D. E., & Zernicka-Goetz, M. (2008). Formation of the embryonic-abembryonic axis of the mouse blastocyst: Relationships between orientation of early cleavage divisions and pattern of symmetric/asymmetric divisions. *Development, 135*(5), 953–962.

Boyer, L. A., Lee, T. I., Cole, M. F., Johnstone, S. E., Levine, S. S., Zucker, J. P., et al. (2005). Core transcriptional regulatory circuitry in human embryonic stem cells. *Cell, 122*(6), 947–956.

Brykczynska, U., Hisano, M., Erkek, S., Ramos, L., Oakeley, E. J., Roloff, T. C., et al. (2010). Repressive and active histone methylation mark distinct promoters in human and mouse spermatozoa. *Nature Structural and Molecular Biology, 17*(6), 679–687.

Burton, A., Muller, J., Tu, S., Padilla-Longoria, P., Guccione, E., & Torres-Padilla, M. E. (2013). Single-cell profiling of epigenetic modifiers identifies PRDM14 as an inducer of cell fate in the mammalian embryo. *Cell Reports, 5*(3), 687–701.

Burton, A., & Torres-Padilla, M. E. (2014). Chromatin dynamics in the regulation of cell fate allocation during early embryogenesis. *Nature Reviews. Molecular Cell Biology, 15*(11), 723–734.

Chambers, I., Colby, D., Robertson, M., Nichols, J., Lee, S., Tweedie, S., et al. (2003). Functional expression cloning of Nanog, a pluripotency sustaining factor in embryonic stem cells. *Cell, 113*(5), 643–655.

Chen, K. H., Boettiger, A. N., Moffitt, J. R., Wang, S., & Zhuang, X. (2015). Spatially resolved, highly multiplexed RNA profiling in single cells. *Science, 348*(6233), 413–426.

Chen, J., Zhang, Z., Li, L., Chen, B. C., Revyakin, A., Hajj, B., et al. (2014). Single-molecule dynamics of enhanceosome assembly in embryonic stem cells. *Cell, 156*, 1274–1285.

Cho, W.-K., Jayanth, N., English, B. P., Inoue, T., Andrews, J. O., Conway, W., et al. (2016). RNA polymerase II cluster dynamics predict mRNA output in living cells. *eLife, 5*, e13617.

Cisse, I. I., Izeddin, I., Causse, S. Z., Boudarene, L., Senecal, A., Muresan, L., et al. (2013). Real-time dynamics of RNA polymerase II clustering in live human cells. *Science, 341*(6146), 664–667.

Cockburn, K., Biechele, S., Garner, J., & Rossant, J. (2013). The Hippo pathway member Nf2 is required for inner cell mass specification. *Current Biology, 23*(13), 1195–1201.

Coskun, A. F., & Cai, L. (2016). Dense transcript profiling in single cells by image correlation decoding. *Nature Methods, 13*(8), 657–660.

Dar, R. D., Razooky, B. S., Singh, A., Trimeloni, T. V., McCollum, J. M., Cox, C. D., et al. (2012). Transcriptional burst frequency and burst size are equally modulated across the human genome. *Proceedings of the National Academy of Sciences of the United States of America, 109*(43), 17454–17459.

Daujat, S., Weiss, T., Mohn, F., Lange, U. C., Ziegler-Birling, C., Zeissler, U., et al. (2009). H3K64 trimethylation marks heterochromatin and is dynamically remodeled during developmental reprogramming. *Nature Structural & Molecular Biology, 16*(7), 777–781.

Deng, Q., Ramskold, D., Reinius, B., & Sandberg, R. (2014). Single-cell RNA-seq reveals dynamic, random monoallelic gene expression in mammalian cells. *Science, 343*(6167), 193–196.

Dietrich, J. E., & Hiiragi, T. (2007). Stochastic patterning in the mouse pre-implantation embryo. *Development, 134*, 4219–4231.

Digman, M. A., & Gratton, E. (2011). Lessons in fluctuation correlation spectroscopy. *Annual Review of Physical Chemistry, 62*, 645–668.

Eldar, A., & Elowitz, M. B. (2010). Functional roles for noise in genetic circuits. *Nature, 467*(7312), 167–173.

Evans, M. J., & Kaufman, M. H. (1981). Establishment in culture of pluripotential cells from mouse embryos. *Nature, 292*(5819), 154–156.

Fierro-Gonzalez, J. C., White, M. R., Silva, J., & Plachta, N. (2013). Cadherin-dependent filopodia control preimplantation embryo compaction. *Nature Cell Biology*, *15*, 1424–1433.
Gaglia, G., Guan, Y., Shah, J. V., & Lahav, G. (2013). Activation and control of p53 tetramerization in individual living cells. *Proceedings of the National Academy of Sciences of the United States of America*, *110*(38), 15497–15501.
Gardner, R. L. (2002). Experimental analysis of second cleavage in the mouse. *Human Reproduction*, *17*(12), 3178–3189.
Gebhardt, J. C., Suter, D. M., Roy, R., Zhao, Z. W., Chapman, A. R., Basu, S., et al. (2013). Single-molecule imaging of transcription factor binding to DNA in live mammalian cells. *Nature Methods*, *10*, 421–426.
Ghamari, A., van de Corput, M. P. C., Thongjuea, S., van Cappellen, W. A., van IJcken, W., van Haren, J., et al. (2013). In vivo live imaging of RNA polymerase II transcription factories in primary cells. *Genes Development*, *27*(7), 767–777.
Goolam, M., Scialdone, A., Graham, S. J., Macaulay, I. C., Jedrusik, A., Hupalowska, A., et al. (2016). Heterogeneity in Oct4 and Sox2 targets biases cell fate in 4-cell mouse embryos. *Cell*, *165*(1), 61–74.
Gorman, J., & Greene, E. C. (2008). Visualizing one-dimensional diffusion of proteins along DNA. *Nature Structural & Molecular Biology*, *15*, 768–774.
Guo, G., Huss, M., Tong, G. Q., Wang, C., Li Sun, L., Clarke, N. D., et al. (2010). Resolution of cell fate decisions revealed by single-cell gene expression analysis from zygote to blastocyst. *Developmental Cell*, *18*, 675–685.
Hager, G. L., McNally, J. G., & Misteli, T. (2009). Transcription dynamics. *Molecular Cell*, *35*(6), 741–753.
Halford, S. E., & Marko, J. F. (2004). How do site-specific DNA-binding proteins find their targets? *Nucleic Acids Research*, *32*, 3040–3052.
Hinde, E., Pandzic, E., Yang, Z., Ng, I. H. W., Jans, D. A., Bogoyevitch, M. A., et al. (2016). Quantifying the dynamics of the oligomeric transcription factor STAT3 by pair correlation of molecular brightness. *Nature Communications*, 7(1104), 1–4.
Hirate, Y., Hirahara, S., Inoue, K., Kiyonari, H., Niwa, H., & Sasaki, H. (2015). Par-aPKC-dependent and -independent mechanisms cooperatively control cell polarity, Hippo signaling, and cell positioning in 16-cell stage mouse embryos. *Development, Growth & Differentiation*, *57*, 544–556.
Hirate, Y., Hirahara, S., Inoue, K., Suzuki, A., Alarcon, V. B., Akimoto, K., et al. (2013). Polarity-dependent distribution of angiomotin localizes Hippo signaling in preimplantation embryos. *Current Biology*, *23*(13), 1181–1194.
Jacob, F., & Monod, J. (1961). Genetic regulatory mechanisms in the synthesis of proteins. *Journal of Molecular Biology*, *3*, 318–356.
Jedrusik, A., Parfitt, D. E., Guo, G., Skamagki, M., Grabarek, J. B., Johnson, M. H., et al. (2008). Role of Cdx2 and cell polarity in cell allocation and specification of trophectoderm and inner cell mass in the mouse embryo. *Genes & Development*, *22*, 2692–2706.
Jungmann, R., Avendano, M. S., Woehrstein, J. B., Dai, M., Shih, W. M., & Yin, P. (2014). Multiplexed 3D cellular super-resolution imaging with DNA-PAINT and exchange-PAINT. *Nature Methods*, *11*(3), 313–318.
Kaur, G., Costa, M. W., Nefzger, C. M., Silva, J., Fierro-Gonzalez, J. C., Polo, J. M., et al. (2013). Probing transcription factor diffusion dynamics in the living mammalian embryo with photoactivatable fluorescence correlation spectroscopy. *Nature Communications*, *4*, 1637.
Kurotaki, Y., Hatta, K., Nakao, K., Nabeshima, Y., & Fujimori, T. (2007). Blastocyst axis is specified independently of early cell lineage but aligns with the ZP shape. *Science*, *316*, 719–723.

Lahav, G., Rosenfeld, N., Sigal, A., Geva-Zatorsky, N., Levine, A. J., Elowitz, M. B., et al. (2004). Dynamics of the p53-Mdm2 feedback loop in individual cells. *Nature Genetics*, *36*(2), 147–150.

Larson, D. R., Fritzsch, C., Sun, L., Meng, X., Lawrence, D. S., & Singer, R. H. (2013). Direct observation of frequency modulated transcription in single cells using light activation. *eLife*, *2*, e00750.

Larue, L., Ohsugi, M., Hirchenhain, J., & Kemler, R. (1994). E-cadherin null mutant embryos fail to form a trophectoderm epithelium. *Proceedings of the National Academy of Sciences of the United States of America*, *91*(17), 8263–8267.

Lepikhov, K., & Walter, J. (2004). Differential dynamics of histone H3 methylation at positions K4 and K9 in the mouse zygote. *BMC Developmental Biology*, *4*, 12.

Leung, C. Y., & Zernicka-Goetz, M. (2013). Angiomotin prevents pluripotent lineage differentiation in mouse embryos via Hippo pathway-dependent and -independent mechanisms. *Nature Communications*, *4*, 2251.

Louvet-Vallee, S., Vinot, S., & Maro, B. (2005). Mitotic spindles and cleavage planes are oriented randomly in the two-cell mouse embryo. *Current Biology*, *15*, 464–469.

Machan, R., & Wohland, T. (2014). Recent applications of fluorescence correlation spectroscopy in live systems. *FEBS Letters*, *588*(19), 3571–3584.

Maitre, J. L., Niwayama, R., Turlier, H., Nedelec, F., & Hiiragi, T. (2015). Pulsatile cell-autonomous contractility drives compaction in the mouse embryo. *Nature Cell Biology*, *17*(7), 849–855.

Marikawa, Y., & Alarcon, V. B. (2012). Creation of trophectoderm, the first epithelium, in mouse preimplantation development. *Results and Problems in Cell Differentiation*, *55*, 165–184.

Martin, G. R. (1981). Isolation of a pluripotent cell line from early mouse embryos cultured in medium conditioned by teratocarcinoma stem cells. *Proceedings of the National Academy of Sciences of the United States of America*, *78*(12), 7634–7638.

Mayer, W., Niveleau, A., Walter, J., Fundele, R., & Haaf, T. (2000). Demethylation of the zygotic paternal genome. *Nature*, *403*(6769), 501–502.

McDole, K., Xiong, Y., Iglesias, P. A., & Zheng, Y. (2011). Lineage mapping the pre-implantation mouse embryo by two-photon microscopy, new insights into the segregation of cell fates. *Developmental Biology*, *355*(2), 239–249.

Mitsui, K., Tokuzawa, Y., Itoh, H., Segawa, K., Murakami, M., Takahashi, K., et al. (2003). The homeoprotein Nanog is required for maintenance of pluripotency in mouse epiblast and ES cells. *Cell*, *113*(5), 631–642.

Miyanari, Y., & Torres-Padilla, M. E. (2012). Control of ground-state pluripotency by allelic regulation of Nanog. *Nature*, *483*(7390), 470–473.

Molina, N., Suter, D. M., Cannavo, R., Zoller, B., Gotic, I., & Naef, F. (2013). Stimulus-induced modulation of transcriptional bursting in a single mammalian gene. *Proceedings of the National Academy of Sciences of the United States of America*, *110*(51), 20563–20568.

Morris, S. A., Guo, Y., & Zernicka-Goetz, M. (2012). Developmental plasticity is bound by pluripotency and the Fgf and Wnt signaling pathways. *Cell Reports*, *2*(4), 756–765.

Mueller, F., Stasevich, T. J., Mazza, D., & McNally, J. G. (2013). Quantifying transcription factor kinetics: At work or at play? *Critical Reviews in Biochemistry and Molecular Biology*, *48*(5), 492–514.

Nelson, D. E., Ihekwaba, A. E., Elliott, M., Johnson, J. R., Gibney, C. A., Foreman, B. E., et al. (2004). Oscillations in NF-kappaB signaling control the dynamics of gene expression. *Science*, *306*(5696), 704–708.

Nichols, J., & Smith, A. (2012). Pluripotency in the embryo and in culture. *Cold Spring Harbor Perspectives in Biology*, *4*(8), a008128.

Nichols, J., Zevnik, B., Anastassiadis, K., Niwa, H., Klewe-Nebenius, D., Chambers, I., et al. (1998). Formation of pluripotent stem cells in the mammalian embryo depends on the POU transcription factor Oct4. *Cell*, *95*(3), 379–391.

Nishioka, N., Inoue, K., Adachi, K., Kiyonari, H., Ota, M., Ralston, A., et al. (2009). The Hippo signaling pathway components Lats and Yap pattern Tead4 activity to distinguish mouse trophectoderm from inner cell mass. *Developmental Cell*, *16*(3), 398–410.

Niwa, H., Toyooka, Y., Shimosato, D., Strumpf, D., Takahashi, K., Yagi, R., et al. (2005). Interaction between Oct3/4 and Cdx2 determines trophectoderm differentiation. *Cell*, *123*, 917–929.

Normanno, D., Boudarène, L., Dugast-Darzacq, C., Chen, J., Richter, C., Proux, F., et al. (2015). Probing the target search of DNA-binding proteins in mammalian cells using TetR as model searcher. *Nature Communications*, *6*, 7357.

Oda, H., Okamoto, I., Murphy, N., Chu, J., Price, S. M., Shen, M. M., et al. (2009). Monomethylation of histone H4-lysine 20 is involved in chromosome structure and stability and is essential for mouse development. *Molecular Cell Biology*, *29*(8), 2278–2295.

Pantazis, P., & Gonzalez-Gaitan, M. (2007). Localized multiphoton photoactivation of paGFP in Drosophila wing imaginal discs. *Journal of Biomedical Optics*, *12*(4), 044004.

Patterson, G. H., & Lippincott-Schwartz, J. (2002). A photoactivatable GFP for selective photolabeling of proteins and cells. *Science*, *297*(5588), 1873–1877.

Perlmann, T., Eriksson, P., & Wrange, O. (1990). Quantitative analysis of the glucocorticoid receptor-DNA interaction at the mouse mammary tumor virus glucocorticoid response element. *The Journal of Biological Chemistry*, *265*, 17222–17229.

Piotrowska-Nitsche, K., Perea-Gomez, A., Haraguchi, S., & Zernicka-Goetz, M. (2005). Four-cell stage mouse blastomeres have different developmental properties. *Development*, *132*(3), 479–490.

Piotrowska-Nitsche, K., & Zernicka-Goetz, M. (2005). Spatial arrangement of individual 4-cell stage blastomeres and the order in which they are generated correlate with blastocyst pattern in the mouse embryo. *Mechanisms of Development*, *122*(4), 487–500.

Plachta, N., Bollenbach, T., Pease, S., Fraser, S. E., & Pantazis, P. (2011). Oct4 kinetics predict cell lineage patterning in the early mammalian embryo. *Nature Cell Biology*, *13*(2), 117–123.

Plusa, B., Frankenberg, S., Chalmers, A., Hadjantonakis, A. K., Moore, C. A., Papalopulu, N., et al. (2005). Downregulation of Par3 and aPKC function directs cells towards the ICM in the preimplantation mouse embryo. *Journal of Cell Science*, *118*(Pt. 3), 505–515.

Puschendorf, M., Terranova, R., Boutsma, E., Mao, X., Isono, K., Brykczynska, U., et al. (2008). PRC1 and Suv39h specify parental asymmetry at constitutive heterochromatin in early mouse embryos. *Nature Genetics*, *40*(4), 411–420.

Raj, A., Peskin, C. S., Tranchina, D., Vargas, D. Y., & Tyagi, S. (2006). Stochastic mRNA synthesis in mammalian cells. *PLoS Biology*, *4*(10), e309.

Raj, A., & van Oudenaarden, A. (2008). Nature, nurture, or chance: Stochastic gene expression and its consequences. *Cell*, *135*(2), 216–226.

Rossant, J. (2004). Lineage development and polar asymmetries in the peri-implantation mouse blastocyst. *Seminars in Cell & Developmental Biology*, *15*(5), 573–581.

Rossant, J., & Tam, P. P. (2009). Blastocyst lineage formation, early embryonic asymmetries and axis patterning in the mouse. *Development*, *136*(5), 701–713.

Samarage, C. R., White, M. D., Alvarez, Y. D., Fierro-Gonzalez, J. C., Henon, Y., Jesudason, E. C., et al. (2015). Cortical tension allocates the first inner cells of the mammalian embryo. *Developmental Cell*, *34*(4), 435–447.

Santenard, A., Ziegler-Birling, C., Koch, M., Tora, L., Bannister, A. J., & Torres-Padilla, M. E. (2010). Heterochromatin formation in the mouse embryo requires critical residues of the histone variant H3.3. *Nature Cell Biology*, *12*(9), 853–862.

Santos, F., Peters, A. H., Otte, A. P., Reik, W., & Dean, W. (2005). Dynamic chromatin modifications characterise the first cell cycle in mouse embryos. *Developmental Biology, 280*(1), 225–236.

Scholer, H. R., Dressler, G. R., Balling, R., Rohdewohld, H., & Gruss, P. (1990). Oct-4: A germline-specific transcription factor mapping to the mouse t-complex. *EMBO Journal, 9*(7), 2185–2195.

Senecal, A., Munsky, B., Proux, F., Ly, N., Braye, F. E., Zimmer, C., et al. (2014). Transcription factors modulate c-Fos transcriptional bursts. *Cell Reports, 8*(1), 75–83.

Shi, J., Chen, Q., Li, X., Zheng, X., Zhang, Y., Qiao, J., et al. (2015). Dynamic transcriptional symmetry-breaking in pre-implantation mammalian embryo development revealed by single-cell RNA-seq. *Development, 142*(20), 3468–3477.

Singer, Z. S., Yong, J., Tischler, J., Hackett, J. A., Altinok, A., Surani, M. A., et al. (2014). Dynamic heterogeneity and DNA methylation in embryonic stem cells. *Molecular Cell, 55*(2), 319–331.

Singh, A., Razooky, B., Cox, C. D., Simpson, M. L., & Weinberger, L. S. (2010). Transcriptional bursting from the HIV-1 promoter is a significant source of stochastic noise in HIV-1 gene expression. *Biophysical Journal, 98*(8), L32–L34.

Smallwood, S. A., Tomizawa, S., Krueger, F., Ruf, N., Carli, N., Segonds-Pichon, A., et al. (2011). Dynamic CpG island methylation landscape in oocytes and preimplantation embryos. *Nature Genetics, 43*(8), 811–814.

Stephenson, R. O., Yamanaka, Y., & Rossant, J. (2010). Disorganized epithelial polarity and excess trophectoderm cell fate in preimplantation embryos lacking E-cadherin. *Development, 137*(20), 3383–3391.

Strnad, P., Gunther, S., Reichmann, J., Krzic, U., Balazs, B., de Medeiros, G., et al. (2015). Inverted light-sheet microscope for imaging mouse pre-implantation development. *Nature Methods, 13*(2), 139–142.

Strumpf, D., Mao, C. A., Yamanaka, Y., Ralston, A., Chawengsaksophak, K., Beck, F., et al. (2005). Cdx2 is required for correct cell fate specification and differentiation of trophectoderm in the mouse blastocyst. *Development, 132*(9), 2093–2102.

Suter, D. M., Molina, N., Gatfield, D., Schneider, K., Schibler, U., & Naef, F. (2011). Mammalian genes are transcribed with widely different bursting kinetics. *Science, 332*(6028), 472–474.

Tabansky, I., Lenarcic, A., Draft, R. W., Loulier, K., Keskin, D. B., Rosains, J., et al. (2013). Developmental bias in cleavage-stage mouse blastomeres. *Current Biology, 23*(1), 21–31.

Takahashi, K., & Yamanaka, S. (2016). A decade of transcription factor-mediated reprogramming to pluripotency. *Nature Reviews. Molecular Cell Biology, 17*(3), 183–193.

Tanaka, S., Kunath, T., Hadjantonakis, A. K., Nagy, A., & Rossant, J. (1998). Promotion of trophoblast stem cell proliferation by FGF4. *Science, 282*(5396), 2072–2075.

Tang, F., Barbacioru, C., Bao, S., Lee, C., Nordman, E., Wang, X., et al. (2010). Tracing the derivation of embryonic stem cells from the inner cell mass by single-cell RNA-Seq analysis. *Cell Stem Cell, 6*(5), 468–478.

Tay, S., Hughey, J. J., Lee, T. K., Lipniacki, T., Quake, S. R., & Covert, M. W. (2010). Single-cell NF-κB dynamics reveal digital activation and analogue information processing. *Nature, 466*(7303), 267–271.

Torres-Padilla, M. E., Parfitt, D. E., Kouzarides, T., & Zernicka-Goetz, M. (2007). Histone arginine methylation regulates pluripotency in the early mouse embryo. *Nature, 445*(7124), 214–218.

Turner, D. A., Paszek, P., Woodcock, D. J., Nelson, D. E., Horton, C. A., Wang, Y., et al. (2010). Physiological levels of TNFα stimulation induce stochastic dynamics of NF-κB responses in single living cells. *Journal of Cell Science, 123*(16), 2834–2843.

von Hippel, P. H., & Berg, O. G. (1989). Facilitated target location in biological systems. *The Journal of Biological Chemistry, 264*, 675–678.

Wang, Z., Gerstein, M., & Snyder, M. (2009). RNA-Seq: A revolutionary tool for transcriptomics. *Nature Reviews Genetics*, *10*(1), 57–63.

White, M. D., Angiolini, J. F., Alvarez, Y. D., Kaur, G., Zhao, Z. W., Mocskos, E., et al. (2016). Long-lived binding of Sox2 to DNA predicts cell fate in the four-cell mouse embryo. *Cell*, *165*(1), 75–87.

White, M. D., Bissiere, S., Alvarez, Y. D., & Plachta, N. (2016). Mouse embryo compaction. *Current Topics in Developmental Biology*, *120*, 235–258.

Wicklow, E., Blij, S., Frum, T., Hirate, Y., Lang, R. A., Sasaki, H., et al. (2014). HIPPO pathway members restrict SOX2 to the inner cell mass where it promotes ICM fates in the mouse blastocyst. *PLoS Genetics*, *10*, e1004618.

Yamanaka, Y., Ralston, A., Stephenson, R. O., & Rossant, J. (2006). Cell and molecular regulation of the mouse blastocyst. *Developmental Dynamics*, *235*(9), 2301–2314.

Yeap, L. S., Hayashi, K., & Surani, M. A. (2009). ERG-associated protein with SET domain (ESET)-Oct4 interaction regulates pluripotency and represses the trophectoderm lineage. *Epigenetics & Chromatin*, *2*(1), 12.

Yuan, P., Han, J., Guo, G., Orlov, Y. L., Huss, M., Loh, Y. H., et al. (2009). Eset partners with Oct4 to restrict extraembryonic trophoblast lineage potential in embryonic stem cells. *Genes and Development*, *23*(21), 2507–2520.

Zernicka-Goetz, M., Morris, S. A., & Bruce, A. W. (2009). Making a firm decision: Multifaceted regulation of cell fate in the early mouse embryo. *Nature Reviews. Genetics*, *10*(7), 467–477.

Zhao, Z. W., Roy, R., Gebhardt, J. C. M., Suter, D. M., Chapman, A. R., & Xie, X. S. (2014). Spatial organization of RNA polymerase II inside a mammalian cell nucleus revealed by reflected light-sheet superresolution microscopy. *Proceedings of the National Academy of Sciences of the United States of America*, *111*(2), 681–686.

Zhao, Z. W., White, M. D., Alvarez, Y. D., Zenker, J., Bissiere, S., & Plachta, N. (2017). Quantifying transcription factor-DNA binding in single cells in vivo with photoactivatable fluorescence correlation spectroscopy. *Nature Protocols*, *12*(7), 1458–1471.

Zhao, Z. W., White, M. D., Bissiere, S., Levi, V., & Plachta, N. (2016). Quantitative imaging of mammalian transcriptional dynamics: From single cells to whole embryos. *BMC Biology*, *14*(1), 115.

CHAPTER FOUR

Our First Choice: Cellular and Genetic Underpinnings of Trophectoderm Identity and Differentiation in the Mammalian Embryo

Sergio Menchero, Julio Sainz de Aja, Miguel Manzanares[1]
Centro Nacional de Investigaciones Cardiovasculares (CNIC), Madrid, Spain
[1]Corresponding author: e-mail address: mmanzanares@cnic.es

Contents

Abstract

The trophectoderm (TE) is the first cell population to appear in the mammalian preimplantation embryo, as the result of the differentiation of totipotent blastomeres located on the outer surface of the late morula. Trophectodermal cells arrange in a monolayer covering the expanding blastocyst and acquire an epithelial phenotype with distinct apicobasal polarity and a basal lamina placed toward the blastocyst interior. During later development through the periimplantation and gastrulation stages, the TE gives rise to extraembryonic membranes and cell types that will eventually form most of the fetal placenta, the specialized organ through which the embryo obtains maternal nourishment necessary for subsequent exponential growth. The specification of the TE is controlled by the combination of morphological cues arising from cell polarity with differential activity of signaling pathways such as Hippo and Notch, and the restriction to outer cells of lineage specifiers such as CDX2. This is possibly the first symmetry-breaking decision undertaken by the uncommitted cells produced by a handful of mitosis divisions from the newly fertilized zygote. Understanding how this cell lineage is

Current Topics in Developmental Biology, Volume 128
ISSN 0070-2153
https://doi.org/10.1016/bs.ctdb.2017.10.009

specified will therefore provide unique information about development, differentiation, and how the interplay between cellular morphology and signaling and regulatory factors results in a correctly 3D-patterned embryo.

During development, a complete organism is built from a unicellular zygote. This single cell must generate not only the embryo itself but also the extra-embryonic structures. The totipotent capacity of the zygote is gradually lost as the cell populations in the embryo differentiate toward committed fates. In mammals, the first symmetry-breaking event occurs during preimplantation development and establishes two cell lineages: the trophectoderm (TE) and the inner cell mass (ICM). The TE gives rise to extraembryonic lineages, such as the extraembryonic ectoderm (ExE) and the ectoplacental cone (EPC), which will later differentiate into the major portion of the fetal placenta. In contrast, the ICM gives rise to the embryo proper, the yolk sac, and the allantois.

1. HETEROGENEITY EMERGENCE IN THE MOUSE EMBRYO

The mouse zygote undergoes a series of divisions that increase the number of cells without altering the size of the embryo. These cells, called blastomeres, are initially equal, and each one can potentially develop into a full embryo (Suwinska, Czolowska, Ozdzenski, & Tarkowski, 2008; Tarkowski, 1959; Tarkowski & Wroblewska, 1967). Heterogeneities among these blastomeres appear gradually, raising questions about the extent to which these emerging differences bias cells toward a particular lineage contribution or reveal a certain variability within an apparently equivalent population. Thus although the two first distinct cell populations are clearly distinguishable at the blastocyst stage, at embryonic (E) day 3.5, the key to understanding how these cells commit to a specific lineage must be sought earlier.

Recent research into gene expression differences between individual cells of four-cell embryos suggests that molecular heterogeneities between blastomeres appear some time before the first lineage decision is established, and that differential gene expression at this stage can bias the contribution of descendant cells to one lineage or another (Goolam et al., 2016; Torres-Padilla, Parfitt, Kouzarides, & Zernicka-Goetz, 2007). However, at these stages cell totipotency is undiminished. Rather, it appears that blastomeres acquire tendencies or preferences to contribute to a particular lineage, while at this stage retaining their capacity to give rise to every population in the embryo (Kelly, 1977; Tabansky et al., 2013; Tarkowski & Wroblewska, 1967). The first morphological evidence of differentiation is seen at the morula stage. After three rounds of division, blastomeres increase their intercellular interactions, and cells on the surface acquire an apical-basal polarity, with the apical domain facing outward (Ziomek & Johnson, 1980). This process, known as compaction, is driven at the eight-cell stage by periodic cortical waves controlled by the actomyosin cortex in cooperation with CDH1 (E-cadherin) (Maitre, Niwayama, Turlier, Nedelec, & Hiiragi, 2015). Moreover, cell shape changes during compaction have been found to depend on CDH1-dependent filopodia (Fierro-Gonzalez, White, Silva, & Plachta, 2013). The increasing number of blastomeres generated during subsequent divisions forces the cells to rearrange their positions, resulting in the emergence of two distinct cell populations: an outer layer of polarized cells enclosing an inner group of apolar cells. The acquisition of polarity and the subsequent cell divisions have been the focus of the two main hypotheses about how the TE and the ICM are segregated. The inside–outside model proposes that the position of a cell in the morula exposes it to specific conditions that are the key to deciding its fate. Thus, cells on the outside of the morula give rise to the TE and the cells on the inside give rise to ICM (Tarkowski & Wroblewska, 1967). Supporting this model, when blastomeres are placed on the outside of another embryo, those cells contribute predominantly to the TE (Hillman, Sherman, & Graham, 1972). In contrast, the polarization model proposes that differentiation is established according to the inheritance of an apical pole in the blastomeres. When a polar cell divides symmetrically, both daughter cells inherit the apical-basal polarity and remain in the outer population. However, if a polarized cell divides asymmetrically, only one daughter cell inherits the apical pole, and the nonpolarized daughter cell will form part of the inner population (Johnson & Ziomek, 1981).

Polarization and inner vs outer locations are still considered the basis of the first lineage choice. However, several recent studies have shown that the division angle and initial position does not robustly predict blastomere fate, because substantial cell relocations force blastomeres to internalize prior to the blastocyst configuration (Anani, Bhat, Honma-Yamanaka, Krawchuk, & Yamanaka, 2014; McDole, Xiong, Iglesias, & Zheng, 2011; Toyooka, Oka, & Fujimori, 2016; Watanabe, Biggins, Tannan, & Srinivas, 2014). Actomyosin cortical networks subject the outer cells to heterogeneous tensile forces, while myosin II accumulates around constricting cells that will become embedded in the inner population (Samarage et al., 2015). Polarized and unpolarized blastomeres have differing contractilities, with unpolarized cells having more cortical myosin and a higher contraction amplitude, triggering their internalization (Maitre et al., 2016).

Since the publication of the first models of TE and ICM segregation, several studies have confirmed the importance of apical-basal polarity in outer cells (Alarcon, 2010; Johnson & Ziomek, 1981; Kondratiuk, Bazydlo, Maleszewski, & Szczepanska, 2012; Plusa et al., 2005). Interestingly, a recent report showed that the apical domain is sufficient to promote TE fate; the apical domains of polarized eight-cell blastomeres induce asymmetric division when transplanted into apolar eight-cell blastomeres, with the daughter cell inheriting the apical domain differentiating toward TE (Korotkevich et al., 2017).

These studies make it clear that the preimplantation embryo is a self-organizing system that is able to interpret morphogenetic cues in order to exit the totipotent state while taking lineage decisions.

2. THE TE GENE REGULATORY NETWORK

The establishment of the first cell populations in the mouse embryo is underpinned by a transcriptional circuitry orchestrating the fate of every cell. The core of these networks is a set of transcription factors (TFs) that drive the commitment of cells to particular lineages. The first lineage choice is controlled by two specific programs: the pluripotency network and the TE network.

The core genes of the pluripotency network are *Oct4* (*Pou5f1*) (Nichols et al., 1998), *Sox2* (Avilion et al., 2003), and *Nanog* (Chambers et al., 2003; Mitsui et al., 2003). This small set of TFs maintains pluripotency both in vivo and in vitro, and at the same time represses differentiation and lineage commitment. Accordingly, the core pluripotency network actively represses the expression of TE-specific genes (Niwa et al., 2005), and the absence of *Oct4* leads to increased differentiation to the trophoblast lineage both in vivo and in vitro (Nichols et al., 1998; Niwa, Miyazaki, & Smith, 2000).

The TE network is controlled by *Cdx2* (Strumpf et al., 2005), a mammalian homologue of the *Drosophila* homeotic gene *caudal* (Mlodzik, Fjose, & Gehring, 1985) that encodes a homeodomain TF. *Cdx2* was reported to be expressed in the blastocyst TE and in the ExE at postimplantation stages (Beck, Erler, Russell, & James, 1995). Analysis of *Cdx2*$^{-/-}$ embryos showed that they were unable to maintain the blastocoel, with suppressed expression of other TE markers and pluripotency factors such as OCT4 and NANOG detected in all cells (Strumpf et al., 2005). Another TE marker, the cytokeratin intermediate filament Krt8 (Chisholm & Houliston, 1987) is also dependent on *Cdx2* (Ralston & Rossant, 2008). These results suggest that *Cdx2* is necessary for the correct formation of the TE and its differentiation, and place *Cdx2* at the core of the regulatory network (Fig. 1).

The zinc finger-coding gene *Gata3* also regulates trophoblast identity (Home et al., 2009; Ralston et al., 2010) (Fig. 1). *Gata3* is coexpressed with *Cdx2* at preimplantation stages, but at postimplantation stages its expression is higher in the EPC than in the ExE (Ralston et al., 2010). *Gata2* was recently found to be coexpressed with *Gata3*, and deletion of both genes impairs the expression of the TE-related genes *Cdx2*, *Eomes*, and *Elf5* (Home et al., 2017). The T-box gene *Eomes* (*Eomesodermin*) is also expressed in the TE at E3.5, in the polar TE at E4.5, and in the ExE at postimplantation

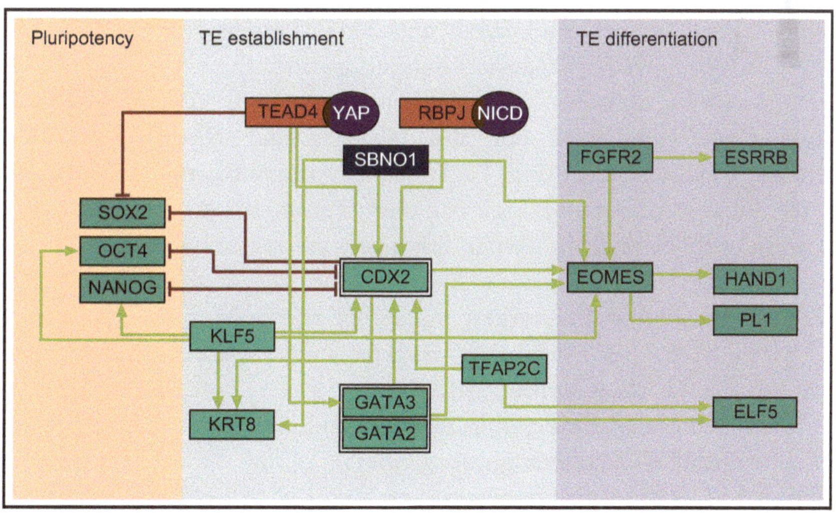

Fig. 1 The gene regulatory network driving trophectoderm identity. The establishment and differentiation of the trophectoderm (TE) in the embryo require the sequential activation of transcription factors and other TE-related genes (*middle* and *right panels*) as well as the repression of the pluripotency network (*left panel*). The transcriptional effectors of the Hippo and Notch pathways are indicated in *light red* (DNA-binding factors) and in *purple* (coactivators). SBNO1 (*dark blue*) is a chromatin remodeler. *Green arrows* indicate activation and *red lines* indicate repression.

stages (Hancock, Agulnik, Silver, & Papaioannou, 1999; Ralston & Rossant, 2008; Strumpf et al., 2005). *Eomes* is required later than *Cdx2*, and while *Eomes* expression is impaired in *Cdx2*$^{-/-}$ embryos, the reverse is not observed (Ralston & Rossant, 2008; Strumpf et al., 2005). In addition, *Eomes* regulates some markers of differentiated trophoblast giant cells (TGCs) such as *Hand1* and *Pl1* (Cross et al., 1995; Faria, Ogren, Talamantes, Linzer, & Soares, 1991; Strumpf et al., 2005) (Fig. 1). The TF AP-2γ (TFAP2C) specifies the TE lineage upstream of *Cdx2* and *Elf5* (Cao et al., 2015) (Fig. 1) and also regulates *Pard6b*, an important apical domain component required for blastocyst formation (Alarcon, 2010). *Fgfr2* is expressed in the blastocyst TE and could be involved in the maintenance of the TE and its derivatives (Haffner-Krausz, Gorivodsky, Chen, & Lonai, 1999; Tanaka, Kunath, Hadjantonakis, Nagy, & Rossant, 1998). A recent work has shown by means of two reporter mice that *Fgfr2* is expressed since the eight-cell stage and *Fgfr1* is also expressed in the TE. However the only TE phenotype that *Fgfr1*$^{-/-}$; *Fgfr2*$^{-/-}$ mutant show is a reduced number of TE cells but no defects in any TE marker has been reported (Kang, Garg, & Hadjantonakis, 2017; Molotkov, Mazot, Brewer, Cinalli, & Soriano, 2017). *Elf5* and *Ets2* are markers of the ExE, and their deletion interferes with proper gastrulation (Donnison et al., 2005; Polydorou & Georgiades, 2013). Intriguingly, *Klf5* regulates markers of both the embryonic and the extraembryonic lineages. *Klf5*$^{-/-}$ blastocysts have reduced levels of *Cdx2*, *Eomes*, and *Krt8* but also of *Oct4* and *Nanog* (Lin, Wani, Whitsett, & Wells, 2010) (Fig. 1). The advent of next-generation sequencing techniques, particularly single cell transcriptomics, has allowed the identification of the earliest molecular differences among cells, and these studies have identified new candidate members of the TE network. For example, *Id2* was identified as the earliest marker of the outer population in the embryo (Guo et al., 2010), and new markers such as *Dab2* or *Lrp2* have been linked to the TE (Posfai et al., 2017).

The TE lineage is thus established and maintained through the progressive activation of TFs that promote the induction of other TE-related markers and their own regulation and at the same time repress the pluripotency network (Fig. 1). Direct repression between CDX2 and the pluripotency factors OCT4, NANOG, and SOX2 appears to be important for maintaining lineage segregation (Huang et al., 2017). In vitro studies suggest that this repression is reciprocal between both lineages (Chen et al., 2009; Niwa et al., 2005).

3. ESTABLISHMENT OF TE FATE

Core factors such as OCT4, NANOG, SOX2, and CDX2 play a well-defined role in determining the first lineage decisions in the embryo. However, surprisingly their initial expression is stochastic among blastomeres up to the late-morula stage, after which they gradually become restricted to their definitive domains (Dietrich & Hiiragi, 2007; Guo et al., 2010; Ohnishi et al., 2014). This dynamic raises the question of how the expression of core lineage factors is controlled during these first stages of development.

Cdx2 is initially expressed at the eight-cell stage in a salt-and-pepper pattern, before becoming restricted to the TE at the blastocyst stage (Dietrich & Hiiragi, 2007; Ralston & Rossant, 2008) (Fig. 2). *Cdx2* is in fact the first factor to be restricted to a specific lineage in the blastocyst. *Cdx2* lies

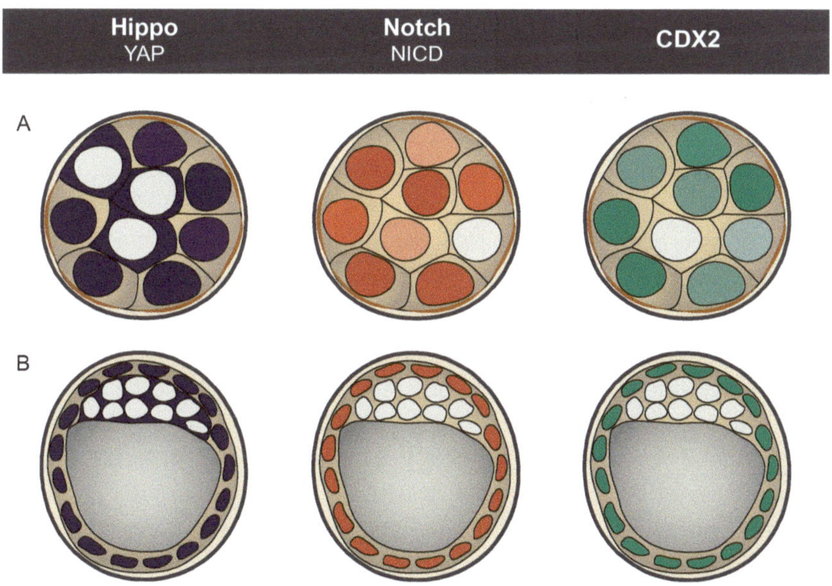

Fig. 2 Hippo and Notch regulate trophectoderm fate. The stochastic expression of CDX2 (*green*) in the early morula cannot be fully explained by the different levels of nuclear YAP (*purple*) and Notch intracellular domain (NICD, *red*) in each blastomere (A). Polarized blastomeres (apical domain in *yellow*) correlate with nuclear YAP. Cytoplasmic YAP is highlighted in unpolarized blastomeres (A). At the blastocyst stage, there is complete overlap of nuclear YAP, NICD, and CDX2, which are all restricted to the trophectoderm (B).

downstream of the Hippo pathway (Figs. 1 and 2), which was first identified in *Drosophila*, where it is involved in tissue growth control (Pan, 2010). However, in the preimplantation embryo, Hippo is not involved in growth regulation but instead provides the link between polarization and cell fate. The differential distribution of pathway components in polarized and unpolarized blastomeres dictates the activity of the pathway (Manzanares & Rodriguez, 2013; Sasaki, 2015).

In apolar inner cells, where the Hippo pathway is active, the junction-associated protein AMOT (angiomotin) localizes at adherent junctions (Hirate et al., 2013; Leung & Zernicka-Goetz, 2013). The kinase LATS1/2 phosphorylates AMOT and stabilizes a complex formed by LATS1/2, AMOT, and NF2 (Hirate et al., 2013). LATS1/2 then phosphorylates the transcriptional coactivator YAP (and the Yap-related protein TAZ), which is retained in the cytoplasm. In polar outer cells, AMOT is sequestered by the apical complex. The kinases are inactive and, consequently, YAP is not phosphorylated. Therefore, YAP is translocated to the nucleus, where it binds the TF TEAD4 and triggers the expression of its target genes (Nishioka et al., 2009), including the TE lineage factors *Cdx2* (Nishioka et al., 2009) and *Gata3* (Ralston et al., 2010). If Hippo pathway activation is blocked due to disruption of any of its components, ectopic *Cdx2* expression can be detected in inner cells (Cockburn, Biechele, Garner, & Rossant, 2013; Hirate et al., 2013; Leung & Zernicka-Goetz, 2013; Nishioka et al., 2009). Most embryos lacking *Tead4*, the Hippo pathway effector, are unable to cavitate and to implant, and have below normal expression of *Cdx2* (Nishioka et al., 2008; Yagi et al., 2007). *Tead4*$^{-/-}$ morulae therefore have low levels of CDX2, and in *Tead4*$^{-/-}$ blastocysts CDX2 are not detected (Nishioka et al., 2008; Rayon et al., 2014). Remarkably, this phenotype can be circumvented if *Tead4*$^{-/-}$ embryos are cultured in low oxygen conditions that reduce oxidative stress (Kaneko & Depamphilis, 2013), suggesting a role for the Hippo pathway in the regulation of energy homeostasis in the blastocyst. The role of the Hippo pathway is not only to promote the TE lineage but also to contribute to the establishment of the ICM by limiting the expression of SOX2, independently of *Cdx2*, in this population (Wicklow et al., 2014).

There is a good overall correlation between blastomere position in the early embryo and polarization: outer cells tend to be polarized and inner cells unpolarized. Nonetheless, since polarization happens progressively, apolar cells can be found in the outer layer of 16-cell embryos, and interestingly the Hippo pathway is active in these cells, with phosphorylated YAP maintained in the cytoplasm (Anani et al., 2014; Hirate et al., 2015). Hippo is thus linked to polarity rather than to cell position (Fig. 2A).

Hippo was the first signaling pathway identified to play a role in lineage determination in the preimplantation embryo (Nishioka et al., 2009). However, recent evidence shows that activation and maintenance of the TE program are ensured not by one pathway but rather by the combination of signals from different inputs. One of these is the Notch pathway, which cooperates with Hippo to regulate *Cdx2* expression (Rayon et al., 2014) (Figs. 1 and 2). Notch is a cell-to-cell signaling pathway involved in many developmental programs (Koch, Lehal, & Radtke, 2013). Disruption of this pathway does not affect embryo survival until postimplantation stages (Conlon, Reaume, & Rossant, 1995; Shi, Stahl, Lu, & Stanley, 2005; Souilhol, Cormier, Tanigaki, Babinet, & Cohen-Tannoudji, 2006), and for this reason its putative role in preimplantation stages was ruled out. However, double mutants for *Tead4* and *Rbpj*, the transcriptional effectors of the Hippo and Notch pathways, respectively, exacerbate their single phenotypes and lead to embryonic death before the blastocyst stage. Notch is active specifically in the TE, and its blockade diminishes *Cdx2* levels. Furthermore, forced expression of the active domain of Notch1 (N1ICD) unbalances the TE/ICM cell ratio (Rayon et al., 2014). Both pathways act in parallel and converge on a regulatory element upstream of *Cdx2*, called the TE enhancer (Rayon et al., 2014), and they do this through interaction with the chromatin remodeler SBNO1 (Strawberry Notch1) (Fig. 1). *Sbno1*$^{-/-}$ embryos fail to cavitate and to express the TE markers *Cdx2*, *Eomes*, and *Krt8*. SBNO1 physically interacts with the YAP/TEAD and NICD/RBPJ transcriptional complexes, and lack of *Sbno1* prevents YAP/TEAD and NICD/RBPJ from properly inducing *Cdx2* (Watanabe et al., 2017).

TE determination may also involve other signaling inputs (Menchero, Rayon, Andreu, & Manzanares, 2017). For example, Rho/Rock signaling has been shown to mediate cell polarization and Hippo regulation. Blockade of Rho/Rock results in disruption of apical-basal polarity and ectopic activation of Hippo (Kono, Tamashiro, & Alarcon, 2014; Mihajlovic & Bruce, 2016). We can therefore conclude that the first differentiation event is driven not by one pathway but by a multitude of transcriptional and signaling players. The Hippo pathway is the main one orchestrating TE fate, but its role is supported by other players like Notch and Rho. This multiplicity may function to ensure the robustness of embryonic development in the event that any branch of the circuitry is blocked. This interplay between pathways shows the need to study combined inputs in order to unravel unsuspected roles during the establishment of the first lineages.

4. PLASTICITY DURING LINEAGE SEGREGATION

Much of the research into mammalian preimplantation development has focused on finding key factors and regulators that drive a cell to a specific population. Nevertheless, it is becoming clear that lineage commitment is a gradual process that is not dependent on a single factor, and that cells retain a certain degree of plasticity.

The relocation of cells during the morula to blastocyst transition raises the question of whether, at the time of internalization, cells are uncommitted to a particular population or if their lineage fate gets altered (Anani et al., 2014; Watanabe et al., 2014). The use of a CDX2-GFP reporter line for live imaging during preimplantation stages has proven particularly useful in addressing this issue. These studies revealed that some GFP-positive outer cells internalized, after which their GFP intensity gradually decreased until it became undetectable (McDole & Zheng, 2012; Toyooka et al., 2016). There is thus patent cell plasticity during this time window, but the precise timing of lineage commitment and its correlation with transcriptional changes is unclear. A recent report used the same CDX2-GFP line to study the potential of single cells during lineage segregation. Transcriptome profiling of single cells from 16-cell to 64-cell embryos revealed a sharp divergence in gene expression between the late 16-cell and early 32-cell stages, with genes related to the TE network being the first to be activated when transcriptional differences arise among blastomeres (Posfai et al., 2017). Nevertheless, the triggering of lineage programs does not immediately restrict cell potency. The ability of a cell to contribute to both populations decreases at the 16–32-cells transition, but interestingly the temporal dynamics of CDX2 expression changes is not the same for the TE and ICM lineages. CDX2-GFP^{high} cells preferentially contribute to the TE, but are still able to contribute to the ICM until the late 32-cell stage. In contrast, CDX2-GFP^{low} cells contribute exclusively to the ICM from the early 32-cell stage. This plasticity is extended in embryos reconstructed exclusively from CDX2-GFP^{high} or CDX2-GFP^{low} cells. Reconstituted embryos composed of late 32-cell stage CDX2-GFP^{low} cells still form both lineages, and it is not until the 64-cell stage that cells lose their ability to form TE. However, embryos made from late 32-cell stage CDX2-GFP^{high} cells do not develop because the cells have already lost their ability to form ICM (Posfai et al., 2017).

These differences in plasticity suggest that TE is the first committed lineage in the embryo, whereas ICM cells retain a broader potential when placed in forced situations.

5. RECAPITULATING TE IDENTITY IN CULTURE

The ability to obtain blastocyst-derived stem cells is a major advance in the analysis of regulatory networks underlying the emergence of the first lineages in the embryo. These populations maintain a high level of plasticity and receptiveness to genetic manipulations that can be used to study molecular mechanisms. Embryonic stem (ES) cells are derived from the ICM (Evans & Kaufman, 1981; Martin, 1981), whereas trophoblast stem (TS) cells are derived from the TE (Tanaka et al., 1998). These stem cell populations mostly preserve their potency and regulatory networks, can be cultured in vitro in stemness conditions, and can also be directed to differentiate and contribute to their target lineage in chimera assays (Beddington & Robertson, 1989; Tanaka et al., 1998).

Transdifferentiation assays allow to change the identity of a stem-cell population have underlined the importance of specific factors for the establishment of early lineages. Overexpression of *Cdx2*, *Gata3*, *Eomes*, *Tfap2c*, or *Tead4* in ES cells triggers their transdifferentiation to TS-like cells (Kuckenberg et al., 2010; Nishioka et al., 2009; Niwa et al., 2005; Ralston et al., 2010). The use of TS cells has obvious advantages, such as the possibility to obtain enough material for biochemical or ChIP-seq assays (Home et al., 2012; Latos et al., 2015), which would be technically challenging with embryonic material obtained directly. However, it is important to recognize that TS cells are not completely equivalent to the in vivo TE. Differences between the TE and TS cells have been reported in epigenetic status (Senner, Krueger, Oxley, Andrews, & Hemberger, 2012) and in the usage of *Cdx2* regulatory elements (Rayon et al., 2016). Furthermore, TS cells comprise a population whose identity is already established, and therefore most of the mechanisms at play are likely related more to maintaining the trophoblast phenotype than to its establishment. For instance, the pivotal role played by cell polarity in establishing TE identity in the embryo cannot be recreated in 2D cultures of TS cells.

This could explain why, although the core TFs such as CDX2 are crucial in both the TE and the TS cells, there are important differences in signaling pathways acting upstream of these factors. For example, YAP is nuclear in the TE and excluded from nuclei in the ICM (Nishioka et al., 2009), but it is located in the nuclei of ES cells (the counterpart of the ICM), where it has been reported to maintain pluripotency and control differentiation (Chung et al., 2016; Lian et al., 2010). Similarly, heterogeneous Notch activity has been detected in ES cells (Nowotschin, Xenopoulos, Schrode, & Hadjantonakis, 2013), whereas Notch is inactive in the ICM (Rayon et al., 2014).

6. TE DERIVATIVES AND THEIR ROLE IN THE POSTIMPLANTATION EMBRYO

After hatching from the zona pellucida, the blastocyst implants in the uterus through the mural TE, formed by the trophoblast population opposite the ICM pole. Implantation requires a complex set of interactions between the embryo and the uterine epithelium, involving several proteases, adhesion molecules, and signaling events (Aplin & Ruane, 2017).

The first definitive cells to differentiate from the TE upon implantation are the TGCs. TGCs cover the entire embryo except for the region covered by the polar TE, which directly overlies the ICM. The polar TE gives rise to two different cone-shaped structures: the ExE, in direct contact with the epiblast; and the EPC, which is located proximal to the ExE and mediates early interactions with the maternal endometrium. By E6.5, both the ExE and the EPC are present and the chorion has formed. The chorion is a bilayered tissue derived from embryo-derived mesoderm that has migrated to the extraembryonic region and from the trophoblast progenitor cells of the ExE. Another structure, the allantois, is also derived from embryonic mesoderm that migrates to the extraembryonic compartment at the posterior end of the embryo; when the allantois reaches the chorion, chorio-allantoic fusion occurs. The resulting chorio-allantoic attachment will form the labyrinth zone of the placenta, the main route for nutrient exchange between mother and embryo (Cross, Nakano, Natale, Simmons, & Watson, 2006). The labyrinth is formed by three trophoblast cell types derived from distinct precursors in the chorion: spongiotrophoblasts, glycogen cells, and secondary TGCs. TGCs give rise to several subtypes, including maternal arterial canal-associated TGCs, parietal TGCs, and spiral artery-associated TGCs (Latos & Hemberger, 2016; Rossant & Cross, 2001). These cells, together with glycogen cells, invade the decidua and associate with maternal tissue (Mould, Morgan, Li, Bikoff, & Robertson, 2012; Simmons et al., 2008). Cells of another TGC subtype, the sinusoidal TGCs, establish themselves in the maternal blood sinusoids. Invasion of maternal tissue occurs with the help of TGC-expressed proteases such as CTS7 and CTS8 (Screen, Dean, Cross, & Hemberger, 2008).

One of the most interesting phases in trophoblast differentiation is ExE differentiation during gastrulation. The ExE establishes a complex crosstalk with the epiblast that is necessary not only for the maintenance and differentiation of ExE trophoblasts and for establishing the anterior–posterior axis, but also for germ-cell formation and gastrulation in the epiblast (Beck et al., 2002; Feldman, Poueymirou, Papaioannou, DeChiara, & Goldfarb, 1995; Yoshimizu, Obinata, & Matsui, 2001) (Fig. 3).

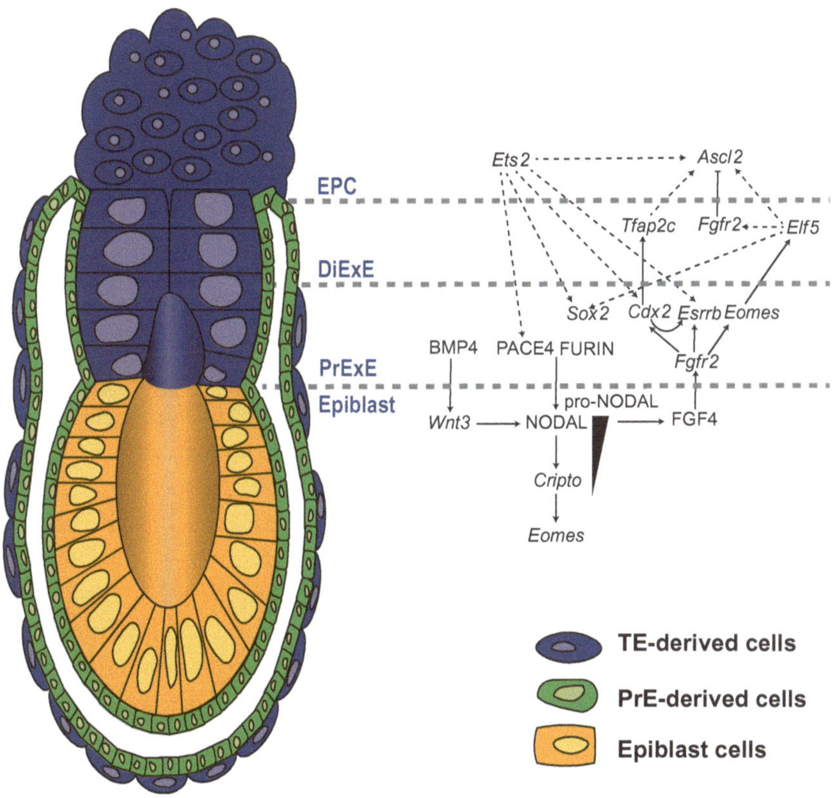

Fig. 3 Role of the extraembryonic ectoderm in patterning the embryonic epiblast. Diagram showing the signaling interactions taking place between the embryonic and extraembryonic portions of the E6.5 embryo. Localized expression of genes (indicated by *italics*) and secreted factors (*upper case*) is indicated in the different regions (epiblast; proximal extraembryonic ectoderm, PrExE; distal extraembryonic ectoderm, DiExE; ectoplacental cone, EPC). Direct interactions are indicated by *solid lines*; *dashed lines* indicate suggested interactions. The gradient of Nodal through the epiblast is indicated by a *black triangle*.

The ExE is formed after implantation at around E5.5 and is the reservoir for TS cells that will proliferate and contribute to different cell types of the placenta during its differentiation and growth. TS cell proliferation depends on diffusible signals from neighboring cells in the epiblast, including FGF4, which acts through FGFR2 expressed on TS cells in the ExE. FGF4 is induced by NODAL in the epiblast and has a proliferative effect on ExE TS cells, activating the expression of genes encoding critical TFs necessary for their maintenance, such as *Cdx2*, *Eomes*, and *Esrrb*, and at the same time blocking specification genes such as *Ascl2*, expressed in the EPC (Guzman-Ayala, Ben-Haim, Beck, & Constam, 2004) (Fig. 3). ExE development is also determined by the levels of *Elf5* and *Ets2*. *Elf5* is important for TS maintenance and its deletion causes TS cells to differentiate to EPC cells. Knockout mice for both genes show loss of ExE upon gastrulation, although a compensatory upregulation of *Ets2* partially rescues the defects in *Elf5* mutants (Donnison et al., 2005; Donnison, Broadhurst, & Pfeffer, 2015; Yamamoto et al., 1998).

One of the major players in the crosstalk between the ExE with the embryonic epiblast is NODAL, a secreted protein of the TGFβ family. *Nodal* is expressed in the epiblast, producing the uncleaved precursor pro-NODAL. Secreted pro-NODAL activates the expression of BMP4 and the convertases PACE4 and FURIN in the ExE through Activin receptor complexes (Fig. 3). PACE4- and FURIN-mediated pro-NODAL cleavage and maturation set up a positive loop in which FOXH1 and SMAD2 activate further *Nodal* expression (Brennan et al., 2001; Norris, Brennan, Bikoff, & Robertson, 2002). BMP4 is able to activate another feedback loop by inducing WNT3, which induces *Nodal* expression. In the posterior epiblast, BMP4 and WNT3 also induce *Brachyury* expression, necessary for correct gastrulation and mesoderm formation (Ben-Haim et al., 2006). BMP4 and its downstream effector SMAD1 are also important for allantois formation and are ultimately responsible for correct placenta development (Tremblay, Dunn, & Robertson, 2001).

The importance of the ExE in primordial germ cell (PGC) differentiation was revealed by the observation that the proximal region of the epiblast gives rise to PGCs while the distal part never does so unless transplanted to a more proximal location (Tam & Zhou, 1996). Later experiments confirmed that the ExE secretes BMP4 and BMP8b, which are essential for PGC generation (Lawson et al., 1999).

These observations demonstrate the critical role of the crosstalk at the embryonic–extraembryonic interface during early mouse development. The possible evolutionary conservation of these mechanisms has not been resolved because of the different geometries of other mammalian postimplantation embryos, such as human or rabbit, which are not cup shaped like the mouse but flat disk shaped (Sheng, 2015), which would require crosstalk without regionalized apposition of embryonic and extraembryonic territories.

The instructive role of extraembryonic cells in patterning noncommitted pluripotent embryonic cells was recently addressed by constructing ES–TS aggregates in 3D (Harrison, Sozen, Christodoulou, Kyprianou, & Zernicka-Goetz, 2017). These structures are capable of expressing *Brachyury* and PGC markers such as *Dppa3* in the ES cell component in a spatially defined fashion and dependent on NODAL signaling. This level of complexity regarding NODAL pathway and PGC formation is not observed in similar aggregates generated exclusively from ES or TS cells. These experiments demonstrate that trophoblast-derived cells can drive pluripotent embryonic cells along spatial and temporal developmental paths similar to those occurring in the embryo. On the other hand, it has been shown that embryonic organoids derived exclusively from ES cells self-organize spatially and acquire axial organization, independently of extraembryonic tissue and, interestingly, of BMP signaling (van den Brink et al., 2014). It can therefore be concluded that the embryo has an intrinsic ability to break its symmetry, and extraembryonic tissues may simply provide the cues needed to bias this intrinsic ability (Turner et al., 2017). This could partly explain how the A–P axis can be established in embryos of species such as humans, in which ExE and epiblast contact is not polarized but takes place over the entire embryo surface.

7. CONCLUSIONS

The emergence of the TE is the first cell-type differentiation event in mammalian embryos. The acquisition of an epithelial phenotype is tightly linked to the differential expression of core TFs, which leads to lineage commitment in the blastocyst. Genetic and morphological properties of the cell, such as apical-basal polarity, are both required for this initial symmetry-breaking event. The study of TE specification provides important clues about how cellular and transcriptional events combine to generate lineage commitment from the naïve state.

ACKNOWLEDGMENTS

We thank Teresa Rayon and members of the Manzanares lab for fruitful discussions and Simon Bartlett (CNIC) for English editing. Work in our lab is supported by the Spanish government (FPI-SO Fellowship to S.M., FPI Fellowship to J.G.S.A., and grant BFU2014-54608-P to M.M.). The CNIC is supported by the Ministry of Economy, Industry and Competitiveness (MEIC) and the Pro CNIC Foundation, and is a Severo Ochoa Center of Excellence (SEV-2015-0505).

REFERENCES

Alarcon, V. B. (2010). Cell polarity regulator PARD6B is essential for trophectoderm formation in the preimplantation mouse embryo. *Biology of Reproduction, 83*, 347–358.

Anani, S., Bhat, S., Honma-Yamanaka, N., Krawchuk, D., & Yamanaka, Y. (2014). Initiation of hippo signaling is linked to polarity rather than to cell position in the preimplantation mouse embryo. *Development, 141*, 2813–2824.

Aplin, J. D., & Ruane, P. T. (2017). Embryo-epithelium interactions during implantation at a glance. *Journal of Cell Science, 130*, 15–22.

Avilion, A. A., Nicolis, S. K., Pevny, L. H., Perez, L., Vivian, N., & Lovell-Badge, R. (2003). Multipotent cell lineages in early mouse development depend on SOX2 function. *Genes & Development, 17*, 126–140.

Beck, F., Erler, T., Russell, A., & James, R. (1995). Expression of Cdx-2 in the mouse embryo and placenta: Possible role in patterning of the extra-embryonic membranes. *Developmental Dynamics, 204*, 219–227.

Beck, S., Le Good, J. A., Guzman, M., Ben Haim, N., Roy, K., Beermann, F., et al. (2002). Extraembryonic proteases regulate nodal signalling during gastrulation. *Nature Cell Biology, 4*, 981–985.

Beddington, R. S., & Robertson, E. J. (1989). An assessment of the developmental potential of embryonic stem cells in the midgestation mouse embryo. *Development, 105*, 733–737.

Ben-Haim, N., Lu, C., Guzman-Ayala, M., Pescatore, L., Mesnard, D., Bischofberger, M., et al. (2006). The nodal precursor acting via activin receptors induces mesoderm by maintaining a source of its convertases and BMP4. *Developmental Cell, 11*, 313–323.

Brennan, J., Lu, C. C., Norris, D. P., Rodriguez, T. A., Beddington, R. S., & Robertson, E. J. (2001). Nodal signalling in the epiblast patterns the early mouse embryo. *Nature, 411*, 965–969.

Cao, Z., Carey, T. S., Ganguly, A., Wilson, C. A., Paul, S., & Knott, J. G. (2015). Transcription factor AP-2gamma induces early Cdx2 expression and represses HIPPO signaling to specify the trophectoderm lineage. *Development, 142*, 1606–1615.

Chambers, I., Colby, D., Robertson, M., Nichols, J., Lee, S., Tweedie, S., et al. (2003). Functional expression cloning of Nanog, a pluripotency sustaining factor in embryonic stem cells. *Cell, 113*, 643–655.

Chen, L., Yabuuchi, A., Eminli, S., Takeuchi, A., Lu, C. W., Hochedlinger, K., et al. (2009). Cross-regulation of the Nanog and Cdx2 promoters. *Cell Research, 19*, 1052–1061.

Chisholm, J. C., & Houliston, E. (1987). Cytokeratin filament assembly in the preimplantation mouse embryo. *Development, 101*, 565–582.

Chung, H., Lee, B. K., Uprety, N., Shen, W., Lee, J., & Kim, J. (2016). Yap1 is dispensable for self-renewal but required for proper differentiation of mouse embryonic stem (ES) cells. *EMBO Reports, 17*, 519–529.

Cockburn, K., Biechele, S., Garner, J., & Rossant, J. (2013). The hippo pathway member Nf2 is required for inner cell mass specification. *Current Biology, 23*, 1195–1201.

Conlon, R. A., Reaume, A. G., & Rossant, J. (1995). Notch1 is required for the coordinate segmentation of somites. *Development*, *121*, 1533–1545.

Cross, J. C., Flannery, M. L., Blanar, M. A., Steingrimsson, E., Jenkins, N. A., Copeland, N. G., et al. (1995). Hxt encodes a basic helix-loop-helix transcription factor that regulates trophoblast cell development. *Development*, *121*, 2513–2523.

Cross, J. C., Nakano, H., Natale, D. R., Simmons, D. G., & Watson, E. D. (2006). Branching morphogenesis during development of placental villi. *Differentiation*, *74*, 393–401.

Dietrich, J. E., & Hiiragi, T. (2007). Stochastic patterning in the mouse pre-implantation embryo. *Development*, *134*, 4219–4231.

Donnison, M., Beaton, A., Davey, H. W., Broadhurst, R., L'Huillier, P., & Pfeffer, P. L. (2005). Loss of the extraembryonic ectoderm in Elf5 mutants leads to defects in embryonic patterning. *Development*, *132*, 2299–2308.

Donnison, M., Broadhurst, R., & Pfeffer, P. L. (2015). Elf5 and Ets2 maintain the mouse extraembryonic ectoderm in a dosage dependent synergistic manner. *Developmental Biology*, *397*, 77–88.

Evans, M. J., & Kaufman, M. H. (1981). Establishment in culture of pluripotential cells from mouse embryos. *Nature*, *292*, 154–156.

Faria, T. N., Ogren, L., Talamantes, F., Linzer, D. I., & Soares, M. J. (1991). Localization of placental lactogen-I in trophoblast giant cells of the mouse placenta. *Biology of Reproduction*, *44*, 327–331.

Feldman, B., Poueymirou, W., Papaioannou, V. E., DeChiara, T. M., & Goldfarb, M. (1995). Requirement of FGF-4 for postimplantation mouse development. *Science*, *267*, 246–249.

Fierro-Gonzalez, J. C., White, M. D., Silva, J. C., & Plachta, N. (2013). Cadherin-dependent filopodia control preimplantation embryo compaction. *Nature Cell Biology*, *15*, 1424–1433.

Goolam, M., Scialdone, A., Graham, S. J., Macaulay, I. C., Jedrusik, A., Hupalowska, A., et al. (2016). Heterogeneity in Oct4 and Sox2 targets biases cell fate in 4-cell mouse embryos. *Cell*, *165*, 61–74.

Guo, G., Huss, M., Tong, G. Q., Wang, C., Li Sun, L., Clarke, N. D., et al. (2010). Resolution of cell fate decisions revealed by single-cell gene expression analysis from zygote to blastocyst. *Developmental Cell*, *18*, 675–685.

Guzman-Ayala, M., Ben-Haim, N., Beck, S., & Constam, D. B. (2004). Nodal protein processing and fibroblast growth factor 4 synergize to maintain a trophoblast stem cell microenvironment. *Proceedings of the National Academy of Sciences of the United States of America*, *101*, 15656–15660.

Haffner-Krausz, R., Gorivodsky, M., Chen, Y., & Lonai, P. (1999). Expression of Fgfr2 in the early mouse embryo indicates its involvement in preimplantation development. *Mechanisms of Development*, *85*, 167–172.

Hancock, S. N., Agulnik, S. I., Silver, L. M., & Papaioannou, V. E. (1999). Mapping and expression analysis of the mouse ortholog of Xenopus Eomesodermin. *Mechanisms of Development*, *81*, 205–208.

Harrison, S. E., Sozen, B., Christodoulou, N., Kyprianou, C., & Zernicka-Goetz, M. (2017). Assembly of embryonic and extraembryonic stem cells to mimic embryogenesis in vitro. *Science*, *356*, eaal1810.

Hillman, N., Sherman, M. I., & Graham, C. (1972). The effect of spatial arrangement on cell determination during mouse development. *Journal of Embryology and Experimental Morphology*, *28*, 263–278.

Hirate, Y., Hirahara, S., Inoue, K., Kiyonari, H., Niwa, H., & Sasaki, H. (2015). Par-aPKC-dependent and -independent mechanisms cooperatively control cell polarity, hippo signaling, and cell positioning in 16-cell stage mouse embryos. *Development, Growth & Differentiation*, *57*, 544–556.

Hirate, Y., Hirahara, S., Inoue, K., Suzuki, A., Alarcon, V. B., Akimoto, K., et al. (2013). Polarity-dependent distribution of angiomotin localizes hippo signaling in preimplantation embryos. *Current Biology*, *23*, 1181–1194.

Home, P., Kumar, R. P., Ganguly, A., Saha, B., Milano-Foster, J., Bhattacharya, B., et al. (2017). Genetic redundancy of GATA factors in the extraembryonic trophoblast lineage ensures the progression of preimplantation and postimplantation mammalian development. *Development*, *144*, 876–888.

Home, P., Ray, S., Dutta, D., Bronshteyn, I., Larson, M., & Paul, S. (2009). GATA3 is selectively expressed in the trophectoderm of peri-implantation embryo and directly regulates Cdx2 gene expression. *The Journal of Biological Chemistry*, *284*, 28729–28737.

Home, P., Saha, B., Ray, S., Dutta, D., Gunewardena, S., Yoo, B., et al. (2012). Altered subcellular localization of transcription factor TEAD4 regulates first mammalian cell lineage commitment. *Proceedings of the National Academy of Sciences of the United States of America*, *109*, 7362–7367.

Huang, D., Guo, G., Yuan, P., Ralston, A., Sun, L., Huss, M., et al. (2017). The role of Cdx2 as a lineage specific transcriptional repressor for pluripotent network during the first developmental cell lineage segregation. *Scientific Report*, *7*, 17156.

Johnson, M. H., & Ziomek, C. A. (1981). The foundation of two distinct cell lineages within the mouse morula. *Cell*, *24*, 71–80.

Kaneko, K. J., & Depamphilis, M. L. (2013). TEAD4 establishes the energy homeostasis essential for blastocoel formation. *Development*, *140*, 3680–3690.

Kang, M., Garg, V., & Hadjantonakis, A. K. (2017). Lineage establishment and progression within the inner cell mass of the mouse blastocyst requires FGFR1 and FGFR2. *Developmental Cell*, *41*, 496–510.e495.

Kelly, S. J. (1977). Studies of the developmental potential of 4- and 8-cell stage mouse blastomeres. *The Journal of Experimental Zoology*, *200*, 365–376.

Koch, U., Lehal, R., & Radtke, F. (2013). Stem cells living with a notch. *Development*, *140*, 689–704.

Kondratiuk, I., Bazydlo, K., Maleszewski, M., & Szczepanska, K. (2012). Delay of polarization event increases the number of Cdx2-positive blastomeres in mouse embryo. *Developmental Biology*, *368*, 54–62.

Kono, K., Tamashiro, D. A., & Alarcon, V. B. (2014). Inhibition of RHO-ROCK signaling enhances ICM and suppresses TE characteristics through activation of hippo signaling in the mouse blastocyst. *Developmental Biology*, *394*, 142–155.

Korotkevich, E., Niwayama, R., Courtois, A., Friese, S., Berger, N., Buchholz, F., et al. (2017). The apical domain is required and sufficient for the first lineage segregation in the mouse embryo. *Developmental Cell*, *40*, 235–247.e237.

Kuckenberg, P., Buhl, S., Woynecki, T., van Furden, B., Tolkunova, E., Seiffe, F., et al. (2010). The transcription factor TCFAP2C/AP-2gamma cooperates with CDX2 to maintain trophectoderm formation. *Molecular and Cellular Biology*, *30*, 3310–3320.

Latos, P. A., & Hemberger, M. (2016). From the stem of the placental tree: Trophoblast stem cells and their progeny. *Development*, *143*, 3650–3660.

Latos, P. A., Sienerth, A. R., Murray, A., Senner, C. E., Muto, M., Ikawa, M., et al. (2015). Elf5-centered transcription factor hub controls trophoblast stem cell self-renewal and differentiation through stoichiometry-sensitive shifts in target gene networks. *Genes & Development*, *29*, 2435–2448.

Lawson, K. A., Dunn, N. R., Roelen, B. A., Zeinstra, L. M., Davis, A. M., Wright, C. V., et al. (1999). Bmp4 is required for the generation of primordial germ cells in the mouse embryo. *Genes & Development*, *13*, 424–436.

Leung, C. Y., & Zernicka-Goetz, M. (2013). Angiomotin prevents pluripotent lineage differentiation in mouse embryos via hippo pathway-dependent and -independent mechanisms. *Nature Communications*, *4*, 2251.

Lian, I., Kim, J., Okazawa, H., Zhao, J., Zhao, B., Yu, J., et al. (2010). The role of YAP transcription coactivator in regulating stem cell self-renewal and differentiation. *Genes & Development*, *24*, 1106–1118.

Lin, S. C., Wani, M. A., Whitsett, J. A., & Wells, J. M. (2010). Klf5 regulates lineage formation in the pre-implantation mouse embryo. *Development*, *137*, 3953–3963.

Maitre, J. L., Niwayama, R., Turlier, H., Nedelec, F., & Hiiragi, T. (2015). Pulsatile cell-autonomous contractility drives compaction in the mouse embryo. *Nature Cell Biology*, *17*, 849–855.

Maitre, J. L., Turlier, H., Illukkumbura, R., Eismann, B., Niwayama, R., Nedelec, F., et al. (2016). Asymmetric division of contractile domains couples cell positioning and fate specification. *Nature*, *536*, 344–348.

Manzanares, M., & Rodriguez, T. A. (2013). Development: Hippo signalling turns the embryo inside out. *Current Biology*, *23*, R559–561.

Martin, G. R. (1981). Isolation of a pluripotent cell line from early mouse embryos cultured in medium conditioned by teratocarcinoma stem cells. *Proceedings of the National Academy of Sciences of the United States of America*, *78*, 7634–7638.

McDole, K., Xiong, Y., Iglesias, P. A., & Zheng, Y. (2011). Lineage mapping the pre-implantation mouse embryo by two-photon microscopy, new insights into the segregation of cell fates. *Developmental Biology*, *355*, 239–249.

McDole, K., & Zheng, Y. (2012). Generation and live imaging of an endogenous Cdx2 reporter mouse line. *Genesis*, *50*, 775–782.

Menchero, S., Rayon, T., Andreu, M. J., & Manzanares, M. (2017). Signaling pathways in mammalian preimplantation development: Linking cellular phenotypes to lineage decisions. *Developmental Dynamics*, *246*, 245–261.

Mihajlovic, A. I., & Bruce, A. W. (2016). Rho-associated protein kinase regulates subcellular localisation of angiomotin and hippo-signalling during preimplantation mouse embryo development. *Reproductive Biomedicine Online*, *33*, 381–390.

Mitsui, K., Tokuzawa, Y., Itoh, H., Segawa, K., Murakami, M., Takahashi, K., et al. (2003). The homeoprotein Nanog is required for maintenance of pluripotency in mouse epiblast and ES cells. *Cell*, *113*, 631–642.

Mlodzik, M., Fjose, A., & Gehring, W. J. (1985). Isolation of caudal, a drosophila homeo box-containing gene with maternal expression, whose transcripts form a concentration gradient at the pre-blastoderm stage. *The EMBO Journal*, *4*, 2961–2969.

Molotkov, A., Mazot, P., Brewer, J. R., Cinalli, R. M., & Soriano, P. (2017). Distinct requirements for FGFR1 and FGFR2 in primitive endoderm development and exit from pluripotency. *Developmental Cell*, *41*, 511–526.e514.

Mould, A., Morgan, M. A., Li, L., Bikoff, E. K., & Robertson, E. J. (2012). Blimp1/Prdm1 governs terminal differentiation of endovascular trophoblast giant cells and defines multipotent progenitors in the developing placenta. *Genes & Development*, *26*, 2063–2074.

Nichols, J., Zevnik, B., Anastassiadis, K., Niwa, H., Klewe-Nebenius, D., Chambers, I., et al. (1998). Formation of pluripotent stem cells in the mammalian embryo depends on the POU transcription factor Oct4. *Cell*, *95*, 379–391.

Nishioka, N., Inoue, K., Adachi, K., Kiyonari, H., Ota, M., Ralston, A., et al. (2009). The hippo signaling pathway components Lats and yap pattern Tead4 activity to distinguish mouse trophectoderm from inner cell mass. *Developmental Cell*, *16*, 398–410.

Nishioka, N., Yamamoto, S., Kiyonari, H., Sato, H., Sawada, A., Ota, M., et al. (2008). Tead4 is required for specification of trophectoderm in pre-implantation mouse embryos. *Mechanisms of Development*, *125*, 270–283.

Niwa, H., Miyazaki, J., & Smith, A. G. (2000). Quantitative expression of Oct-3/4 defines differentiation, dedifferentiation or self-renewal of ES cells. *Nature Genetics*, *24*, 372–376.

Niwa, H., Toyooka, Y., Shimosato, D., Strumpf, D., Takahashi, K., Yagi, R., et al. (2005). Interaction between Oct3/4 and Cdx2 determines trophectoderm differentiation. *Cell, 123*, 917–929.

Norris, D. P., Brennan, J., Bikoff, E. K., & Robertson, E. J. (2002). The Foxh1-dependent autoregulatory enhancer controls the level of nodal signals in the mouse embryo. *Development (Cambridge, England), 129*, 3455–3468.

Nowotschin, S., Xenopoulos, P., Schrode, N., & Hadjantonakis, A. K. (2013). A bright single-cell resolution live imaging reporter of notch signaling in the mouse. *BMC Developmental Biology, 13*, 15.

Ohnishi, Y., Huber, W., Tsumura, A., Kang, M., Xenopoulos, P., Kurimoto, K., et al. (2014). Cell-to-cell expression variability followed by signal reinforcement progressively segregates early mouse lineages. *Nature Cell Biology, 16*, 27–37.

Pan, D. (2010). The Hippo signaling pathway in development and cancer. *Developmental Cell, 19*, 491–505.

Plusa, B., Frankenberg, S., Chalmers, A., Hadjantonakis, A. K., Moore, C. A., Papalopulu, N., et al. (2005). Downregulation of Par3 and aPKC function directs cells towards the ICM in the preimplantation mouse embryo. *Journal of Cell Science, 118*, 505–515.

Polydorou, C., & Georgiades, P. (2013). Ets2-dependent trophoblast signalling is required for gastrulation progression after primitive streak initiation. *Nature Communications, 4*, 1658.

Posfai, E., Petropoulos, S., de Barros, F. R., Schell, J. P., Jurisica, I., Sandberg, R., et al. (2017). Position- and Hippo signaling-dependent plasticity during lineage segregation in the early mouse embryo. *eLife, 6*, pii: e22906.

Ralston, A., Cox, B. J., Nishioka, N., Sasaki, H., Chea, E., Rugg-Gunn, P., et al. (2010). Gata3 regulates trophoblast development downstream of Tead4 and in parallel to Cdx2. *Development, 137*, 395–403.

Ralston, A., & Rossant, J. (2008). Cdx2 acts downstream of cell polarization to cell-autonomously promote trophectoderm fate in the early mouse embryo. *Developmental Biology, 313*, 614–629.

Rayon, T., Menchero, S., Nieto, A., Xenopoulos, P., Crespo, M., Cockburn, K., et al. (2014). Notch and Hippo converge on cdx2 to specify the trophectoderm lineage in the mouse blastocyst. *Developmental Cell, 30*, 410–422.

Rayon, T., Menchero, S., Rollan, I., Ors, I., Helness, A., Crespo, M., et al. (2016). Distinct mechanisms regulate Cdx2 expression in the blastocyst and in trophoblast stem cells. *Scientific Reports, 6*, 27139.

Rossant, J., & Cross, J. C. (2001). Placental development: Lessons from mouse mutants. *Nature Reviews. Genetics, 2*, 538–548.

Samarage, C. R., White, M. D., Alvarez, Y. D., Fierro-Gonzalez, J. C., Henon, Y., Jesudason, E. C., et al. (2015). Cortical tension allocates the first inner cells of the mammalian embryo. *Developmental Cell, 34*, 435–447.

Sasaki, H. (2015). Position- and polarity-dependent Hippo signaling regulates cell fates in preimplantation mouse embryos. *Seminars in Cell & Developmental Biology, 47–48*, 80–87.

Screen, M., Dean, W., Cross, J. C., & Hemberger, M. (2008). Cathepsin proteases have distinct roles in trophoblast function and vascular remodelling. *Development, 135*, 3311–3320.

Senner, C. E., Krueger, F., Oxley, D., Andrews, S., & Hemberger, M. (2012). DNA methylation profiles define stem cell identity and reveal a tight embryonic–extraembryonic lineage boundary. *Stem Cells, 30*, 2732–2745.

Sheng, G. (2015). Epiblast morphogenesis before gastrulation. *Developmental Biology, 401*, 17–24.

Shi, S., Stahl, M., Lu, L., & Stanley, P. (2005). Canonical notch signaling is dispensable for early cell fate specifications in mammals. *Molecular and Cellular Biology*, *25*, 9503–9508.

Simmons, D. G., Natale, D. R., Begay, V., Hughes, M., Leutz, A., & Cross, J. C. (2008). Early patterning of the chorion leads to the trilaminar trophoblast cell structure in the placental labyrinth. *Development*, *135*, 2083–2091.

Souilhol, C., Cormier, S., Tanigaki, K., Babinet, C., & Cohen-Tannoudji, M. (2006). RBP-Jkappa-dependent notch signaling is dispensable for mouse early embryonic development. *Molecular and Cellular Biology*, *26*, 4769–4774.

Strumpf, D., Mao, C. A., Yamanaka, Y., Ralston, A., Chawengsaksophak, K., Beck, F., et al. (2005). Cdx2 is required for correct cell fate specification and differentiation of trophectoderm in the mouse blastocyst. *Development*, *132*, 2093–2102.

Suwinska, A., Czolowska, R., Ozdzenski, W., & Tarkowski, A. K. (2008). Blastomeres of the mouse embryo lose totipotency after the fifth cleavage division: Expression of Cdx2 and Oct4 and developmental potential of inner and outer blastomeres of 16- and 32-cell embryos. *Developmental Biology*, *322*, 133–144.

Tabansky, I., Lenarcic, A., Draft, R. W., Loulier, K., Keskin, D. B., Rosains, J., et al. (2013). Developmental bias in cleavage-stage mouse blastomeres. *Current Biology*, *23*, 21–31.

Tam, P. P., & Zhou, S. X. (1996). The allocation of epiblast cells to ectodermal and germ-line lineages is influenced by the position of the cells in the gastrulating mouse embryo. *Developmental Biology*, *178*, 124–132.

Tanaka, S., Kunath, T., Hadjantonakis, A. K., Nagy, A., & Rossant, J. (1998). Promotion of trophoblast stem cell proliferation by FGF4. *Science*, *282*, 2072–2075.

Tarkowski, A. K. (1959). Experiments on the development of isolated blastomers of mouse eggs. *Nature*, *184*, 1286–1287.

Tarkowski, A. K., & Wroblewska, J. (1967). Development of blastomeres of mouse eggs isolated at the 4- and 8-cell stage. *Journal of Embryology and Experimental Morphology*, *18*, 155–180.

Torres-Padilla, M. E., Parfitt, D. E., Kouzarides, T., & Zernicka-Goetz, M. (2007). Histone arginine methylation regulates pluripotency in the early mouse embryo. *Nature*, *445*, 214–218.

Toyooka, Y., Oka, S., & Fujimori, T. (2016). Early preimplantation cells expressing Cdx2 exhibit plasticity of specification to TE and ICM lineages through positional changes. *Developmental Biology*, *411*, 50–60.

Tremblay, K. D., Dunn, N. R., & Robertson, E. J. (2001). Mouse embryos lacking Smad1 signals display defects in extra-embryonic tissues and germ cell formation. *Development*, *128*, 3609–3621.

Turner, D. A., Girgin, M., Alonso-Crisostomo, L., Trivedi, V., Baillie-Johnson, P., Glodowski, C. R., et al. (2017). Anteroposterior polarity and elongation in the absence of extra-embryonic tissues and of spatially localisedsignalling in gastruloids: Mammalian embryonic organoids. *Development*, *144*, 3894–3906.

van den Brink, S. C., Baillie-Johnson, P., Balayo, T., Hadjantonakis, A. K., Nowotschin, S., Turner, D. A., et al. (2014). Symmetry breaking, germ layer specification and axial organisation in aggregates of mouse embryonic stem cells. *Development*, *141*, 4231–4242.

Watanabe, T., Biggins, J. S., Tannan, N. B., & Srinivas, S. (2014). Limited predictive value of blastomere angle of division in trophectoderm and inner cell mass specification. *Development*, *141*, 2279–2288.

Watanabe, Y., Miyasaka, K. Y., Kubo, A., Kida, Y. S., Nakagawa, O., Hirate, Y., et al. (2017). Notch and Hippo signaling converge on strawberry notch 1 (Sbno1) to synergistically activate Cdx2 during specification of the trophectoderm. *Scientific Reports*, *7*, 46135.

Wicklow, E., Blij, S., Frum, T., Hirate, Y., Lang, R. A., Sasaki, H., et al. (2014). HIPPO pathway members restrict SOX2 to the inner cell mass where it promotes ICM fates in the mouse blastocyst. *PLoS Genetics*, *10*, e1004618.

Yagi, R., Kohn, M. J., Karavanova, I., Kaneko, K. J., Vullhorst, D., DePamphilis, M. L., et al. (2007). Transcription factor TEAD4 specifies the trophectoderm lineage at the beginning of mammalian development. *Development*, *134*, 3827–3836.

Yamamoto, H., Flannery, M. L., Kupriyanov, S., Pearce, J., McKercher, S. R., Henkel, G. W., et al. (1998). Defective trophoblast function in mice with a targeted mutation of Ets2. *Genes & Development*, *12*, 1315–1326.

Yoshimizu, T., Obinata, M., & Matsui, Y. (2001). Stage-specific tissue and cell interactions play key roles in mouse germ cell specification. *Development*, *128*, 481–490.

Ziomek, C. A., & Johnson, M. H. (1980). Cell surface interaction induces polarization of mouse 8-cell blastomeres at compaction. *Cell*, *21*, 935–942.

CHAPTER FIVE

Primitive Endoderm Differentiation: From Specification to Epithelialization

Cécilia Bassalert[2], Lorena Valverde-Estrella[2], Claire Chazaud[1]
GReD, Université Clermont Auvergne, CNRS, INSERM, Clermont-Ferrand, France
[1]Corresponding author: e-mail address: claire.chazaud@uca.fr

Contents

Abstract

At the time of implantation, the mouse blastocyst has developed three cell lineages: the epiblast (Epi), the primitive endoderm (PrE), and the trophectoderm (TE). The PrE and TE are extraembryonic tissues but their interactions with the Epi are critical to sustain embryonic growth, as well as to pattern the embryo. We review here the cellular and molecular events that lead to the production of PrE and Epi lineages and discuss the different hypotheses that are proposed for the induction of these cell types. In the second part, we report the current knowledge about the epithelialization of the PrE.

[2] Equal contribution.

Current Topics in Developmental Biology, Volume 128
ISSN 0070-2153
https://doi.org/10.1016/bs.ctdb.2017.12.001

1. INTRODUCTION

To sustain nutritional needs, mammalian embryos have evolved by developing extraembryonic tissues. These are not only providing nutrient requirements but are also involved in the patterning of embryonic tissues by controlling key events such as the antero–posterior axis patterning or the induction of germ cells. Two extraembryonic lineages are segregated before implantation: the trophectoderm (TE) lineage, which will later participate to the placenta, and the primitive endoderm (PrE) lineage that will be one of the constituents of the yolk sac. In the meanwhile, the embryonic lineage, the epiblast (Epi), has to maintain its capacity to produce all the cells of the future body, protecting itself from the surrounding differentiation cues and preparing for later gastrulation.

This review will focus on how the PrE lineage is induced and then differentiates. Interestingly this tissue goes through many morphogenetic changes including cell movements, epithelialization, and epithelium to mesenchyme transition (EMT), making it an attractive model for dynamic changes.

2. GENERATION OF TROPHECTODERM AND INNER CELL MASS CELL LINEAGES

After fertilization, the zygote divides and reaches the 8-cell stage, where the cells polarize and increase their contacts, leading to a compacted embryo (reviewed in White, Bissiere, Alvarez, & Plachta, 2016). Due to the compaction, the two subsequent division rounds (from 8- to 16- and 16- to 32-cell stage) will be differentiative, generating inner and outer cells due to complex mechanisms (reviewed in Maître, 2017). Outer cells remain polarized and will later constitute the TE, whereas inner apolarized cells compose the inner cell mass (ICM). The ICM lineage is then produced from two sequential divisions. Around E3.0–E3.25 (20- to 32-cell stage), the blastocoel cavity appears, constraining the ICM to one pole of the embryo that has then become a blastocyst (Fig. 1).

The acquisition of the TE or ICM fate essentially depends on the position-dependent Hippo pathway (Nishioka et al., 2009). Combined to the Notch activity, it restricts the expression of the transcription factors (TFs), CDX2 and GATA3, to outside cells (Rayon et al., 2014). These TFs are required for the acquisition and maintenance of the TE identity. While CDX2 and OCT4 are initially coexpressed in all blastomeres, the restricted expression of CDX2 in outside cells blocks the transcription of

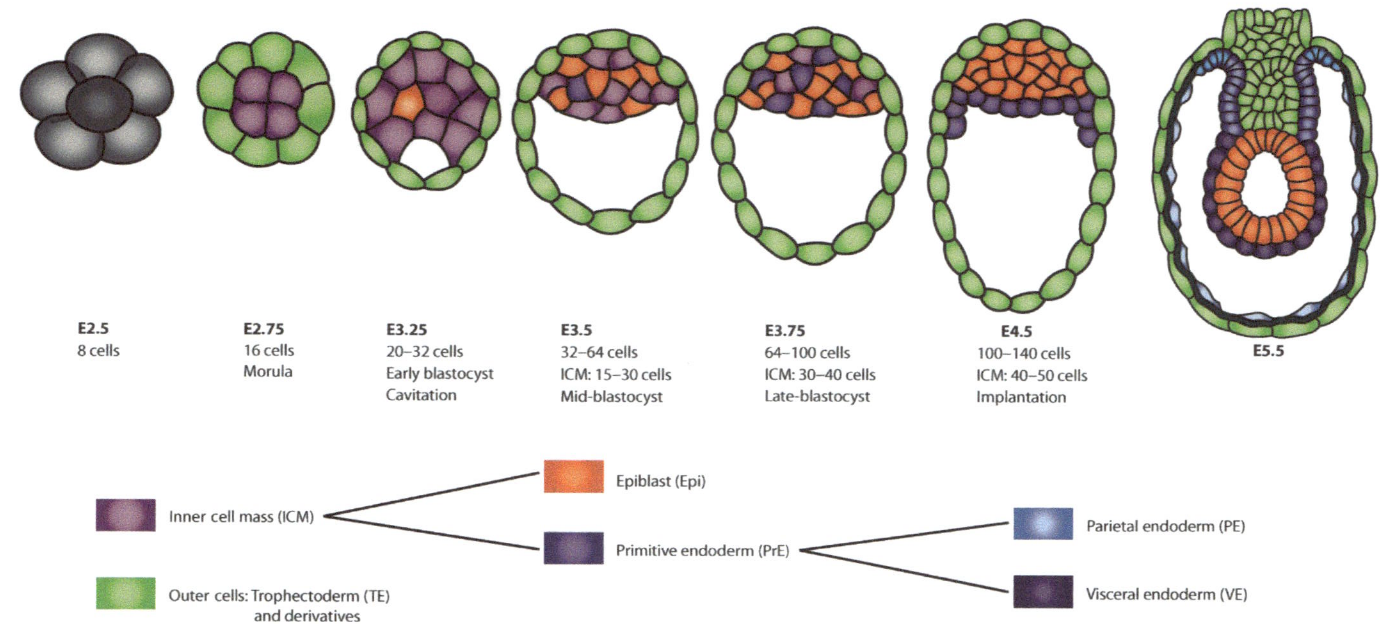

Fig. 1 Mouse development from the 8-cell stage to early implantation and its associated cell lineages.

Oct4 in these cells, which expression is then confined to ICM cells (Niwa et al., 2005). Another ICM marker, SOX2, starts to be expressed in inner cells from the 16-cell stage, also in link with the Hippo pathway but independently of CDX2 (Wicklow et al., 2014). Both OCT4 and SOX2 are essential for the maintenance of ICM derivatives (Avilion et al., 2003; Le Bin et al., 2014; Nichols et al., 1998; Wicklow et al., 2014).

3. REGULATION OF THE Epi/PrE SPECIFICATION

Starting from E3.0–E3.25 (20- to 32-cell stage) individual ICM cells start to specify into either Epi or PrE. This is evidenced by the reciprocal and exclusive expression of the TF NANOG and GATA6, in a "salt and pepper" pattern (Fig. 1), which is completed by E3.75–E4.0 stage. In fact, NANOG and GATA6 are coexpressed in all blastomeres from the 8-cell stage then, during the specification process, one of these markers is lost individually in each cell, both at RNA and protein levels, in an apparently random pattern (Bessonnard et al., 2014; Guo et al., 2010; Kurimoto et al., 2006; Plusa, Piliszek, Frankenberg, Artus, & Hadjantonakis, 2008; Saiz, Williams, Seshan, & Hadjantonakis, 2016). Importantly, the specification is asynchronous among ICM cells (see Figs. 1 and 2). Indeed, the first cells start to specify around E3.0–E3.25, whereas the last ones do so at E3.75–E4.0, stage at which no coexpressing cells are found any more (Bessonnard et al., 2014; Gerbe, Cox, Rossant, & Chazaud, 2008; Plusa et al., 2008; Saiz et al., 2016). Interestingly, even after cell specification (around E3.75) there are great variations in protein levels of NANOG in Epi and of GATA6 in PrE cells, suggesting that cell identity is acquired by the loss of one of the markers rather than the increase of the reciprocal factor (Bessonnard et al., 2014; Dietrich & Hiiragi, 2007; Saiz et al., 2016). It is thus difficult to assess cell identity looking at only one of the markers. Therefore, analyzing relative levels of NANOG and GATA6 expression in each cell better defines cell identity (Bessonnard et al., 2014; Guo et al., 2010; Kurimoto et al., 2006; Saiz et al., 2016): precursor cells are double positive for the markers, while specified cells have downregulated one of the markers: NANOG^{high}/GATA6^{low} for Epi cells and NANOG^{low}/GATA6^{high} for PrE cells.

The deletion of *Nanog* and *Gata6* genes has revealed their driving function for cell fate specification. In *Gata6* mutants, all ICM cells express NANOG without any PrE markers, indicating the acquisition of an Epi fate (Bessonnard et al., 2014; Schrode, Saiz, Di Talia, & Hadjantonakis, 2014), whereas in *Nanog* mutants the opposite phenotype is observed, as all ICM cells express GATA6 (Frankenberg et al., 2011). Thus, NANOG and

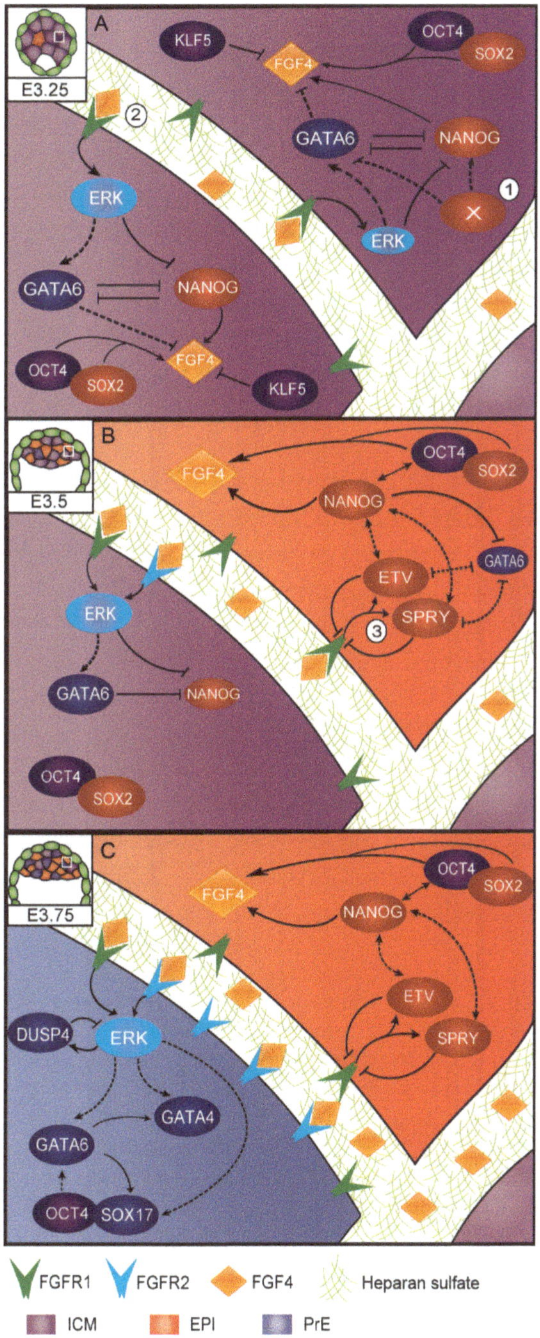

Fig. 2 See legend on next page.

GATA6 are master regulators driving a cell toward an Epi or a PrE fate, respectively. *Gata6* heterozygote embryos exhibit an imbalance in ICM composition with less PrE cells (Bessonnard et al., 2014; Schrode et al., 2014). In addition, lowering the dose of GATA6 by removing one allele leads to an earlier specification of Epi cells, visualized by more NANOG^{high}/GATA6^{low} at E3.25 (Bessonnard et al., 2014; Schrode et al., 2014) and by a higher number of cells expressing the Epi marker *Fgf4* (Bessonnard et al., 2014). In parallel, PrE cell specification is delayed in these embryos, as seen by a higher proportion of double-positive cells around E3.75 (Schrode et al., 2014) that finally specify into Epi and PrE by the late blastocyst stage (around E4.0) but in skewed proportions with less PrE cells (Bessonnard et al., 2014; Schrode et al., 2014).This demonstrates that probably due to the antagonism between NANOG and GATA6, the dose of GATA6 is directly converted into cell numbers and also defines the timing of cell differentiation. However, Epi and PrE cells specify in correct proportions in *Nanog* heterozygotes (Miyanari & Torres-Padilla, 2012), denoting a probable compensation mechanism.

The initial coexpression of NANOG and GATA6 in precursor cells implies that other factor(s) is (are) required to tip the scales into one lineage fate. Blocking the MEK1/ERK pathway using pharmacological inhibitors or the KO of FGF pathway members such as *Grb2* (an adaptor for RTK receptors), *Fgf4*, or *Fgfr1/2* prevents PrE cell differentiation, leading to a fully epiblastic ICM (Chazaud, Yamanaka, Pawson, & Rossant, 2006; Kang, Garg, & Hadjantonakis, 2017; Kang, Piliszek, Artus, & Hadjantonakis, 2013; Krawchuk, Honma-Yamanaka, Anani, & Yamanaka, 2013; Molotkov, Mazot, Brewer, Cinalli, & Soriano, 2017; Nichols, Silva, Roode, & Smith,

Fig. 2 Specification of Epi and PrE lineages. (A) At E3.0, unspecified ICM cells are similar and coexpress NANOG and GATA6. Low levels of FGF4 are produced. Concomitantly to a low ERK signaling, an unknown factor (X) may activate NANOG expression or inactivate GATA6's ①. Alternatively, in the intercellular space, inhomogeneous FGF4 could create differences between two neighboring cells ②, starting to imbalance ERK signaling. (B) At E3.5, some cells have started to differentiate. Epi cells specify first with a high NANOG/GATA6 ratio. This leads to an increased expression and then secretion of FGF4. ETVs and Sproutys reinforce the Epi state. Their expression is either induced by a high NANOG/GATA6 ratio or, alternatively, their initial presence ③ could create an imbalance in the transmission of the FGF signaling. The neighboring cell, still in a precursor state, receives a high dose of FGF4 and will start to differentiate into PrE by activating the FGFR1/ERK pathway. (C) At E3.75, most of Epi and PrE cells are specified. PrE cells express a low NANOG/GATA6 ratio. They receive high doses of FGF4 signal, which activates the FGFR/ERK pathway and leads, in coordination with GATA6, to the expression of several mature PrE factors such as GATA4 or SOX17.

2009). On the reverse, administration of recombinant FGF4 (recFGF4) to WT embryos biases the cells toward a PrE identity (Yamanaka, Lanner, & Rossant, 2010). These results show that FGF4/FGFRs are the RTK pathway required for PrE differentiation. Interestingly, deleting *Fgf4* or *Fgfrs* does not prevent GATA6 expression up to 32-cell stage, demonstrating an independent induction from the 4/8-cell stage (Kang et al., 2013; Molotkov et al., 2017). *Fgf4* is expressed by ICM cells (Niswander & Martin, 1992) and is specific to Epi cells (Frankenberg et al., 2011; Guo et al., 2010 ; Kurimoto et al., 2006; Ohnishi et al., 2014). Of note, *Fgf4* is the earliest factor to be differentially expressed in the ICM from E3.25 (Guo et al., 2010; Ohnishi et al., 2014). It was initially thought that the FGF4 signal was transduced by the FGFR2 receptor, notably because this receptor is specifically expressed by PrE cells (Guo et al., 2010; Kurimoto et al., 2006; Ohnishi et al., 2014). However, recent analyses of *Fgfr1* and/or *Fgfr2* mutants have revealed that FGFR1 is the principal receptor driving cell heterogeneity by inducing PrE cells, while only the absence of both receptors leads to a full loss of PrE (Kang et al., 2017; Molotkov et al., 2017). Interestingly, *Fgfr1* is equally expressed in all three cell lineages of the blastocyst, TE, PrE, and Epi as soon as E3.25 (Guo et al., 2010; Kang et al., 2017; Molotkov et al., 2017; Ohnishi et al., 2014), strongly suggesting that all the cells have an equal capacity to receive the FGF signal. Thus, the differential FGF signaling activity could be due to different amount of FGF4 surrounding the cells.

However, *Etv4–5* as well as *Spry4*, that are negative regulators of the RTK pathway (Hayashi et al., 2009; Kramer, Okabe, Hacohen, Krasnow, & Hiromi, 1999; Reich, Sapir, & Shilo, 1999; Taniguchi et al., 2007), are preferentially expressed in Epi cells and are under the dependence of FGFR1 activity (Kang et al., 2017). On the other hand, in PrE cells, FGFR1 activation induces the expression of Dusp4, a phosphatase protein that inhibits ERK. These results show that FGFR1 seems to be activated in both Epi and PrE cells, but the downstream signaling pathways are different between the two cell types suggesting distinct downstream activities. ETV4 and SPRY4 would block FGFRs activity only in Epi cells, most likely to desensitize the cells from their own secretion (Kang et al., 2017), whereas in PrE cells, Dusp4 possibly allows a fine-tune regulation, or enables a resensitization, so that the cells can perceive novel pulses of ERK activation.

The dosage and timing of the FGF4/FGFR signaling are critical. Rescue experiments of *Fgf4* mutants treated by increasing doses of recFGF4 were carried out. One article describes an ICM exhibiting all Nanog-positive or all Gata6-positive cells (Kang et al., 2013), suggesting that FGF4 must be distributed unevenly to recapitulate a proper salt and pepper pattern.

Another report shows a dose-dependent number of differentiated PrE cells (Krawchuk et al., 2013). What causes this difference is not clear. Anyhow, decreasing the number of FGFRs WT alleles gradually impairs the PrE/Epi ratio (Kang et al., 2017; Molotkov et al., 2017). As well, FGF4 expression is increased in *Klf5* mutants (Azami et al., 2017), leading to a higher number of PrE cells (Azami et al., 2017; Lin, Wani, Whitsett, & Wells, 2010). Thus altogether, these results show that precisely tuned FGF signaling controls the proportions of PrE and Epi populations. Whether extracellular FGF4 must be heterogeneously distributed is still in debate.

In addition, the use of time-window treatments with recFGF4 or MEK inhibitors revealed that ICM cells are plastic, i.e., can switch from Epi to PrE or vice versa upon cell grafts and Fgf4 pathway activation or inhibition (Grabarek et al., 2012; Saiz et al., 2016; Yamanaka et al., 2010), meaning they can adopt either identity depending on the environment. This plasticity is present before E4.0 as, subsequently, specified cells become insensitive to these treatments in an asynchronous manner (Grabarek et al., 2012; Saiz et al., 2016; Yamanaka et al., 2010). Actually, a recent analysis of NANOG and GATA6 expression in individual cells shows that the proportion of sensitive cells correlates with the number of precursors, unspecified cells that coexpress both TFs. Thus, the plasticity of the ICM as a whole would be linked mainly to the presence of these unspecified bipotential cells (Saiz et al., 2016). Still, in such experiments, a low number of NANOG^{high}/GATA6^{low} or NANOG^{low}/GATA6^{high} cells remained plastic upon environmental changes by switching identity. This suggests that this ability exists in a very short time window, probably when the cells are in transition toward a stable identity, shortly after leaving the precursor state.

Interestingly, loss of cell plasticity occurs earlier for Epi cells than for PrE cells. This was observed with cell transplantation experiments (Grabarek et al., 2012) as well as with insensitivity to Fgf4 or MEK inhibitors treatments (Bessonnard et al., 2017).

4. MECHANISM FOR SALT AND PEPPER PATTERN PROPAGATION

Altogether, work from different labs shows that around E3.0–E3.25, by elevating NANOG/GATA6 ratio, Epi cells are the first to specify (Bessonnard et al., 2017, 2014). The primacy of Epi cell specification is also supported by mathematical models (Bessonnard et al., 2014) and strongly correlates with an earlier loss of plasticity of these cells compared to PrE cells (Bessonnard et al., 2017; Grabarek et al., 2012; Saiz et al., 2016) (Fig. 2).

The Epi state causes an increase of FGF4 expression and secretion (Frankenberg et al., 2011; Kang et al., 2017; Ohnishi et al., 2014), which, in sufficient amounts, can convert naïve neighboring cells into PrE. All precursor cells are competent to transduce FGF4 from early time points as the administration of recFGF4 between E2.5 and E3.25 rapidly downregulates NANOG expression (Bessonnard et al., 2017; Saiz et al., 2016) and induces a precocious SOX17 expression in most ICM cells (Bessonnard et al., 2014). *Klf5* KO embryos also exhibit an earlier differentiation due to a precociously elevated FGF4 expression (Azami et al., 2017).

Mathematical models based on in vivo or in vitro data show that NANOG, GATA6, and the FGFR/ERK pathway constitute a gene regulatory network, which self-organization is sufficient to lead to the binary cell fate specification (Bessonnard et al., 2014; Schröter, Rué, Mackenzie, & Martinez Arias, 2015). These are based on the reciprocal inhibition between NANOG and GATA6 expression, balanced by the FGF4/FGFR pathway. Experiments on ES cells converted into extraembryonic endoderm by ectopic expression of GATA6 or active c-RAF, which is upstream of ERK, propose that the sole repression of NANOG expression by ERK activity could be sufficient, with no direct impact on GATA6 expression (Hamilton & Brickman, 2014; Schröter et al., 2015).

5. INDUCTION OF Epi/PrE SPECIFICATION

Finding what triggers the conversion of a precursor cell into an Epi state will probably be crucial to understand how the salt and pepper pattern is induced. Therefore, any factor elevating the NANOG/GATA6 ratio could behave as an inducer and the loss of GATA6 expression seems to be key in the system. It was hypothesized that a diminution of ERK signaling could trigger the mechanism (Bessonnard et al., 2014). However, recent data suggest that having low ERK signaling is necessary but it is not sufficient to change a precursor identity into an Epi identity. Indeed, when embryos are cultured with MEK inhibitors from E2.5 to E3.75 or to E4.5 all ICM cells have an Epi identity (Bessonnard et al., 2017; Saiz et al., 2016; Yamanaka et al., 2010). However, a culture from E2.5 to E3.5 in the same conditions leads to an ICM composed of Epi cells as well as of cells coexpressing NANOG and GATA6 (Bessonnard et al., 2017; Saiz et al., 2016). Thus despite the MEK inhibitor treatment, a subpopulation of cells remains in a precursor state. These cells are still bipotential, as shown by a further culture in control medium (Saiz et al., 2016). Therefore, to be competent to become Epi, a precursor cell requires not only a low ERK activity but also unknown factor(s) to elevate the NANOG/GATA6 ratio (Fig. 2A). Whether it depends on a clock or on other factor(s) is currently unknown. Interestingly, this competency is acquired asynchronously and is thus probably driving the asynchrony of Epi/PrE differentiation.

ETV4–5 or SPRY2–4 could play a role in the initiation of the Epi/Pre specification, by preventing ERK activity. Knowing how and when they are induced before E3.5, as well as their role, will be interesting. It will be significant to know whether they are expressed once Epi cells have started to specify, i.e., whether their expression depends on a NANOG^{high}/GATA6^{low} activity. In that case, they would only reinforce the Epi state (Fig. 2B) while an earlier independent expression may enroll them in the initiation mechanism of the salt and pepper.

ICM cells are resulting from both the fourth (8–16 cells) and fifth (16–32 cells) divisions that are differentiative. It has been proposed that heritage of FGFR2 could be different between the cells produced from the two divisions, potentially inducing an Epi/PrE bias (Morris, Graham, Jedrusik, & Zernicka-Goetz, 2013). However, *Fgfr2* mutants have a minor phenotype in Epi/PrE specification (Kang et al., 2017; Molotkov et al., 2017) so FGFR2 expression would unlikely bias cell fate. Nevertheless, other factors could be differentially inherited from outer cells.

Inner cells express *Fgf4* at moderate levels between the 16- and 32-cell stage (Kang et al., 2017; Krupa et al., 2014; Ohnishi et al., 2014). Chimera experiments have suggested that the accumulation of inner cells, produced by inner and outer cell division, could lead to an elevation of global FGF4 secretion (Krupa et al., 2014), sufficient to induce the differentiation of some cells into PrE. Still, FGF4 concentration would have to be distributed unevenly to lead to a "salt and pepper" pattern, and this could be possibly due to differences in diffusion through the extracellular matrix (Fig. 2A). In that case, PrE cells would be actively differentiating, while Epi cells would be a default state. Alternatively, the accumulation of inner cells would allow a greater exposure to FGF4 due to a higher number of neighbors that may induce differentiation into PrE (Tosenberger et al., 2017).

Nevertheless, cell accumulation might not be enough to activate FGFR1, and indeed an increase of *Fgf4* expression is observed in some individual cells around the 32-cell stage (Guo et al., 2010; Ohnishi et al., 2014). It is not known whether this increase depends on a NANOG^{high}/GATA6^{low} activity in a few cells. At least, it seems to be due to the release of KLF5 repression, as *Klf5* KO embryos have an elevated *Fgf4* expression already at E3.0 (Azami et al., 2017). KLF factors promote pluripotency and can cooccupy several sites with NANOG in ES cells, including *Nanog* locus itself as well as *Fgf4*'s (Azami et al., 2017; Jiang et al., 2008). However, promoting Nanog/Sox2/Oct4 expression and repressing Fgf4 expression at the same time seems contradictory and more analyses will be required.

Chromatin modifiers could be potential initiators of Epi/PrE specification by modulating *Nanog* or *Gata6* expression, directly or indirectly. A few factors such as Prdm14, Carm1, and Satb1 have been identified as potential actors due to their differential expression between Epi and PrE cells and/or their effect during additive transgenesis. Satb1, which is enriched in PrE cells, controls *Nanog* expression through a direct repression in ES cells (Savarese et al., 2009). In embryos, it was shown that Satb1 expression is downstream of the FGF pathway. However, loss or gain of function of Satb1 only partially biases the Epi/PrE ratio (Goolam & Zernicka-Goetz, 2017), and no phenotype during preimplantation stages has been described so far with mutant embryos (Alvarez et al., 2000). Prdm14 also limits the FGF pathway activity, at least in ES cells (Grabole et al., 2013). Despite a specific expression in Epi cells and a promotion of the Epi state by forced expression (Burton et al., 2013) no phenotype concerning Epi vs PrE specification was observed in mutants embryos (Payer et al., 2013; Yamaji et al., 2008). In *Carm1* mutant embryos, the Epi cell number is moderately impaired at E4.5, and whether more PrE cells are produced was not indicated (Panamarova et al., 2016). Thus, the chromatin modifiers examined so far seem to be downstream of the Epi/PrE specification rather than acting as inducers. The functional analysis of other differentially expressed modifiers (Burton et al., 2013; Nestorov, Hotz, Liu, & Peters, 2015) might bring out upstream activators.

Due to the cell-to-cell propagation of the salt and pepper mechanism mediated by FGF4 secretion, only a few Epi cells would be necessary to engage and spread the Epi and PrE specification. Mathematical simulations indicate that only one specified Epi cell could be sufficient to set the mechanism on (Bessonnard et al., 2014). Small stochastic variations due to transcriptional noise in one or a few cells have been proposed to elevate the NANOG/GATA6 ratio (Ohnishi et al., 2014). Differences in transcriptional noise have been observed throughout preimplantation (Piras, Tomita, & Selvarajoo, 2014), and it would be particularly interesting to focus around the 16- to 32-cell stage. Such mechanisms for cell differentiation have been observed in other tissues or organisms (Chang, Hemberg, Barahona, Ingber, & Huang, 2008; Maamar, Raj, & Dubnau, 2007). However, a mathematical analysis shows that if transcriptional noise of *Nanog* and

Gata6 is considered as a source of randomness, simulations predict unrealistic cell behaviors (De Mot et al., 2016). Still, noise-driven induction may depend on other unknown factors not considered in the mathematical model.

6. PrE MATURATION THROUGH THE FGF PATHWAY

Once the cells have specified into Epi or PrE, their differentiation continues by the adoption of novel markers necessary for further morphogenetic changes, owing to their future tissue functions. In the PrE, subsequently to the downregulation of NANOG and maintenance of GATA6 levels, several proteins expression such as SOX17, SOX7, GATA4, LRP2, and Dab2 (Gerbe et al., 2008; Guo et al., 2010; Kurimoto et al., 2006; Morris et al., 2010; Niakan et al., 2010; Plusa et al., 2008) are induced in a temporal manner slightly before implantation (Ohnishi et al., 2014). Whereas none are individually required at that stage, the TF (SOX17, SOX7, GATA4, and FOXQ1) could collectively act in reinforcing the salt and pepper pattern. In the meanwhile, other proteins are preparing the PrE cell for the following epithelialization steps (collagens, laminins, LRP2, etc.) (Gerbe et al., 2008; Ohnishi et al., 2014).

Surprisingly, these later PrE markers are weakly expressed or not found in *Nanog* mutants (Silva et al., 2009). Further work showed that NANOG acts on the PrE in a non cell-autonomous manner through FGF4 secretion to allow the induction of SOX17, GATA4, and PDGFRa expression (Frankenberg et al., 2011; Messerschmidt & Kemler, 2010). In *Nanog*$^{-/-}$ embryos, GATA6 is expressed but it is not sufficient to induce its downstream PrE targets, meaning that they require also the activity of the FGFR/MEK pathway. The deletion of *Gata6* leads to an absence of the later PrE markers expression (Bessonnard et al., 2014; Cai, Capo-Chichi, Rula, Yang, & Xu, 2008; Schrode et al., 2014), and this phenotype cannot be rescued by recFGF4 administration (Bessonnard et al., 2014; Schrode et al., 2014). Thus, GATA6 and the FGF pathway act jointly to differentiate the PrE, in a cell-autonomous and non cell-autonomous way, respectively.

OCT4 and SOX2 are also expressed in the Epi and their individual ablation leads to a decrease of FGF4 secretion by Epi cells (Nichols et al., 1998; Wicklow et al., 2014; Yuan, Corbi, Basilico, & Dailey, 1995), concomitant to a lower expression of GATA6 in PrE cells, and impairs PrE maturation (Le Bin et al., 2014; Wicklow et al., 2014). Another study shows that *Oct4* is required cell-autonomously in the PrE (Frum et al., 2013) where it is expressed until early E4.5 (Grabarek et al., 2012). Indeed, it was shown that OCT4 associates with SOX2 in the Epi and with SOX17 in the PrE (Aksoy et al., 2013; Stefanovic et al., 2009), where they activate target genes specific of each tissue (Aksoy et al., 2013; Frum et al., 2013). This mechanism probably reinforces the Epi/PrE specification and might be involved in the loss of plasticity.

7. OTHER SIGNALING PATHWAYS INVOLVED IN PrE MATURATION AND MAINTENANCE

Pdgfra is expressed in all ICM cells of the early blastocyst and its expression becomes restricted and enhanced in PrE cells after E3.75 (Artus, Panthier, & Hadjantonakis, 2010; Guo et al., 2010; Plusa et al., 2008). Its depletion does not impact PrE specification. However, mutant embryos exhibit an increased cell death within the PrE lineage, which is rescued by caspases inhibitor. Moreover, extraembryonic endoderm stem cells (XEN) cannot be derived from mutant embryos, further supporting a role for the PDGF signaling in the survival of PrE cells (Artus, Kang, Cohen-Tannoudji, & Hadjantonakis, 2013; Artus et al., 2010). A similar role has also been found for the BMP pathway as impairing its signaling also leads to an increased cell death in the PrE lineage (Graham et al., 2014). The Jak/Stat pathway is required for the maintenance of both lineages of the ICM by the late blastocyst stage, as observed in *Stat3* mutants and in embryo cultures with a JAK inhibitor (Do et al., 2013; Morgani & Brickman, 2015). STAT3 is important for ES cell maintenance and renewal (Boeuf, Hauss, Graeve, Baran, & Kedinger, 1997), suggesting a direct effect also on the Epi lineage within the embryo. Whether the role of *Stat3* in PrE cells is cell-autonomous or due to the defective Epi is not known.

8. CELL SORTING AND PrE EPITHELIALIZATION AT THE LATE BLASTOCYST STAGE

After specification around E3.75–E4.0, ICM cells acquire different physical properties leading to sorting and formation of two distinct tissues, with PrE cells separating the Epi from the blastocoel cavity (Fig. 1). Lineage tracing has shown that cells acquire their identity before sorting and maintain it (Chazaud et al., 2006; Meilhac et al., 2009; Plusa et al., 2008). The mechanisms involved in this cell-type segregation are not understood yet, but differential adhesion and cortical tension have been proposed to promote cell rearrangements and cell sorting in cell aggregates from diverse systems (Foty & Steinberg, 2005; Krieg et al., 2008). Embryoid bodies are convenient tools to recapitulate cell sorting and PrE epithelium formation, as PrE cells specify in a salt and pepper pattern like in the ICM (Rula et al., 2007). Using such models, it was shown that the absence of *E*-cadherin does not prevent sorting (Moore, Cai, Escudero, & Xu, 2009, Moore, Tao, Meng, et al., 2014). However, other cell adhesion molecules could be at play. Moreover, a mathematical modeling suggests that differential adhesion cannot be the sole mechanism and that adding directional force pushing PrE cells on the surface can improve the sorting pattern (Krupinski, Chickarmane, & Peterson, 2011). Noteworthy, actin-based cell movements seem to be involved (Meilhac et al., 2009). Still, it is currently unknown whether these actin-based cell displacements are the consequence of cell protrusions and/or of differences in cortical tensions.

Around E3.75–E4.0, 10%–20% (depending on studies) of ICM cells die by apoptosis (Copp, 1978). Epi and PrE fluorescent reporters indicate that both *Nanog* and *Pdgfra* expressing cells are affected (Plusa et al., 2008; Xenopoulos, Kang, Puliafito, Di Talia, & Hadjantonakis, 2015), demonstrating that both Epi and PrE cells are concerned. The current hypotheses to explain this cell death are either the consequence of misplaced cells after sorting (Plusa et al., 2008) or the impossibility for a cell to choose between an Epi and PrE identity (Bessonnard et al., 2014).

During sorting, PrE cells reach the surface of the ICM in an asynchronous manner (Gerbe et al., 2008; Plusa et al., 2008). Surprisingly, as soon as they reach the surface they polarize, attested by the asymmetric localization of DAB2 and LRP2 on the future apical side of the epithelium (Fig. 3B) (Gerbe et al., 2008). DAB2 is an endocytic adaptor that can bind to different receptors such as LRP2 and is implicated in many different pathways (Tao, Moore, Smith, & Xu, 2016). These PrE cells also start to express the TF SOX7 (Artus, Piliszek, & Hadjantonakis, 2011). It is only once all PrE cells have reached the surface that the polarization marker aPKC relocates to the apical border (Saiz, Grabarek, Sabherwal, Papalopulu, & Plusa, 2013). Interestingly, live imaging has revealed that aPKC function is to anchor PrE cells at the surface to avoid cell remixing (Saiz et al., 2013). This is reminiscent of *Dab2* mutant embryos and embryoid bodies that fail to polarize their PrE cells, and therefore can be found intermingled with Epi cells (Moore, Cai, Tao, Smith, & Xu, 2013; Yang et al., 2002). Thus, cell polarization seems to occur only when the cells have reached the surface and is required to maintain them there.

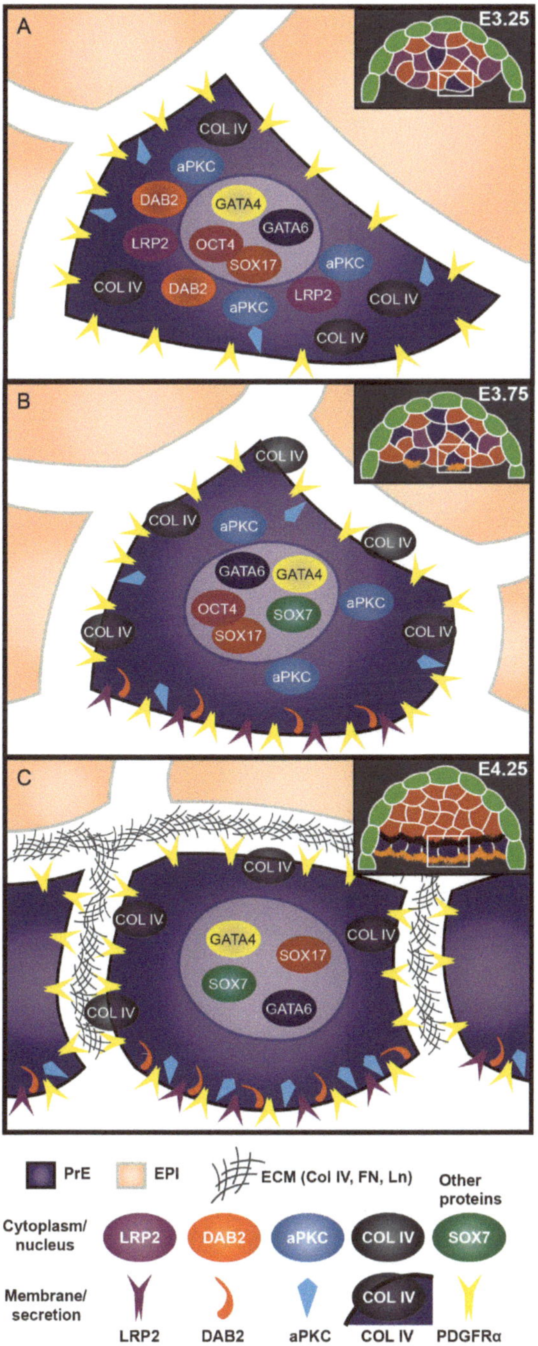

Fig. 3 See legend on next page.

Cell polarization becomes also evident with the formation of the basement membrane, which is visible at E4.5. Collagen IV and Laminin 1 start to be produced before cell sorting (Fig. 3B–C) (Gerbe et al., 2008; Guo et al., 2010; Niakan et al., 2010; Ohnishi et al., 2014), but it is difficult to know when they are secreted and if they are playing a role in cell sorting. *Laminin-c1* and *Integrin-b1* mutants fail to form a proper epithelium (Fässler & Meyer, 1995; Smyth et al., 1998; Stephens et al., 1995). In *Integrin-b1* mutant embryoid bodies, cell sorting occurs but extraembryonic endoderm cells detach from the Epi afterward, due to a failure to form focal adhesions on laminin and fibronectin (Liu et al., 2009; Moore, Tao, Smith, & Xu, 2014). Interestingly, the PrE cells of these mutants express GATA4 only in the cytoplasm. Such a phenotype has also been found when blocking aPKC signaling pathway (Saiz et al., 2013), indicating that correct nuclear localization of GATA4 depends on the polarization of PrE cells. Integrin-b1 is also expressed in the Epi and its specific deletion either in Epi or in PrE shows that the attachment and polarization of PrE cells require its expression in the Epi (Moore, Tao, Smith, et al., 2014). Thus, Epi and PrE are interdependent as PrE secretes the major components of the basement membrane and Epi produces Integrin-b1 to organize the polarization of both Epi and PrE cells in cell-autonomous and non cell-autonomous manners, respectively (Liu et al., 2009).

9. THE PrE DERIVATIVES: THE VISCERAL AND THE PARIETAL ENDODERM

The PrE epithelium covers the Epi and is in contact with the trophectoderm at its periphery (Fig. 1). From E4.75, PrE cells start to migrate along the inner surface of the trophectoderm. This migration

Fig. 3 (A) Before E3.75, some PrE cells are already situated at the ICM surface and express specific markers but are not polarized yet. (B) Around E3.75, PrE cells that have emerged to the surface start to be polarized, evidenced by DAB2 and LRP2 localized in the future apical membrane that faces the blastocoel cavity. SOX7 is now expressed within the nucleus of these cells. It is not clear if COLIV is already secreted by then or at the next step. (C) At E4.25, sorting is completed as all PrE cells have arisen at the surface, and start to form a structured epithelium with a basolateral extracellular matrix (ECM), and an apical pole labeled by LRP2, DAB2, and now aPKC.

follows an EMT, the first in the embryo, producing PE cells. The cells remaining in contact with the Epi and also with the developing extraembryonic ectoderm, a trophectoderm derivative, remain polarized and constitute the visceral endoderm (VE) around E5.5, identified by the beginning of expression of alpha-fetoprotein (Dziadek & Andrews, 1983; Kwon et al., 2006). The VE cells are specialized in nutrient absorption. The role of the PE has not been really identified yet but its absence, for example, in *Lama1* mutants, leads to a rapid embryo death (Alpy et al., 2005; Miner, Li, Mudd, Go, & Sutherland, 2004).

Little is known about the in vivo development of these tissues due to the extremely small size and the poor accessibility of the embryo that has just implanted. However, in vitro tools have been developed such as F9 embryonal carcinoma cells or plated ES embryoid bodies and have unraveled some mechanisms. Differences in the substratum can stimulate or prevent cell detachment and migration (Behrendtsen, Alexander, & Werb, 1995; Paca et al., 2012), showing the importance of the extracellular matrix. PrE and then VE cells can produce PE cells upon contact with trophectoderm cells. Indeed, trophectoderm cells produce high quantities of parathyroid hormone-related protein (PTHRP), which is sufficient to transform VE cells into PE cells (Behrendtsen et al., 1995). The PTHRP induces the AMPc pathway, which was shown to differentiate retinoic acid treated F9 cells into PE (Strickland & Mahdavi, 1978).

More recently, XEN cells have been isolated from blastocysts (Kunath et al., 2005). These cells share characteristics of both VE and PE at the level of markers identity as well as morphology (Brown et al., 2010; Kunath et al., 2005), which is changing from round to epithelioid. In chimera experiments, XEN cells contribute mainly to PE and very rarely to VE, indicating that XEN cells are closer to PE than VE (Kunath et al., 2005). Interestingly, administration of BMP or of Nodal/Cripto can induce a VE differentiation of XEN cells (Artus et al., 2012; Kruithof-de Julio et al., 2011; Paca et al., 2012) as well as of freshly isolated embryonic PE (Paca et al., 2012), with a better contribution to VE in chimera (see Moerkamp et al., 2013 for a comprehensive review on XEN cells). These experiments were showing for the first time that PE cells could revert to a VE identity. In rat XEN cells, the WNT pathway has been shown to have a similar effect, which is reversed by activating the cAMP pathway (Chuykin, Schulz, Guan, & Bader, 2013). Still, the mechanisms of PE differentiation remain largely unknown, especially in vivo.

10. CONCLUSION

In recent years, the complementarity of different technological advances such as single cell transcriptomics or live imaging, coupled to gain- and loss-of-function experiments have greatly increased our knowledge on PrE differentiation. Nevertheless, important questions like the initiation of the salt and pepper pattern or the mechanism of cell sorting remain unsolved, and we are only starting to understand the process of epithelialization or of PE formation. These mechanisms are poorly understood, due to the difficulty to access the embryo. The newly developed micropattern or 3D culture setups with embryos or stem cells (Bedzhov & Zernicka-Goetz, 2014; Deglincerti et al., 2016; Harrison, Sozen, Christodoulou, Kyprianou, & Zernicka-Goetz, 2017; Warmflash, Sorre, Etoc, Siggia, & Brivanlou, 2014) allow analyses out of the maternal womb and will certainly help to decipher these mechanisms.

REFERENCES

Aksoy, I., Jauch, R., Chen, J., Dyla, M., Divakar, U., Bogu, G. K., et al. (2013). Oct4 switches partnering from Sox2 to Sox17 to reinterpret the enhancer code and specify endoderm. *The EMBO Journal*, *32*, 938–953.

Alpy, F., Jivkov, I., Sorokin, L., Klein, A., Arnold, C., Huss, Y., et al. (2005). Generation of a conditionally null allele of the laminin alpha1 gene. *Genesis*, *2000*(43), 59–70.

Alvarez, J. D., Yasui, D. H., Niida, H., Joh, T., Loh, D. Y., & Kohwi-Shigematsu, T. (2000). The MAR-binding protein SATB1 orchestrates temporal and spatial expression of multiple genes during T-cell development. *Genes & Development*, *14*, 521–535.

Artus, J., Douvaras, P., Piliszek, A., Isern, J., Baron, M. H., & Hadjantonakis, A.-K. (2012). BMP4 signalling directs primitive endoderm-derived XEN cells to an extraembryonic visceral endoderm identity. *Developmental Biology*, *361*, 245.

Artus, J., Kang, M., Cohen-Tannoudji, M., & Hadjantonakis, A.-K. (2013). PDGF signaling is required for primitive endoderm cell survival in the inner cell mass of the mouse blastocyst. *Stem Cells (Dayton, Ohio)*, *31*, 1932.

Artus, J., Panthier, J.-J., & Hadjantonakis, A.-K. (2010). A role for PDGF signaling in expansion of the extra-embryonic endoderm lineage of the mouse blastocyst. *Development (Cambridge, England)*, *137*, 3361.

Artus, J., Piliszek, A., & Hadjantonakis, A.-K. (2011). The primitive endoderm lineage of the mouse blastocyst: Sequential transcription factor activation and regulation of differentiation by Sox17. *Developmental Biology*, *350*, 393–404.

Avilion, A. A., Nicolis, S. K., Pevny, L. H., Perez, L., Vivian, N., & Lovell-Badge, R. (2003). Multipotent cell lineages in early mouse development depend on SOX2 function. *Genes & Development*, *17*, 126–140.

Azami, T., Waku, T., Matsumoto, K., Jeon, H., Muratani, M., Kawashima, A., et al. (2017). Klf5 maintains the balance of primitive endoderm to epiblast specification during mouse embryonic development by suppression of Fgf4. *Development (Cambridge, England)*, *144*, 3706–3718.

Bedzhov, I., & Zernicka-Goetz, M. (2014). Self-organizing properties of mouse pluripotent cells initiate morphogenesis upon implantation. *Cell*, *156*, 1032–1044.
Behrendtsen, O., Alexander, C. M., & Werb, Z. (1995). Cooperative interactions between extracellular matrix, integrins and parathyroid hormone-related peptide regulate parietal endoderm differentiation in mouse embryos. *Development (Cambridge, England)*, *121*, 4137–4148.
Bessonnard, S., Coqueran, S., Vandormael-Pournin, S., Dufour, A., Artus, J., & Cohen-Tannoudji, M. (2017). ICM conversion to epiblast by FGF/ERK inhibition is limited in time and requires transcription and protein degradation. *Scientific Reports*, *7*, 12285.
Bessonnard, S., De Mot, L., Gonze, D., Barriol, M., Dennis, C., Goldbeter, A., et al. (2014). Gata6, Nanog and Erk signaling control cell fate in the inner cell mass through a tristable regulatory network. *Development (Cambridge, England)*, *141*, 3637–3648.
Boeuf, H., Hauss, C., Graeve, F. D., Baran, N., & Kedinger, C. (1997). Leukemia inhibitory factor–dependent transcriptional activation in embryonic stem cells. *The Journal of Cell Biology*, *138*, 1207.
Brown, K., Legros, S., Artus, J., Doss, M. X., Khanin, R., Hadjantonakis, A.-K., et al. (2010). A comparative analysis of extra-embryonic endoderm cell lines. *PLoS One*, *5*, e12016.
Burton, A., Muller, J., Tu, S., Padilla-Longoria, P., Guccione, E., & Torres-Padilla, M.-E. (2013). Single-cell profiling of epigenetic modifiers identifies PRDM14 as an inducer of cell fate in the mammalian embryo. *Cell Reports*, *5*, 687–701.
Cai, K. Q., Capo-Chichi, C. D., Rula, M. E., Yang, D.-H., & Xu, X.-X. (2008). Dynamic GATA6 expression in primitive endoderm formation and maturation in early mouse embryogenesis. *Developmental Dynamics: An Official Publication of the American Association of Anatomists*, *237*, 2820–2829.
Chang, H. H., Hemberg, M., Barahona, M., Ingber, D. E., & Huang, S. (2008). Transcriptome-wide noise controls lineage choice in mammalian progenitor cells. *Nature*, *453*, 544–547.
Chazaud, C., Yamanaka, Y., Pawson, T., & Rossant, J. (2006). Early lineage segregation between epiblast and primitive endoderm in mouse blastocysts through the Grb2-MAPK pathway. *Developmental Cell*, *10*, 615–624.
Chuykin, I., Schulz, H., Guan, K., & Bader, M. (2013). Activation of the PTHRP/adenylate cyclase pathway promotes differentiation of rat XEN cells into parietal endoderm, whereas Wnt/β-catenin signaling promotes differentiation into visceral endoderm. *Journal of Cell Science*, *126*, 128–138.
Copp, A. J. (1978). Interaction between inner cell mass and trophectoderm of the mouse blastocyst. *Development*, *48*, 109–125.
De Mot, L., Gonze, D., Bessonnard, S., Chazaud, C., Goldbeter, A., & Dupont, G. (2016). Cell fate specification based on tristability in the inner cell mass of mouse blastocysts. *Biophysical Journal*, *110*, 710–722.
Deglincerti, A., Croft, G. F., Pietila, L. N., Zernicka-Goetz, M., Siggia, E. D., & Brivanlou, A. H. (2016). Self-organization of the in vitro attached human embryo. *Nature*, *533*, 251–254.
Dietrich, J.-E., & Hiiragi, T. (2007). Stochastic patterning in the mouse pre-implantation embryo. *Development*, *134*, 4219–4231.
Do, D. V., Ueda, J., Messerschmidt, D. M., Lorthongpanich, C., Zhou, Y., Feng, B., et al. (2013). A genetic and developmental pathway from STAT3 to the OCT4–NANOG circuit is essential for maintenance of ICM lineages in vivo. *Genes & Development*, *27*, 1378.
Dziadek, M. A., & Andrews, G. K. (1983). Tissue specificity of alpha-fetoprotein messenger RNA expression during mouse embryogenesis. *The EMBO Journal*, *2*, 549.
Fässler, R., & Meyer, M. (1995). Consequences of lack of beta 1 integrin gene expression in mice. *Genes & Development*, *9*, 1896–1908.

Foty, R. A., & Steinberg, M. S. (2005). The differential adhesion hypothesis: A direct evaluation. *Developmental Biology*, *278*, 255–263.

Frankenberg, S., Gerbe, F., Bessonnard, S., Belville, C., Pouchin, P., Bardot, O., et al. (2011). Primitive endoderm differentiates via a three-step mechanism involving Nanog and RTK signaling. *Developmental Cell*, *21*, 1005–1013.

Frum, T., Halbisen, M. A., Wang, C., Amiri, H., Robson, P., & Ralston, A. (2013). Oct4 cell-autonomously promotes primitive endoderm development in the mouse blastocyst. *Developmental Cell*, *25*, 610–622.

Gerbe, F., Cox, B., Rossant, J., & Chazaud, C. (2008). Dynamic expression of Lrp2 pathway members reveals progressive epithelial differentiation of primitive endoderm in mouse blastocyst. *Developmental Biology*, *313*, 594–602.

Goolam, M., & Zernicka-Goetz, M. (2017). The chromatin modifier Satb1 regulates cell fate through Fgf signalling in the early mouse embryo. *Development (Cambridge, England)*, *144*, 1450–1461.

Grabarek, J. B., Żyżyńska, K., Saiz, N., Piliszek, A., Frankenberg, S., Nichols, J., et al. (2012). Differential plasticity of epiblast and primitive endoderm precursors within the ICM of the early mouse embryo. *Development*, *139*, 129–139.

Grabole, N., Tischler, J., Hackett, J. A., Kim, S., Tang, F., Leitch, H. G., et al. (2013). *Prdm14* promotes germline fate and naive pluripotency by repressing FGF signalling and DNA methylation. *EMBO Reports*, *14*, 629–637.

Graham, S. J. L., Wicher, K. B., Jedrusik, A., Guo, G., Herath, W., Robson, P., et al. (2014). BMP signalling regulates the pre-implantation development of extra-embryonic cell lineages in the mouse embryo. *Nature Communications*, *5*, 5667.

Guo, G., Huss, M., Tong, G. Q., Wang, C., Li Sun, L., Clarke, N. D., et al. (2010). Resolution of cell fate decisions revealed by single-cell gene expression analysis from zygote to blastocyst. *Developmental Cell*, *18*, 675–685.

Hamilton, W. B., & Brickman, J. M. (2014). Erk signaling suppresses embryonic stem cell self-renewal to specify endoderm. *Cell Reports*, *9*, 2056–2070.

Harrison, S. E., Sozen, B., Christodoulou, N., Kyprianou, C., & Zernicka-Goetz, M. (2017). Assembly of embryonic and extraembryonic stem cells to mimic embryogenesis in vitro. *Science*, *356*, pii: eaal1810.

Hayashi, S., Shimoda, T., Nakajima, M., Tsukada, Y., Sakumura, Y., Dale, J. K., et al. (2009). Sprouty4, an FGF inhibitor, displays cyclic gene expression under the control of the notch segmentation clock in the mouse PSM. *PLoS One*, *4*, e5603.

Jiang, J., Chan, Y.-S., Loh, Y.-H., Cai, J., Tong, G.-Q., Lim, C.-A., et al. (2008). A core Klf circuitry regulates self-renewal of embryonic stem cells. *Nature Cell Biology*, *10*, 353–360.

Kang, M., Garg, V., & Hadjantonakis, A.-K. (2017). Lineage establishment and progression within the inner cell mass of the mouse blastocyst requires FGFR1 and FGFR2. *Developmental Cell*, *41*, 496–510.e5.

Kang, M., Piliszek, A., Artus, J., & Hadjantonakis, A.-K. (2013). FGF4 is required for lineage restriction and salt-and-pepper distribution of primitive endoderm factors but not their initial expression in the mouse. *Development (Cambridge, England)*, *140*, 267–279.

Kramer, S., Okabe, M., Hacohen, N., Krasnow, M. A., & Hiromi, Y. (1999). Sprouty: A common antagonist of FGF and EGF signaling pathways in drosophila. *Development (Cambridge, England)*, *126*, 2515–2525.

Krawchuk, D., Honma-Yamanaka, N., Anani, S., & Yamanaka, Y. (2013). FGF4 is a limiting factor controlling the proportions of primitive endoderm and epiblast in the ICM of the mouse blastocyst. *Developmental Biology*, *384*, 65–71.

Krieg, M., Arboleda-Estudillo, Y., Puech, P.-H., Käfer, J., Graner, F., Müller, D. J., et al. (2008). Tensile forces govern germ-layer organization in zebrafish. *Nature Cell Biology*, *10*, 429–436.

Kruithof-de Julio, M., Alvarez, M. J., Galli, A., Chu, J., Price, S. M., Califano, A., et al. (2011). Regulation of extra-embryonic endoderm stem cell differentiation by nodal and cripto signaling. *Development (Cambridge, England), 138*, 3885–3895.

Krupa, M., Mazur, E., Szczepańska, K., Filimonow, K., Maleszewski, M., & Suwińska, A. (2014). Allocation of inner cells to epiblast vs primitive endoderm in the mouse embryo is biased but not determined by the round of asymmetric divisions (8→16- and 16→32-cells). *Developmental Biology, 385*, 136–148.

Krupinski, P., Chickarmane, V., & Peterson, C. (2011). Simulating the mammalian blastocyst—Molecular and mechanical interactions pattern the embryo. *PLoS Computational Biology*, 7, e1001128.

Kunath, T., Arnaud, D., Uy, G. D., Okamoto, I., Chureau, C., Yamanaka, Y., et al. (2005). Imprinted X-inactivation in extra-embryonic endoderm cell lines from mouse blastocysts. *Development (Cambridge, England), 132*, 1649–1661.

Kurimoto, K., Yabuta, Y., Ohinata, Y., Ono, Y., Uno, K. D., Yamada, R. G., et al. (2006). An improved single-cell cDNA amplification method for efficient high-density oligonucleotide microarray analysis. *Nucleic Acids Research, 34*, e42.

Kwon, G. S., Fraser, S. T., Eakin, G. S., Mangano, M., Isern, J., Sahr, K. E., et al. (2006). Tg(Afp-GFP) expression marks primitive and definitive endoderm lineages during mouse development. *Developmental Dynamics: An Official Publication of the American Association of Anatomists, 235*, 2549.

Le Bin, G. C., Muñoz-Descalzo, S., Kurowski, A., Leitch, H., Lou, X., Mansfield, W., et al. (2014). Oct4 is required for lineage priming in the developing inner cell mass of the mouse blastocyst. *Development (Cambridge, England), 141*, 1001–1010.

Lin, S.-C. J., Wani, M. A., Whitsett, J. A., & Wells, J. M. (2010). Klf5 regulates lineage formation in the pre-implantation mouse embryo. *Development (Cambridge, England), 137*, 3953–3963.

Liu, J., He, X., Corbett, S. A., Lowry, S. F., Graham, A. M., Fässler, R., et al. (2009). Integrins are required for the differentiation of visceral endoderm. *Journal of Cell Science, 122*, 233–242.

Maamar, H., Raj, A., & Dubnau, D. (2007). Noise in gene expression determines cell fate in Bacillus subtilis. *Science, 317*, 526–529.

Maître, J.-L. (2017). Mechanics of blastocyst morphogenesis: Mechanics of blastocyst morphogenesis. *Biology of the Cell, 109*, 323–338.

Meilhac, S. M., Adams, R. J., Morris, S. A., Danckaert, A., Le Garrec, J.-F., & Zernicka-Goetz, M. (2009). Active cell movements coupled to positional induction are involved in lineage segregation in the mouse blastocyst. *Developmental Biology, 331*, 210–221.

Messerschmidt, D. M., & Kemler, R. (2010). Nanog is required for primitive endoderm formation through a non-cell autonomous mechanism. *Developmental Biology, 344*, 129–137.

Miner, J. H., Li, C., Mudd, J. L., Go, G., & Sutherland, A. E. (2004). Compositional and structural requirements for laminin and basement membranes during mouse embryo implantation and gastrulation. *Development (Cambridge, England), 131*, 2247–2256.

Miyanari, Y., & Torres-Padilla, M.-E. (2012). Control of ground-state pluripotency by allelic regulation of Nanog. *Nature, 483*, 470–473.

Moerkamp, A. T., Paca, A., Goumans, M.-J., Kunath, T., Kruithof, B. P. T., & Kruithof-de Julio, M. (2013). Extraembryonic endoderm cells as a model of endoderm development. *Development, Growth & Differentiation, 55*, 301–308.

Molotkov, A., Mazot, P., Brewer, J. R., Cinalli, R. M., & Soriano, P. (2017). Distinct requirements for FGFR1 and FGFR2 in primitive endoderm development and exit from Pluripotency. *Developmental Cell, 41*, 511.e4-526.e4.

Moore, R., Cai, K. Q., Escudero, D. O., & Xu, X.-X. (2009). Cell adhesive affinity does not dictate primitive endoderm segregation and positioning during murine embryoid body formation. *Genesis, 2000*(47), 579–589.
Moore, R., Cai, K. Q., Tao, W., Smith, E. R., & Xu, X.-X. (2013). Differential requirement for Dab2 in the development of embryonic and extra-embryonic tissues. *BMC Developmental Biology, 13*, 39.
Moore, R., Tao, W., Meng, Y., Smith, E. R., & Xu, X.-X. (2014). Cell adhesion and sorting in embryoid bodies derived from N- or E-cadherin deficient murine embryonic stem cells. *Biology Open, 3*, 121–128.
Moore, R., Tao, W., Smith, E. R., & Xu, X.-X. (2014). The primitive endoderm segregates from the epiblast in β1 integrin-deficient early mouse embryos. *Molecular and Cellular Biology, 34*, 560.
Morgani, S. M., & Brickman, J. M. (2015). LIF supports primitive endoderm expansion during pre-implantation development. *Development (Cambridge, England), 142*, 3488–3499.
Morris, S. A., Graham, S. J. L., Jedrusik, A., & Zernicka-Goetz, M. (2013). The differential response to Fgf signalling in cells internalized at different times influences lineage segregation in preimplantation mouse embryos. *Open Biology, 3*, 130104.
Morris, S. A., Teo, R. T. Y., Li, H., Robson, P., Glover, D. M., & Zernicka-Goetz, M. (2010). Origin and formation of the first two distinct cell types of the inner cell mass in the mouse embryo. *Proceedings of the National Academy of Sciences of the United States of America, 107*, 6364–6369.
Nestorov, P., Hotz, H.-R., Liu, Z., & Peters, A. H. F. M. (2015). Dynamic expression of chromatin modifiers during developmental transitions in mouse preimplantation embryos. *Scientific Reports, 5*, 14347.
Niakan, K. K., Ji, H., Maehr, R., Vokes, S. A., Rodolfa, K. T., Sherwood, R. I., et al. (2010). Sox17 promotes differentiation in mouse embryonic stem cells by directly regulating extraembryonic gene expression and indirectly antagonizing self-renewal. *Genes & Development, 24*, 312–326.
Nichols, J., Silva, J., Roode, M., & Smith, A. (2009). Suppression of Erk signalling promotes ground state pluripotency in the mouse embryo. *Development (Cambridge, England), 136*, 3215–3222.
Nichols, J., Zevnik, B., Anastassiadis, K., Niwa, H., Klewe-Nebenius, D., Chambers, I., et al. (1998). Formation of pluripotent stem cells in the mammalian embryo depends on the POU transcription factor Oct4. *Cell, 95*, 379–391.
Nishioka, N., Inoue, K., Adachi, K., Kiyonari, H., Ota, M., Ralston, A., et al. (2009). The Hippo signaling pathway components Lats and Yap pattern Tead4 activity to distinguish mouse trophectoderm from inner cell mass. *Developmental Cell, 16*, 398–410.
Niswander, L., & Martin, G. R. (1992). Fgf-4 expression during gastrulation, myogenesis, limb and tooth development in the mouse. *Development (Cambridge, England), 114*, 755–768.
Niwa, H., Toyooka, Y., Shimosato, D., Strumpf, D., Takahashi, K., Yagi, R., et al. (2005). Interaction between Oct3/4 and Cdx2 determines trophectoderm differentiation. *Cell, 123*, 917–929.
Ohnishi, Y., Huber, W., Tsumura, A., Kang, M., Xenopoulos, P., Kurimoto, K., et al. (2014). Cell-to-cell expression variability followed by signal reinforcement progressively segregates early mouse lineages. *Nature Cell Biology, 16*, 27–37.
Paca, A., Séguin, C. A., Clements, M., Ryczko, M., Rossant, J., Rodriguez, T. A., et al. (2012). BMP signaling induces visceral endoderm differentiation of XEN cells and parietal endoderm. *Developmental Biology, 361*, 90–102.
Panamarova, M., Cox, A., Wicher, K. B., Butler, R., Bulgakova, N., Jeon, S., et al. (2016). The BAF chromatin remodelling complex is an epigenetic regulator of lineage specification in the early mouse embryo. *Development (Cambridge, England), 143*, 1271–1283.

Payer, B., Rosenberg, M., Yamaji, M., Yabuta, Y., Koyanagi-Aoi, M., Hayashi, K., et al. (2013). Tsix RNA and the germline factor, PRDM14, link X reactivation and stem cell reprogramming. *Molecular Cell, 52*, 805–818.

Piras, V., Tomita, M., & Selvarajoo, K. (2014). Transcriptome-wide variability in single embryonic development cells. *Scientific Reports, 4*, 7137.

Plusa, B., Piliszek, A., Frankenberg, S., Artus, J., & Hadjantonakis, A.-K. (2008). Distinct sequential cell behaviours direct primitive endoderm formation in the mouse blastocyst. *Development (Cambridge, England), 135*, 3081–3091.

Rayon, T., Menchero, S., Nieto, A., Xenopoulos, P., Crespo, M., Cockburn, K., et al. (2014). Notch and hippo converge on Cdx2 to specify the trophectoderm lineage in the mouse blastocyst. *Developmental Cell, 30*, 410–422.

Reich, A., Sapir, A., & Shilo, B. (1999). Sprouty is a general inhibitor of receptor tyrosine kinase signaling. *Development (Cambridge, England), 126*, 4139–4147.

Rula, M. E., Cai, K. Q., Moore, R., Yang, D.-H., Staub, C. M., Capo-Chichi, C. D., et al. (2007). Cell autonomous sorting and surface positioning in the formation of primitive endoderm in embryoid bodies. *Genesis, 2000*(45), 327–338.

Saiz, N., Grabarek, J. B., Sabherwal, N., Papalopulu, N., & Plusa, B. (2013). Atypical protein kinase C couples cell sorting with primitive endoderm maturation in the mouse blastocyst. *Development (Cambridge, England), 140*, 4311–4322.

Saiz, N., Williams, K. M., Seshan, V. E., & Hadjantonakis, A.-K. (2016). Asynchronous fate decisions by single cells collectively ensure consistent lineage composition in the mouse blastocyst. *Nature Communications*, 7, 13463.

Savarese, F., Dávila, A., Nechanitzky, R., Rosa-Velazquez, I. D. L., Pereira, C. F., Engelke, R., et al. (2009). Satb1 and Satb2 regulate embryonic stem cell differentiation and Nanog expression. *Genes & Development, 23*, 2625–2638.

Schrode, N., Saiz, N., Di Talia, S., & Hadjantonakis, A.-K. (2014). GATA6 levels modulate primitive endoderm cell fate choice and timing in the mouse blastocyst. *Developmental Cell, 29*, 454–467.

Schröter, C., Rué, P., Mackenzie, J. P., & Martinez Arias, A. (2015). FGF/MAPK signaling sets the switching threshold of a bistable circuit controlling cell fate decisions in embryonic stem cells. *Development (Cambridge, England), 142*, 4205–4216.

Silva, J., Nichols, J., Theunissen, T. W., Guo, G., van Oosten, A. L., Barrandon, O., et al. (2009). Nanog is the gateway to the pluripotent ground state. *Cell, 138*, 722–737.

Smyth, N., Vatansever, H. S., Meyer, M., Frie, C., Paulsson, M., & Edgar, D. (1998). The targeted deletion of the LAMC1 gene. *Annals of the New York Academy of Sciences, 857*, 283–286.

Stefanovic, S., Abboud, N., Désilets, S., Nury, D., Cowan, C., & Pucéat, M. (2009). Interplay of Oct4 with Sox2 and Sox17: A molecular switch from stem cell pluripotency to specifying a cardiac fate. *The Journal of Cell Biology, 186*, 665–673.

Stephens, L. E., Sutherland, A. E., Klimanskaya, I. V., Andrieux, A., Meneses, J., Pedersen, R. A., et al. (1995). Deletion of beta 1 integrins in mice results in inner cell mass failure and peri-implantation lethality. *Genes & Development, 9*, 1883–1895.

Strickland, S., & Mahdavi, V. (1978). The induction of differentiation in teratocarcinoma stem cells by retinoic acid. *Cell, 15*, 393–403.

Taniguchi, K., Ayada, T., Ichiyama, K., Kohno, R.-I., Yonemitsu, Y., Minami, Y., et al. (2007). Sprouty2 and Sprouty4 are essential for embryonic morphogenesis and regulation of FGF signaling. *Biochemical and Biophysical Research Communications, 352*, 896–902.

Tao, W., Moore, R., Smith, E. R., & Xu, X.-X. (2016). Endocytosis and physiology: Insights from Disabled-2 deficient mice. *Frontiers in Cell and Development Biology, 4*, 129.

Tosenberger, A., Gonze, D., Bessonnard, S., Cohen-Tannoudji, M., Chazaud, C., & Dupont, G. (2017). A multiscale model of early cell lineage specification including cell division. *NPJ Systems Biology and Applications, 3*, 16.

Warmflash, A., Sorre, B., Etoc, F., Siggia, E. D., & Brivanlou, A. H. (2014). A method to recapitulate early embryonic spatial patterning in human embryonic stem cells. *Nature Methods*, *11*, 847–854.
White, M. D., Bissiere, S., Alvarez, Y. D., & Plachta, N. (2016). Mouse embryo compaction. *Current Topics in Developmental Biology*, *120*, 235–258.
Wicklow, E., Blij, S., Frum, T., Hirate, Y., Lang, R. A., Sasaki, H., et al. (2014). HIPPO pathway members restrict SOX2 to the inner cell mass where it promotes ICM fates in the mouse blastocyst. *PLoS Genetics*, *10*, e1004618.
Xenopoulos, P., Kang, M., Puliafito, A., Di Talia, S., & Hadjantonakis, A.-K. (2015). Heterogeneities in Nanog expression drive stable commitment to pluripotency in the mouse blastocyst. *Cell Reports*, *10*, 1508–1520.
Yamaji, M., Seki, Y., Kurimoto, K., Yabuta, Y., Yuasa, M., Shigeta, M., et al. (2008). Critical function of Prdm14 for the establishment of the germ cell lineage in mice. *Nature Genetics*, *40*, 1016–1022.
Yamanaka, Y., Lanner, F., & Rossant, J. (2010). FGF signal-dependent segregation of primitive endoderm and epiblast in the mouse blastocyst. *Development (Cambridge, England)*, *137*, 715–724.
Yang, D.-H., Smith, E. R., Roland, I. H., Sheng, Z., He, J., Martin, W. D., et al. (2002). Disabled-2 is essential for endodermal cell positioning and structure formation during mouse embryogenesis. *Developmental Biology*, *251*, 27–44.
Yuan, H., Corbi, N., Basilico, C., & Dailey, L. (1995). Developmental-specific activity of the FGF-4 enhancer requires the synergistic action of Sox2 and Oct-3. *Genes & Development*, *9*, 2635–2645.

CHAPTER SIX

The Regulative Nature of Mammalian Embryos

Katarzyna Klimczewska, Anna Kasperczuk, Aneta Suwińska[1]
Faculty of Biology, University of Warsaw, Warsaw, Poland
[1]Corresponding author: e-mail address: asuwinska@biol.uw.edu.pl

Contents

Abstract

The striking developmental plasticity of early mammalian embryos has been known since the classical experiments performed in the 1950s and 1960s. There are many lines of evidence that the mammalian embryo is able to continue normal development even

Current Topics in Developmental Biology, Volume 128
ISSN 0070-2153
https://doi.org/10.1016/bs.ctdb.2017.10.010

when exposed to severe experimental manipulations of the number and position of cells within the embryo. These observations have raised the question about the mechanisms involved in emergence, maintenance, and progressive restriction of this plasticity. Only recently, we have begun to understand these mechanisms. In this review, in order to explain the molecular and cellular events underlying the remarkable plasticity of the early mammalian embryo, we discuss results of classical experiments demonstrating developmental potential of mammalian embryos and link them with the novel data provided by contemporary experimental approaches. We also show how developmental flexibility of mammalian embryos is manifested in nature, and discuss its implications for basic research and medicine.

1. INTRODUCTION

Embryogenesis of mouse, which is considered as a model mammal, takes place in the female reproductive tract. The preimplantation period of mammalian development is characterized by two key cell fate decisions that allow the transformation of the totipotent mother cell, the zygote, into the blastocyst, ready for implantation into the maternal uterus. By the moment of implantation, two crucial and complementary events must have occurred: the generation of material for an embryo proper and the formation of extraembryonic cell lineages supporting its development in the uterus, by mediating nutrient and metabolite exchange between the embryo and the mother. These lineage specification events are accompanied by gradual restriction and ultimately a loss of developmental potential of the resulting cells—blastomeres. Initial totipotency of blastomeres, defined as their capability of developing into a complete organism, becomes restricted with developmental time, first to pluripotency (i.e., the ability to differentiate into only embryonic tissues) and subsequently to a definitive cell lineage commitment. Between these periods of totipotency and the ultimate determination, cells exist in a state of plasticity (i.e., they have a developmental preference toward cell lineage, but this preference can still be reversible, for example in response to an external stimulus).

Until the 8-cell stage, blastomeres are morphologically indistinguishable, but as a result of compaction and polarization, two types of cells arise as an embryo divides from 8- to 16- and from 16- to 32-cell stage: polar cells (usually placed outside) and apolar cells (predominantly located inside the embryo). They ultimately have distinct destinations and form the first two primary cell lineages of the blastocyst. Thus, the first dilemma to be resolved by initially totipotent cells is whether to join the extraembryonic trophectoderm cell lineage (TE, that is the precursor of the future placenta) or the pluripotent cell line—the inner cell mass (ICM). The second decision that must be undertaken is the choice made by cells of the ICM between populating the epiblast (EPI) or differentiating to form the primitive endoderm (PE). The epiblast develops primarily into the body of the fetus proper, whereas the PE contributes predominantly to the yolk sac, and serves as a source of signals for establishment of anteroposterior axis of the embryo (Beddington & Robertson, 1999). According to our current knowledge, such early cell fate decisions in mammalian embryo are made based on a combination of factors, including intercellular interactions, cell polarity, spatial positioning, and signals from cellular microenvironment, rather than partitioning of maternal determinants within the fertilized egg. Thus, this developmental strategy is translated into a high degree of plasticity and self-organization of early mammalian embryo development.

In this chapter, we demonstrate the various aspects of the regulative abilities of mammalian embryos, with a special emphasis on the mouse, that can be observed either in nature or in experimental research. However, the results obtained for this model species might not be applicable to all mammals and thus should be extrapolated with special caution. Additionally, we show how the knowledge on the plasticity and developmental potential of mammalian embryos can be applied in practice, in veterinary and biomedical research.

2. BLASTOMERE POTENCY TO GIVE RISE TO THE WHOLE ORGANISM

Totipotency sits atop of the cell potency hierarchy. According to the most common and stringent definition, totipotency is the ability of a single cell to produce all the cells of the body, as well as the transient structures of the embryo (i.e., placenta and fetal membranes that function only during the prenatal life). Totipotency defined this way can be proven only when a single blastomere isolated from an embryo is able to compensate for the loss of other cells, complete normal development, and give rise to a fertile adult individual. Such capability is obviously normal for a zygote (Fig. 1A), but the occurrence of monozygotic multiplets, either after experimental intervention or spontaneously in nature, proves that totipotency is maintained also at further developmental stages.

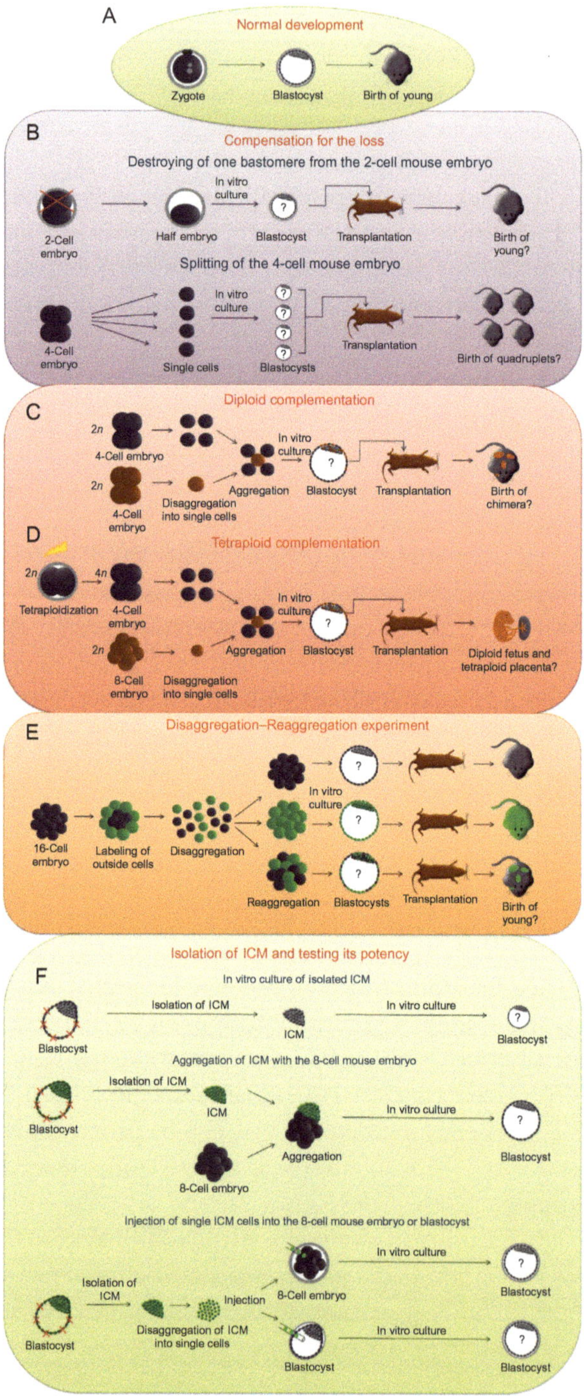

Fig. 1 See legend on opposite page.

2.1 Compensation for the Loss: Embryo Splitting as a Proof of Totipotency and Plasticity of Blastomeres

In the 1950s, Tarkowski showed that the destruction of one blastomere of the 2-cell mouse embryo with a thin glass needle results in the formation of a morphologically normal blastocyst, that after transplantation to a recipient mouse, could generate a healthy adult individual (Tarkowski, 1959; Fig. 1B). This pioneering experiment demonstrated that at least one blastomere of the 2-cell mouse embryo, like the zygote, is totipotent. Similarly in rats, the mechanical separation of blastomeres of 2-cell embryos has been used in the production of identical twins (Matsumoto, Miyake, Utsumi, & Iritani, 1989). However, attempts to experimentally obtain twins, triplets, and quadruplets from all blastomeres of individual 2- to 4-cell stage mouse embryos were rarely demonstrated to be successful (Mullen, Whitten, & Carter, 1970; Papaioannou & Ebert, 1995; Papaioannou, Mkandawire, & Biggers, 1989; Tsunoda & McLaren, 1983).

Totipotency of mouse blastomeres in still later stages of development has never been demonstrated. Indeed, Tarkowski and Wróblewska (1967) showed that isolated blastomeres of the 4- and 8-cell mouse embryos usually form small blastocysts with ICMs consisting of only a few cells or completely devoid of an ICM altogether (Tarkowski & Wróblewska, 1967) and are thus incapable of forming normal fetuses after transplantation to pseudopregnant females (Rossant, 1976; Fig. 1B). These observations suggest that blastocysts

Fig. 1 Various approaches to study developmental potency of mouse blastomeres. (A) Undisturbed zygote develops into a blastocyst that gives rise to a mouse. (B) Destruction of one blastomere of the 2-cell mouse embryo results in development of the remaining blastomere, that after in vitro culture and transfer to pseudopregnant recipient, gives rise to a viable animal. However, if blastomeres of the 4-cell mouse embryo are separated, the individual cells cannot undergo complete embryogenesis until term. (C) Aggregation of a single diploid blastomere with genetically distinct diploid blastomeres results in formation of a chimera with contribution of both types of cells to all embryonic and extraembryonic lineages. (D) In contrast, the descendants of a single diploid blastomere, that was aggregated with carrier tetraploid cells, contribute to the tissues of the pure diploid fetus, whereas tetraploid blastomeres are restricted to the trophectoderm derivatives. (E) Disaggregation of the labeled cleaving embryos into single cells and subsequent reaggregation into clusters composed exclusively of inner, outer, or randomly mixed cells. After transplantation to recipient mice such aggregates can develop into viable animals. (F) Isolation of ICM by immunosurgery or microsurgical manipulation and verifying its potency to regenerate TE. Contribution of ICM cells in chimeras can be assessed after aggregation of ICM with the 8-cell embryo or injection of single ICM cells into the 8-cell embryo, as well as into the cavity of the blastocyst.

developing from blastomeres separated at the 4-cell, or later stages of development, comprise an insufficient number of cells to ensure successful embryogenesis, culminating in the birth of offspring. It has been shown that the key events of preimplantation development, such as cell compaction, polarization, cavitation, and lineage commitment, are subjected to a strict developmental clock, and always take place after a defined number of cell divisions (Morris, Guo, & Zernicka-Goetz, 2012). Therefore, the later one splits an embryo to separate blastomeres, the lower the number of cells that are available to build the blastocyst, and, consequently, such blastocysts have insufficient number of cells to form a functionally sufficient ICM. Thus, the developmental failure of such single isolated mouse blastomeres is probably due to the length of the preimplantation period that in mice lasts approximately 4.5 days.

In other mammalian species, including rabbit, sheep, cattle, goat, horse, and rat, the preimplantation period is longer, and thus at the moment of implantation, their embryos are composed of a higher number of cells than the equivalently staged mouse embryo. In rabbits, uterine implantation begins on the 7th day of embryonic development and single blastomeres of 2-, 4- and 8-cell embryos are capable of developing into normal and viable young (Moore, Adams, & Rowson, 1968; Seidel, 1952, 1960). Similarly, ovine blastomeres isolated at the 2- to 4-cell stage, and occasionally even at 8-cell stage, are totipotent and can give rise to diminutive blastocysts able to develop into healthy individuals after transplantation to recipient females (Willadsen, 1980, 1981). In horses, monozygotic foals can be produced from blastomeres isolated from the embryo at the 2- to 8-cell stage (Allen & Pashen, 1984). In cattle, all four blastomeres of a single 4-cell stage embryo have been shown to develop into monozygotic quadruplets (Johnson, Loskutoff, Plante, & Betteridge, 1995).

The potency of blastomeres to create entire organisms can be also verified by bisection of embryos at later stages of development. In goats and pigs, embryos at the morula or blastocyst stage subjected to bisection, followed by transfer to recipients, also were capable of full-term embryogenesis giving rise to healthy offspring (Reichelt & Niemann, 1994) and even identical twins (Tsunoda, Tokunaga, Sugie, & Katsumata, 1985).

Although the splitting of mammalian embryos has been successfully and efficiently demonstrated in several livestock species, in nonhuman primates this strategy has given inferior results. In Rhesus monkeys, embryo twinning has been attempted by splitting the 8-cell embryo into a set of identical quadruplets, each consisting of two of the original eight cells. However, their transfer to the uteri of surrogates only led to the birth of one monkey (Chan et al., 2000). Despite this, the splitting of Rhesus monkey embryos when performed at the 2- or 4-cell stage, followed by the transplantation, resulted in two twin pregnancies. However, they each failed to give rise to the birth of monozygotic twins. Similar results were achieved using the bisection method, as all pregnancies only led to singletons (Mitalipov, Yeoman, Kuo, & Wolf, 2002).

In contrast to animal models, splitting experiments using human embryos raise ethical and legal concerns. These restrictions, in combination with the scarcity of the research material, cause that data on the potencies of human blastomeres are significantly limited. Moreover, manipulated human embryos are not allowed to be transferred to a uterus and thus totipotency defined by the strictest definition cannot be proven. Nevertheless, it has been shown that single blastomeres of a 4-cell human embryo have the capacity to develop in vitro into blastocysts with ICM and TE (Van de Velde, Cauffman, Tournaye, Devroey, & Liebaers, 2008). Such embryos are considered to be potentially totipotent, supported by the fact that the transfer of a 4-cell human embryo with $<50\%$ intact blastomeres can result in the birth of a healthy infant (Veiga, Calderon, Barri, & Coroleu, 1987). This observation suggests that at least one blastomere of a 4-cell human embryo is totipotent. Moreover, genetically abnormal (triploid) embryos, split into twin embryos at the 2- to 5- and 6- to 8-cell stages, are able to develop into blastocysts (Illmensee, Levanduski, Vidali, Husami, & Goudas, 2010). Interestingly, splitting at the 6- to 8-cell stage resulted in an increased rate of successful embryo twinning when compared to the procedure performed at the earlier stages (Illmensee et al., 2010). However, Noli and coworkers demonstrated that irrespective of the stage of the embryo at the moment of splitting (2–5-cell or 6–10-cell), blastomeres develop into blastocyst-like structures with absent or poor quality ICMs (Noli et al., 2015).

2.2 Monozygotic Multiplets: Proof of Totipotency of Blastomeres in Nature

Splitting experiments mimic, although imperfectly, the natural phenomenon of monozygotic multiplets. Except for the nine-banded armadillo *Dasypus novemcinctus*, which consistently gives birth to genetically identical quadruplets generated from a single zygote (Loughry, Avise, McDonough, & Prodohl, 1998), multiple monozygotic pregnancy is very rare in mammals. In human, monozygotic twins are born all over the world with equally low frequencies: only 4 gestations per 1000 result in birth of identical twins (Hall, 2003; MacGillivray, 1986). Monozygotic multiple pregnancy arises from a single embryo, which divides within 14 days after conception to form two or more embryos. Although the etiology underlying human monozygotic twinning remains largely unexplained, some observations, including the existence and frequency of various forms of monozygotic pregnancies (dichorionic/diamniotic vs diamniotic/monochorionic vs monoamniotic/monochorionic), point to several hypothetical mechanisms (Herranz, 2015).

According to the first theory, the formation of monozygotic twin pregnancy is the result of a genetic mutation (Hamamy, Ajlouni, & Ajlouni, 2004; Harvey, Huntley, & Smith, 1977; Machin, 2009). Due to this mutation, embryonic cells do not adhere tightly to each other, but remain loosely associated, and as a consequence, two individual embryos start to develop. Monozygotic twinning may also be caused by an inherited defect in one of the *zona pellucida* proteins, which could result in formation of a fragile *zona pellucida* leading to a complete embryo bisection or the disruption of ICM cells during abnormal hatching of the blastocyst (Hall, 2003; Verpoest et al., 2009). Additionally, there may be also other reasons of monozygotic twinning. For instance, fertilization of postovulatory aged oocyte surrounded by a *zona pellucida*, that is susceptible to mechanical damage, may lead, in the case of *zona* fracture, to the splitting of the blastocyst into two parts (Hall, 2003).

Within the last 30 years the frequency of monozygotic twin pregnancies has increased 2–12 times due to the practice of assisted reproductive technology (ART) procedures (Alikani, Cekleniak, Walters, & Cohen, 2003; Blickstein, Jones, & Keith, 2003; Blickstein, Verhoeven, & Keith, 1999; Edwards, Mettler, & Walters, 1986; Sills, Tucker, & Palermo, 2000). One of the most important factors facilitating this increased frequency is the artificial induction of ovulation. It has been shown that women undergoing ovarian stimulation deliver monozygotic twins nearly twice as likely as women who have not used this therapy (Källén, Olausson, & Nygren, 2002). The higher twinning frequency could be caused by a hardening of the *zona* induced by ovulatory hormones that increase the likelihood of accidental embryo bisection (Derom, Vlietinck, Derom, Van den Berghe, & Thiery, 1987; Edwards et al., 1986). The second important factor favoring the incidence of twin pregnancies is *zona pellucida* manipulation commonly utilized in ART. It has been proposed that an artificial hole in the *zona*, made during the procedure of intracytoplasmic sperm injection or assisted hatching (i.e., chemical, mechanical, or laser manipulation of the *zona pellucida*), might result in mechanical splitting of the embryo and the formation of monozygotic twins during blastocyst expansion (Hershlag et al., 1999; Saito, Tsutsumi, Noda, Ibuki, & Hiroi, 2000; Schieve et al., 2000; Skiadas, Missmer, Benson, Gee, & Racowsky, 2008; Tarlatzis et al., 2002). Last but not least, ART often involves in vitro culture of embryos that does not perfectly mimic the female uterine environment and may contribute to enhanced apoptosis of cells and impairment of intercellular connections (Ménézo & Sakkas, 2002). Moreover, hardening of the *zona pellucida* following prolonged embryo culture has been also proposed as a possible mechanism contributing to the increased rates of monozygotic twinning (Behr et al., 2000).

Although multiplets occur spontaneously in nature and the regulative abilities of single blastomeres of mammalian embryos have been proven, the attempts to experimentally obtain multiplets from all blastomeres of cleavage-stage mammalian embryo, as previously mentioned, often failed. This suggests the existence of some differences between the blastomeres of the early cleavage embryo that could be reflected in their unequal developmental potential.

3. BLASTOMERES OF THE CLEAVING EMBRYO—MORPHOLOGICALLY IDENTICAL, YET MOLECULARLY DISTINCT

Despite morphological uniformity of mammalian blastomeres up to the 8-cell stage, evidence is accumulating that early blastomeres exhibit differences at the molecular level, starting from the 2-cell stage onward. Recent studies in human and mouse embryos have shown that sister blastomeres in 2-cell stage embryos differ in their gene expression profiles (Shi et al., 2015; Tang et al., 2011). Molecular asymmetries, referring to, for example, X inactive-specific transcript (Xist) RNA that plays a key role in X chromosome inactivation in females, are also present among presumed identical 4- and 8-cell human blastomeres. Indeed, most of the Xist RNA is accumulated preferentially in two out of eight cells (Hartshorn, Rice, & Wangh, 2003).

There are also numerous studies confirming that expression of several transcription regulators, crucial for the establishment of the first embryonic cell lineages, differs between blastomeres in early cleavage embryos. It is therefore probable that the embryo acquires interblastomere differences that might influence the future fate of subsequent daughter cells at the blastocyst stage (Biase, Cao, & Zhong, 2014). Indeed, using a noninvasive cell labeling method, Fujimori and coworkers have shown that in some of the analyzed

4-cell stage mouse embryos, one out of the four blastomeres is biased to contribute to the extraembryonic tissues (Fujimori, Kurotaki, Miyazaki, & Nabeshima, 2003). The probability that the future developmental fate decisions might be already taken at the 4-cell stage has been recently confirmed by a pioneering method of multicolor lineage tracing. Tabansky and coworkers have found that the progeny of some 4-cell stage mouse blastomeres tend to exclusively contribute to either the ICM or TE lineages (Tabansky et al., 2013).

A correlation between blastomere gene expression profiles and subsequent fate has been provided by many reports. For example, it has been shown that the distribution of mRNA for *Cdx2* (caudal-related homeobox 2 gene), a key transcription factor involved in TE differentiation, is asymmetric between 8-cell stage mouse blastomeres and most likely biases future cell fate decisions. Specifically, the low amounts of Cdx2 mRNA in some blastomeres predispose them to preserve their unrestricted potential and create the future ICM lineage, whereas cells with high *Cdx2* expression tend to become the future TE (Jedrusik et al., 2008). Moreover, *Sox21* (sex determining region Y-box 21), that was found to regulate the transcription of the *Cdx2* gene, is also expressed heterogeneously in mouse embryos. Interestingly, the downregulation of *Sox21* gene expression in individual blastomeres is able to enhance the expression of the *Cdx2* gene and as a consequence reduce their pluripotency (Goolam et al., 2016). It is also noteworthy that human blastomeres of 2- to 8-cell embryos differ significantly in the level of ICM-specific gene expression, such as *OCT4* (octamer-binding transcription factor 4, Chavez et al., 2014), and also genes specific for TE, such as *CDX2* (Chavez et al., 2014) and *BHCG* (human chorionic gonadotropin subunit beta) (Hansis, Grifo, & Krey, 2004).

Epigenetic modifications and the kinetics of transcription factor DNA/chromatin binding are yet another source of heterogeneity in early blastomeres. For example, 4-cell stage mouse blastomeres manifest an unequal level of histone H3 methylation that, as has been suggested by Torres-Padilla, Parfitt, Kouzarides, and Zernicka-Goetz (2007), translates to their developmental fate. Progeny of cells with the lowest level of H3R26me (methylation at the arginine 26 residue of histone H3, catalyzed by the enzymatic activity of the epigenetic modifier Carm1) tended to differentiate into mural TE, whereas the maximal level of histone H3 methylation was characteristic for blastomeres biased toward an inner cell fate (Torres-Padilla et al., 2007). Additionally, *Prdm14* (PR domain-containing protein 14, a known interaction partner of Carm1, as well as an epigenetic chromatin modifier) has also been found to be asymmetrically expressed in 4-cell mouse embryos. Two out of four blastomeres express *Prdm14* at detectable levels, whereas the other two display only basal or undetectable levels of *Prdm14*. Moreover, the overexpression of *Prdm14* by the microinjection of specific mRNA increases the level of H3R26me in injected blastomeres, predisposing their progeny to contribute to the ICM (Burton et al., 2013). In human embryos, blastomeres of the 4- and 8-cell stage also show differences in the expression of DNA methyltransferases and histone modification enzymes (Chavez et al., 2014). Such differences proved to be important in preimplantation development, since the knockdown of one of the histone-modifying enzymes, *MSK2* (mitogen- and stress-activated protein kinase 2) in human embryos, has been reported to significantly reduce the rate of appropriate blastocyst formation (Chavez et al., 2014).

Plachta, Bollenbach, Pease, Fraser, and Pantazis (2011), employing an innovative method of live-imaging fluorescence decay visualization after photoactivation, demonstrated that in 8-cell stage mouse embryos, individual blastomere nuclei exhibit two distinct profiles of DNA/chromatin-binding affinity, for the pluripotency-related transcription factor Oct4. Moreover, such distinctive binding kinetics might be considered as a marker of future cell fate, as the blastomere nuclei exhibiting slower rates of Oct4 kinetic exchange (associated with greater retention of Oct4 in the nucleus and increased DNA/chromatin-binding affinities) were found to preferentially contribute to the pluripotent ICM lineage, while those with rapid kinetics (typified by low nuclear retention rates and lower DNA/chromatin-binding affinity) predominantly populated the TE lineage (Plachta et al., 2011). Similarly, the kinetics of Sox2 (sex determining region Y-box2) binding, another transcription factor controlling pluripotency, has also been shown to be nonidentical among early sister blastomeres, thus predisposing them to different fates. Specifically, the blastomeres manifesting long-lasting binding of Sox2 to DNA/chromatin tended to contribute to the ICM rather than the TE (White et al., 2016).

The cause of such early heterogeneity among blastomeres of the same embryo remains unclear. One possible explanation is that some mRNAs encoding transcription factors or other enzymes regulating gene expression may be asymmetrically distributed within the oocyte/zygote. Such initial asymmetries could be then enhanced during the temporally long embryonic cleavage divisions that could eventually translate into distinct developmental potentials for the resulting daughter cells (Hadjantonakis & Arias, 2016; Shi et al., 2015). Indeed, it was shown that 4-cell stage blastomeres differ in their developmental potential depending on the type of cleavage division they originated from. Progeny of blastomeres that inherited predominantly the vegetal part of the oocyte/zygote have been reported to form the mural TE. Consistently, these blastomeres, when combined, fail to develop into healthy chimeric offspring. Conversely, chimeras made of blastomeres that inherited both the vegetal- and animal-associated material from the oocyte/zygote did not manifest any abnormalities in their development (Piotrowska-Nitsche, Perea-Gomez, Haraguchi, & Zernicka-Goetz, 2005). Correlation between the fate of blastomeres and their origin, with regard to the inheritance of the vegetal/animal component of the oocyte/zygote, has also been proven at the 8-cell stage. Indeed, the blastomeres originating from the vegetal part were characterized by the higher *Cdx2* expression than their sister blastomeres (Jedrusik et al., 2008).

Concluding, the heterogeneity with regard to the molecular differences observed among blastomeres of the cleaving embryos seems to bias rather than ultimately determine their future fate, as evidenced by "compensation" experiments described earlier. Bearing that in mind it is plausible that developmental fate of blastomeres depends on the experimental approach that triggers a specified microenvironment, i.e., cell–cell contact and signaling.

4. POTENCY OF BLASTOMERES TO MULTILINEAGE DIFFERENTIATION

Although, as discussed earlier, many scientists use the term "totipotent" in the most strictest sense (i.e., to refer to an ability of isolated cell of the embryo to produce a fertile individual), others interpret it less rigorously, as a capability of a single cell to produce all cell types without organizing them into a coherent body plan.

4.1 Complementation Experiments and Blastomere Potency

In mouse, only the zygote and the blastomeres of the 2-cell stage embryo are totipotent in the strictest sense. One method to compensate for the insufficient number of ICM cells, that is considered to preclude complete development of blastomeres isolated at later stages, is to aggregate them with genetically distinct cells, so-called carrier blastomeres. As such, diploid or tetraploid carriers can be used as a source of supporting cells, resulting in the formation of a chimera (i.e., an aggregate of two or more embryos or parts of embryos that can integrate and create a proper blastocyst). For example, it has been shown that the aggregation of single 4- or 8-cell stage mouse blastomeres with carrier diploid blastomeres results in the formation of normal blastocysts that, following the transplantation to recipient foster mice, can undergo complete embryogenesis (Kelly, 1975, 1977; Fig. 1C). The descendants of such single tested blastomeres have been identified in both the ICM and TE derivatives of both postimplantation chimeric fetuses and neonates. Tetraploid cells, which can be obtained by electrofusion of blastomeres at the 2-cell stage (Kubiak & Tarkowski, 1985), are also often used as the carrier blastomeres (Fig. 1D). It has been reported that in experimentally obtained diploid–tetraploid mouse embryo chimeras, tetraploid cells show a restricted tissue distribution: 4*n* cells are commonly present in both the fetal part of placenta and some fetal membranes, but they rarely contribute to fetus itself (James, Klerkx, Keighren, Flockhart, & West, 1995; Mackay & West, 2005; Tarkowski, Witkowska, & Opas, 1977; Fig. 1D). This method, termed "tetraploid complementation," permits the derivation of a diploid and nonchimeric animal that exclusively originates from the single isolated experimental blastomere. For instance, it has been demonstrated that the aggregation of single blastomeres isolated from 4-, 8-, or even 16-cell embryos with tetraploid cells can result in the birth of "pure" diploid individuals, and in the case of blastomeres isolated from the same embryo, genetically identical multiplets (twins, triplets, and quadruplets) (Tarkowski, Ozdzenski, & Czołowska, 2001; Tarkowski, Ozdzenski, & Czolowska, 2005; Tarkowski, Suwińska, Czołowska, & Ożdżeński, 2010; Fig. 1D). Furthermore, when single inner or outer 32-cell stage blastomeres were surrounded by tetraploid cells, their descendants were able to populate the first three cell lineages of the chimeric blastocyst (i.e., TE, PE, and EPI) and even sporadically produced fetuses after transplantation to recipients (Tarkowski et al., 2010). In such complementation experiments the tested experimental blastomeres participated in the developmental processes, but they did not initiate them on their own. Therefore, it is important to note that they would not be capable of generating a full organism without the assistance of either diploid or tetraploid cells.

4.2 The Disaggregation–Reaggregation Approach and Blastomere Potency

The potency of blastomeres to multilineage differentiation has also been investigated by disaggregation–reaggregation approaches. Ziomek and coworkers examined the developmental potential of 16-cell stage mouse blastomeres by creating aggregates of solely outer, inner, or mixed groups of cells (Fig. 1E) (Ziomek, Johnson, & Handyside, 1982). Their experiments revealed that such clusters could form blastocysts and following their transfer to pseudopregnant mice could develop to 9.5-day-old fetuses (Ziomek et al., 1982). The fate of such embryos was further and thoroughly studied by Suwińska and coworkers, who characterized the mechanisms enabling the successful development of aggregates composed of inner, outer, or randomly mixed 16-cell stage blastomeres, to the blastocyst stage (Suwińska, Czołowska, Ozdzeński, & Tarkowski, 2008; Fig. 1E). In embryos reconstructed exclusively from inner or outer blastomeres, a molecular reprogramming was observed: blastomeres located on the outside of the solely "inner" aggregates, although initially negative for *Cdx2* expression, upregulated *Cdx2*, while in the solely "outer" aggregates, some of the blastomeres located internally downregulated their *Cdx2* expression status to become Cdx2 negative. However, the blastomeres in mixed aggregates relocated within the embryo, to restore their original, prerearrangement position in a manner consistent with their initial *Cdx2* expression status (Suwińska et al., 2008). Irrespective of the applied regulative strategy, the postimplantation development of all blastocysts was found to be normal and resulted in the birth of fertile mice (Suwińska et al., 2008).

Attempts to obtain animals from aggregates of 32-cell stage blastomeres have been less successful (Suwińska et al., 2008). Aggregates of outer 32-cell stage cells typically formed only trophoblastic vesicles devoid of ICMs, and thus they were incapable of developing into fetuses. In contrast, the aggregation of inner 32-cell stage blastomeres or mixed inner and outer blastomeres resulted in the formation of morphologically normal blastocysts. Nevertheless, only the blastocysts originating from the "mixed" aggregates were able to implant into recipient mice and give rise to pups, implying that molecular reprogramming in the "inner" aggregates was incomplete (Suwińska et al., 2008). Recently, these experiments have been revisited using a *Cdx2-eGFP* fluorescent genetic marker reporter construct (Posfai et al., 2017) that allowed the correlation between the developmental potential of a given cell of the mouse embryo with the dynamics of its intracellular *Cdx2-eGFP* expression. This study confirmed that when a cell is placed in a competitive environment (as in the "mixed" aggregates), it contributes to the lineage compatible with its intrinsic identity, as characterized by the reporter gene expression profile. In contrast, when there is no competition (i.e., in the aggregates composed of only ICM or only TE progenitor cells), some cells, when forced into different positions, can change their fate (Posfai et al., 2017; Suwińska et al., 2008). Moreover, Posfai and coworkers confirmed that *Cdx2*-expressing cells are committed to TE as soon as at the 32-cell stage (i.e., just after blastocyst formation) that is manifested by their inability to give rise to ICM. In contrast, ICM cells delay their ultimate commitment for an additional cell cycle (i.e., up to 64-cell stage) (Posfai et al., 2017). These results would appear to contradict earlier studies performed by Rossant and Vijh (1980), who demonstrated that outer 30-cell stage blastomeres injected under *zona pellucida* of 8-cell embryos contributed to both ICM and TE derivatives in the chimeric embryos by the 8th day of pregnancy (Rossant & Vijh, 1980). Further evidence for the totipotency (in this sense defined solely as the ability of the cell to contribute progeny to all lineages, not support full development) of outer 30-cell stage blastomeres was provided by their aggregation followed by injection into empty *zona pellucida*. The majority of such aggregates developed into structures morphologically resembling normal blastocysts with ICMs, that after transplantation to pseudopregnant recipients, were able to develop to E6.5 stage egg cylinders (Rossant & Vijh, 1980). Experiments performed by Rossant and Vijh (1980) and Posfai et al. (2017) were conducted nearly 40 years apart. Thus, the discrepancies in their outcomes may result from differences in developmental stages of embryos that they used.

4.3 Isolation of ICM as a Test of Its Potency to Regenerate TE

The potency of the inner cells of the blastocyst can also be characterized by the ability of isolated ICMs to regenerate TE (Fig. 1F). ICMs can be isolated either by immunosurgery, involving the sequential treatment of blastocysts with antiserum and active complement that initiates an immune reaction causing the lysis of the trophectoderm cells (Solter & Knowles, 1975), or by microsurgical manipulation (Cortés et al., 2007; Gardner & Johnson, 1972; Tanaka, Takeuchi, Neri, Sills, & Palermo, 2006).

Several reports have suggested that the potential of ICM cells to differentiate into TE is lost by the 32-cell stage (Gardner & Johnson, 1972; Rossant, 1975a, 1975b). Other studies, however, show that some ICM cells obtained immunosurgically from early 32- to 50-cell stage blastocyst still have the potential to reconstitute TE (Handyside, 1978; Hogan & Tilly, 1978; Nichols & Gardner, 1984; Pierce, Arechaga, Muro, & Wells, 1988; Spindle, 1978), and are even able to form small blastocyst-like vesicles that can give rise to trophoblast giant cells during in vitro culture or following ectopic transfer. Moreover, Rossant and Lis (1979), who aggregated ICMs derived from 32- to 50-cell blastocysts with 8-cell embryos, reported that the progeny of the ICM cells could contribute to postimplantation TE derivatives (Rossant & Lis, 1979; Fig. 1F), and that the aggregation of two isolated ICMs could reconstitute morphologically appropriate E5.5 egg cylinders. However, when 8-cell embryos were aggregated with ICMs obtained from 50- to 70-cell stage blastocysts, the ICM progeny rarely contributed to the TE or its derivative in vivo (Rossant & Lis, 1979). These results were also confirmed by an analysis performed by Chazaud, Yamanaka, Pawson, and Rossant (2006) whereby GFP-expressing cells isolated from the ICM of 64-cell blastocysts were aggregated with 8-cell embryos and shown to contribute to the TE in only 8% of any obtained blastocysts. Moreover, when single ICM cells from 64-cell embryos were injected into recipient blastocysts, they only colonized the PE or EPI lineages (Chazaud et al., 2006; Fig. 1F). Thus, the potential to differentiate into TE appears to be gradually lost as development progresses, most likely at the transition from 32- to 64-cell stage.

The discrepancies between the outcomes of each of the described experiments could be a result of inefficiency in the immunosurgery method. Indeed, Szczepanska and coworkers have shown that immunosurgery sometimes leaves single TE cells on the surface of ICMs that could be responsible for the regeneration of the TE layer observed in the isolated ICMs of the different developmental stages referred (Szczepanska, Stanczuk, & Maleszewski, 2011).

It has been suggested that complete loss of plasticity and final commitment to one of the three cell lineages (TE, EPI, or PE) in murine embryo occurs between the late (>100 cells) and the periimplantation (>140 cells) blastocyst stage (Grabarek et al., 2012). Specifically, Grabarek et al. (2012) injected epiblast and PE precursors, selectively isolated from early (<64 cells), mid (>64 cells), and late (>100 cells) blastocysts into recipient 8-cell embryos (Fig. 1F), and showed that individual ICM cells can form all three lineages of the blastocyst. Even the donor cells from the late blastocysts (>100 cells), in which EPI and PE precursors were already morphologically sorted into their respective cell layers, maintained their capacity to contribute to multiple lineages. Complete loss of plasticity was only observed in the case of inner cells derived from the periimplantation embryos (>140 cells), in which the ICM cells contributed exclusively to the lineage predicted by their gene expression profile (Grabarek et al., 2012).

4.4 Various Ways to Study Potency of Human Blastomeres

The ability of cells to undergo multilineage differentiation has also been tested in human embryos, although less extensively than in the mouse model. Although, as stated in previous paragraphs, 2- to 8-cell stage human blastomeres differentially express potent differentiation regulators, such as *OCT4* or *CDX2* (Chavez et al., 2014; Hansis et al., 2004), they do not seem to be lineage committed at that developmental stage. When individual blastomeres of 2- to 8-cell stage human embryos are randomly injected with a fluorescent lineage tracer and cultured to the blastocyst stage, the descendants of the injected blastomeres are able to form both the TE and ICM of blastocysts (Mottla et al., 1995).

To investigate whether TE and ICM cells of the human embryo are irreversibly committed or still able to change their fate, De Paepe and coworkers isolated outer and inner cells from 5-day-old human blastocysts and reconstituted "outer-cells-only" and "inner-cells-only" aggregates (De Paepe et al., 2013). Unexpectedly, both types of aggregates resulted in the formation of blastocysts with ICM cells expressing the EPI-specific marker NANOG. In addition, when TE cells were placed inside of host embryos, they could integrate within the host ICM and initiate the expression of *NANOG*. These results suggest that final commitment of TE and ICM cells in the human embryo takes place at a developmentally later time-point equivalent than in mouse embryos.

5. POTENCY OF BLASTOMERES TO GENERATE CHIMERAS: BREAKING INTERSPECIES BARRIERS

As stated earlier, the generation of chimeric embryos, composed of two or more populations of genetically distinct cells, but still able to successfully develop to term, proves the mammalian embryo's ability to regulate the number and arrangement of its cells. The first intraspecies chimeras were obtained independently by Mintz (1962) and Tarkowski (1961) via the aggregation of two 8-cell embryos. Apart from mice (Markert & Petters, 1978; Mintz, 1962, 1964; Ozdzeński, Szczesny, & Tarkowski, 1997; Tarkowski, 1961; Tarkowski, Jagiello, Czolowska, & Ozdzenski, 2005; Tarkowski & Wojewodzka, 1982), intraspecies chimeras have also been produced in several other mammalian species, including rat (Mayer & Fritz, 1974), sheep (Tucker, Moor, & Rowson, 1974), rabbit (Gardner & Munro, 1974), cattle (Brem, Tenhumberg, & Kräußlich, 1984), pig

(Matsunari et al., 2013), and nonhuman primates (Tachibana et al., 2012). However, it quickly emerged that the remarkable plasticity of mammalian embryos allows them to overcome the interspecies barrier, permitting the creation of chimeras made by combining cells of two different species. The first interspecies chimeras were generated by Gardner and Johnson (1973, 1975), who constructed mouse–rat chimeras via the injection of rat ICMs into mouse blastocysts (Gardner & Johnson, 1973, 1975). However, the resulting offspring contained very few rat cells. Recently, these classical results have been confirmed by another approach, whereby 8-cell stage mouse and rat embryos were aggregated and able to complete embryogenesis in mouse foster mothers and to reach adulthood (Bożyk et al., 2017). However, Bożyk and coworkers noticed that the proportion of mouse cells dominated over that of rat cells in all the cell lineages of chimeric blastocysts, apparently due to the selective elimination of the rat cells. In similarity to the chimeras obtained by Gardner and Johnson, such mouse–rat aggregation chimeras, that were able to live in a good health for at least 2.5 months, had very few remaining rat cells in adulthood (Bożyk et al., 2017). Indeed, individual animals containing higher percentages of rat cells died, primarily due to severe developmental malformations. The reason for this interspecies incompatibility is not clear. However, despite the common belief that mice and rats are closely related, these two species evolutionarily diverged about 20.9 million years ago (mya). It is possible that this early genetic diversification could have resulted in ligand–receptor incompatibilities and differences in the affinity of adhesion molecules (for a review, see Wu et al., 2016). Mammalian species differ considerably in various aspects, including the time and manner of embryo implantation, embryo size, pattern of postimplantation epiblast development, speed of development, and the duration of pregnancy (Mascetti & Pedersen, 2016; Wu et al., 2016). Consequently, these and other factors may negatively affect the efficiency of interspecies chimera formation. However, it has been shown that the generation of viable chimeras between two closely related mouse species, *Mus musculus* and *Mus caroli* (evolutionarily diverged only 7.8 mya), is possible. Such chimeras were produced either by injection of *M. caroli* ICMs into *M. musculus* blastocysts or by aggregation of two 8-cell embryos (one of each species) followed by their transfer to *M. musculus* foster mothers (Rossant & Chapman, 1983; Rossant, Croy, Chapman, Siracusa, & Clark, 1982; Rossant & Frels, 1980; Rossant, Mauro, & Croy, 1982). These experiments have additionally shown that *M. caroli* blastocysts are not able to survive in the *M. musculus* uterus, as injection of *M. musculus* ICMs into *M. caroli* blastocysts, followed by the transplantation to the *M. musculus* recipient females, did not result in any live births. These results clearly indicate that the successful production of interspecies chimeras is possible, but only when trophoblast cells and maternal uterus originate from the same species. Based on these findings, other viable chimeras have been generated: *Ovis aries–Capra hircus* (Fehilly, Willadsen, & Tucker, 1984; Jaszczak, Członkowska, Guszkiewicz, & Parada, 1991; MacLaren, Anderson, BonDurant, & Edmondson, 1993; Polzin et al., 1987) and *Bos taurus–Bos indicus* (Summers, Shelton, & Bell, 1983; Williams, Munro, & Shelton, 1990).

Interspecies chimeras can also be formed by the injection of embryonic stem cells (ESCs) or induced pluripotent stem cells (iPSCs) into the blastocysts of another species. In this case, donor cells contribute predominantly to the EPI (Beddington & Robertson, 1989; Isotani, Hatayama, Kaseda, Ikawa, & Okabe, 2011; Kobayashi et al., 2010). The first ESC-derived interspecies chimeras were produced by Xiang and coworkers, who injected *Apodemus sylvaticus* (the common wood mouse) ESCs into *M. musculus* blastocysts (Xiang et al., 2008). They obtained viable chimeras, despite the considerable evolutionary distance between these two species (14.59 mya divergence). In addition, the obtained adult chimeric animals contained a substantial contribution of *A. sylvaticus* cells that differentiated into a wide range of cell types, including the germ line. ESCs and iPSCs have also been successfully utilized for generation of mouse–rat chimeras (Isotani et al., 2011; Isotani, Yamagata, Okabe, & Ikawa, 2016; Kobayashi et al., 2010; Wu et al., 2017; Yamaguchi et al., 2017). These experiments provided the first evidence that the formation of interspecies chimeras between pluripotent cells and the embryos of more distantly related species is possible. However, despite the availability of human and nonhuman primate ESCs, none of these lines have yet been demonstrated to be efficient in generating chimeras in combination with blastocyst of different species (James, Noggle, Swigut, & Brivanlou, 2006; Simerly et al., 2011), probably due to the mismatch between the donor cells and developmental stage of the host embryo.

In recent years, renewed interest in generating interspecies chimeras has emerged due to the potential possibility to "grow" replacement human organs in genetically engineered "organ niches" in large animals. Indeed, the first attempts to inject human pluripotent stem cells into pig blastocysts, followed by transplantation to surrogate sows, have been performed, albeit with a limited success: the contribution of human cells in the resulting human–pig chimeras was very low (Wu et al., 2017).

5.1 Naturally Occurring Mammalian Chimeras

The existence of chimerism is not limited to the research laboratory. Spontaneous chimeras occur also in nature, taking as an example humans. Among various possible mechanisms leading to the emergence of human chimerism is dizygotic twinning. In such an instance, dispermic chimeras, also called tetragametic or whole body chimeras, result as a consequence of the fertilization of two oocytes (or the oocyte and the second polar body) by two spermatozoa, and their subsequent fusion into one single embryo (Malan, Vekemans, & Turleau, 2006; Rinkevich, 2001). This type of chimerism may cause various abnormalities in the resulting organism. The majority of such chimeras are identified due to gonadal abnormalities (e.g., hermaphroditism, when the chimeric individual originates from fused male and female embryos), or other unusual phenotypic features, such as heterochromia (differently colored eyes), mixture of skin colors, and developmental abnormalities (Berger-Zaslav et al., 2009; Draper, Conley, Smith, & Benson, 2008; Hong et al., 2013; Moores, 1973; Mosebach, Parkner, Jakubiczka, Wieacker, & Heim, 2006; Strain, Warner, Johnston, & Bonthron, 1995; Tippet, 1984; Watkins et al., 1981; Yunis, Zuniga, Romero, & Yunis, 2007). Other human chimeras show normal phenotypes and because of a lack of visible symptoms they may never be diagnosed, leading to underestimation of this phenomenon (Boklage, 2006).

6. ESCs—AN IN VITRO REFLECTION OF EMBRYONIC PLASTICITY AND POTENCY

Developmental potency and plasticity of early mammalian blastomeres can also be reflected in their ability to give rise to ESCs. The first mammalian ESCs were independently established by two groups in the mouse model in 1981 (Evans & Kaufman, 1981; Martin, 1981). The successful isolation of murine ESCs opened a new chapter in developmental research and resulted in numerous attempts to derive ESCs from other mammalian species. However, to date, other ESCs, whose pluripotency has been verified via the common and accepted tests, have only been isolated from monkeys (Thomson et al., 1995, 1996), humans (Thomson et al., 1998), and rats (Buehr et al., 2008; Li et al., 2008; Ueda et al., 2008). Since the isolation of human ESCs from blastocysts raises ethical concerns due to the necessity to destroy the embryo, much attention has been placed on deriving these cells from single blastomeres and even polar bodies (Becker & Chung, 2006; Chung et al., 2006, 2008; Delhaise, Bralion, Schuurbiers, & Dessy,

1996; Geens et al., 2009; González, Ibáñez, & Santaló, 2010; Klimanskaya, Chung, Becker, Lu, & Lanza, 2006; Lorthongpanich, Yang, Piotrowska-Nitsche, Parnpai, & Chan, 2008; Wakayama et al., 2007; Yang, Mai, Li, & Zhou, 2013; Zdravkovic et al., 2015). However, the efficiency of establishing such novel human ESCs lines has been shown to vary and seems to depend on numerous distinct parameters, some of which are discussed later.

Apart from the genetic background of an embryo (Brook & Gardner, 1997), its developmental stage seems to be the most important factor affecting the efficiency of ESC derivation. Since 2-cell stage mouse blastomeres can be totipotent (Tarkowski, 1959), one could assume that the individual blastomeres of 2-cell embryos should be equally capable of establishing both ESCs and live pups. Indeed, Teramura and coworkers have managed to obtain functional ESCs from blastocysts that are themselves derived from single blastomeres of the 2-cell stage mouse embryos (Teramura et al., 2007). Moreover, González and coworkers have shown that it is possible to isolate ESCs lines from all 4-cell stage mouse blastomeres (González, Ibáñez, & Santaló, 2011a). However, many reports contradict these results. Lorthongpanich and coworkers have reported that not all 2- and 4-cell stage mouse blastomeres have the potential to form ICM cells and, in turn, ESCs (Lorthongpanich et al., 2008). Similarly, in human embryos only one of the four 4-cell stage blastomeres has been reported to be able to give rise to human ESCs lines (Geens et al., 2009). Also, Wakayama et al. (2007) have reported that the efficiency of mouse ESCs derivation from 2-cell stage embryos is around 70%. They have additionally noted that the ESCs isolation success rate further decreases with embryonic age, from 40% at the early 4-cell stage, to 22% at the late 4-cell stage and 14% at the 8-cell stage (Wakayama et al., 2007).

Despite certain discrepancies, it seems to be a common consensus that individual blastomere potency to form ESCs decreases as embryonic development progresses. Indeed, it has been found that the mRNA transcription profile of 8-cell stage mouse blastomeres more closely resembles that of TE cells than the corresponding profile of the ICM, suggesting that such blastomeres may be primarily designated to contribute to the TE (Tang et al., 2011). This finding might partially account for their lower efficiency in establishing ESCs in comparison to the blastomeres of the earlier 2- and 4-cell stages. Moreover, reports that point to an unequal developmental potency between the early sister blastomeres at the 4-cell stage (see previous paragraphs for more details) could in some measure explain why not all 4-cell stage blastomeres can generate ICM.

Another factor that seems to affect the efficiency of establishing new ESCs lines from single blastomeres is the potential multifactorial nature of the cell milieu. For example, the low success rate of ESCs derivation from mouse and human single 8-cell stage blastomeres can be improved by coculturing them with supporting ESCs colonies (Chung et al., 2006; Klimanskaya et al., 2006). The efficiency rate for ESCs derivation from single human 8-cell stage blastomeres has also been shown to be enhanced when a biopsied single 8-cell stage blastomere was cocultured with its embryo of origin (Chung et al., 2008). ESC derivation is also improved by a 24-h-long culture in media containing E-cadherin, thus mimicking the original adhesion and cell-to-cell interaction parameters (González, Ibáñez, & Santaló, 2011b). Moreover, adrenocorticotropic hormone, a peptide pituitary gland hormone, has also been found to improve the derivation of ESCs lines from single blastomeres. However, its presence in the culture medium has been insufficient to ensure that all sister blastomeres can generate ESCs lines (Wakayama et al., 2007), again indicating that early blastomeres are heterogeneous in terms of their developmental potency.

Another intriguing and yet not fully understood aspect of ESC identity is the developmental potential of established ESCs colonies. Since ESCs stem from the ICM, that is known to be comprised of a heterogeneous population of cells (Chazaud et al., 2006), it should not come as a surprise that the state of pluripotency fluctuates within the ESCs culture itself. Indeed, single-cell analyses of gene expression in mouse ESCs revealed that the expression profiles of pluripotency factors like *Ssea1* (stage-specific embryonic antigen 1), *Rex1* (ring-exported protein 1), *Stella* a.k.a. *Dppa3* (developmental pluripotency associated 3), or *Nanog* (Nanog homeobox) can vary significantly between cells of the same colony, thus highlighting that they in fact form distinct subpopulations (Carter et al., 2008; Chambers et al., 2007; Cui et al., 2004; Graf & Stadtfeld, 2008; Hayashi, Lopes, Tang, & Surani, 2008; Singh, Hamazaki, Hankowski, & Terada, 2007; Toyooka, Shimosato, Murakami, Takahashi, & Niwa, 2008). Moreover, a subfraction of cells in ESCs culture can express markers of differentiation, such as *Sox17* (sex determining region Y-box 17) (Niakan et al., 2010) and *Gata6* (GATA-binding protein 6) (Singh et al., 2007). Interestingly, even in strict culture conditions only around 0.5% of mouse ESCs resemble 2-cell stage embryo blastomeres (and their associated and presumed totipotency). Remarkably, these cells, termed 2C-like (2-cell like) ESCs, have been shown to generate both embryonic and extraembryonic tissues, in spite of the fact that they did not express common pluripotency markers, such as *Oct4*, *Nanog*, and *Sox2*

(Macfarlan et al., 2012). It is therefore tempting to speculate that factors, other than those commonly known, are pivotal for maintaining potency and plasticity in ESCs culture. Thus, the above studies clearly demonstrate that there are many shades of embryonic potency and plasticity, which can be captured in vitro.

7. MECHANISMS UNDERLYING PLASTICITY OF BLASTOMERES AND EMBRYOS

Despite many years of research investigating the regulative nature of mammalian embryos, interpretation of the obtained experimental data is still difficult and mechanisms underlying this unique feature have only just begun to be elucidated.

On one hand, it is clear that individual cells exhibit variation in the expression levels of key transcription factors and epigenetic marks involved in embryogenesis prior to the 8-cell stage, and that this molecular heterogeneity between blastomeres may bias their future fate (Biase et al., 2014; Burton et al., 2013; Goolam et al., 2016; Plachta et al., 2011; Torres-Padilla et al., 2007; White et al., 2016). However, there are many lines of evidence that these cell-intrinsic differences most probably guide rather than dictate the cell fate under unperturbed conditions. According to such a new concept, the preimplantation mammalian embryo can be regarded as a self-organizing system, based on a multifaceted network of interrelations that must be strictly and timely coordinated to guarantee successful development (Wennekamp, Mesecke, Nédélec, & Hiiragi, 2013).

7.1 The First Cell Fate Decision: The Central Role of Hippo Signaling Pathway

According to current knowledge, the first cell fate decision (i.e., the establishment of TE and ICM in the preimplantation embryo) depends on many interconnected and interdependent factors, such as cell polarity, cell position, physical parameters of cells (cortical tension and cell adhesion), cellular interaction and intercellular communication, and the activity of signaling pathways. The Hippo signaling pathway, as elucidated in the mouse model, seems to be, at least in part, a clasp fastening most of these factors. It integrates information on cell polarity, cell–cell contact, and cell position and translates this information into appropriate and cell lineage-specific transcriptional programs (Nishioka et al., 2009; Sasaki, 2017). It is the interblastomeres differences in cell polarity and adhesion (Hirate et al., 2013), as well as contractility (Anani, Bhat, Honma-Yamanaka, Krawchuk, & Yamanaka, 2014; Maître et al., 2016) that result in the differential activation of the Hippo pathway. Active Hippo signaling in apolar, highly contractile cells, sequesters Yap1 (Yes Associated Protein 1) in the cytoplasm, while inactive Hippo pathway in polar, less contractile cells leads to nuclear accumulation of Yap1 and its interaction with the transcription factor Tead4 (TEA-domain protein 4) and the induction of TE-specifying *Cdx2* expression (Hirate et al., 2013; Nishioka et al., 2009). Such activation of *Cdx2* in TE progenitors is also conducted in cooperation with Notch signaling (Rayon et al., 2014) and leads to the subsequent downregulation of the pluripotency related *Oct4* and *Nanog* genes (Chen et al., 2009; Niwa et al., 2005). In ICM progenitors, active Hippo signaling leads to the induction of *Sox2* gene expression (Wicklow et al., 2014). Interestingly, the importance of continuous cell–cell interactions (required to activate the Hippo-signaling pathway) during blastomere fate regulation has been demonstrated by the fact that blastomeres, resulting from constant disaggregation following cell division, in in vitro culture adopt a uniform molecular identity that is distinct from both the TE and ICM (Lorthongpanich, Doris, Limviphuvadh, Knowles, & Solter, 2012).

The first cell fate decision in the mouse embryo takes place during the 8- to 16- and 16- to 32-cell stage transition and is defined by the type of blastomere division (either conservative or differentiative in relation to the inheritance of the apical domain). According to the classical "inside–outside" and "polarity" models, both daughter cells generated by a conservative division inherit apical domains and remain located on the embryo surface, finally contributing to the TE (Johnson & Ziomek, 1981; Tarkowski & Wróblewska, 1967). In contrast, the allocation of cells to the ICM typically occurs through differentiative cell divisions, in which polar cells divide and result in one of the daughter cells that stays outside and gives rise to TE, and the second one, lacking the apical domain, and being deposited inside the embryo as a precursor of ICM. There are two major waves of such cell internalization, at the 8- to 16-cell (first wave of differentiative divisions) and 16- to 32-cell transitions (second wave). However, recent reports have shown that the mechanism of cell internalization is much more dynamic and complex, and depends not only on differentiative divisions but also involves processes such as active cell engulfment, bulging, or displacement (Anani et al., 2014; Maître et al., 2016; McDole, Xiong, Iglesias, & Zheng, 2011; Samarage et al., 2015; Watanabe, Biggins, Tannan, & Srinivas, 2014; Yamanaka, Lanner, & Rossant, 2010). Given these observations, cell cleavage divisions can be considered as dynamic processes involving extensive cell rearrangements resulting from cell polarity and tensile forces generated by the cortical cytoskeleton. Additionally, in order to ensure appropriate proportions of outer and inner cells (i.e., the precursor populations of the TE and ICM, respectively), cleaving embryos appear to use an ill-defined regulative, compensatory mechanism, controlling the frequency of cellular internalization (Humięcka, Krupa, Guzewska, Maleszewski, & Suwińska, 2016; Krupa et al., 2014). There is a wide variation in the reported number of cells internalized during the 8- to 16-cell transition, ranging from 1–2 to even 8 (Fleming, 1987; Handyside, 1981; Krupa et al., 2014; Morris et al., 2010; Suwińska et al., 2008). However, ultimately, each embryo ends up with similar number of ICM cells by the blastocyst stage (Fleming, 1987; Krupa et al., 2014). Those presenting a low number of inner cells internalized during 8- to 16-cell transition compensate for this deficiency by an increased frequency of internalization during the next round of divisions. Moreover, in extreme cases, when the number of inner cells is still insufficient at 32-cell stage, this pool can be supplemented as a result of the third round of internalization (Humięcka et al., 2016; Morris et al., 2010). Some additional mechanisms can also contribute to this regulation. At the 16-cell stage, some outer cells have been observed to be apolar (Anani et al., 2014; Hirate et al., 2015). Their higher surface tension and higher contractility usually induces them to occupy an inner position (Maître et al., 2016), but sometimes apolar outer cells undergo repolarization and remain on the outside (Anani et al., 2014; Korotkevich et al., 2017). Conversely, until the early blastocyst stage, there are some Cdx2-positive cells that can translocate from the outer position inward, downregulate *Cdx2* expression, and finally contribute to the ICM (McDole & Zheng, 2012; Toyooka, Oka, & Fujimori, 2016). How the cells choose such alternative fates remains an open question, but adjusting the proper number of inner and outer cells seems to be of special importance. This is underlined by the observation that successful development to birth requires the establishment of at least four inner, pluripotent cells before implantation, and otherwise development will arrest (Morris et al., 2012).

7.2 The Second Cell Fate Decision: The Role of Fgf4/Map Kinase Signaling Pathway

As the blastocyst matures, the second cell fate decision results in the partitioning of the ICM into EPI and PE. Until the 64-cell stage, the expression of lineage-specific genes, such as *Gata6* and *Nanog*, overlaps in some cells of the ICM (Plusa, Piliszek, Frankenberg, Artus, & Hadjantonakis, 2008). As development progresses, PE and EPI precursor cells, each expressing lineage-specific factors (i.e., *Gata6* and *Nanog*, respectively), in a mutually exclusive manner arise (Chazaud et al., 2006; Kurimoto et al., 2006) and coexist in a randomized pattern with double-positive, bipotential cells, across the ICM (i.e., the so-called salt and pepper pattern) (Saiz, Williams, Seshan, & Hadjantonakis, 2016). These precursor cells are then sorted into the correct positions, either inside or to the cavity-facing surface of ICM, via an integrated combination of positional information, active cell migration, and programmed cell death of incorrectly positioned cells (Meilhac et al., 2009; Plusa et al., 2008). Additionally, it has recently been shown that the bipotential cells are allocated to either the PE and EPI lineages asynchronously, thus allowing the embryo to control its ICM composition (i.e., to reflect appropriate proportions of PE and EPI cells) (Saiz et al., 2016).

Fgf (fibroblast growth factor)/Mapk (mitogen-activated protein kinase) signaling has been shown to play an essential role during the formation of the PE lineage in the mouse model (Kang, Piliszek, Artus, & Hadjantonakis, 2013; Krawchuk, Honma-Yamanaka, Anani, & Yamanaka, 2013). Fgf4, produced by *Nanog*-expressing EPI-biased cells, has been proposed to activate Fgf/Mapk pathway in neighboring cells, thus increasing the expression of *Gata6*, and repressing the expression of *Nanog* (Schrode, Saiz, Di Talia, & Hadjantonakis, 2014). Pharmacological inhibition of Fgf signaling (via Fgf receptor or downstream signaling Erk1/2 and p38–Mapk14/11 inhibition) is sufficient for ICM cells to adopt the EPI fate (Nichols, Silva, Roode, & Smith, 2009; Thamodaran & Bruce, 2016; Yamanaka et al., 2010), whereas Fgf4 administration shifts all ICM cells toward adopting a PE fate (Yamanaka et al., 2010).

The question of whether the heterogeneity of the ICM cell population, as it matures, is the consequence of stochastic fluctuations in gene expression, results from the previous cleavage history of blastomeres and/or depends on the activity of the Fgf signaling pathway has been widely addressed and has given rise to contradictory conclusions (Morris et al., 2010; Yamanaka et al., 2010). Studies by Morris, Graham, Jedrusik, and Zernicka-Goetz (2013) and Krupa et al. (2014) together link the time of ICM founder cell internalization with intrinsic cell differences in the expression of the *Fgf4* and *Fgfr2* genes that act to bias them toward either EPI or PE precursors. Based on their results, the fate of ICM cells would appear to be biased by timing of cell internalization and the number of cells generated in the first and the second waves of internalization. Inner cells derived from the first internalization wave express higher levels of *Fgf4* and are predisposed to form EPI, and those internalized during the second round upregulate *Fgfr2*, thereby becoming more sensitive to Fgf signaling essential for PE formation (Krupa et al., 2014; Morris et al., 2013). These data are consistent with the findings of Mihajlović, Thamodaran, and Bruce (2015), who created embryos containing some blastomeres unable to initiate TE differentiation, through RNAi-mediated *Tead4* knockdown, and found that the resulting descendants of TE-inhibited cells preferentially contributed to EPI, due to reduced *Fgfr2* and other PE marker gene expression. Thus, the longer outer cells are exposed to TE-differentiation cues, provided by the inhibition of Hippo signaling, the greater probability that the resulting ICM founder cells will contribute to the PE (Mihajlović et al., 2015).

Interestingly, it has been shown that the significance of the Fgf4/Mapk signaling pathway in PE specification is not universal for all mammals. It does not impact on ICM cell specification in human embryos, despite the presence of a salt and pepper pattern of *NANOG* and *GATA6* gene expression (Kuijk et al., 2012; Roode et al., 2012). Similarly, inhibition of this pathway in bovine embryos fails to prevent PE formation, indicating the involvement of some other regulative mechanisms (Kuijk et al., 2012). In contrast, in nonhuman primates, the specification of the PE lineage is regulated by both the Fgf and Wnt signaling pathways (Boroviak et al., 2015).

7.3 The Transition From Totipotency to Pluripotency

When comparing the dynamics of fate restriction shown by the cells of the blastocyst, it was observed that TE progenitors lose their totipotency, defined as the ability to differentiation to all cell lineages, by the 32-cell stage (i.e., one cleavage cycle earlier than ICM progenitors). ICM cells lose the ability to differentiate to TE when the PE and EPI progenitors are established and this transition from totipotency to the lineage priming seems to be regulated by the activity of Fgf/Mapk pathway (Wigger et al., 2017). However, the restricted gene expression pattern in the ICM does not preclude ICM cells from switching their fates (between PE and EPI) if their environment is experimentally disturbed, either through the modulation of Fgf signaling (Nichols et al., 2009; Yamanaka et al., 2010) or by changing their position in a "chimera assay" (Grabarek et al., 2012). This asynchronous lineage commitment and the ability of ICM cells to choose alternative cell fate (PE or EPI) until the periimplantation stage (>140 cell)

(Grabarek et al., 2012), together with the incremental lineage allocation (Saiz et al., 2016) and lower plasticity of EPI cells in comparison with PE cells (Grabarek et al., 2012), may collectively form a mechanism ensuring that the correct number of cells is allocated to the embryonic and extraembryonic cell lineage compartments.

8. FROM THEORY TO PRACTICE: POTENTIAL AND REAL BENEFITS OF EMBRYO PLASTICITY

The high plasticity and unrestricted developmental potential of the early mammalian embryo, as well as the fact that these abilities can be captured in vitro by the generation of ESCs can have numerous practical applications. The plasticity of mammalian blastomeres and ESCs lines is not only an interesting research subject for developmental biologists but also a powerful tool that may be useful in all kinds of veterinary and biomedical studies (Fig. 2).

Nowadays, embryo splitting can be used in veterinary medicine and animal breeding to maintain high quality and healthy livestock with desirable genetic characteristics (Yang et al., 2007). Moreover, as mammalian embryos are able to continue normal development after removing single blastomeres, it is possible to subject them to preimplantation genetic diagnosis (PGD) in order to identify potential harmful mutations and chromosomal aberrations (Simpson & Rechitsky, 2017). PGD is applied to embryos generated via in vitro fertilization and allows for selection of healthy embryos for transfer (Fig. 2A). Biopsied blastomeres may also serve as an ethically acceptable source of cells for creating ESCs (Chung et al., 2008; Geens et al., 2009; Klimanskaya et al., 2006; Yang et al., 2013; Zdravkovic et al., 2015).

Moreover, the fact that sister blastomeres of the early embryo are not identical has an important implication for the existence of discordant symptoms of many diseases observed in monozygotic twins. Such twins are thus invaluable tools for research on the genetic and environmental background of human diseases (Fig. 2B). Among conditions which were found to manifest different phenotypes in monozygotic twins are chromosomal abnormalities, such as trisomy 13, Patau syndrome (Naor, Amir, Cohen, & Davidson, 1987), chromosomal structural aberrations, such as the DiGeorge syndrome (Hillebrand, Siebert, Simeoni, & Santer, 2000), as well as autosomal monogenetic diseases, such as sickle cell anemia (Amin et al., 1991) and X chromosome-linked conditions (Buchbinder, Nadeau, & Nugent, 2011).

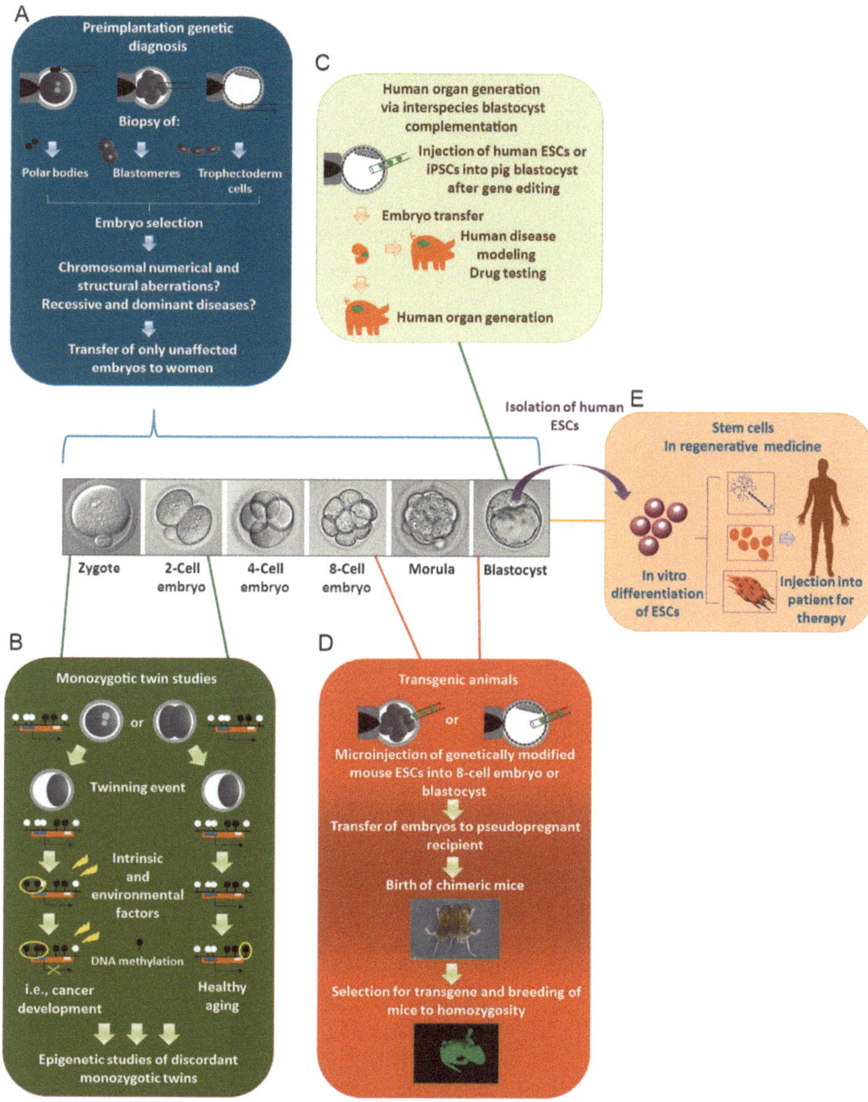

Fig. 2 Potential applications of mammalian embryos and embryonic stem cells. (A) An embryo biopsy sample for preimplantation genetic testing. After diagnosis, only the unaffected embryos are selected for transfer to the uterus. (B) Usage of discordant monozygotic twins in molecular-genetic studies. The *black* and *white circles* indicate methylated and unmethylated sites of the tumor suppressor gene. After the twinning event, both monozygotic twins have an identical DNA methylation profile. During the lifetime, this pattern becomes different because of the intrinsic and environmental factors. One of the twins (on the *left*) displays an increase in methylation in the promoter site and suppression of gene expression in comparison with the healthy twin (on the *right*), who acquired just minor changes. Consequently, this epigenetic discordance

(Continued)

Moreover, many genetic abnormalities were reported to affect only one of the twins (Bennett, Boye, & Neufeld, 2008; Bourthoumieu et al., 2005; Cheng, Shaw, Shih, & Soong, 2006). The exact nature of this discrepancy is still unknown; however, it may be related to differences in epigenetic modifications between monozygotic twins (Castillo-Fernandez, Spector, & Bell, 2014).

The remarkable plasticity and potency of mammalian embryos are, as thoroughly discussed in the previous paragraphs, reflected in their ability to form chimeras. Such animals provide an important resource to study cell fate and the mechanisms that govern mammalian embryogenesis and organogenesis, as well as to serve as a tool for investigating the potential of various pluripotent stem cell states. Another possible application of chimeras is the generation of human organs for transplantations in animals with similar to human organ size and physiology (Fig. 2C). Indeed, the recent work of Wu and coworkers has demonstrated that human pluripotent stem cells are able to incorporate into postimplantation pig embryos and form chimeras, which raise the possibility that in future we will be able to generate transplantable human organs in farm animals (Wu et al., 2017).

Chimeric animals generated using ESCs (Gossler, Doetschman, Korn, Serfling, & Kemler, 1986) may also serve as a tool for testing the pluripotency of newly established ESC lines. Moreover, transgenic animals, which can be obtained using genetically modified ESCs, are a vitally important model in studies on gene function (Fig. 2D; Bouabe & Okkenhaug, 2013), and the molecular background of numerous human diseases, including atherosclerosis, myotonic dystrophy, or cancers (reviewed in Gomes-Pereira, Cooper, & Gourdon, 2011; Lee et al., 2017; Zhang, Moore, & Ji, 2011). They can also

Fig. 2—Cont'd may account for the difference in the cancer disease susceptibility between monozygotic twins. (C) Human organ generation via interspecies blastocyst complementation. Pig embryos after gene editing may potentially be complemented with human stem cells (ESCs or iPSCs) to generate the functional replacements for damaged human organs. This approach also offers an attractive platform for the study of human diseases and testing for drug efficacy and toxicity under in vivo conditions. (D) Formation of transgenic animals through injection of mouse ESCs into mouse embryos. The obtained animals can serve as a model in studies on gene function, the etiology, onset, and progression of many human diseases, as well as models in drug discovery research. (E) Stem cell applications in regenerative medicine. Embryonic stem cells can differentiate into various types of cells, including neural, blood, and smooth muscle cells. Thus, they are considered as the hope of establishing a successful regenerative therapy.

serve as models in drug discovery research (Zambrowicz & Sands, 2003). Furthermore, the ESCs themselves play an important role in medical studies, since their proliferative and developmental potential holds a promise for future therapeutic applications (Fig. 2E). For example, Schwartz and coworkers have recently reported that macular degeneration patients regained sight after the injection of human ESC-derived cells into the space behind the retina (Schwartz et al., 2012, 2015). This successful case of ESCs therapy in humans gives hope for further improvements in this field, despite many possible obstacles.

9. SUMMARY

Advances in experimental approaches and technologies, such as live-cell imaging, use of fluorescent reporter proteins, genetic manipulations, and computational image analysis, have led in recent years to the revealing of the remarkable complexity and dynamics of mammalian development. However, researchers are just beginning to solve, piece by piece, this fascinating developmental puzzle. Until recently, segregation of cell lineages, TE vs ICM and PE vs EPI, has been regarded as two separate cell fate decisions. It is becoming clear, however, that they are interlinked, for example, through the involvement of bone morphogenetic protein (Graham et al., 2014) and Sox2-signaling pathways (Wicklow et al., 2014). Moreover, experiments performed by Morris and coworkers, Krupa and coworkers, and Mihajlović and coworkers indicate that all primary cell lineages (TE, PE, and EPI) begin to arise already at the 8- to 16-cell stage transition (Krupa et al., 2014; Mihajlović et al., 2015; Morris et al., 2010). Thus, the integration of our present knowledge regarding these two specification events is the next great challenge for developmental biologists, and most likely it will shed new light on the concepts of blastomere potency and the remarkable regulative abilities of the mammalian embryos.

ACKNOWLEDGMENTS

The authors would like to apologize for any potential omission of the relevant literature. It would have been unintended. The authors thank Dr. Anna Ajduk, Dr. Katarzyna Szczepańska, and Professor Marek Maleszewski for critical comments. During the preparation of this work, A.S. and A.K. were supported by the Grant SONATA 2014/15/D/NZ3/02435 and A.S. and K.K. by the Grant OPUS 2013/09/B/NZ3/02404 from the National Science Center (Poland).

REFERENCES

Alikani, M., Cekleniak, N. A., Walters, E., & Cohen, J. (2003). Monozygotic twinning following assisted conception: An analysis of 81 consecutive cases. *Human Reproduction (Oxford, England)*, *18*(9), 1937–1943.

Allen, W. R., & Pashen, R. L. (1984). Production of monozygotic (identical) horse twins by embryo micromanipulation. *Journal of Reproduction and Fertility*, *71*(2), 607–613.

Amin, B. R., Bauersachs, R. M., Meiselman, H. J., Mohandas, N., Hebbel, R. P., Bowen, P. E., et al. (1991). Monozygotic twins with sickle cell anemia and discordant clinical courses: Clinical and laboratory studies. *Hemoglobin*, *15*(4), 247–256.

Anani, S., Bhat, S., Honma-Yamanaka, N., Krawchuk, D., & Yamanaka, Y. (2014). Initiation of Hippo signaling is linked to polarity rather than to cell position in the pre-implantation mouse embryo. *Development (Cambridge, England)*, *141*(14), 2813–2824. https://doi.org/10.1242/dev.107276.

Becker, S., & Chung, Y. (2006). Embryonic stem cells from single blastomeres. *Methods in Enzymology*, *418*, 108–116. https://doi.org/10.1016/S0076-6879(06)18007-6.

Beddington, R. S., & Robertson, E. J. (1989). An assessment of the developmental potential of embryonic stem cells in the midgestation mouse embryo. *Development (Cambridge, England)*, *105*(4), 733–737.

Beddington, R. S., & Robertson, E. J. (1999). Axis development and early asymmetry in mammals. *Cell*, *96*(2), 195–209.

Behr, B., Fisch, J. D., Racowsky, C., Miller, K., Pool, T. B., & Milki, A. A. (2000). Blastocyst-ET and monozygotic twinning. *Journal of Assisted Reproduction and Genetics*, *17*(6), 349–351.

Bennett, C. M., Boye, E., & Neufeld, E. J. (2008). Female monozygotic twins discordant for hemophilia A due to nonrandom X-chromosome inactivation. *American Journal of Hematology*, *83*(10), 778–780. https://doi.org/10.1002/ajh.21219.

Berger-Zaslav, A.-L., Mehta, L., Jacob, J., Mercado, T., Gadi, I., Tepperberg, J. H., et al. (2009). Ovotesticular disorder of sexual development (true hermaphroditism). *Urology*, *73*(2), 293–296. https://doi.org/10.1016/j.urology.2008.08.428.

Biase, F. H., Cao, X., & Zhong, S. (2014). Cell fate inclination within 2-cell and 4-cell mouse embryos revealed by single-cell RNA sequencing. *Genome Research*, *24*(11), 1787–1796. https://doi.org/10.1101/gr.177725.114.

Blickstein, I., Jones, C., & Keith, L. G. (2003). Zygotic-splitting rates after single-embryo transfers in in vitro fertilization. *The New England Journal of Medicine*, *348*(23), 2366–2367. https://doi.org/10.1056/NEJMc026724.

Blickstein, I., Verhoeven, H. C., & Keith, L. G. (1999). Zygotic splitting after assisted reproduction. *The New England Journal of Medicine*, *340*(9), 738–739. https://doi.org/10.1056/NEJM199903043400916.

Boklage, C. E. (2006). Embryogenesis of chimeras, twins and anterior midline asymmetries. *Human Reproduction (Oxford, England)*, *21*(3), 579–591. https://doi.org/10.1093/humrep/dei370.

Boroviak, T., Loos, R., Lombard, P., Okahara, J., Behr, R., Sasaki, E., et al. (2015). Lineage-specific profiling delineates the emergence and progression of naive pluripotency in mammalian embryogenesis. *Developmental Cell*, *35*(3), 366–382. https://doi.org/10.1016/j.devcel.2015.10.011.

Bouabe, H., & Okkenhaug, K. (2013). Gene targeting in mice: A review. *Methods in Molecular Biology (Clifton, NJ)*, *1064*, 315–336. https://doi.org/10.1007/978-1-62703-601-6_23.

Bourthoumieu, S., Yardin, C., Terro, F., Gilbert, B., Laroche, C., Saura, R., et al. (2005). Monozygotic twins concordant for blood karyotype, but phenotypically discordant: A case of "mosaic chimerism" *American Journal of Medical Genetics. Part A*, *135*(2), 190–194. https://doi.org/10.1002/ajmg.a.30674.

Bożyk, K., Gilecka, K., Humięcka, M., Szpila, M., Suwińska, A., & Tarkowski, A. K. (2017). Mouse↔rat aggregation chimaeras can develop to adulthood. *Developmental Biology, 427*(1), 106–120. https://doi.org/10.1016/j.ydbio.2017.05.002.

Brem, G., Tenhumberg, H., & Kräußlich, H. (1984). Chimerism in cattle through microsurgical aggregation of morulae. *Theriogenology, 22*(5), 609–613.

Brook, F. A., & Gardner, R. L. (1997). The origin and efficient derivation of embryonic stem cells in the mouse. *Proceedings of the National Academy of Sciences of the United States of America, 94*(11), 5709–5712.

Buchbinder, D., Nadeau, K., & Nugent, D. (2011). Monozygotic twin pair showing discordant phenotype for X-linked thrombocytopenia and Wiskott-Aldrich syndrome: A role for epigenetics? *Journal of Clinical Immunology, 31*(5), 773–777. https://doi.org/10.1007/s10875-011-9561-3.

Buehr, M., Meek, S., Blair, K., Yang, J., Ure, J., Silva, J., et al. (2008). Capture of authentic embryonic stem cells from rat blastocysts. *Cell, 135*(7), 1287–1298. https://doi.org/10.1016/j.cell.2008.12.007.

Burton, A., Muller, J., Tu, S., Padilla-Longoria, P., Guccione, E., & Torres-Padilla, M.-E. (2013). Single-cell profiling of epigenetic modifiers identifies PRDM14 as an inducer of cell fate in the mammalian embryo. *Cell Reports, 5*(3), 687–701. https://doi.org/10.1016/j.celrep.2013.09.044.

Carter, M. G., Stagg, C. A., Falco, G., Yoshikawa, T., Bassey, U. C., Aiba, K., et al. (2008). An in situ hybridization-based screen for heterogeneously expressed genes in mouse ES cells. *Gene Expression Patterns: GEP, 8*(3), 181–198. https://doi.org/10.1016/j.gep.2007.10.009.

Castillo-Fernandez, J. E., Spector, T. D., & Bell, J. T. (2014). Epigenetics of discordant monozygotic twins: Implications for disease. *Genome Medicine, 6*(7), 60. https://doi.org/10.1186/s13073-014-0060-z.

Chambers, I., Silva, J., Colby, D., Nichols, J., Nijmeijer, B., Robertson, M., et al. (2007). Nanog safeguards pluripotency and mediates germline development. *Nature, 450*(7173), 1230–1234. https://doi.org/10.1038/nature06403.

Chan, A. W., Dominko, T., Luetjens, C. M., Neuber, E., Martinovich, C., Hewitson, L., et al. (2000). Clonal propagation of primate offspring by embryo splitting. *Science (New York, NY), 287*(5451), 317–319.

Chavez, S. L., McElroy, S. L., Bossert, N. L., De Jonge, C. J., Rodriguez, M. V., Leong, D. E., et al. (2014). Comparison of epigenetic mediator expression and function in mouse and human embryonic blastomeres. *Human Molecular Genetics, 23*(18), 4970–4984. https://doi.org/10.1093/hmg/ddu212.

Chazaud, C., Yamanaka, Y., Pawson, T., & Rossant, J. (2006). Early lineage segregation between epiblast and primitive endoderm in mouse blastocysts through the Grb2-MAPK pathway. *Developmental Cell, 10*(5), 615–624. https://doi.org/10.1016/j.devcel.2006.02.020.

Chen, L., Yabuuchi, A., Eminli, S., Takeuchi, A., Lu, C.-W., Hochedlinger, K., et al. (2009). Cross-regulation of the Nanog and Cdx2 promoters. *Cell Research, 19*(9), 1052–1061. https://doi.org/10.1038/cr.2009.79.

Cheng, P.-J., Shaw, S.-W., Shih, J.-C., & Soong, Y.-K. (2006). Monozygotic twins discordant for monosomy 21 detected by first-trimester nuchal translucency screening. *Obstetrics and Gynecology, 107*(2 Pt 2), 538–541. https://doi.org/10.1097/01.AOG.0000190218.24610.ce.

Chung, Y., Klimanskaya, I., Becker, S., Li, T., Maserati, M., Lu, S.-J., et al. (2008). Human embryonic stem cell lines generated without embryo destruction. *Cell Stem Cell, 2*(2), 113–117. https://doi.org/10.1016/j.stem.2007.12.013.

Chung, Y., Klimanskaya, I., Becker, S., Marh, J., Lu, S.-J., Johnson, J., et al. (2006). Embryonic and extraembryonic stem cell lines derived from single mouse blastomeres. *Nature, 439*(7073), 216–219. https://doi.org/10.1038/nature04277.

Cortés, J. L., Cobo, F., Catalina, P., Nieto, A., Cabrera, C., Montes, R., et al. (2007). Evaluation of a laser technique to isolate the inner cell mass of murine blastocysts. *Biotechnology and Applied Biochemistry*, *46*(Pt 4), 205–209.

Cui, L., Johkura, K., Yue, F., Ogiwara, N., Okouchi, Y., Asanuma, K., et al. (2004). Spatial distribution and initial changes of SSEA-1 and other cell adhesion-related molecules on mouse embryonic stem cells before and during differentiation. *The Journal of Histochemistry and Cytochemistry*, *52*(11), 1447–1457. https://doi.org/10.1369/jhc.3A6241.2004.

Delhaise, F., Bralion, V., Schuurbiers, N., & Dessy, F. (1996). Establishment of an embryonic stem cell line from 8-cell stage mouse embryos. *European Journal of Morphology*, *34*(4), 237–243.

De Paepe, C., Cauffman, G., Verloes, A., Sterckx, J., Devroey, P., Tournaye, H., et al. (2013). Human trophectoderm cells are not yet committed. *Human Reproduction (Oxford, England)*, *28*(3), 740–749. https://doi.org/10.1093/humrep/des432.

Derom, C., Vlietinck, R., Derom, R., Van den Berghe, H., & Thiery, M. (1987). Increased monozygotic twinning rate after ovulation induction. *Lancet (London, England)*, *1*(8544), 1236–1238.

Draper, N. L., Conley, C., Smith, C., & Benson, K. (2008). Dispermic chimerism identified during HLA typing for stem cell transplantation. *Transfusion*, *48*(7), 1398–1402. https://doi.org/10.1111/j.1537-2995.2008.01711.x.

Edwards, R. G., Mettler, L., & Walters, D. E. (1986). Identical twins and in vitro fertilization. *Journal of In Vitro Fertilization and Embryo Transfer: IVF*, *3*(2), 114–117.

Evans, M. J., & Kaufman, M. H. (1981). Establishment in culture of pluripotential cells from mouse embryos. *Nature*, *292*(5819), 154–156.

Fehilly, C. B., Willadsen, S. M., & Tucker, E. M. (1984). Interspecific chimaerism between sheep and goat. *Nature*, *307*(5952), 634–636.

Fleming, T. P. (1987). A quantitative analysis of cell allocation to trophectoderm and inner cell mass in the mouse blastocyst. *Developmental Biology*, *119*(2), 520–531.

Fujimori, T., Kurotaki, Y., Miyazaki, J.-I., & Nabeshima, Y.-I. (2003). Analysis of cell lineage in two- and four-cell mouse embryos. *Development (Cambridge, England)*, *130*(21), 5113–5122. https://doi.org/10.1242/dev.00725.

Gardner, R. L., & Johnson, M. H. (1972). An investigation of inner cell mass and trophoblast tissues following their isolation from the mouse blastocyst. *Journal of Embryology and Experimental Morphology*, *28*(2), 279–312.

Gardner, R. L., & Johnson, M. H. (1973). Investigation of early mammalian development using interspecific chimaeras between rat and mouse. *Nature: New Biology*, *246*(151), 86–89.

Gardner, R. L., & Johnson, M. H. (1975). Investigation of cellular interaction and deployment in the early mammalian embryo using interspecific chimaeras between the rat and mouse. *Ciba Foundation Symposium*, (29), 183–200.

Gardner, R. L., & Munro, A. J. (1974). Successful construction of chimaeric rabbit. *Nature*, *250*(462), 146–147.

Geens, M., Mateizel, I., Sermon, K., De Rycke, M., Spits, C., Cauffman, G., et al. (2009). Human embryonic stem cell lines derived from single blastomeres of two 4-cell stage embryos. *Human Reproduction (Oxford, England)*, *24*(11), 2709–2717. https://doi.org/10.1093/humrep/dep262.

Gomes-Pereira, M., Cooper, T. A., & Gourdon, G. (2011). Myotonic dystrophy mouse models: Towards rational therapy development. *Trends in Molecular Medicine*, *17*(9), 506–517. https://doi.org/10.1016/j.molmed.2011.05.004.

González, S., Ibáñez, E., & Santaló, J. (2010). Establishment of mouse embryonic stem cells from isolated blastomeres and whole embryos using three derivation methods. *Journal of Assisted Reproduction and Genetics*, *27*(12), 671–682. https://doi.org/10.1007/s10815-010-9473-9.

González, S., Ibáñez, E., & Santaló, J. (2011a). Influence of early fate decisions at the two-cell stage on the derivation of mouse embryonic stem cell lines. *Stem Cell Research*, 7(1), 54–65. https://doi.org/10.1016/j.scr.2011.03.005.

González, S., Ibáñez, E., & Santaló, J. (2011b). Influence of E-cadherin-mediated cell adhesion on mouse embryonic stem cells derivation from isolated blastomeres. *Stem Cell Reviews*, 7(3), 494–505. https://doi.org/10.1007/s12015-010-9221-7.

Goolam, M., Scialdone, A., Graham, S. J. L., Macaulay, I. C., Jedrusik, A., Hupalowska, A., et al. (2016). Heterogeneity in Oct4 and Sox2 targets biases cell fate in 4-cell mouse embryos. *Cell*, *165*(1), 61–74. https://doi.org/10.1016/j.cell.2016.01.047.

Gossler, A., Doetschman, T., Korn, R., Serfling, E., & Kemler, R. (1986). Transgenesis by means of blastocyst-derived embryonic stem cell lines. *Proceedings of the National Academy of Sciences of the United States of America*, *83*(23), 9065–9069.

Grabarek, J. B., Zyzyńska, K., Saiz, N., Piliszek, A., Frankenberg, S., Nichols, J., et al. (2012). Differential plasticity of epiblast and primitive endoderm precursors within the ICM of the early mouse embryo. *Development (Cambridge, England)*, *139*(1), 129–139. https://doi.org/10.1242/dev.067702.

Graf, T., & Stadtfeld, M. (2008). Heterogeneity of embryonic and adult stem cells. *Cell Stem Cell*, *3*(5), 480–483. https://doi.org/10.1016/j.stem.2008.10.007.

Graham, S. J. L., Wicher, K. B., Jedrusik, A., Guo, G., Herath, W., Robson, P., et al. (2014). BMP signalling regulates the pre-implantation development of extra-embryonic cell lineages in the mouse embryo. *Nature Communications*, *5*, 5667. https://doi.org/10.1038/ncomms6667.

Hadjantonakis, A.-K., & Arias, A. M. (2016). Single-cell approaches: Pandora's box of developmental mechanisms. *Developmental Cell*, *38*(6), 574–578. https://doi.org/10.1016/j.devcel.2016.09.012.

Hall, J. G. (2003). Twinning. *Lancet (London, England)*, *362*(9385), 735–743. https://doi.org/10.1016/S0140-6736(03)14237-7.

Hamamy, H. A., Ajlouni, H. K., & Ajlouni, K. M. (2004). Familial monozygotic twinning: Report of an extended multi-generation family. *Twin Research: The Official Journal of the International Society for Twin Studies*, 7(3), 219–222.

Handyside, A. H. (1978). Time of commitment of inside cells isolated from preimplantation mouse embryos. *Journal of Embryology and Experimental Morphology*, *45*, 37–53.

Handyside, A. H. (1981). Immunofluorescence techniques for determining the numbers of inner and outer blastomeres in mouse morulae. *Journal of Reproductive Immunology*, *2*(6), 339–350.

Hansis, C., Grifo, J. A., & Krey, L. C. (2004). Candidate lineage marker genes in human preimplantation embryos. *Reproductive Biomedicine Online*, *8*(5), 577–583.

Hartshorn, C., Rice, J. E., & Wangh, L. J. (2003). Differential pattern of Xist RNA accumulation in single blastomeres isolated from 8-cell stage mouse embryos following laser zona drilling. *Molecular Reproduction and Development*, *64*(1), 41–51. https://doi.org/10.1002/mrd.10223.

Harvey, M. A., Huntley, R. M., & Smith, D. W. (1977). Familial monozygotic twinning. *The Journal of Pediatrics*, *90*(2), 246–247.

Hayashi, K., Lopes, S. M. C. d. S., Tang, F., & Surani, M. A. (2008). Dynamic equilibrium and heterogeneity of mouse pluripotent stem cells with distinct functional and epigenetic states. *Cell Stem Cell*, *3*(4), 391–401. https://doi.org/10.1016/j.stem.2008.07.027.

Herranz, G. (2015). The timing of monozygotic twinning: A criticism of the common model. *Zygote (Cambridge, England)*, *23*(1), 27–40. https://doi.org/10.1017/S0967199413000257.

Hershlag, A., Paine, T., Cooper, G. W., Scholl, G. M., Rawlinson, K., & Kvapil, G. (1999). Monozygotic twinning associated with mechanical assisted hatching. *Fertility and Sterility*, *71*(1), 144–146.

Hillebrand, G., Siebert, R., Simeoni, E., & Santer, R. (2000). DiGeorge syndrome with discordant phenotype in monozygotic twins. *Journal of Medical Genetics*, *37*(9) . E23.

Hirate, Y., Hirahara, S., Inoue, K.-I., Kiyonari, H., Niwa, H., & Sasaki, H. (2015). Par-aPKC-dependent and -independent mechanisms cooperatively control cell polarity, Hippo signaling, and cell positioning in 16-cell stage mouse embryos. *Development, Growth & Differentiation*, *57*(8), 544–556. https://doi.org/10.1111/dgd.12235.

Hirate, Y., Hirahara, S., Inoue, K.-I., Suzuki, A., Alarcon, V. B., Akimoto, K., et al. (2013). Polarity-dependent distribution of angiomotin localizes Hippo signaling in preimplantation embryos. *Current Biology: CB*, *23*(13), 1181–1194. https://doi.org/10.1016/j.cub.2013.05.014.

Hogan, B., & Tilly, R. (1978). In vitro development of inner cell masses isolated immunosurgically from mouse blastocysts. II. Inner cell masses from 3.5- to 4.0-day p.c. blastocysts. *Journal of Embryology and Experimental Morphology*, *45*, 107–121.

Hong, X., Ying, Y., Xu, X., Liu, Y., Chen, Z., Lan, X., et al. (2013). A dispermic chimera was identified in a healthy man with mixed field agglutination reaction in ABO blood grouping and mosaic 46, XY/46, XX karyotype. *Transfusion and Apheresis Science*, *48*(2), 223–228. https://doi.org/10.1016/j.transci.2012.10.002.

Humięcka, M., Krupa, M., Guzewska, M. M., Maleszewski, M., & Suwińska, A. (2016). ESCs injected into the 8-cell stage mouse embryo modify pattern of cleavage and cell lineage specification. *Mechanisms of Development*, *141*, 40–50. https://doi.org/10.1016/j.mod.2016.06.002.

Illmensee, K., Levanduski, M., Vidali, A., Husami, N., & Goudas, V. T. (2010). Human embryo twinning with applications in reproductive medicine. *Fertility and Sterility*, *93*(2), 423–427. https://doi.org/10.1016/j.fertnstert.2008.12.098.

Isotani, A., Hatayama, H., Kaseda, K., Ikawa, M., & Okabe, M. (2011). Formation of a thymus from rat ES cells in xenogeneic nude mouse↔rat ES chimeras. *Genes to Cells: Devoted to Molecular & Cellular Mechanisms*, *16*(4), 397–405. https://doi.org/10.1111/j.1365-2443.2011.01495.x.

Isotani, A., Yamagata, K., Okabe, M., & Ikawa, M. (2016). Generation of Hprt-disrupted rat through mouse←rat ES chimeras. *Scientific Reports*, *6*, 24215. https://doi.org/10.1038/srep24215.

James, R. M., Klerkx, A. H., Keighren, M., Flockhart, J. H., & West, J. D. (1995). Restricted distribution of tetraploid cells in mouse tetraploid<==>diploid chimaeras. *Developmental Biology*, *167*(1), 213–226. https://doi.org/10.1006/dbio.1995.1018.

James, D., Noggle, S. A., Swigut, T., & Brivanlou, A. H. (2006). Contribution of human embryonic stem cells to mouse blastocysts. *Developmental Biology*, *295*(1), 90–102. https://doi.org/10.1016/j.ydbio.2006.03.026.

Jaszczak, K., Członkowska, M., Guszkiewicz, A., & Parada, R. (1991). Cytogenetic analysis of experimental interspecies goat-sheep chimera. *The Journal of Heredity*, *82*(3), 244–245.

Jedrusik, A., Parfitt, D.-E., Guo, G., Skamagki, M., Grabarek, J. B., Johnson, M. H., et al. (2008). Role of Cdx2 and cell polarity in cell allocation and specification of trophectoderm and inner cell mass in the mouse embryo. *Genes & Development*, *22*(19), 2692–2706. https://doi.org/10.1101/gad.486108.

Johnson, W. H., Loskutoff, N. M., Plante, Y., & Betteridge, K. J. (1995). Production of four identical calves by the separation of blastomeres from an in vitro derived four-cell embryo. *The Veterinary Record*, *137*(1), 15–16.

Johnson, M. H., & Ziomek, C. A. (1981). The foundation of two distinct cell lineages within the mouse morula. *Cell*, *24*(1), 71–80.

Källén, B., Olausson, P. O., & Nygren, K. G. (2002). Neonatal outcome in pregnancies from ovarian stimulation. *Obstetrics and Gynecology*, *100*(3), 414–419.

Kang, M., Piliszek, A., Artus, J., & Hadjantonakis, A.-K. (2013). FGF4 is required for lineage restriction and salt-and-pepper distribution of primitive endoderm factors but not their initial expression in the mouse. *Development (Cambridge, England)*, *140*(2), 267–279. https://doi.org/10.1242/dev.084996.

Kelly, S. J. (1975). Studies of the potency of the early cleavage blastomeres of the mouse. In M. Balls & A. E. Wild (Eds.), *The early development of mammals* (pp. 97–105). Cambridge: Cambridge University Press.

Kelly, S. J. (1977). Studies of the developmental potential of 4- and 8-cell stage mouse blastomeres. *The Journal of Experimental Zoology*, *200*(3), 365–376. https://doi.org/10.1002/jez.1402000307.

Klimanskaya, I., Chung, Y., Becker, S., Lu, S.-J., & Lanza, R. (2006). Human embryonic stem cell lines derived from single blastomeres. *Nature*, *444*(7118), 481–485. https://doi.org/10.1038/nature05142.

Kobayashi, T., Yamaguchi, T., Hamanaka, S., Kato-Itoh, M., Yamazaki, Y., Ibata, M., et al. (2010). Generation of rat pancreas in mouse by interspecific blastocyst injection of pluripotent stem cells. *Cell*, *142*(5), 787–799. https://doi.org/10.1016/j.cell.2010.07.039.

Korotkevich, E., Niwayama, R., Courtois, A., Friese, S., Berger, N., Buchholz, F., et al. (2017). The apical domain is required and sufficient for the first lineage segregation in the mouse embryo. *Developmental Cell*, *40*(3), 235–247.

Krawchuk, D., Honma-Yamanaka, N., Anani, S., & Yamanaka, Y. (2013). FGF4 is a limiting factor controlling the proportions of primitive endoderm and epiblast in the ICM of the mouse blastocyst. *Developmental Biology*, *384*(1), 65–71. https://doi.org/10.1016/j.ydbio.2013.09.023.

Krupa, M., Mazur, E., Szczepańska, K., Filimonow, K., Maleszewski, M., & Suwińska, A. (2014). Allocation of inner cells to epiblast vs primitive endoderm in the mouse embryo is biased but not determined by the round of asymmetric divisions (8→16- and 16→32-cells). *Developmental Biology*, *385*(1), 136–148. https://doi.org/10.1016/j.ydbio.2013.09.008.

Kubiak, J. Z., & Tarkowski, A. K. (1985). Electrofusion of mouse blastomeres. *Experimental Cell Research*, *157*(2), 561–566.

Kuijk, E. W., van Tol, L. T. A., Van de Velde, H., Wubbolts, R., Welling, M., Geijsen, N., et al. (2012). The roles of FGF and MAP kinase signaling in the segregation of the epiblast and hypoblast cell lineages in bovine and human embryos. *Development (Cambridge, England)*, *139*(5), 871–882. https://doi.org/10.1242/dev.071688.

Kurimoto, K., Yabuta, Y., Ohinata, Y., Ono, Y., Uno, K. D., Yamada, R. G., et al. (2006). An improved single-cell cDNA amplification method for efficient high-density oligonucleotide microarray analysis. *Nucleic Acids Research*, *34*(5), e42. https://doi.org/10.1093/nar/gkl050.

Lee, Y. T., Lin, H. Y., Chan, Y. W. F., Li, K. H. C., To, O. T. L., Yan, B. P., et al. (2017). Mouse models of atherosclerosis: A historical perspective and recent advances. *Lipids in Health and Disease*, *16*(1), 12. https://doi.org/10.1186/s12944-016-0402-5.

Li, P., Tong, C., Mehrian-Shai, R., Jia, L., Wu, N., Yan, Y., et al. (2008). Germline competent embryonic stem cells derived from rat blastocysts. *Cell*, *135*(7), 1299–1310. https://doi.org/10.1016/j.cell.2008.12.006.

Lorthongpanich, C., Doris, T. P. Y., Limviphuvadh, V., Knowles, B. B., & Solter, D. (2012). Developmental fate and lineage commitment of singled mouse blastomeres. *Development (Cambridge, England)*, *139*(20), 3722–3731. https://doi.org/10.1242/dev.086454.

Lorthongpanich, C., Yang, S.-H., Piotrowska-Nitsche, K., Parnpai, R., & Chan, A. W. S. (2008). Development of single mouse blastomeres into blastocysts, outgrowths and the establishment of embryonic stem cells. *Reproduction (Cambridge, England)*, *135*(6), 805–813. https://doi.org/10.1530/REP-07-0478.

Loughry, W. J., Avise, J., McDonough, C., & Prodohl, P. (1998). Polyembryony in armadillos. *American Scientist*, *86*(3), 274–279. https://doi.org/10.1511/1998.3.274.

Macfarlan, T. S., Gifford, W. D., Driscoll, S., Lettieri, K., Rowe, H. M., Bonanomi, D., et al. (2012). Embryonic stem cell potency fluctuates with endogenous retrovirus activity. *Nature*, *487*(7405), 57–63. https://doi.org/10.1038/nature11244.

MacGillivray, I. (1986). Epidemiology of twin pregnancy. *Seminars in Perinatology*, *10*(1), 4–8.

Machin, G. (2009). Familial monozygotic twinning: A report of seven pedigrees. *American Journal of Medical Genetics. Part C, Seminars in Medical Genetics*, *151C*(2), 152–154. https://doi.org/10.1002/ajmg.c.30211.

Mackay, G. E., & West, J. D. (2005). Fate of tetraploid cells in 4n<-->2n chimeric mouse blastocysts. *Mechanisms of Development*, *122*(12), 1266–1281. https://doi.org/10.1016/j.mod.2005.09.001.

MacLaren, L. A., Anderson, G. B., BonDurant, R. H., & Edmondson, A. J. (1993). Reproductive cycles and pregnancy in interspecific sheep<==>goat chimaeras. *Reproduction, Fertility, and Development*, *5*(3), 261–270.

Maître, J.-L., Turlier, H., Illukkumbura, R., Eismann, B., Niwayama, R., Nédélec, F., et al. (2016). Asymmetric division of contractile domains couples cell positioning and fate specification. *Nature*, *536*(7616), 344–348.

Malan, V., Vekemans, M., & Turleau, C. (2006). Chimera and other fertilization errors. *Clinical Genetics*, *70*(5), 363–373. https://doi.org/10.1111/j.1399-0004.2006.00689.x.

Markert, C. L., & Petters, R. M. (1978). Manufactured hexaparental mice show that adults are derived from three embryonic cells. *Science (New York, N.Y.)*, *202*(4363), 56–58.

Martin, G. R. (1981). Isolation of a pluripotent cell line from early mouse embryos cultured in medium conditioned by teratocarcinoma stem cells. *Proceedings of the National Academy of Sciences of the United States of America*, *78*(12), 7634–7638.

Mascetti, V. L., & Pedersen, R. A. (2016). Contributions of mammalian chimeras to pluripotent stem cell research. *Cell Stem Cell*, *19*(2), 163–175. https://doi.org/10.1016/j.stem.2016.07.018.

Matsumoto, K., Miyake, M., Utsumi, K., & Iritani, A. (1989). Production of identical twins by separating two-cell rat embryos. *Gamete Research*, *22*(3), 257–263. https://doi.org/10.1002/mrd.1120220303.

Matsunari, H., Nagashima, H., Watanabe, M., Umeyama, K., Nakano, K., Nagaya, M., et al. (2013). Blastocyst complementation generates exogenic pancreas in vivo in apancreatic cloned pigs. *Proceedings of the National Academy of Sciences of the United States of America*, *110*(12), 4557–4562. https://doi.org/10.1073/pnas.1222902110.

Mayer, J. F., & Fritz, H. I. (1974). The culture of preimplantation rat embryos and the production of allophenic rats. *Journal of Reproduction and Fertility*, *39*(1), 1–9.

McDole, K., Xiong, Y., Iglesias, P. A., & Zheng, Y. (2011). Lineage mapping the preimplantation mouse embryo by two-photon microscopy, new insights into the segregation of cell fates. *Developmental Biology*, *355*(2), 239–249. https://doi.org/10.1016/j.ydbio.2011.04.024.

McDole, K., & Zheng, Y. (2012). Generation and live imaging of an endogenous Cdx2 reporter mouse line. *Genesis (New York, NY: 2000)*, *50*(10), 775–782. https://doi.org/10.1002/dvg.22049.

Meilhac, S. M., Adams, R. J., Morris, S. A., Danckaert, A., Le Garrec, J.-F., & Zernicka-Goetz, M. (2009). Active cell movements coupled to positional induction are involved in lineage segregation in the mouse blastocyst. *Developmental Biology*, *331*(2), 210–221. https://doi.org/10.1016/j.ydbio.2009.04.036.

Ménézo, Y. J. R., & Sakkas, D. (2002). Monozygotic twinning: Is it related to apoptosis in the embryo? *Human Reproduction (Oxford, England)*, *17*(1), 247–248.

Mihajlović, A. I., Thamodaran, V., & Bruce, A. W. (2015). The first two cell-fate decisions of preimplantation mouse embryo development are not functionally independent. *Scientific Reports*, *5*, 15034. https://doi.org/10.1038/srep15034.

Mintz, B. (1962). Formation of genotypically mosaic mouse embryos. *American Zoologist*, *2*, 432.

Mintz, B. (1964). Formation of genetically mosaic mouse embryos and early development of "lethal (t12/t12)-normal" mosaics. *The Journal of Experimental Zoology*, *157*, 273–292.

Mitalipov, S. M., Yeoman, R. R., Kuo, H.-C., & Wolf, D. P. (2002). Monozygotic twinning in rhesus monkeys by manipulation of in vitro-derived embryos. *Biology of Reproduction*, *66*(5), 1449–1455.

Moore, N. W., Adams, C. E., & Rowson, L. E. (1968). Developmental potential of single blastomeres of the rabbit egg. *Journal of Reproduction and Fertility*, *17*(3), 527–531.

Moores, P. (1973). Blood group and tissue mosaicism in a Natal Indian woman. *Acta Haematologica*, *50*(5), 299–304.

Morris, S. A., Graham, S. J. L., Jedrusik, A., & Zernicka-Goetz, M. (2013). The differential response to Fgf signalling in cells internalized at different times influences lineage segregation in preimplantation mouse embryos. *Open Biology*, *3*(11), 130104. https://doi.org/10.1098/rsob.130104.

Morris, S. A., Guo, Y., & Zernicka-Goetz, M. (2012). Developmental plasticity is bound by pluripotency and the Fgf and Wnt signaling pathways. *Cell Reports*, *2*(4), 756–765. https://doi.org/10.1016/j.celrep.2012.08.029.

Morris, S. A., Teo, R. T. Y., Li, H., Robson, P., Glover, D. M., & Zernicka-Goetz, M. (2010). Origin and formation of the first two distinct cell types of the inner cell mass in the mouse embryo. *Proceedings of the National Academy of Sciences of the United States of America*, *107*(14), 6364–6369. https://doi.org/10.1073/pnas.0915063107.

Mosebach, M., Parkner, A., Jakubiczka, S., Wieacker, P., & Heim, M. U. (2006). Dispermic chimerism identified during blood group determination and HLA typing. *Transfusion*, *46*(11), 1978–1981. https://doi.org/10.1111/j.1537-2995.2006.01005.x.

Mottla, G. L., Adelman, M. R., Hall, J. L., Gindoff, P. R., Stillman, R. J., & Johnson, K. E. (1995). Lineage tracing demonstrates that blastomeres of early cleavage-stage human pre-embryos contribute to both trophectoderm and inner cell mass. *Human Reproduction (Oxford, England)*, *10*(2), 384–391.

Mullen, R., Whitten, W., & Carter, S. (1970). Studies on chimeric mice and half-embryos. In *Annual report of the Jackson Laboratory* (pp. 67–68). Bar Harbor, Maine: Jackson Laboratory.

Naor, N., Amir, Y., Cohen, T., & Davidson, S. (1987). Trisomy 13 in monozygotic twins discordant for major congenital anomalies. *Journal of Medical Genetics*, *24*(8), 500–502.

Niakan, K. K., Ji, H., Maehr, R., Vokes, S. A., Rodolfa, K. T., Sherwood, R. I., et al. (2010). Sox17 promotes differentiation in mouse embryonic stem cells by directly regulating extraembryonic gene expression and indirectly antagonizing self-renewal. *Genes & Development*, *24*(3), 312–326. https://doi.org/10.1101/gad.1833510.

Nichols, J., & Gardner, R. L. (1984). Heterogeneous differentiation of external cells in individual isolated early mouse inner cell masses in culture. *Journal of Embryology and Experimental Morphology*, *80*, 225–240.

Nichols, J., Silva, J., Roode, M., & Smith, A. (2009). Suppression of Erk signalling promotes ground state pluripotency in the mouse embryo. *Development (Cambridge, England)*, *136*(19), 3215–3222. https://doi.org/10.1242/dev.038893.

Nishioka, N., Inoue, K., Adachi, K., Kiyonari, H., Ota, M., Ralston, A., et al. (2009). The Hippo signaling pathway components Lats and Yap pattern Tead4 activity to distinguish mouse trophectoderm from inner cell mass. *Developmental Cell*, *16*(3), 398–410. https://doi.org/10.1016/j.devcel.2009.02.003.

Niwa, H., Toyooka, Y., Shimosato, D., Strumpf, D., Takahashi, K., Yagi, R., et al. (2005). Interaction between Oct3/4 and Cdx2 determines trophectoderm differentiation. *Cell, 123*(5), 917–929. https://doi.org/10.1016/j.cell.2005.08.040.

Noli, L., Dajani, Y., Capalbo, A., Bvumbe, J., Rienzi, L., Ubaldi, F. M., et al. (2015). Developmental clock compromises human twin model created by embryo splitting. *Human Reproduction (Oxford, England), 30*(12), 2774–2784. https://doi.org/10.1093/humrep/dev252.

Ozdzeński, W., Szczesny, E., & Tarkowski, A. K. (1997). Postimplantation development of mouse blastocysts with two separate inner cell masses. *Anatomy and Embryology, 195*(5), 467–471.

Papaioannou, V. E., & Ebert, K. M. (1995). Mouse half embryos: Viability and allocation of cells in the blastocyst. *Developmental Dynamics: An Official Publication of the American Association of Anatomists, 203*(4), 393–398. https://doi.org/10.1002/aja.1002030402.

Papaioannou, V. E., Mkandawire, J., & Biggers, J. D. (1989). Development and phenotypic variability of genetically identical half mouse embryos. *Development (Cambridge, England), 106*(4), 817–827.

Pierce, G. B., Arechaga, J., Muro, C., & Wells, R. S. (1988). Differentiation of ICM cells into trophectoderm. *The American Journal of Pathology, 132*(2), 356–364.

Piotrowska-Nitsche, K., Perea-Gomez, A., Haraguchi, S., & Zernicka-Goetz, M. (2005). Four-cell stage mouse blastomeres have different developmental properties. *Development (Cambridge, England), 132*(3), 479–490. https://doi.org/10.1242/dev.01602.

Plachta, N., Bollenbach, T., Pease, S., Fraser, S. E., & Pantazis, P. (2011). Oct4 kinetics predict cell lineage patterning in the early mammalian embryo. *Nature Cell Biology, 13*(2), 117–123. https://doi.org/10.1038/ncb2154.

Plusa, B., Piliszek, A., Frankenberg, S., Artus, J., & Hadjantonakis, A.-K. (2008). Distinct sequential cell behaviours direct primitive endoderm formation in the mouse blastocyst. *Development (Cambridge, England), 135*(18), 3081–3091. https://doi.org/10.1242/dev.021519.

Polzin, V. J., Anderson, D. L., Anderson, G. B., BonDurant, R. H., Butler, J. E., Pashen, R. L., et al. (1987). Production of sheep-goat chimeras by inner cell mass transplantation. *Journal of Animal Science, 65*(1), 325–330.

Posfai, E., Petropoulos, S., de Barros, F. R. O., Schell, J. P., Jurisica, I., Sandberg, R., et al. (2017). Position- and Hippo signaling-dependent plasticity during lineage segregation in the early mouse embryo. *eLife, 6*. https://doi.org/10.7554/eLife.22906.

Rayon, T., Menchero, S., Nieto, A., Xenopoulos, P., Crespo, M., Cockburn, K., et al. (2014). Notch and hippo converge on Cdx2 to specify the trophectoderm lineage in the mouse blastocyst. *Developmental Cell, 30*(4), 410–422. https://doi.org/10.1016/j.devcel.2014.06.019.

Reichelt, B., & Niemann, H. (1994). Generation of identical twin piglets following bisection of embryos at the morula and blastocyst stage. *Journal of Reproduction and Fertility, 100*(1), 163–172.

Rinkevich, B. (2001). Human natural chimerism: An acquired character or a vestige of evolution? *Human Immunology, 62*(6), 651–657.

Roode, M., Blair, K., Snell, P., Elder, K., Marchant, S., Smith, A., et al. (2012). Human hypoblast formation is not dependent on FGF signalling. *Developmental Biology, 361*(2), 358–363. https://doi.org/10.1016/j.ydbio.2011.10.030.

Rossant, J. (1975a). Investigation of the determinative state of the mouse inner cell mass. I. Aggregation of isolated inner cell masses with morulae. *Journal of Embryology and Experimental Morphology, 33*(4), 979–990.

Rossant, J. (1975b). Investigation of the determinative state of the mouse inner cell mass. II. The fate of isolated inner cell masses transferred to the oviduct. *Journal of Embryology and Experimental Morphology, 33*(4), 991–1001.

Rossant, J. (1976). Postimplantation development of blastomeres isolated from 4- and 8-cell mouse eggs. *Journal of Embryology and Experimental Morphology*, *36*(2), 283–290.

Rossant, J., & Chapman, V. M. (1983). Somatic and germline mosaicism in interspecific chimaeras between Mus musculus and Mus caroli. *Journal of Embryology and Experimental Morphology*, *73*, 193–205.

Rossant, J., Croy, B. A., Chapman, V. M., Siracusa, L., & Clark, D. A. (1982). Interspecific chimeras in mammals: A new experimental system. *Journal of Animal Science*, *55*(5), 1241–1248.

Rossant, J., & Frels, W. I. (1980). Interspecific chimeras in mammals: Successful production of live chimeras between Mus musculus and Mus caroli. *Science (New York, N.Y.)*, *208*(4442), 419–421.

Rossant, J., & Lis, W. T. (1979). Potential of isolated mouse inner cell masses to form trophectoderm derivatives in vivo. *Developmental Biology*, *70*(1), 255–261.

Rossant, J., Mauro, V. M., & Croy, B. A. (1982). Importance of trophoblast genotype for survival of interspecific murine chimaeras. *Journal of Embryology and Experimental Morphology*, *69*, 141–149.

Rossant, J., & Vijh, K. M. (1980). Ability of outside cells from preimplantation mouse embryos to form inner cell mass derivatives. *Developmental Biology*, *76*(2), 475–482.

Saito, H., Tsutsumi, O., Noda, Y., Ibuki, Y., & Hiroi, M. (2000). Do assisted reproductive technologies have effects on the demography of monozygotic twinning? *Fertility and Sterility*, *74*(1), 178–179.

Saiz, N., Williams, K. M., Seshan, V. E., & Hadjantonakis, A.-K. (2016). Asynchronous fate decisions by single cells collectively ensure consistent lineage composition in the mouse blastocyst. *Nature Communications*, 7, 13463. https://doi.org/10.1038/ncomms13463.

Samarage, C. R., White, M. D., Álvarez, Y. D., Fierro-González, J. C., Henon, Y., Jesudason, E. C., et al. (2015). Cortical tension allocates the first inner cells of the mammalian embryo. *Developmental Cell*, *34*(4), 435–447. https://doi.org/10.1016/j.devcel.2015.07.004.

Sasaki, H. (2017). Roles and regulations of Hippo signaling during preimplantation mouse development. *Development, Growth & Differentiation*, *59*(1), 12–20. https://doi.org/10.1111/dgd.12335.

Schieve, L. A., Meikle, S. F., Peterson, H. B., Jeng, G., Burnett, N. M., & Wilcox, L. S. (2000). Does assisted hatching pose a risk for monozygotic twinning in pregnancies conceived through in vitro fertilization? *Fertility and Sterility*, *74*(2), 288–294.

Schrode, N., Saiz, N., Di Talia, S., & Hadjantonakis, A.-K. (2014). GATA6 levels modulate primitive endoderm cell fate choice and timing in the mouse blastocyst. *Developmental Cell*, *29*(4), 454–467. https://doi.org/10.1016/j.devcel.2014.04.011.

Schwartz, S. D., Hubschman, J.-P., Heilwell, G., Franco-Cardenas, V., Pan, C. K., Ostrick, R. M., et al. (2012). Embryonic stem cell trials for macular degeneration: A preliminary report. *Lancet (London, England)*, *379*(9817), 713–720. https://doi.org/10.1016/S0140-6736(12)60028-2.

Schwartz, S. D., Regillo, C. D., Lam, B. L., Eliott, D., Rosenfeld, P. J., Gregori, N. Z., et al. (2015). Human embryonic stem cell-derived retinal pigment epithelium in patients with age-related macular degeneration and Stargardt's macular dystrophy: Follow-up of two open-label phase 1/2 studies. *Lancet (London, England)*, *385*(9967), 509–516. https://doi.org/10.1016/S0140-6736(14)61376-3.

Seidel, F. (1952). Die entwicklugspotenzen einen isolierten blastomere des zweizellenstadiums im saugetierei. *Naturwissenschaften*, *39*, 355–356.

Seidel, F. (1960). Die entwicklungsfahigkeiten isolierter furchungszellen aus dem ei des kaninchens Oryctolagus cuniculus. *Roux Archiv fur Entwicklungsmechanik*, *152*, 43–130.

Shi, J., Chen, Q., Li, X., Zheng, X., Zhang, Y., Qiao, J., et al. (2015). Dynamic transcriptional symmetry-breaking in pre-implantation mammalian embryo development revealed by single-cell RNA-seq. *Development (Cambridge, England)*, *142*(20), 3468–3477. https://doi.org/10.1242/dev.123950.

Sills, E. S., Tucker, M. J., & Palermo, G. D. (2000). Assisted reproductive technologies and monozygous twins: Implications for future study and clinical practice. *Twin Research: The Official Journal of the International Society for Twin Studies*, *3*(4), 217–223.

Simerly, C., McFarland, D., Castro, C., Lin, C.-C., Redinger, C., Jacoby, E., et al. (2011). Interspecies chimera between primate embryonic stem cells and mouse embryos: Monkey ESCs engraft into mouse embryos, but not post-implantation fetuses. *Stem Cell Research*, 7(1), 28–40. https://doi.org/10.1016/j.scr.2011.03.002.

Simpson, J. L., & Rechitsky, S. (2017). Preimplantation diagnosis and other modern methods for prenatal diagnosis. *The Journal of Steroid Biochemistry and Molecular Biology*, *165*, 124–130.

Singh, A. M., Hamazaki, T., Hankowski, K. E., & Terada, N. (2007). A heterogeneous expression pattern for Nanog in embryonic stem cells. *Stem Cells (Dayton, Ohio)*, *25*(10), 2534–2542. https://doi.org/10.1634/stemcells.2007-0126.

Skiadas, C. C., Missmer, S. A., Benson, C. B., Gee, R. E., & Racowsky, C. (2008). Risk factors associated with pregnancies containing a monochorionic pair following assisted reproductive technologies. *Human Reproduction (Oxford, England)*, *23*(6), 1366–1371. https://doi.org/10.1093/humrep/den045.

Solter, D., & Knowles, B. B. (1975). Immunosurgery of mouse blastocyst. *Proceedings of the National Academy of Sciences of the United States of America*, *72*(12), 5099–5102.

Spindle, A. I. (1978). Trophoblast regeneration by inner cell masses isolated from cultured mouse embryos. *The Journal of Experimental Zoology*, *203*(3), 483–489. https://doi.org/10.1002/jez.1402030315.

Strain, L., Warner, J. P., Johnston, T., & Bonthron, D. T. (1995). A human parthenogenetic chimaera. *Nature Genetics*, *11*(2), 164–169. https://doi.org/10.1038/ng1095-164.

Summers, P. M., Shelton, J. N., & Bell, K. (1983). Synthesis of primary Bos taurus–Bos indicus chimaeric calves. *Animal Reproduction Science*, *6*(2), 91–102. https://doi.org/10.1016/0378-4320(83)90013-1.

Suwińska, A., Czołowska, R., Ozdzeński, W., & Tarkowski, A. K. (2008). Blastomeres of the mouse embryo lose totipotency after the fifth cleavage division: Expression of Cdx2 and Oct4 and developmental potential of inner and outer blastomeres of 16- and 32-cell embryos. *Developmental Biology*, *322*(1), 133–144. https://doi.org/10.1016/j.ydbio.2008.07.019.

Szczepanska, K., Stanczuk, L., & Maleszewski, M. (2011). Isolated mouse inner cell mass is unable to reconstruct trophectoderm. *Differentiation; Research in Biological Diversity*, *82*(1), 1–8. https://doi.org/10.1016/j.diff.2011.04.001.

Tabansky, I., Lenarcic, A., Draft, R. W., Loulier, K., Keskin, D. B., Rosains, J., et al. (2013). Developmental bias in cleavage-stage mouse blastomeres. *Current Biology: CB*, *23*(1), 21–31. https://doi.org/10.1016/j.cub.2012.10.054.

Tachibana, M., Sparman, M., Ramsey, C., Ma, H., Lee, H.-S., Penedo, M. C. T., et al. (2012). Generation of chimeric rhesus monkeys. *Cell*, *148*(1–2), 285–295. https://doi.org/10.1016/j.cell.2011.12.007.

Tanaka, N., Takeuchi, T., Neri, Q. V., Sills, E. S., & Palermo, G. D. (2006). Laser-assisted blastocyst dissection and subsequent cultivation of embryonic stem cells in a serum/cell free culture system: Applications and preliminary results in a murine model. *Journal of Translational Medicine*, *4*, 20. https://doi.org/10.1186/1479-5876-4-20.

Tang, F., Barbacioru, C., Nordman, E., Bao, S., Lee, C., Wang, X., et al. (2011). Deterministic and stochastic allele specific gene expression in single mouse blastomeres. *PLoS One*, *6*(6), e21208. https://doi.org/10.1371/journal.pone.0021208.

Tarkowski, A. K. (1959). Experiments on the development of isolated blastomers of mouse eggs. *Nature, 184*, 1286–1287.
Tarkowski, A. K. (1961). Mouse chimaeras developed from fused eggs. *Nature, 190*, 857–860.
Tarkowski, A. K., Jagiello, K., Czolowska, R., & Ozdzenski, W. (2005). Mouse chimaeras developed from electrofused blastocysts: New evidence for developmental plasticity of the inner cell mass. *The International Journal of Developmental Biology, 49*(8), 909–914. https://doi.org/10.1387/ijdb.052017at.
Tarkowski, A. K., Ozdzenski, W., & Czołowska, R. (2001). Mouse singletons and twins developed from isolated diploid blastomeres supported with tetraploid blastomeres. *The International Journal of Developmental Biology, 45*(3), 591–596.
Tarkowski, A. K., Ozdzenski, W., & Czolowska, R. (2005). Identical triplets and twins developed from isolated blastomeres of 8- and 16-cell mouse embryos supported with tetraploid blastomeres. *The International Journal of Developmental Biology, 49*(7), 825–832. https://doi.org/10.1387/ijdb.052018at.
Tarkowski, A. K., Suwińska, A., Czołowska, R., & Ożdżeński, W. (2010). Individual blastomeres of 16- and 32-cell mouse embryos are able to develop into foetuses and mice. *Developmental Biology, 348*(2), 190–198. https://doi.org/10.1016/j.ydbio.2010.09.022.
Tarkowski, A. K., Witkowska, A., & Opas, J. (1977). Development of cytochalasin in B-induced tetraploid and diploid/tetraploid mosaic mouse embryos. *Journal of Embryology and Experimental Morphology, 41*, 47–64.
Tarkowski, A. K., & Wojewodzka, M. (1982). A method for obtaining chimaeric mouse blastocysts with two separate inner cell masses: A preliminary report. *Journal of Embryology and Experimental Morphology, 71*, 215–221.
Tarkowski, A. K., & Wróblewska, J. (1967). Development of blastomeres of mouse eggs isolated at the 4- and 8-cell stage. *Journal of Embryology and Experimental Morphology, 18*(1), 155–180.
Tarlatzis, B. C., Qublan, H. S., Sanopoulou, T., Zepiridis, L., Grimbizis, G., & Bontis, J. (2002). Increase in the monozygotic twinning rate after intracytoplasmic sperm injection and blastocyst stage embryo transfer. *Fertility and Sterility*, 77(1), 196–198.
Teramura, T., Takehara, T., Kishi, N., Mihara, T., Kawata, N., Takeuchi, H., et al. (2007). A mouse and embryonic stem cell derived from a single embryo. *Cloning and Stem Cells, 9*(4), 485–494. https://doi.org/10.1089/clo.2006.0076.
Thamodaran, V., & Bruce, A. W. (2016). p38 (Mapk14/11) occupies a regulatory node governing entry into primitive endoderm differentiation during preimplantation mouse embryo development. *Open Biology, 6*(9), 160190.
Thomson, J. A., Itskovitz-Eldor, J., Shapiro, S. S., Waknitz, M. A., Swiergiel, J. J., Marshall, V. S., et al. (1998). Embryonic stem cell lines derived from human blastocysts. *Science (New York, NY), 282*(5391), 1145–1147.
Thomson, J. A., Kalishman, J., Golos, T. G., Durning, M., Harris, C. P., Becker, R. A., et al. (1995). Isolation of a primate embryonic stem cell line. *Proceedings of the National Academy of Sciences of the United States of America, 92*(17), 7844–7848.
Thomson, J. A., Kalishman, J., Golos, T. G., Durning, M., Harris, C. P., & Hearn, J. P. (1996). Pluripotent cell lines derived from common marmoset (Callithrix jacchus) blastocysts. *Biology of Reproduction, 55*(2), 254–259.
Tippet, P. (1984). Human chimeras. In N. Le Douarin & A. McLaren (Eds.), *Chimeras in development biology* (p. 165). London: Academic Press.
Torres-Padilla, M.-E., Parfitt, D.-E., Kouzarides, T., & Zernicka-Goetz, M. (2007). Histone arginine methylation regulates pluripotency in the early mouse embryo. *Nature, 445*(7124), 214–218. https://doi.org/10.1038/nature05458.
Toyooka, Y., Oka, S., & Fujimori, T. (2016). Early preimplantation cells expressing Cdx2 exhibit plasticity of specification to TE and ICM lineages through positional changes. *Developmental Biology, 411*(1), 50–60. https://doi.org/10.1016/j.ydbio.2016.01.011.

Toyooka, Y., Shimosato, D., Murakami, K., Takahashi, K., & Niwa, H. (2008). Identification and characterization of subpopulations in undifferentiated ES cell culture. *Development (Cambridge, England)*, *135*(5), 909–918. https://doi.org/10.1242/dev.017400.
Tsunoda, Y., & McLaren, A. (1983). Effect of various procedures on the viability of mouse embryos containing half the normal number of blastomeres. *Journal of Reproduction and Fertility*, *69*(1), 315–322.
Tsunoda, Y., Tokunaga, T., Sugie, T., & Katsumata, M. (1985). Production of monozygotic twins following the transfer of bisected embryos in the goats. *Theriogenology*, *24*(3), 337–343.
Tucker, E. M., Moor, R. M., & Rowson, L. E. (1974). Tetraparental sheep chimaeras induced by blastomere transplantation. Changes in blood type with age. *Immunology*, *26*(3), 613–621.
Ueda, S., Kawamata, M., Teratani, T., Shimizu, T., Tamai, Y., Ogawa, H., et al. (2008). Establishment of rat embryonic stem cells and making of chimera rats. *PLoS One*, *3*(7), e2800. https://doi.org/10.1371/journal.pone.0002800.
Van de Velde, H., Cauffman, G., Tournaye, H., Devroey, P., & Liebaers, I. (2008). The four blastomeres of a 4-cell stage human embryo are able to develop individually into blastocysts with inner cell mass and trophectoderm. *Human Reproduction (Oxford, England)*, *23*(8), 1742–1747. https://doi.org/10.1093/humrep/den190.
Veiga, A., Calderon, G., Barri, P. N., & Coroleu, B. (1987). Pregnancy after the replacement of a frozen-thawed embryo with less than 50% intact blastomeres. *Human Reproduction (Oxford, England)*, *2*(4), 321–323.
Verpoest, W., Van Landuyt, L., Desmyttere, S., Cremers, A., Devroey, P., & Liebaers, I. (2009). The incidence of monozygotic twinning following PGD is not increased. *Human Reproduction (Oxford, England)*, *24*(11), 2945–2950. https://doi.org/10.1093/humrep/dep280.
Wakayama, S., Hikichi, T., Suetsugu, R., Sakaide, Y., Bui, H.-T., Mizutani, E., et al. (2007). Efficient establishment of mouse embryonic stem cell lines from single blastomeres and polar bodies. *Stem Cells*, *25*(4), 986–993. https://doi.org/10.1634/stemcells.2006-0615.
Watanabe, T., Biggins, J. S., Tannan, N. B., & Srinivas, S. (2014). Limited predictive value of blastomere angle of division in trophectoderm and inner cell mass specification. *Development (Cambridge, England)*, *141*(11), 2279–2288. https://doi.org/10.1242/dev.103267.
Watkins, W. M., Yates, A. D., Greenwell, P., Bird, G. W., Gibson, M., Roy, T. C., et al. (1981). A human dispermic chimaera first suspected from analyses of the blood group gene-specified glycosyltransferases. *Journal of Immunogenetics*, *8*(2), 113–128.
Wennekamp, S., Mesecke, S., Nédélec, F., & Hiiragi, T. (2013). A self-organization framework for symmetry breaking in the mammalian embryo. *Nature Reviews. Molecular Cell Biology*, *14*(7), 452–459. https://doi.org/10.1038/nrm3602.
White, M. D., Angiolini, J. F., Alvarez, Y. D., Kaur, G., Zhao, Z. W., Mocskos, E., et al. (2016). Long-lived binding of Sox2 to DNA predicts cell fate in the four-cell mouse embryo. *Cell*, *165*(1), 75–87. https://doi.org/10.1016/j.cell.2016.02.032.
Wicklow, E., Blij, S., Frum, T., Hirate, Y., Lang, R. A., Sasaki, H., et al. (2014). HIPPO pathway members restrict SOX2 to the inner cell mass where it promotes ICM fates in the mouse blastocyst. *PLoS Genetics*, *10*(10), e1004618. https://doi.org/10.1371/journal.pgen.1004618.
Wigger, M., Kisielewska, K., Filimonow, K., Plusa, B., Maleszewski, M., & Suwińska, A. (2017). Plasticity of the inner cell mass in mouse blastocysts is restricted by the activity of FGF/MAPK pathway. *Scientific Reports*, 7(1), 15136.
Willadsen, S. M. (1980). The viability of early cleavage stages containing half the normal number of blastomeres in the sheep. *Journal of Reproduction and Fertility*, *59*(2), 357–362.
Willadsen, S. M. (1981). The development capacity of blastomeres from 4- and 8-cell sheep embryos. *Journal of Embryology and Experimental Morphology*, *65*, 165–172.

Williams, T. J., Munro, R. K., & Shelton, J. N. (1990). Production of interspecies chimeric calves by aggregation of Bos indicus and Bos taurus demi-embryos. *Reproduction, Fertility, and Development*, *2*(4), 385–394.
Wu, J., Greely, H. T., Jaenisch, R., Nakauchi, H., Rossant, J., & Belmonte, J. C. I. (2016). Stem cells and interspecies chimaeras. *Nature*, *540*(7631), 51–59. https://doi.org/10.1038/nature20573.
Wu, J., Platero-Luengo, A., Sakurai, M., Sugawara, A., Gil, M. A., Yamauchi, T., et al. (2017). Interspecies chimerism with mammalian pluripotent stem cells. *Cell*, *168*(3), 473–486. https://doi.org/10.1016/j.cell.2016.12.036.
Xiang, A. P., Mao, F. F., Li, W.-Q., Park, D., Ma, B.-F., Wang, T., et al. (2008). Extensive contribution of embryonic stem cells to the development of an evolutionarily divergent host. *Human Molecular Genetics*, *17*(1), 27–37. https://doi.org/10.1093/hmg/ddm282.
Yamaguchi, T., Sato, H., Kato-Itoh, M., Goto, T., Hara, H., Sanbo, M., et al. (2017). Interspecies organogenesis generates autologous functional islets. *Nature*, *542*(7640), 191–196. https://doi.org/10.1038/nature21070.
Yamanaka, Y., Lanner, F., & Rossant, J. (2010). FGF signal-dependent segregation of primitive endoderm and epiblast in the mouse blastocyst. *Development (Cambridge, England)*, *137*(5), 715–724. https://doi.org/10.1242/dev.043471.
Yang, G., Mai, Q., Li, T., & Zhou, C. (2013). Derivation of human embryonic stem cell lines from single blastomeres of low-quality embryos by direct plating. *Journal of Assisted Reproduction and Genetics*, *30*(7), 953–961. https://doi.org/10.1007/s10815-013-0042-x.
Yang, X., Tian, X. C., Kubota, C., Page, R., Xu, J., Cibelli, J., et al. (2007). Risk assessment of meat and milk from cloned animals. *Nature Biotechnology*, *25*(1), 77–83.
Yunis, E. J., Zuniga, J., Romero, V., & Yunis, E. J. (2007). Chimerism and tetragametic chimerism in humans: Implications in autoimmunity, allorecognition and tolerance. *Immunologic Research*, *38*(1–3), 213–236.
Zambrowicz, B. P., & Sands, A. T. (2003). Knockouts model the 100 best-selling drugs—Will they model the next 100? *Nature Reviews. Drug Discovery*, *2*(1), 38–51. https://doi.org/10.1038/nrd987.
Zdravkovic, T., Nazor, K. L., Larocque, N., Gormley, M., Donne, M., Hunkapillar, N., et al. (2015). Human stem cells from single blastomeres reveal pathways of embryonic or trophoblast fate specification. *Development (Cambridge, England)*, *142*(23), 4010–4025. https://doi.org/10.1242/dev.122846.
Zhang, W., Moore, L., & Ji, P. (2011). Mouse models for cancer research. *Chinese Journal of Cancer*, *30*(3), 149–152.
Ziomek, C. A., Johnson, M. H., & Handyside, A. H. (1982). The developmental potential of mouse 16-cell blastomeres. *The Journal of Experimental Zoology*, *221*(3), 345–355. https://doi.org/10.1002/jez.1402210310.

FURTHER READING

Conn, C. M., Cozzi, J., Harper, J. C., Winston, R. M., & Delhanty, J. D. (1999). Preimplantation genetic diagnosis for couples at high risk of Down syndrome pregnancy owing to parental translocation or mosaicism. *Journal of Medical Genetics*, *36*(1), 45–50.
Khosrotehrani, K., & Bianchi, D. W. (2005). Multi-lineage potential of fetal cells in maternal tissue: A legacy in reverse. *Journal of Cell Science*, *118*(Pt 8), 1559–1563. https://doi.org/10.1242/jcs.02332.

CHAPTER SEVEN

States and Origins of Mammalian Embryonic Pluripotency In Vivo and in a Dish

Priscila Ramos-Ibeas*, Jennifer Nichols†,‡,1, Ramiro Alberio*,1

*School of Biosciences, University of Nottingham, Nottingham, United Kingdom
†Wellcome Trust - Medical Research Council Stem Cell Institute, University of Cambridge, Cambridge, United Kingdom
‡University of Cambridge, Cambridge, United Kingdom
[1]Corresponding authors: e-mail address: jn270@cam.ac.uk; ramiro.alberio@nottingham.ac.uk

Contents

Abstract

Mouse embryonic stem cells (ESC), derived from preimplantation embryos in 1981, defined mammalian pluripotency for many decades. However, after the derivation of human ESC in 1998, comparative studies showed that different types of pluripotency exist in early embryos and that these can be captured in vitro under various culture conditions. Over the past decade much has been learned about the key signaling pathways, growth factor requirements, and transcription factor profiles of pluripotent cells in embryos, allowing improvement of derivation and culture conditions for novel pluripotent stem cell types. More recently, studies using single-cell transcriptomics of embryos from different species provided an unprecedented level of resolution of cellular interactions and cell fate decisions that are informing new ways to understand the emergence of pluripotency in different organisms. These new approaches enhance knowledge of species differences during early embryogenesis and will be instrumental

Current Topics in Developmental Biology, Volume 128
ISSN 0070-2153
https://doi.org/10.1016/bs.ctdb.2017.11.002

for improving methodologies for generating intra- and interspecies chimeric animals using pluripotent stem cells. Here, we discuss the recent developments in our understanding of early embryogenesis in different mammalian species.

1. DERIVATION AND CHARACTERIZATION OF MURINE EMBRYONIC STEM CELL LINES

Interest in states of pluripotency in the context of in vitro models representative of embryonic cells in vivo was initially prompted by the spontaneous appearance of germline teratocarcinomas in the 129 strain of mouse (Stevens, 1962). Testicular tumors were found to contain differentiated cells organized into recognizable structures of tissues representative of all three embryonic germ cell (EGC) layers, and populations of undifferentiated cells that could be propagated under certain culture regimes. The resulting embryonal carcinoma cell (ECC) lines exhibited properties reminiscent of early embryonic cells (undifferentiated cells with multilineage differentiation potential), inspiring attempts to induce a similar response from early embryos grafted into appropriate sites in a congenic adult mouse (Stevens, 1970). The tumorigenesis route to self-renewing embryonal stem cells incurred a substantial risk of spontaneously arising karyotypic abnormalities. Could this defect be eliminated by direct explantation of early mouse embryos into the culture conditions developed for optimal propagation of ECC? Seminal work simultaneously published in the United Kingdom and the United States, utilizing the conditions optimized for ECC propagation, heralded the arrival of the embryonic stem cell (ESC) research era (Evans & Kaufman, 1981; Martin, 1981). That the identity of the resulting cell lines resembled the in vivo counterparts from whence they were derived was demonstrated by their ability to resume normal development and contribute to functional adult tissues, including the germ cells, when integrated into host preimplantation stage embryos (Bradley, Evans, Kaufman, & Robertson, 1984).

Fortuitously, the embryo explantation studies also utilized the 129 strain of mice. Other genetic backgrounds were found to exhibit various levels of recalcitrance to ESC derivation when exposed to the original ECC/ESC culture conditions, consisting of a feeder layer of mitotically inactivated fibroblasts and medium containing serum. Identification of the active cytokine produced by the feeders imparting self-renewal to ESC as leukemia inhibitory factor (LIF) in 1988 enabled dispensation with feeder cells for isolation and culture of 129 ESC (Smith et al., 1988; Williams et al., 1988). Efficiency of derivation remained low, however (~30%), until Brook and Gardner manually dissected the component tissues of periimplantation blastocysts to demonstrate that isolation of the nascent epiblast cells from the prodifferentiative signals of the extraembryonic lineages greatly enhanced the efficiency of ESC derivation (Brook & Gardner, 1997). During normal developmental progression, this periimplantation phase is transient, but can be prolonged in embryonic diapause, inducible by removal or blockade of estrogen activity. The use of diapause blastocysts for provision of epiblasts further enhanced ESC derivation efficiency from both 129 and the more recalcitrant strains, such as C57BL/6 and CBA (Brook & Gardner, 1997). Interestingly, the original ESC derived by Evans and Kaufman employed diapause blastocysts (Evans & Kaufman, 1981).

As well as interstrain variability for ESC derivation, the requirement for serum in the medium imposed a further complexity to hamper derivation efficiency and culture consistency. Substitution of serum with BMP4 enabled reliable culture of ESC from multiple mouse strains (Ying, Nichols, Chambers, & Smith, 2003). When added to serum-free culture medium, BMP4 was able to inhibit neural differentiation, which, in combination with blockade of mesoderm/endoderm formation by LIF, potentiated an undifferentiated state in the cells. The advantage of this regime was that hitherto existing feeder-dependent ESC strains could finally be propagated directly on gelatin. Disappointingly, however, non-129 strains remained recalcitrant to ESC derivation in this serum-free medium. This failure could be overcome by invoking a combination of epiblast isolation and supplementation of the serum-free medium with an inhibitor of the mitogen-activated protein kinase (MAPK) pathway, consisting of Mapk kinase 1/2 (Mek) and Mapk1/2 (Erk), in addition to LIF and BMP4. Germline-competent feeder-free ESC could then be isolated directly from C57BL/6 and CBA embryos (Batlle-Morera, Smith, & Nichols, 2008). The greatest advance was achieved by combined inhibition of the Mek/Erk pathway and glycogen synthase kinase-3 (GSK3) with addition of LIF, which dramatically increased derivation efficiency of multiple strains (Ying et al., 2008), including nonobese diabetic (Nichols, Jones, et al., 2009), which had previously proved completely recalcitrant. Most significantly, derivation of the long-awaited rat ESC became possible (Buehr et al., 2008; Li et al., 2008). Culturing murine embryos from the mid blastocyst stage or earlier in the presence of a Mek/Erk inhibitor, could direct all inner cell mass (ICM) cells to epiblast, which both enhanced the number of cells per embryo competent for ESC derivation and eliminated the primitive endoderm (PrE) and its prodifferentiation activity (Nichols, Silva, Roode, Smith, 2009). Inhibition of GSK3β reduces T-cell factor 3 (Tcf3) repressive influence on the core pluripotency network through stimulation of Wnt signaling by β-catenin (Wray et al., 2011). The two inhibitor-based regime, known as "2i," has thus revolutionized ESC derivation from murine embryos. When propagated in 2i/LIF, ESC exhibit relatively homogeneous expression of factors associated with pluripotency, as seen in diapause epiblasts, compared with those grown in serum/LIF (Boroviak et al., 2015; Nichols, Silva, et al., 2009). Recent studies suggest that long-term Mek1/2 inhibition compromises ESC developmental potential and can result in loss of DNA methylation in female ESC, especially if the passaging regime is prolonged (Choi et al., 2017; Yagi et al., 2017). In serum/LIF, ESC cultures are heterogeneous, exhibiting properties of earlier and later stages of development and show low level promiscuous expression of lineage-specific markers (Canham, Sharov, Ko, & Brickman, 2010; Chambers et al., 2007; Hayashi, Lopes, Tang, & Surani, 2008; Toyooka, Shimosato, Murakami, Takahashi, & Niwa, 2008; van den Berg et al., 2008).

In addition to highlighting the prodifferentiation activity of the extraembryonic lineages during ESC derivation, Brook and Gardner demonstrated that single epiblast cells at the periimplantation stage of development are competent to generate self-renewing ESC colonies (Brook & Gardner, 1997). In their experiments a maximum of three ESC lines were produced per embryo. The advent of 2i/LIF and its enhanced ability to maximize the capture of ESC facilitated identification of cells competent to form ESC from embryos of various stages. From each of 12 periimplantation epiblasts, assumed to comprise around 20 cells each, between 2 and 12 ESC clones were derived (Nichols, Silva, et al., 2009). Interestingly, when single ICM cells were isolated from early blastocysts they failed to generate ESC clones in 2i/LIF on gelatin unless they were precultured on a substrate of basement lamina comprising Fibronectin and/or Laminin 511, known to be abundant at that stage of development (Boroviak, Loos, Bertone, Smith, & Nichols, 2014). As expected, the capacity for clonal ESC colony formation rapidly diminished after implantation and epithelialization of the epiblast. Transcriptional profiling of early embryos and ESC confirmed the equivalence of the periimplantation epiblast to ESC at the molecular level. The defining features of this state are summarized in Table 1.

Table 1 Defining Features of Naïve Pluripotency

Qualifiers for Naïve Pluripotency	Properties of Primed Pluripotency
Contribution to preimplantation chimeras	Unable to contribute to preimplantation chimeras
Competence to form all tissues in the body, including germ cells	Not able to contribute to the germline
High clonogenicity (ability to generate colonies from isolated single cells)	Requirement for cell–cell communication
Preferential organization as three-dimensional, dome-shaped colonies	Proliferation as single layered polarized epithelium
Global hypomethylation	High incidence of methylation
Self-renewal without glycolysis	Glycogen-dependent metabolism
High oxygen consumption	Preference for low oxygen atmosphere
High mitochondrial membrane potential	Low mitochondrial membrane potential
Efficient proliferation with Erk pathway inhibition	Requirement for FGF signaling
Activity of species-specific naïve pluripotency network	Expression of early postimplantation embryonic markers

2. ALTERNATIVE STEM CELL STATES

Prior to segregation of the first extraembryonic lineage, the trophectoderm, from the ICM, all cells of the morula are assumed to be totipotent, since those tested were shown to contribute progeny to derivative tissues of both embryonic and extraembryonic origin (Kelly, 1977). In serum/LIF culture conditions, a proportion of cells at any one time was seen to display molecular and functional properties of totipotency reminiscent of the 2-cell stage embryo (Macfarlan et al., 2012). Cells with the capacity to contribute to both embryonic and extraembryonic tissues have also been purified from 2i/LIF cultures (Canham et al., 2010; Morgani et al., 2013). So far, factors have been identified whose ectopic expression can enrich the "2C" totipotent state, but constant self-renewal in this state has not been proven (Choi et al., 2017) (Table 2).

Murine EpiSC can be derived from embryos soon after implantation until just before the appearance of the first somites (Brons et al., 2007;

Table 2 Toti/Pluripotent States in vivo and In Vitro

			Pluripotent					
	Totipotent		Naïve		Primed		Alternative	
	In Vitro	In Vivo Counterpart	In Vitro	In Vivo Counterpart	In Vitro	In Vivo Counterpart	In Vitro	In Vivo Counterpart
Mouse	mESC[a]	2-cell	mESC	e3.5–e4.5 ICM	mEpiSC	e7 EPI	rsPSC	e7.5 EPI
	EPS	?	mEGC	PGCs				
Human	EPS	?	hESC	e6	hESC	?	rsPSC	?
Pig					pESC	e6–e7 ICM		
					pEpiSC	e10–e11 epiblast		

[a]Only a subset of mESC with these properties has been identified.

Osorno et al., 2012; Tesar et al., 2007). Interestingly, regardless of their origin, during expansion in culture all EpiSC lines examined exhibit a molecular profile representative of the anterior primitive streak of late gastrula stage embryos (Kojima et al., 2014). This observation may help explain why EpiSC cannot contribute to the germline efficiently, since primordial germ cells (PGCs) originate from the posterior proximal epiblast cells in mice. Interestingly, ESC can be readily converted into EpiSC (Guo et al., 2009), and by capturing the cells during this process, the so-called Epiblast-like cell (EpiLC) stage, PGCs can be readily induced (Hayashi, Ohta, Kurimoto, Aramaki, & Saitou, 2011). This intermediate phase, representative of the preprimed, nascent epithelial early postimplantation epiblast, has been termed the "formative" state (Smith, 2017).

Since their initial derivation, the application of murine ESC for biomedical research has expanded massively. An equivalent system for human development and disease modeling was therefore actively pursued. Using embryos left over from assisted conception programs, the first human ESC were derived (Thomson et al., 1998). Although they arose from blastocysts, the properties of these cell lines were found to be more similar to the epiblast stem cell (EpiSC) lines derived from murine postimplantation embryos (Brons et al., 2007; Tesar et al., 2007). The properties of murine ESC were suggested to be a rodent-specific phenomenon. It is still not known why epiblasts from most nonrodent mammalian species progress to the polarized epithelial state so readily in culture, but the potential to intercept this progression and thereby capture self-renewing cell lines in the "naïve" state of pluripotency from human tissue was explored avidly (Gafni et al., 2013; Guo et al., 2016; Hanna et al., 2010; Takashima et al., 2014; Theunissen et al., 2014; Wang et al., 2011). Various conditions for their culture have been established; these are reviewed extensively elsewhere (Boroviak & Nichols, 2017; Zimmerlin, Park, & Zambidis, 2017). Interestingly, a different type of pluripotent cells, designated extended pluripotent stem (EPS) cells, has been recently described in mouse and human that is able to colonize embryonic and extraembryonic lineages in mouse chimeras (Yang et al., 2017).

3. PLURIPOTENT STEM CELL LINES FROM NONRODENT MAMMALS: COMMON AND DIVERGENT PROPERTIES

The initial reports of rhesus macaque and human ESC derived from preimplantation embryos (Thomson et al., 1998, 1995) highlighted the different requirements of primate ESC compared to those of mouse, namely, the need for feeder layers but independence from LIF. They also highlighted the differences in early development between rodents and primates, and the requirement for fibroblast growth factor (FGF) and Activin A for human ESC propagation (Vallier, Alexander, & Pedersen, 2005). These conditions were the basis for the establishment of mouse EpiSC, which can grow under feeder-free conditions (Brons et al., 2007; Tesar et al., 2007). The similarities between mouse EpiSC, derived from postimplantation embryos, and primate ESC, which are derived from preimplantation blastocysts, suggested that during the early stages of derivation those cells continue to advance in development and are captured in a self-renewing state when they reach the postimplantation epiblast equivalent. Support for this idea comes from single-cell transcriptome analysis of Macaque embryos showing that they align with late postimplantation epiblast (Nakamura et al., 2016). The reports showing that hESC can be converted to a naïve pluripotent state in vitro, closer to epiblast cells of mouse blastocysts (Chan et al., 2013; Gafni et al., 2013; Takashima et al., 2014; Theunissen et al., 2014) motivated attempts at direct derivation of naïve ESC from single human ICM cells (Guo et al., 2016). This achievement showed that finely balanced conditions are needed to capture this state from the human blastocyst.

The interest in generating pluripotent stem cells (PSC) from other animals began in the late 1980s with the purpose of improving animal breeding and for the generation of transgenic animals for the production of therapeutic proteins in animal bioreactors (Wall & Seidel, 1992). This was also followed by increased interest in using transgenic animals for biomedical applications and for cell banking from endangered species and breeds. Despite multiple attempts in deriving germline competent farm animal stem cell lines, few, if any, have been sufficiently well characterized and can qualify as *bona fide* ESC. Even if rodents, humans, and domestic animals share similarities in embryo development and pluripotency networks, there are key differences that preclude derivation of PSC from domestic mammals based on culture conditions used for mice and humans (Roberts, Yuan, & Ezashi, 2016). Several reviews have discussed the limited progress in PSC derivation from domestic animals (Blomberg & Telugu, 2012; Ezashi, Yuan, & Roberts, 2016; Goncalves, Ambrosio, & Piedrahita, 2014; Roberts et al., 2016).

The pig has attracted significant attention by virtue of its potential uses in biomedical research; thus, in the last few years many studies reported the derivation of porcine (p) ESC using diverse culture systems. These included supplementation with LIF (Kim et al., 2007; Telugu et al., 2011), bFGF

(Hall & Hyttel, 2014; Hou et al., 2016; Park et al., 2013; Tan et al., 2012), or a combination of both (Brevini et al., 2010; Xue et al., 2016) (Table 3), although it still remains unclear whether LIF is necessary and whether or not its receptor is present in pig embryos (Blomberg, Schreier, & Talbot, 2008; Guo et al., 2009; Hall & Hyttel, 2014). The LIF pathway has been detected by mRNA expression in pESC, although phosphorylated STAT3 protein has not been found, indicating the LIF pathway may not be active in established lines (Hou et al., 2016). Other studies suggested that LIF is important for pESC self-renewal but that it operates through the PI3K/AKT pathway (Brevini et al., 2010), or by activating STAT3

Table 3 Culture Conditions Used to Derive ESC From Domestic Mammals

Species	Culture Conditions	References
Pig	LIF	Telugu et al. (2011)
	bFGF	Hall and Hyttel (2014), Hou et al. (2016), Park et al. (2013), and Tan et al. (2012)
	LIF and bFGF	Brevini et al. (2010) and Xue et al. (2016)
	2i (MEK and GSK3β inhibitors)	Haraguchi, Kikuchi, Nakai, and Tokunaga (2012)
Cow	LIF and bFGF	Cong, Cao, and Liu (2014) and Jin et al. (2012)
	2i (MEK and GSK3β inhibitors)	Furusawa et al. (2013) and Verma, Huang, Kallingappa, and Oback (2013)
	3i (MEK, GSK3β, and FGFR inhibitors)	Kim, Park, Jung, and Roh (2015) and Park, Kim, Jung, and Roh (2015)
Sheep	bFGF and GSK3β inhibitor	Zhao et al. (2011)
Goat	LIF	Behboodi et al. (2011)
Rabbit	bFGF and Activin A/Nodal/TGFβ	Honda, Hirose, and Ogura (2009), Intawicha et al. (2009), and Schmaltz-Panneau et al. (2014)
	bFGF/LIF/no growth factors	Osteil et al. (2016)
Horse	bFGF	Guest and Allen (2007) and Li, Zhou, Imreh, Ahrlund-Richter, and Allen (2006)
Dog	LIF and bFGF	Vaags et al. (2009)
Cat	LIF and bFGF	Gomez et al. (2010)

(Thomson et al., 2012) even if LIFR is absent (Hall & Hyttel, 2014). Thus, there is still controversy regarding which signaling pathways are involved and what cytokines are needed for pESC derivation. The use of GSK3β and Mek inhibitors (2i medium) has also been reported (Haraguchi et al., 2012) (Table 3). However, all pESC lines have shown limited self-renewal ability over many passages. Only one of them was able to survive for over 75 passages, although the cells grew slowly (doubling time of 45.5 h) and it was not clear at which passage pluripotency markers were analyzed (Xue et al., 2016). Regarding their differentiation capacity, some recent papers report the formation of teratomas in immunodeficient mice (Hou et al., 2016; Jung et al., 2014; Xue et al., 2016), and surprisingly one of them with a cell line derived from parthenogenetic embryos cultured with bFGF, LIF, and stem cell factor (SCF) (Jung et al., 2014). Xue and colleagues also generated chimeric fetuses from pESC, although contribution was very low and the cells mainly integrated into the trophoblast (Xue et al., 2016). Thus, germline transmission still remains to be demonstrated from pESC.

The timing for ESC derivation appears to be a determining factor since pluripotent cells emerge at different times depending on the species. Expression of pluripotency factors (OCT4, NANOG, SOX2, and c-MYC) in the pig embryo increases from embryonic day (e) 5 to 7 blastocysts, and decreases again from e7 to e9. Thus, the optimal stage for pESC derivation would be around e6 or e7 (Brevini et al., 2010; Hou et al., 2016; Tan et al., 2012), but other authors have described more efficient derivation from e5.5 blastocysts (Xue et al., 2016). Thus, this issue is still not clear and a better understanding of the emergence of pluripotency in the pig preimplantation embryo is needed.

The protracted preimplantation development of the pig embryo offers a broad window for PSC derivation. It was recently shown for the first time that pig epiblast stem cells (pEpiSC) can be cultured in vitro from late epiblasts. Like human ESC, these cells are Activin A and bFGF dependent, and they are able to give rise to all somatic layers and germ cell precursors in vitro (Alberio, Croxall, & Allegrucci, 2010).

PSC have also been derived from bovine embryos, although most of them showed limited proliferation and cannot generate teratomas in immunodeficient mice. Combined LIF and bFGF (Cong et al., 2014; Jin et al., 2012) and also 2i (Furusawa et al., 2013; Verma et al., 2013) have been used as culture systems. Although one article reported contribution to chimeric fetuses, evidence was based on polymerase chain reaction (PCR) only (Furusawa et al., 2013). More recently, bovine (b) ESC have also been derived under 3i conditions (MEK, GSK3β, and FGFR inhibitors) (Kim et al., 2015; Park et al., 2015) (Table 3). Furthermore, since CDX2, a trophoblast marker in mouse, is still expressed in the bovine ICM and could be a barrier for maintaining pluripotency in vitro, Wu and coworkers generated bESC from CDX2-knockdown blastocysts, although efficiency was low (Wu et al., 2016). As in the case of pig embryos, a more detailed understanding of PSC emergence during bovine embryo development is needed so that better ESC derivation protocols can be developed.

Ovine ESC-like cells have been derived in N2B27 medium supplemented with GSK3β inhibitor and bFGF. Colonies of these cells showed a dome-shaped naïve pluripotent morphology, but were bFGF dependent, and although they formed teratomas they failed to produce chimeric animals (Zhao et al., 2011). ESC-like cells from goat have also been derived under LIF supplementation, showing long-term proliferation (>120 passages) and were the first ungulate species to form teratomas (Behboodi et al., 2011). Rabbit (r) ESC, relying on bFGF and Activin A/Nodal/TGFβ signaling have been derived by various groups (Honda et al., 2009; Intawicha et al., 2009; Schmaltz-Panneau et al., 2014), but remarkably one report described derivation of rESC from single cells in bFGF-free medium that colonized the ICM in rabbit embryos. The chimeric contribution in vivo was not tested (Osteil et al., 2016). Primed PSC from horse embryos have been derived under standard bFGF supplementation, however, they failed to form teratomas (Guest & Allen, 2007; Li et al., 2006). Canine ESC, able to form teratomas and to grow beyond 30 passages, have been obtained under combined LIF and bFGF supplementation (Vaags et al., 2009). Very limited self-renewal and differentiation capacity has been reported for feline ESC-like cell lines (Gomez et al., 2010) (Table 3). Furthermore, none has been used for further studies.

The development of iPSC (induced pluripotent stem cells) technology opened new avenues in PSC generation from domestic animals (Fujishiro et al., 2012; Sartori et al., 2012; West et al., 2010, 2011; Zhang et al., 2015). These cells are obtained by reprogramming through the introduction of specific transcription factors (Takahashi & Yamanaka, 2006). However, chimera contribution remains controversial and germline transmission has not been demonstrated using iPSC. One article reported germline contribution from pig iPSC; however, the efficiency was very low (2 out of 43 piglets) and the evidence was based on PCR only (West et al., 2011). As for ESC derivation, porcine iPSC culture systems have been based on those developed for primed hESC with bFGF supplementation, or on those used for mouse naïve ESC adding LIF, sometimes together with 2i, or a combination of both LIF and bFGF (Cheng et al., 2012; Telugu, Ezashi, & Roberts, 2010; Wang et al., 2013). However, persistent expression of ectopic reprogramming factors is a common problem for iPSC derivation in domestic species and probably a barrier to differentiation (Rodriguez, Allegrucci, & Alberio, 2012). Another alternative source of PSC is EGCs, derived from PGCs. EGC, which can be efficiently derived from mouse and rat pregonadal PGCs, have similar molecular features to ESC and can generate germline chimeras (Leitch et al., 2010). However, equivalent stable cell lines have not been established in domestic animals or humans. It is possible that this is due to differences in PGCs specification mechanisms between rodents and nonrodent mammals recently reported (Kobayashi et al., 2017). For example, early mouse PGCs express *Nanog*, *Oct-4*, and *Sox2*, which may facilitate the establishment of EGCs, however, human/nonhuman primate and pig PGC do not express *SOX2*, but instead express *SOX17* (Kobayashi et al., 2017; Sasaki et al., 2016).

In conclusion, PSC from nonrodent mammals express pluripotency markers and can differentiate in vitro into multiple lineages, however, these cell lines exhibit a low self-renewal capacity and generally cannot be maintained long term.

4. USE OF CHIMERAS TO DETERMINE NAÏVE STATE: SUCCESSES AND LIMITATIONS

A chimera is an organism made of cells from different individuals. The capacity of a pluripotent cell to contribute to the formation of different tissues when introduced into a host embryo is considered the gold standard for assessing pluripotency. Naïve mouse ESC are able to contribute to all tissues in the chimera assay, including the germline (Bradley et al., 1984; Nagy, Rossant, Nagy, Abramow-Newerly, & Roder, 1993). In contrast, primed mouse EpiSC are barely able to contribute to preimplantation embryo chimeras (Brons et al., 2007; Tesar et al., 2007).

Interspecies chimeras using distinct animal host embryos can be employed to assess pluripotency of hESC. Different studies have shown that primed hESC have a very limited capacity to contribute to preimplantation chimeras (James, Noggle, Swigut, & Brivanlou, 2006; Masaki et al., 2015); however, they can contribute to postimplantation chimeras (Mascetti & Pedersen, 2016). Remarkably, human–mouse interspecies chimeras have been generated with high efficiency by transplanting hESC and hiPSC to the epiblast and primitive streak of gastrulating mouse embryos (99.2% and 71.8%, respectively). Thus, it seems the success of the technique relies on matching in vitro pluripotent cells with their in vivo counterpart (Mascetti & Pedersen, 2016). However, these findings are not in agreement with a previous paper in which hESC did not integrate into the postimplantation mouse embryo (Wu et al., 2015).

Naïve hPSC have been matched to mouse morulae/early blastocysts; however, it still remains controversial whether they can contribute to the development of mouse preimplantation chimeric embryos. Although Gafni et al. reported hiPSC integration into mouse embryos (Gafni et al., 2013), two studies testing multiple hESC lines cultured under different conditions failed to detect significant differences in chimera contribution (Takashima et al., 2014; Theunissen et al., 2016) as chimera formation was too low. Moreover, the PCR assay used to detect the presence of human DNA cannot determine if these cells are dead or alive (Theunissen et al., 2016). Other authors have captured an alternative state of pluripotency, termed "region-selective" PSC from both primate and mouse embryos by addition of bFGF and a Wnt pathway inhibitor. These cells could be aligned with their in vivo counterpart as they integrate in the posterior epiblast of the late gastrulation mouse embryo (60.69% chimera formation rate) (Wu et al., 2015).

Chimera assays have also been used to test pluripotency in other mammalian cell lines. Although primed monkey iPSC have been injected into mouse embryos with no success, naïve monkey iPSC and ESC were able to form fetal chimeras (Chen et al., 2015; Fang et al., 2014). In the pig, germline transmission from one iPSC line was reported, but evidence was based on PCR (West et al., 2011). Contribution to chimeras from pESC was very low and mainly localized in the trophoblast (Xue et al., 2016). In cattle, evidence of chimera contribution from ESC was very weak and based on PCR only in the fetus (Furusawa et al., 2013). Thus, the authentication process needs improving because most evidence is based on PCR detection, a sensitive method prone to artifact and DNA contamination (Ezashi et al., 2016).

Interest in the application of interspecies chimeras as a means to generate bespoke organs for potential transplantation purposes is increasing. The first example, rat-mouse chimeras created by injecting rat PSC into mouse blastocysts, showed moderate levels of full-term chimerism (rat ESC, 47%; rat iPSC, 41%) but with extensive contribution to specific organs, achieved by targeted genetic ablation of the host tissue, such as pancreas (Kobayashi et al., 2010), thymus (Isotani, Hatayama, Kaseda, Ikawa, & Okabe, 2011), or kidney (Usui et al., 2012). These proof-of-concept studies, combined with revolutionary gene-editing technologies, could be developed to ablate the host organ, for example, a porcine pancreas, allowing human PSC to fill the vacant niche. This strategy can transform regenerative medicine by creating human–animal chimeras for organ production, or be used for developing new human disease models and drug screening (Wu et al., 2017). Due to the evolutionary distance between mice and humans, size, length of gestation, and early embryonic development, large animals are being considered as more suitable hosts for generating human organs. Recently, it was shown that human iPSC in different states of pluripotency can contribute to chimeric pig fetuses, albeit with low efficiency and occasional interference with normal development (Wu et al., 2017). Analysis of the equivalence of properties between different types of PSC captured in vitro and those present in the embryo could be critical for ensuring successful integration of donor cells into the host embryo. To this end, single-cell transcriptomics represent a precious tool for detailed analysis of the emergence and development of pluripotency in different species and for determining the in vivo counterpart of PSC lines. Other alternatives to improve chimerism could be to provide selective advantage to donor PSC by genetic engineering through CRISPR–Cas9 technology or to capture distinct PSC states resembling the epiblast of the embryo at each developmental stage.

5. PROPERTIES OF BLASTOCYSTS FROM DIFFERENT SPECIES: MORPHOLOGICAL AND TRANSCRIPTIONAL OVERLAP AND DIVERGENCE

The first lineage segregation in mammals takes place at the blastocyst stage, where cells differentiate into the epiblast that will form the fetus proper, and two extraembryonic lineages: trophectoderm (TE) and hypoblast, also known as PrE. Understanding the underlying mechanisms of lineage segregation is crucial for the generation of PSC and for assisted reproductive technologies. Although this process is conserved in mammals, there are several differences between species. For example, specification of the PrE takes place before implantation in all species; however, together with the epiblast they form a cup-shaped embryo in rodents, while in other species they develop as a discoid laminar embryo. In humans, an amniotic epithelium segregates from the epiblast, forming a nascent amniotic cavity (Rossant & Tam, 2017). Furthermore, the process of implantation is delayed in most species compared to rodents, but it is especially extended in large domestic species such as cow, sheep, horse, and pig, where there is an extensive proliferation and remodeling of extraembryonic tissue before attachment.

The emergence of pluripotent cells in early mammalian embryos is characterized by the sequential activation of key genes, starting from the early morula through to the blastocyst stage. This process shows variable kinetics between species. For example in mice, Nanog, Oct4, and Sox2 proteins are coexpressed in compact morulae and the ICM of blastocysts. In human, pig, and cattle embryos, however, NANOG protein expression is delayed, and coexpression of all three factors is only evident in expanded blastocysts (Cauffman, De Rycke, Sermon, Liebaers, & Van de Velde, 2009; Khan et al., 2012; Kuijk et al., 2008) (Fig. 1). Interestingly, the epiblast in *Cynomolgus* monkey maintains stable expression of pluripotency-related genes over 1 week from early epiblast to gastrulation (Nakamura et al., 2016). Similarly, the embryonic disc in pig embryos shows expression of OCT4, NANOG, and SOX2 (Alberio et al., 2010; Hall & Hyttel, 2014).

High-throughput single-cell RNA sequencing technology has revealed the transcriptional programs of early embryogenesis at an exceptional level of

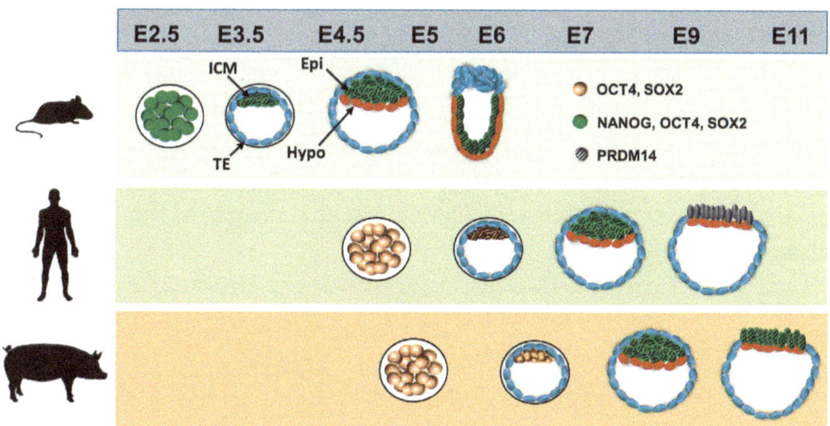

Fig. 1 Comparative developmental timeline between species. Genes expressed at different stages are indicated. Note that in human and pig embryos expression of key pluripotency genes is delayed compared to mouse embryos, suggesting that acquisition of pluripotency is delayed in these species. *Epi*, epiblast; *Hypo*: hypoblast; *ICM*, inner cell mass; *TE*, trophectoderm.

resolution and detail (Blakeley et al., 2015; Boroviak et al., 2015; Nakamura et al., 2016; Petropoulos et al., 2016; Yan et al., 2013). Cross-species comparisons have shown that although mouse and primates share transcriptional programs, there are important differences in timing and specific transcription factors (Xue et al., 2013). For example, several genes are specifically expressed in the primate epiblast such as *KLF17*, while other mouse pluripotency markers are absent (*Klf2*) or expressed in extraembryonic lineages (*ESRRB, BMP4*). *PRDM14* is expressed in mouse and human early blastocysts, but is not detected until the expanded blastocyst stage in pig (Blakeley et al., 2015; Cao et al., 2014) and robustly expressed in the late epiblast in spherical and ovoid embryos (Kobayashi et al., 2017). These examples show that although many key markers of pluripotency are conserved between mammals, a distinct profile of gene expression is evident in the pig embryo (although it is largely shared with the human), consistent with their protracted development compared to mice (Table 4). The gradual emergence of pluripotency may explain the difficulties in establishing suitable ESC derivation conditions in domestic animals.

Cdx2, a key gene for trophectoderm specification in mice, where it is expressed from the morula stage, is not detectable until the blastocyst stage in humans, and only in a proportion of trophectoderm cells (Niakan & Eggan, 2013). *Oct4,* restricted to the epiblast in the mouse, is expressed in all lineages until implantation in other mammals (Blakeley et al., 2015; Boroviak et al., 2015; Liu et al., 2015; Rossant, 2011). Other important genes for TE development such as *Elf5* and *Eomes* are only expressed in mouse embryos, while *CLDN10*, *PLAC8*, and *TRIML* are specific to humans (Blakeley et al., 2015). Furthermore, *Laminin* is crucial for PrE specification in mice to separate the ICM into epiblast and PrE (Chazaud, Yamanaka, Pawson, & Rossant, 2006), but it is absent in the human PrE (Niakan & Eggan, 2013).

One of the largest RNA sequencing studies of human embryos analyzed 1529 individual cells from 88 embryos between e3 and e7 (Petropoulos et al., 2016). This study shows that zygotic genome activation occurs around e3/e4, in agreement to previous studies (Ko, 2016). Moreover, and in contrast to mice, *XIST* and X-linked genes are biallelically expressed at e7 and there is a progressive gene downregulation in both X chromosomes until dosage compensation between female and male embryos is reached (Petropoulos et al., 2016). It will be interesting to determine whether this is a primate-specific phenomenon, or whether it also applies to other mammalian species.

Table 4 Differences in Embryo Development Between Mouse, Primate, and Pig

	Mouse	**Primate**	**Pig**
Embryo shape	Cup shape	Bilaminar disc	Bilaminar disc
Preimplantation period	e4.5	~e9	~e14
Timing of embryonic genome activation	2-cell	8-cell	4-cell
Stable expression of pluripotency genes	2 days (e3.5–e5.5)	1 week in monkey	1 week (e5–e11)
Pluripotency markers in the epiblast	*Klf2*, *Esrrb*, *Bmp4*	*KLF4*, *KLF17*	*KLF4*, *KLF17*
PRDM14 expression	From early blastocyst	From early blastocyst	From late epiblast (e8–e11)
NANOG expression	From morula	From expanded blastocyst	From expanded blastocyst
CDX2/OCT4 expression	*Cdx2* restricted to outside cells prior to blastocyst formation, *Oct4* in ICM in early blastocyst	No *CDX2* expression prior to blastocyst formation, *OCT4* not restricted to ICM until implantation	No *CDX2* expression prior to blastocyst formation, *OCT4* not restricted to ICM until implantation
X inactivation status	Paternal X inactivation from 4-cell to late blastocyst	Reduced expression of both X chromosomes until random X-inactivation at late blastocyst	Not determined
TE/ICM/PrE specification	TE restricted by late morula, ICM by early blastocyst, and PrE by late blastocyst	TE, EPI, and PrE segregate at the same time in early blastocysts	TE specified by early blastocyst and PrE by late blastocyst
FGF/ERK signaling in the ICM	Determines EPI vs PrE	ERK inhibition does not block PrE formation	ERK inhibition blocks PrE formation

6. OVERCOMING DEVELOPMENTAL PROGRESSION TO PROMOTE EPIBLAST SELF-RENEWAL

Several signaling pathways are involved in the emergence of the pluripotent epiblast and its segregation from the PrE. Although some of them are conserved in mammals, there are important differences between species. FGF engages the MAPK pathway, consisting of Mapk kinase 1/2 (MEK) and Mapk1/2 (ERK), inducing PrE segregation (Kunath et al., 2007). Thus, blocking MAPK signaling protects epiblast cells from differentiating into extraembryonic lineages in the mouse (Nichols, Silva, et al., 2009). On the other hand, WNT signaling can be stimulated by inhibition of GSK3β to antagonize indirectly FGF/MAPK (ten Berge et al., 2011). Double inhibition of ERK and Gsk3β (2i) in the mouse promotes epiblast specification over PrE and allows derivation of naïve ESCs that maintain germline competence and contribute to chimeras efficiently (Nichols, Silva, et al., 2009; Ying et al., 2008). Conversely, increasing FGF/MAPK signaling can convert the entire ICM into PrE (Kuijk et al., 2012; Yamanaka, Lanner, & Rossant, 2010). However, suppression of ERK alone or as part of the 2i treatment only partially prevents PrE segregation in marmoset (Boroviak et al., 2015), pig (Rodriguez et al., 2012), and bovine embryos (Kuijk et al., 2012) and does not interfere with PrE formation in human (Kuijk et al., 2012; Roode et al., 2012; Van der Jeught et al., 2013), suggesting additional mechanisms participating in PrE segregation in these species.

hESC and mouse EpiSC show rapid differentiation in response to WNT activation (Dravid et al., 2005; Greber et al., 2010), whereas WNT inhibition enhances their pluripotency (Sumi, Oki, Kitajima, & Meno, 2013; Wu et al., 2015). While WNT signaling is dispensable for lineage segregation in mice (Biechele, Cox, & Rossant, 2011), WNT inhibition in marmoset embryos significantly increased *NANOG* expression in the epiblast, and when combined with ERK inhibition induces a significant reduction of PrE segregation (Boroviak et al., 2015). Interestingly, in human embryos GSK3β inhibition, with a comparable effect to WNT pathway activation, does not reduce the number of *NANOG* positive cells (Roode et al., 2012); WNT signaling was shown to play a role in TE development and not in the ICM (Krivega, Essahib, & Van de Velde, 2015). WNT does not play a major role in the segregation of PrE and epiblast in bovine embryos (Kuijk et al., 2012). Thus, the effects of modulating ERK and WNT signaling can have different effects depending upon the species and type of pluripotent cells being targeted.

Janus kinase/signal transducer and activator of transcription (JAK/STAT) is directly activated by LIF, leading to STAT3 phosphorylation and transcription of ground state pluripotency genes such as *Klf4* (Stuart et al., 2014) and *Tfcp2l1* (Martello, Bertone, & Smith, 2013; Ye, Li, Tong, & Ying, 2013). However, despite the fact that LIF receptor may be present in hESC, LIF is apparently not necessary to maintain pluripotency (Humphrey et al., 2004). Other important signaling pathways implicated in lineage segregation in human embryos are NODAL, PDGF, and TGFβ, although strong evidence is still missing (Rossant & Tam, 2017). For example, TGFβ inhibition in one study abrogated *NANOG* and *SOX17* and reduced *OCT4* expression (Blakeley et al., 2015), but in another study it increased the number of cells expressing NANOG while GATA6 remained constant (Van der Jeught et al., 2014). These differences could be explained by the different doses of inhibitor used. However, TGFβ inhibition in mice and cattle affects neither *NANOG*/*OCT4* expression (Blakeley et al., 2015; Kuijk et al., 2012) nor lineage segregation in marmoset embryos, based on NANOG and GATA6 expression (Boroviak et al., 2015). In pigs, experiments on explant cultures derived from e5 and e10 embryos using inhibitors of FGF, JAK/STAT, BMP, WNT, and NODAL pathways revealed the importance of FGF, JAK/STAT, and BMP signaling for capturing pluripotency (Hall & Hyttel, 2014).

In recent years, several studies have analyzed the effect of different inhibitors on lineage segregation in bovine embryos. Kuijk and colleagues found that GSK3β inhibitor alone increases the percentage of NANOG-positive cells but there was no synergistic effect as part of the 2i treatment (Kuijk et al., 2012). However, another study using qPCR of mechanically separated ICM and TE showed that 2i treatment increased *NANOG* and *SOX2* and decreased *GATA4* transcription. However, expression of other pluripotency-related genes (*POU5F1*, *KLF4*, *DPPA3*) was not affected (Harris, Huang, & Oback, 2013). It was also found that increasing the concentration of ERK inhibitor (PD0325901) from 0.4 to 10 μM improved blastocyst morphology and increased *NANOG* and *FGF4* expression while reducing *SOX17* and *PDGFRα* (McLean, Meng, Henderson, Turner, & Oback, 2014). After screening nine principal signaling pathways, including TGFβ, BMP, EGF, VEGF, PDGF, FGF, cAMP, PI3K, and JAK, it was concluded that PI3K was important for ICM growth. JAK inhibition abolished ICM formation, repressing epiblast, PrE, and pluripotency-related genes (Meng, Forrester-Gauntlett, Turner, Henderson, & Oback, 2015).

7. THE PIG AS A MODEL SYSTEM FOR HUMAN DEVELOPMENT

When choosing animal models to study basic biology and for biomedical research, scientists need to consider many factors to ensure that the biological questions asked can be adequately addressed. Mice have been traditionally used to study mammalian developmental biology because of their small size, prolificacy, easy handling, and limited space requirements. Furthermore, the availability of ESC and a full genome sequence simplified the generation of genetically modified mice. However, rodents have important differences in embryology, physiology, and anatomy from humans, which limits their utility. The pig as a model in biomedical research has acquired significant relevance more recently. This is primarily due to the close similarities between pigs and humans. Although phylogenetically pigs and humans diverged over 90 million years ago, they have more nucleotide sequence conservation with humans than do mice (Archibald et al., 2010). From a developmental perspective, pig and human embryos follow similar kinetics during the first ~10 days, except for embryo implantation which occurs on ~e9 in humans, whereas in pigs attachment of the embryo occurs from ~e14. Thus, pig embryos can be recovered from the uterus at late epiblast stages and during the early stages of gastrulation. From the point of view of physiology and size, pigs are considered excellent models for xenotransplantation, with organs such as kidneys and heart being of similar dimension and function to those of humans. Other organs of interest for transplantation are pancreas and cornea. More research is needed to understand the compatibility between human and pig cells before this technology can offer a solution; however, studies of early lineage segregation are beginning to show a clearer picture of how similar human and pig embryos are at these early stages (Kobayashi et al., 2017). Furthermore, genetically engineered pigs are excellent models for many human diseases, for example, cystic fibrosis, retinitis pigmentosa, cancer, immunodeficiency, neurodegenerative diseases, cardiovascular disease, and Crohn's disease (Holm, Alstrup, & Luo, 2016; Walters, Wells, Bryda, Schommer, & Prather, 2017; Yao, Huang, & Zhao, 2016). Because of a similar physiology, pigs also represent excellent models of metabolic syndrome, osteoarthritis, and for toxicological studies, which combined with gene editing can offer excellent platforms for modeling human conditions more reliably.

Recent advances in molecular embryology and genetic engineering using CRISPR/Cas9 system offer a great prospect for developing the toolkit needed to establish the pig as an ethically acceptable, cost effective, and reliable model for biomedical research. Key priorities for the near future are to gain deeper knowledge of pig developmental biology and cultivate more robust tools for precise genetic modification. We can foresee the generation of pigs carrying multiple genetic modifications eliminating the genetic barriers that prevented a wider use of this model previously.

ACKNOWLEDGMENTS

P.R.I. is funded by a Marie Sklodowska Curie Fellowship. Work in R.A. laboratory is funded by BBSRC and MRC. J.N. is funded by the University of Cambridge, Wellcome Trust, MRC, and BBSRC.

REFERENCES

Alberio, R., Croxall, N., & Allegrucci, C. (2010). Pig epiblast stem cells depend on activin/nodal signaling for pluripotency and self-renewal. *Stem Cells and Development*, *19*, 1627–1636.

Archibald, A. L., Bolund, L., Churcher, C., Fredholm, M., Groenen, M. A., Harlizius, B., et al. (2010). Pig genome sequence—Analysis and publication strategy. *BMC Genomics*, *11*, 438.

Batlle-Morera, L., Smith, A., & Nichols, J. (2008). Parameters influencing derivation of embryonic stem cells from murine embryos. *Genesis*, *46*, 758–767.

Behboodi, E., Bondareva, A., Begin, I., Rao, K., Neveu, N., Pierson, J. T., et al. (2011). Establishment of goat embryonic stem cells from in vivo produced blastocyst-stage embryos. *Molecular Reproduction and Development*, *78*, 202–211.

Biechele, S., Cox, B. J., & Rossant, J. (2011). Porcupine homolog is required for canonical Wnt signaling and gastrulation in mouse embryos. *Developmental Biology*, *355*, 275–285.

Blakeley, P., Fogarty, N. M., Del Valle, I., Wamaitha, S. E., Hu, T. X., Elder, K., et al. (2015). Defining the three cell lineages of the human blastocyst by single-cell RNA-Seq. *Development*, *142*, 3151–3165.

Blomberg, L. A., Schreier, L. L., & Talbot, N. C. (2008). Expression analysis of pluripotency factors in the undifferentiated porcine inner cell mass and epiblast during in vitro culture. *Molecular Reproduction and Development*, *75*, 450–463.

Blomberg, L. A., & Telugu, B. P. (2012). Twenty years of embryonic stem cell research in farm animals. *Reproduction in Domestic Animals*, *47*(Suppl. 4), 80–85.

Boroviak, T., Loos, R., Bertone, P., Smith, A., & Nichols, J. (2014). The ability of inner-cell-mass cells to self-renew as embryonic stem cells is acquired following epiblast specification. *Nature Cell Biology*, *16*, 516–528.

Boroviak, T., Loos, R., Lombard, P., Okahara, J., Behr, R., Sasaki, E., et al. (2015). Lineage-specific profiling delineates the emergence and progression of naive pluripotency in mammalian embryogenesis. *Developmental Cell*, *35*, 366–382.

Boroviak, T., & Nichols, J. (2017). Primate embryogenesis predicts the hallmarks of human naive pluripotency. *Development*, *144*, 175–186.

Bradley, A., Evans, M., Kaufman, M. H., & Robertson, E. (1984). Formation of germ-line chimaeras from embryo-derived teratocarcinoma cell lines. *Nature*, *309*, 255–256.

Brevini, T. A., Pennarossa, G., Attanasio, L., Vanelli, A., Gasparrini, B., & Gandolfi, F. (2010). Culture conditions and signalling networks promoting the establishment of cell lines from parthenogenetic and biparental pig embryos. *Stem Cell Reviews, 6*, 484–495.

Brons, I. G., Smithers, L. E., Trotter, M. W., Rugg-Gunn, P., Sun, B., Chuva de Sousa Lopes, S. M., et al. (2007). Derivation of pluripotent epiblast stem cells from mammalian embryos. *Nature, 448*, 191–195.

Brook, F. A., & Gardner, R. L. (1997). The origin and efficient derivation of embryonic stem cells in the mouse. *Proceedings of the National Academy of Sciences of the United States of America, 94*, 5709–5712.

Buehr, M., Meek, S., Blair, K., Yang, J., Ure, J., Silva, J., et al. (2008). Capture of authentic embryonic stem cells from rat blastocysts. *Cell, 135*, 1287–1298.

Canham, M. A., Sharov, A. A., Ko, M. S., & Brickman, J. M. (2010). Functional heterogeneity of embryonic stem cells revealed through translational amplification of an early endodermal transcript. *PLoS Biology, 8*, e1000379.

Cao, S., Han, J., Wu, J., Li, Q., Liu, S., Zhang, W., et al. (2014). Specific gene-regulation networks during the pre-implantation development of the pig embryo as revealed by deep sequencing. *BMC Genomics, 15*, 4.

Cauffman, G., De Rycke, M., Sermon, K., Liebaers, I., & Van de Velde, H. (2009). Markers that define stemness in ESC are unable to identify the totipotent cells in human preimplantation embryos. *Human Reproduction, 24*, 63–70.

Chambers, I., Silva, J., Colby, D., Nichols, J., Nijmeijer, B., Robertson, M., et al. (2007). Nanog safeguards pluripotency and mediates germline development. *Nature, 450*, 1230–1234.

Chan, Y. S., Goke, J., Ng, J. H., Lu, X., Gonzales, K. A., Tan, C. P., et al. (2013). Induction of a human pluripotent state with distinct regulatory circuitry that resembles preimplantation epiblast. *Cell Stem Cell, 13*, 663–675.

Chazaud, C., Yamanaka, Y., Pawson, T., & Rossant, J. (2006). Early lineage segregation between epiblast and primitive endoderm in mouse blastocysts through the Grb2-MAPK pathway. *Developmental Cell, 10*, 615–624.

Chen, Y., Niu, Y., Li, Y., Ai, Z., Kang, Y., Shi, H., et al. (2015). Generation of Cynomolgus monkey chimeric fetuses using embryonic stem cells. *Cell Stem Cell, 17*, 116–124.

Cheng, D., Guo, Y., Li, Z., Liu, Y., Gao, X., Gao, Y., et al. (2012). Porcine induced pluripotent stem cells require LIF and maintain their developmental potential in early stage of embryos. *PLoS One*, 7, e51778.

Choi, Y. J., Lin, C. P., Risso, D., Chen, S., Kim, T. A., Tan, M. H., et al. (2017). Deficiency of microRNA miR-34a expands cell fate potential in pluripotent stem cells. *Science, 355*, pii: eaag1927.

Cong, S., Cao, G., & Liu, D. (2014). Effects of different feeder layers on culture of bovine embryonic stem cell-like cells in vitro. *Cytotechnology, 66*, 995–1005.

Dravid, G., Ye, Z., Hammond, H., Chen, G., Pyle, A., Donovan, P., et al. (2005). Defining the role of Wnt/beta-catenin signaling in the survival, proliferation, and self-renewal of human embryonic stem cells. *Stem Cells, 23*, 1489–1501.

Evans, M. J., & Kaufman, M. (1981). Establishment in culture of pluripotential cells from mouse embryos. *Nature, 292*, 154–156.

Ezashi, T., Yuan, Y., & Roberts, R. M. (2016). Pluripotent stem cells from domesticated mammals. *Annual Review of Animal Biosciences, 4*, 223–253.

Fang, R., Liu, K., Zhao, Y., Li, H., Zhu, D., Du, Y., et al. (2014). Generation of naive induced pluripotent stem cells from rhesus monkey fibroblasts. *Cell Stem Cell, 15*, 488–496.

Fujishiro, S. H., Nakano, K., Mizukami, Y., Azami, T., Arai, Y., Matsunari, H., et al. (2012). Generation of naive-like porcine-induced pluripotent stem cells capable of contributing to embryonic and fetal development. *Stem Cells and Development, 22*, 473–482.

Furusawa, T., Ohkoshi, K., Kimura, K., Matsuyama, S., Akagi, S., Kaneda, M., et al. (2013). Characteristics of bovine inner cell mass-derived cell lines and their fate in chimeric conceptuses. *Biology of Reproduction, 89*, 28.

Gafni, O., Weinberger, L., Mansour, A. A., Manor, Y. S., Chomsky, E., Ben-Yosef, D., et al. (2013). Derivation of novel human ground state naive pluripotent stem cells. *Nature, 504*, 282–286.

Gomez, M. C., Serrano, M. A., Pope, C. E., Jenkins, J. A., Biancardi, M. N., Lopez, M., et al. (2010). Derivation of cat embryonic stem-like cells from in vitro-produced blastocysts on homologous and heterologous feeder cells. *Theriogenology, 74*, 498–515.

Goncalves, N. N., Ambrosio, C. E., & Piedrahita, J. A. (2014). Stem cells and regenerative medicine in domestic and companion animals: A multispecies perspective. *Reproduction in Domestic Animals, 49*(Suppl. 4), 2–10.

Greber, B., Wu, G., Bernemann, C., Joo, J. Y., Han, D. W., Ko, K., et al. (2010). Conserved and divergent roles of FGF signaling in mouse epiblast stem cells and human embryonic stem cells. *Cell Stem Cell, 6*, 215–226.

Guest, D. J., & Allen, W. R. (2007). Expression of cell-surface antigens and embryonic stem cell pluripotency genes in equine blastocysts. *Stem Cells and Development, 16*, 789–796.

Guo, G., von Meyenn, F., Santos, F., Chen, Y., Reik, W., Bertone, P., et al. (2016). Naive pluripotent stem cells derived directly from isolated cells of the human inner cell mass. *Stem Cell Reports, 6*, 437–446.

Guo, G., Yang, J., Nichols, J., Hall, J. S., Eyres, I., Mansfield, W., et al. (2009). Klf4 reverts developmentally programmed restriction of ground state pluripotency. *Development, 136*, 1063–1069.

Hall, V. J., & Hyttel, P. (2014). Breaking down pluripotency in the porcine embryo reveals both a premature and reticent stem cell state in the inner cell mass and unique expression profiles of the naive and primed stem cell states. *Stem Cells and Development, 23*, 2030–2045.

Hanna, J., Cheng, A. W., Saha, K., Kim, J., Lengner, C. J., Soldner, F., et al. (2010). Human embryonic stem cells with biological and epigenetic characteristics similar to those of mouse ESCs. *Proceedings of the National Academy of Sciences of the United States of America, 107*, 9222–9227.

Haraguchi, S., Kikuchi, K., Nakai, M., & Tokunaga, T. (2012). Establishment of self-renewing porcine embryonic stem cell-like cells by signal inhibition. *The Journal of Reproduction and Development, 58*, 707–716.

Harris, D., Huang, B., & Oback, B. (2013). Inhibition of MAP2K and GSK3 signaling promotes bovine blastocyst development and epiblast-associated expression of pluripotency factors. *Biology of Reproduction, 88*, 74.

Hayashi, K., Lopes, S. M., Tang, F., & Surani, M. A. (2008). Dynamic equilibrium and heterogeneity of mouse pluripotent stem cells with distinct functional and epigenetic states. *Cell Stem Cell, 3*, 391–401.

Hayashi, K., Ohta, H., Kurimoto, K., Aramaki, S., & Saitou, M. (2011). Reconstitution of the mouse germ cell specification pathway in culture by pluripotent stem cells. *Cell, 146*, 519–532.

Holm, I. E., Alstrup, A. K., & Luo, Y. (2016). Genetically modified pig models for neurodegenerative disorders. *The Journal of Pathology, 238*, 267–287.

Honda, A., Hirose, M., & Ogura, A. (2009). Basic FGF and Activin/Nodal but not LIF signaling sustain undifferentiated status of rabbit embryonic stem cells. *Experimental Cell Research, 315*, 2033–2042.

Hou, D. R., Jin, Y., Nie, X. W., Zhang, M. L., Ta, N., Zhao, L. H., et al. (2016). Derivation of porcine embryonic stem-like cells from in vitro-produced blastocyst-stage embryos. *Scientific Reports, 6*, 25838.

Humphrey, R. K., Beattie, G. M., Lopez, A. D., Bucay, N., King, C. C., Firpo, M. T., et al. (2004). Maintenance of pluripotency in human embryonic stem cells is STAT3 independent. *Stem Cells, 22*, 522–530.

Intawicha, P., Ou, Y. W., Lo, N. W., Zhang, S. C., Chen, Y. Z., Lin, T. A., et al. (2009). Characterization of embryonic stem cell lines derived from New Zealand white rabbit embryos. *Cloning and Stem Cells, 11*, 27–38.

Isotani, A., Hatayama, H., Kaseda, K., Ikawa, M., & Okabe, M. (2011). Formation of a thymus from rat ES cells in xenogeneic nude mouse ↔ rat ES chimeras. *Genes to Cells, 16*, 397–405.

James, D., Noggle, S. A., Swigut, T., & Brivanlou, A. H. (2006). Contribution of human embryonic stem cells to mouse blastocysts. *Developmental Biology, 295*, 90–102.

Jin, M., Wu, A., Dorzhin, S., Yue, Q., Ma, Y., & Liu, D. (2012). Culture conditions for bovine embryonic stem cell-like cells isolated from blastocysts after external fertilization. *Cytotechnology, 64*, 379–389.

Jung, S. K., Kim, H. J., Kim, C. L., Lee, J. H., You, J. Y., Lee, E. S., et al. (2014). Enhancing effects of serum-rich and cytokine-supplemented culture conditions on developing blastocysts and deriving porcine parthenogenetic embryonic stem cells. *Journal of Veterinary Science, 15*, 519–528.

Kelly, S. J. (1977). Studies of the developmental potential of 4- and 8-cell stage mouse blastomeres. *The Journal of Experimental Zoology, 200*, 365–376.

Khan, D. R., Dube, D., Gall, L., Peynot, N., Ruffini, S., Laffont, L., et al. (2012). Expression of pluripotency master regulators during two key developmental transitions: EGA and early lineage specification in the bovine embryo. *PLoS One*, 7, e34110.

Kim, D., Park, S., Jung, Y. G., & Roh, S. (2015). In vitro culture of stem-like cells derived from somatic cell nuclear transfer bovine embryos of the Korean beef cattle species, HanWoo. *Reproduction, Fertility, and Development, 28*, 1762–1780.

Kim, H. S., Son, H. Y., Kim, S., Lee, G. S., Park, C. H., Kang, S. K., et al. (2007). Isolation and initial culture of porcine inner cell masses derived from in vitro-produced blastocysts. *Zygote, 15*, 55–63.

Ko, M. S. (2016). Zygotic genome activation revisited: Looking through the expression and function of Zscan4. *Current Topics in Developmental Biology, 120*, 103–124.

Kobayashi, T., Yamaguchi, T., Hamanaka, S., Kato-Itoh, M., Yamazaki, Y., Ibata, M., et al. (2010). Generation of rat pancreas in mouse by interspecific blastocyst injection of pluripotent stem cells. *Cell, 142*, 787–799.

Kobayashi, T., Zhang, H., Tang, W. W. C., Irie, N., Withey, S., Klisch, D., et al. (2017). Principles of early human development and germ cell program from conserved model systems. *Nature, 546*, 416–420.

Kojima, Y., Kaufman-Francis, K., Studdert, J. B., Steiner, K. A., Power, M. D., Loebel, D. A., et al. (2014). The transcriptional and functional properties of mouse epiblast stem cells resemble the anterior primitive streak. *Cell Stem Cell, 14*, 107–120.

Krivega, M., Essahib, W., & Van de Velde, H. (2015). WNT3 and membrane-associated beta-catenin regulate trophectoderm lineage differentiation in human blastocysts. *Molecular Human Reproduction, 21*, 711–722.

Kuijk, E. W., Du Puy, L., Van Tol, H. T., Oei, C. H., Haagsman, H. P., Colenbrander, B., et al. (2008). Differences in early lineage segregation between mammals. *Developmental Dynamics, 237*, 918–927.

Kuijk, E. W., van Tol, L. T., Van de Velde, H., Wubbolts, R., Welling, M., Geijsen, N., et al. (2012). The roles of FGF and MAP kinase signaling in the segregation of the epiblast and hypoblast cell lineages in bovine and human embryos. *Development, 139*, 871–882.

Kunath, T., Saba-El-Leil, M. K., Almousailleakh, M., Wray, J., Meloche, S., & Smith, A. (2007). FGF stimulation of the Erk1/2 signalling cascade triggers transition of pluripotent embryonic stem cells from self-renewal to lineage commitment. *Development, 134*, 2895–2902.

Leitch, H. G., Blair, K., Mansfield, W., Ayetey, H., Humphreys, P., Nichols, J., et al. (2010). Embryonic germ cells from mice and rats exhibit properties consistent with a generic pluripotent ground state. *Development, 137*, 2279–2287.

Li, P., Tong, C., Mehrian-Shai, R., Jia, L., Wu, N., Yan, Y., et al. (2008). Germline competent embryonic stem cells derived from rat blastocysts. *Cell, 135*, 1299–1310.

Li, X., Zhou, S. G., Imreh, M. P., Ahrlund-Richter, L., & Allen, W. R. (2006). Horse embryonic stem cell lines from the proliferation of inner cell mass cells. *Stem Cells and Development, 15*, 523–531.

Liu, S., Bou, G., Sun, R., Guo, S., Xue, B., Wei, R., et al. (2015). Sox2 is the faithful marker for pluripotency in pig: Evidence from embryonic studies. *Developmental Dynamics, 244*, 619–627.

Macfarlan, T. S., Gifford, W. D., Driscoll, S., Lettieri, K., Rowe, H. M., Bonanomi, D., et al. (2012). Embryonic stem cell potency fluctuates with endogenous retrovirus activity. *Nature, 487*, 57–63.

Martello, G., Bertone, P., & Smith, A. (2013). Identification of the missing pluripotency mediator downstream of leukaemia inhibitory factor. *The EMBO Journal, 32*, 2561–2574.

Martin, G. R. (1981). Isolation of a pluripotent cell line from early mouse embryos cultured in medium conditioned by teratocarcinoma stem cells. *Proceedings of the National Academy of Sciences of the United States of America, 78*, 7634–7638.

Masaki, H., Kato-Itoh, M., Umino, A., Sato, H., Hamanaka, S., Kobayashi, T., et al. (2015). Interspecific in vitro assay for the chimera-forming ability of human pluripotent stem cells. *Development, 142*, 3222–3230.

Mascetti, V. L., & Pedersen, R. A. (2016). Human-mouse chimerism validates human stem cell pluripotency. *Cell Stem Cell, 18*, 67–72.

McLean, Z., Meng, F., Henderson, H., Turner, P., & Oback, B. (2014). Increased MAP kinase inhibition enhances epiblast-specific gene expression in bovine blastocysts. *Biology of Reproduction, 91*, 49.

Meng, F., Forrester-Gauntlett, B., Turner, P., Henderson, H., & Oback, B. (2015). Signal inhibition reveals JAK/STAT3 pathway as critical for bovine inner cell mass development. *Biology of Reproduction, 93*, 132.

Morgani, S. M., Canham, M. A., Nichols, J., Sharov, A. A., Migueles, R. P., Ko, M. S., et al. (2013). Totipotent embryonic stem cells arise in ground-state culture conditions. *Cell Reports, 3*, 1945–1957.

Nagy, A., Rossant, J., Nagy, R., Abramow-Newerly, W., & Roder, J. C. (1993). Derivation of completely cell culture-derived mice from early-passage embryonic stem cells. *Proceedings of the National Academy of Sciences of the United States of America, 90*, 8424–8428.

Nakamura, T., Okamoto, I., Sasaki, K., Yabuta, Y., Iwatani, C., Tsuchiya, H., et al. (2016). A developmental coordinate of pluripotency among mice, monkeys and humans. *Nature, 537*, 57–62.

Niakan, K. K., & Eggan, K. (2013). Analysis of human embryos from zygote to blastocyst reveals distinct gene expression patterns relative to the mouse. *Developmental Biology, 375*, 54–64.

Nichols, J., Jones, K., Phillips, J. M., Newland, S. A., Roode, M., Mansfield, W., et al. (2009). Validated germline-competent embryonic stem cell lines from nonobese diabetic mice. *Nature Medicine, 15*, 814–818.

Nichols, J., Silva, J., Roode, M., & Smith, A. (2009). Suppression of Erk signalling promotes ground state pluripotency in the mouse embryo. *Development, 136*, 3215–3222.

Osorno, R., Tsakiridis, A., Wong, F., Cambray, N., Economou, C., Wilkie, R., et al. (2012). The developmental dismantling of pluripotency is reversed by ectopic Oct4 expression. *Development, 139*, 2288–2298.

Osteil, P., Moulin, A., Santamaria, C., Joly, T., Jouneau, L., Aubry, M., et al. (2016). A panel of embryonic stem cell lines reveals the variety and dynamic of pluripotent states in rabbits. *Stem Cell Reports*, *7*, 383–398.

Park, S., Kim, D., Jung, Y. G., & Roh, S. (2015). Thiazovivin, a Rho kinase inhibitor, improves stemness maintenance of embryo-derived stem-like cells under chemically defined culture conditions in cattle. *Animal Reproduction Science*, *161*, 47–57.

Park, J. K., Kim, H. S., Uh, K. J., Choi, K. H., Kim, H. M., Lee, T., et al. (2013). Primed pluripotent cell lines derived from various embryonic origins and somatic cells in pig. *PLoS One*, *8*, e52481.

Petropoulos, S., Edsgard, D., Reinius, B., Deng, Q., Panula, S. P., Codeluppi, S., et al. (2016). Single-cell RNA-Seq reveals lineage and X chromosome dynamics in human preimplantation embryos. *Cell*, *165*, 1012–1026.

Roberts, R. M., Yuan, Y., & Ezashi, T. (2016). Exploring early differentiation and pluripotency in domestic animals. *Reproduction, Fertility, and Development*, *29*, 101–107.

Rodriguez, A., Allegrucci, C., & Alberio, R. (2012). Modulation of pluripotency in the porcine embryo and iPS cells. *PLoS One*, *7*, e49079.

Roode, M., Blair, K., Snell, P., Elder, K., Marchant, S., Smith, A., et al. (2012). Human hypoblast formation is not dependent on FGF signalling. *Developmental Biology*, *361*, 358–363.

Rossant, J. (2011). Developmental biology: A mouse is not a cow. *Nature*, *471*, 457–458.

Rossant, J., & Tam, P. P. (2017). New insights into early human development: Lessons for stem cell derivation and differentiation. *Cell Stem Cell*, *20*, 18–28.

Sartori, C., DiDomenico, A. I., Thomson, A. J., Milne, E., Lillico, S. G., Burdon, T. G., et al. (2012). Ovine-induced pluripotent stem cells can contribute to chimeric lambs. *Cellular Reprogramming*, *14*, 8–19.

Sasaki, K., Nakamura, T., Okamoto, I., Yabuta, Y., Iwatani, C., Tsuchiya, H., et al. (2016). The germ cell fate of cynomolgus monkeys is specified in the nascent amnion. *Developmental Cell*, *39*, 169–185.

Schmaltz-Panneau, B., Jouneau, L., Osteil, P., Tapponnier, Y., Afanassieff, M., Moroldo, M., et al. (2014). Contrasting transcriptome landscapes of rabbit pluripotent stem cells in vitro and in vivo. *Animal Reproduction Science*, *149*, 67–79.

Smith, A. (2017). Formative pluripotency: The executive phase in a developmental continuum. *Development*, *144*, 365–373.

Smith, A. G., Heath, J. K., Donaldson, D. D., Wong, G. G., Moreau, J., Stahl, M., et al. (1988). Inhibition of pluripotential embryonic stem cell differentiation by purified polypeptides. *Nature*, *336*, 688–690.

Stevens, L. C. (1962). The biology of teratomas including evidence indicating their origin form primordial germ cells. *L'Annèe Biologique*, *1*, 585–610.

Stevens, L. C. (1970). The development of transplantable teratocarcinomas from intratesticular grafts of pre- and postimplantation mouse embryos. *Developmental Biology*, *21*, 364–382.

Stuart, H. T., van Oosten, A. L., Radzisheuskaya, A., Martello, G., Miller, A., Dietmann, S., et al. (2014). NANOG amplifies STAT3 activation and they synergistically induce the naive pluripotent program. *Current Biology*, *24*, 340–346.

Sumi, T., Oki, S., Kitajima, K., & Meno, C. (2013). Epiblast ground state is controlled by canonical Wnt/beta-catenin signaling in the postimplantation mouse embryo and epiblast stem cells. *PLoS One*, *8*, e63378.

Takahashi, K., & Yamanaka, S. (2006). Induction of pluripotent stem cells from mouse embryonic and adult fibroblast cultures by defined factors. *Cell*, *126*, 663–676.

Takashima, Y., Guo, G., Loos, R., Nichols, J., Ficz, G., Krueger, F., et al. (2014). Resetting transcription factor control circuitry toward ground-state pluripotency in human. *Cell*, *158*, 1254–1269.

Tan, G., Ren, L., Huang, Y., Tang, X., Zhou, Y., Zhou, Y., et al. (2012). Isolation and culture of embryonic stem-like cells from pig nuclear transfer blastocysts of different days. *Zygote, 20*, 347–352.
Telugu, B. P., Ezashi, T., & Roberts, R. M. (2010). Porcine induced pluripotent stem cells analogous to naive and primed embryonic stem cells of the mouse. *The International Journal of Developmental Biology, 54*, 1703–1711.
Telugu, B. P., Ezashi, T., Sinha, S., Alexenko, A. P., Spate, L., Prather, R. S., et al. (2011). Leukemia inhibitory factor (LIF)-dependent, pluripotent stem cells established from inner cell mass of porcine embryos. *The Journal of Biological Chemistry, 286*, 28,948–28,953.
ten Berge, D., Kurek, D., Blauwkamp, T., Koole, W., Maas, A., Eroglu, E., et al. (2011). Embryonic stem cells require Wnt proteins to prevent differentiation to epiblast stem cells. *Nature Cell Biology, 13*, 1070–1075.
Tesar, P. J., Chenoweth, J. G., Brook, F. A., Davies, T. J., Evans, E. P., Mack, D. L., et al. (2007). New cell lines from mouse epiblast share defining features with human embryonic stem cells. *Nature, 448*, 196–199.
Theunissen, T. W., Friedli, M., He, Y., Planet, E., O'Neil, R. C., Markoulaki, S., et al. (2016). Molecular criteria for defining the naive human pluripotent state. *Cell Stem Cell, 19*, 502–515.
Theunissen, T. W., Powell, B. E., Wang, H., Mitalipova, M., Faddah, D. A., Reddy, J., et al. (2014). Systematic identification of culture conditions for induction and maintenance of naive human pluripotency. *Cell Stem Cell, 15*, 471–487.
Thomson, J. A., Itskovitz-Eldor, J., Shapiro, S. S., Waknitz, M. A., Swiergiel, J. J., Marshall, V. S., et al. (1998). Embryonic stem cell lines derived from human blastocysts. *Science, 282*, 1145–1147.
Thomson, J. A., Kalishman, J., Golos, T. G., Durning, M., Harris, C. P., Becker, R. A., et al. (1995). Isolation of a primate embryonic stem cell line. *Proceedings of the National Academy of Sciences of the United States of America, 92*, 7844–7848.
Thomson, A. J., Pierart, H., Meek, S., Bogerman, A., Sutherland, L., Murray, H., et al. (2012). Reprogramming pig fetal fibroblasts reveals a functional LIF signaling pathway. *Cellular Reprogramming, 14*, 112–122.
Toyooka, Y., Shimosato, D., Murakami, K., Takahashi, K., & Niwa, H. (2008). Identification and characterization of subpopulations in undifferentiated ES cell culture. *Development, 135*, 909–918.
Usui, J., Kobayashi, T., Yamaguchi, T., Knisely, A. S., Nishinakamura, R., & Nakauchi, H. (2012). Generation of kidney from pluripotent stem cells via blastocyst complementation. *The American Journal of Pathology, 180*, 2417–2426.
Vaags, A. K., Rosic-Kablar, S., Gartley, C. J., Zheng, Y. Z., Chesney, A., Villagomez, D. A., et al. (2009). Derivation and characterization of canine embryonic stem cell lines with in vitro and in vivo differentiation potential. *Stem Cells, 27*, 329–340.
Vallier, L., Alexander, M., & Pedersen, R. A. (2005). Activin/Nodal and FGF pathways cooperate to maintain pluripotency of human embryonic stem cells. *Journal of Cell Science, 118*, 4495–4509.
van den Berg, D. L., Zhang, W., Yates, A., Engelen, E., Takacs, K., Bezstarosti, K., et al. (2008). Estrogen-related receptor beta interacts with Oct4 to positively regulate Nanog gene expression. *Molecular and Cellular Biology, 28*, 5986–5995.
Van der Jeught, M., Heindryckx, B., O'Leary, T., Duggal, G., Ghimire, S., Lierman, S., et al. (2014). Treatment of human embryos with the TGFbeta inhibitor SB431542 increases epiblast proliferation and permits successful human embryonic stem cell derivation. *Human Reproduction, 29*, 41–48.
Van der Jeught, M., O'Leary, T., Ghimire, S., Lierman, S., Duggal, G., Versieren, K., et al. (2013). The combination of inhibitors of FGF/MEK/Erk and GSK3beta signaling

increases the number of OCT3/4- and NANOG-positive cells in the human inner cell mass, but does not improve stem cell derivation. *Stem Cells and Development*, *22*, 296–306.

Verma, V., Huang, B., Kallingappa, P. K., & Oback, B. (2013). Dual kinase inhibition promotes pluripotency in finite bovine embryonic cell lines. *Stem Cells and Development*, *22*, 1728–1742.

Wall, R. J., & Seidel, G. E., Jr. (1992). Transgenic farm animals—A critical analysis. *Theriogenology*, *38*, 337–357.

Walters, E. M., Wells, K. D., Bryda, E. C., Schommer, S., & Prather, R. S. (2017). Swine models, genomic tools and services to enhance our understanding of human health and diseases. *Lab Animal*, *46*, 167–172.

Wang, J., Gu, Q., Hao, J., Jia, Y., Xue, B., Jin, H., et al. (2013). Tbx3 and Nr5alpha2 play important roles in pig pluripotent stem cells. *Stem Cell Reviews*, *9*, 700–708.

Wang, W., Yang, J., Liu, H., Lu, D., Chen, X., Zenonos, Z., et al. (2011). Rapid and efficient reprogramming of somatic cells to induced pluripotent stem cells by retinoic acid receptor gamma and liver receptor homolog 1. *Proceedings of the National Academy of Sciences of the United States of America*, *108*, 18283–18288.

West, F. D., Terlouw, S. L., Kwon, D. J., Mumaw, J. L., Dhara, S. K., Hasneen, K., et al. (2010). Porcine induced pluripotent stem cells produce chimeric offspring. *Stem Cells and Development*, *19*, 1211–1220.

West, F. D., Uhl, E. W., Liu, Y., Stowe, H., Lu, Y., Yu, P., et al. (2011). Brief report: Chimeric pigs produced from induced pluripotent stem cells demonstrate germline transmission and no evidence of tumor formation in young pigs. *Stem Cells*, *29*, 1640–1643.

Williams, R. L., Hilton, D. J., Pease, S., Willson, T. A., Stewart, C. L., Gearing, D. P., et al. (1988). Myeloid leukaemia inhibitory factor maintains the developmental potential of embryonic stem cells. *Nature*, *336*, 684–687.

Wray, J., Kalkan, T., Gomez-Lopez, S., Eckardt, D., Cook, A., Kemler, R., et al. (2011). Inhibition of glycogen synthase kinase-3 alleviates Tcf3 repression of the pluripotency network and increases embryonic stem cell resistance to differentiation. *Nature Cell Biology*, *13*, 838–845.

Wu, J., Okamura, D., Li, M., Suzuki, K., Luo, C., Ma, L., et al. (2015). An alternative pluripotent state confers interspecies chimaeric competency. *Nature*, *521*, 316–321.

Wu, J., Platero-Luengo, A., Sakurai, M., Sugawara, A., Gil, M. A., Yamauchi, T., et al. (2017). Interspecies chimerism with mammalian pluripotent stem cells. *Cell*, *168*, 473-486 e415.

Wu, X., Song, M., Yang, X., Liu, X., Liu, K., Jiao, C., et al. (2016). Establishment of bovine embryonic stem cells after knockdown of CDX2. *Scientific Reports*, *6*, 28343.

Xue, Z., Huang, K., Cai, C., Cai, L., Jiang, C. Y., Feng, Y., et al. (2013). Genetic programs in human and mouse early embryos revealed by single-cell RNA sequencing. *Nature*, *500*, 593–597.

Xue, B., Li, Y., He, Y., Wei, R., Sun, R., Yin, Z., et al. (2016). Porcine pluripotent stem cells derived from IVF embryos contribute to chimeric development in vivo. *PLoS One*, *11*, e0151737.

Yagi, M., Kishigami, S., Tanaka, A., Semi, K., Mizutani, E., Wakayama, S., et al. (2017). Derivation of ground-state female ES cells maintaining gamete-derived DNA methylation. *Nature*, *548*, 224–227.

Yamanaka, Y., Lanner, F., & Rossant, J. (2010). FGF signal-dependent segregation of primitive endoderm and epiblast in the mouse blastocyst. *Development*, *137*, 715–724.

Yan, L., Yang, M., Guo, H., Yang, L., Wu, J., Li, R., et al. (2013). Single-cell RNA-Seq profiling of human preimplantation embryos and embryonic stem cells. *Nature Structural & Molecular Biology*, *20*, 1131–1139.

Yang, Y., Liu, B., Xu, J., Wang, J., Wu, J., Shi, C., et al. (2017). Derivation of pluripotent stem cells with in vivo embryonic and extraembryonic potency. *Cell*, *169*, 243-257 e225.

Yao, J., Huang, J., & Zhao, J. (2016). Genome editing revolutionize the creation of genetically modified pigs for modeling human diseases. *Human Genetics*, *135*, 1093–1105.
Ye, S., Li, P., Tong, C., & Ying, Q. L. (2013). Embryonic stem cell self-renewal pathways converge on the transcription factor Tfcp2l1. *The EMBO Journal*, *32*, 2548–2560.
Ying, Q. L., Nichols, J., Chambers, I., & Smith, A. (2003). BMP induction of Id proteins suppresses differentiation and sustains embryonic stem cell self-renewal in collaboration with STAT3. *Cell*, *115*, 281–292.
Ying, Q. L., Wray, J., Nichols, J., Batlle-Morera, L., Doble, B., Woodgett, J., et al. (2008). The ground state of embryonic stem cell self-renewal. *Nature*, *453*, 519–523.
Zhang, W., Pei, Y., Zhong, L., Wen, B., Cao, S., & Han, J. (2015). Pluripotent and metabolic features of two types of porcine iPSCs derived from defined mouse and human ES cell culture conditions. *PLoS One*, *10*, e0124562.
Zhao, Y., Lin, J., Wang, L., Chen, B., Zhou, C., Chen, T., et al. (2011). Derivation and characterization of ovine embryonic stem-like cell lines in semi-defined medium without feeder cells. *Journal of Experimental Zoology. Part A, Ecological Genetics and Physiology*, *315*, 639–648.
Zimmerlin, L., Park, T. S., & Zambidis, E. T. (2017). Capturing human naive pluripotency in the embryo and in the dish. *Stem Cells and Development*, *26*, 1141–1161.

CHAPTER EIGHT

Capturing and Interconverting Embryonic Cell Fates in a Dish

Jennifer Watts*,†,‡,2, Alyson Lokken*,2, Alexandra Moauro*,‡, Amy Ralston*,†,1

*Michigan State University, East Lansing, MI, United States
†Program in Reproductive and Developmental Sciences, Michigan State University, East Lansing, MI, United States
‡Graduate Program in Physiology, Michigan State University, East Lansing, MI, United States
[1]Corresponding author: e-mail address: aralston@msu.edu

Contents

Abstract

Cells of the early embryo are totipotent because they will differentiate to produce the fetus and its surrounding extraembryonic tissues. By contrast, embryonic stem (ES) cells are considered to be merely pluripotent because they lack the ability to efficiently produce extraembryonic cell types. The relatively limited developmental potential of ES cells can be explained by the observation that ES cells are derived from the embryo after its cells have already begun to specialize and lose totipotency. Meanwhile, at the time that pluripotent ES cell progenitors are specified, so are the multipotent progenitors of two extraembryonic stem cell types: trophoblast stem (TS) cells and extraembryonic endoderm stem (XEN) cells. Notably, all three embryo-derived stem cell types are capable of either self-renewing or differentiating in a lineage-appropriate manner. These three types of embryo-derived stem cell serve as paradigms for defining the genes and pathways that define and maintain unique stem cell identities. Remarkably, some

[2] Equal contribution.

Current Topics in Developmental Biology, Volume 128
ISSN 0070-2153
https://doi.org/10.1016/bs.ctdb.2017.11.008

of the mechanisms that maintain the specific developmental potential of each stem cell line do so by preventing conversion to another stem cell fate. This chapter highlights noteworthy studies that have identified the genes and pathways that normally limit the interconversion of stem cell identities.

1. INTRODUCTION

Very early in mammalian embryogenesis, cells make decisions to produce either the fetus or the extraembryonic tissues of the placenta and yolk sac. Failure to properly execute cell fate decisions can result in miscarriage, birth defects, and can even lead to long-term health issues in the adult. It is now widely appreciated that cell fates must be actively maintained, and that a failure to maintain cell fate can lead cells to adopt aberrant phenotypes. Progress toward elucidating genes important for directing cell fate decisions and maintaining cellular phenotypes has been provided by analyses in embryo models. Additionally, our understanding of the molecular underpinnings of cell fate has been substantially advanced by the study of stem cell lines that represent the fetal and extraembryonic lineages in vitro. As an experimental system, stem cell lines provide many of the benefits of studying embryos because they can be differentiated to a variety of mature endpoints. Yet, stem cells provide an advantage over embryo models because they can be expanded to provide massive cellular quantities, which are more limited in embryos. Accordingly, stem cell lines have been used to identify factors that normally initiate and maintain cell identities during embryogenesis.

Some of the first paradigms for capturing and preserving specific developmental cell fates in vitro included pluripotent stem cell lines, such as embryonal carcinoma (EC) (Kelly & Gatie, 2017) and embryonic stem (ES) cell lines (Evans & Kaufman, 1981; Martin, 1981). These pluripotent cell lines made possible the expansion of largely pure populations with which to perform controlled studies of differentiation. Additionally, pluripotent cell lines provided precedent that specific embryonic cell states could indeed be captured and preserved in vitro. The subsequent derivation of epiblast stem cells (EpiSCs) from later-stage embryos (Brons et al., 2007; Tesar et al., 2007) demonstrated that pluripotent stem cell progenitors could be propagated from multiple developmental stages.

While pluripotent stem cells can differentiate into any mature cell type of the body, they are incapable of efficiently producing extraembryonic cell types of the trophoblast and extraembryonic endoderm lineages (Beddington & Robertson, 1989). Nevertheless, this limitation is mitigated by the existence of extraembryonic stem cell lines, including trophoblast stem (TS) and extraembryonic endoderm stem (XEN) cells, which have been derived from pre- and postimplantation stage embryos (Kunath et al., 2005; Lin, Khan, Zapiec, & Mombaerts, 2016; Tanaka, Kunath, Hadjantonakis, Nagy, & Rossant, 1998). Like ES cells, TS and XEN stem cells are capable of either self-renewing or differentiating to more mature, lineage-appropriate endpoints in response to extrinsic cues. Extraembryonic stem cell lines have enabled researchers to learn critical lessons regarding how extraembryonic cell fates are specified and maintained during development, and provide key insight into the mechanisms that enable stem cells to maintain their lineage-specific developmental potential.

2. MECHANISMS REPRESSING TS CELL FATE IN ES CELLS

During embryonic development, the trophoblast lineage is the first lineage to be specified, beginning as the trophectoderm of the blastocyst, and then gradually differentiating to produce multiple types of differentiated cell. The ultimate goal of the trophoblast lineage is to connect with extraembryonic mesoderm-derived umbilical cord and produce a functioning placenta (Fig. 1). Given the fundamental importance of the placenta in fetal health, understanding the origins of the trophoblast lineage has been a central goal in reproductive and developmental biology. In this regard, the mouse has an advantage over many mammalian models because self-renewing, multipotent TS cells can be derived from mouse embryos (Tanaka et al., 1998). TS cell lines are considered *bona fide* stem cell lines because they can either self-renew in the presence of fibroblast growth factor 4 (FGF4) and Activin or TGFβ (Erlebacher, Price, & Glimcher, 2004; Kubaczka et al., 2014), or differentiate on withdrawal of self-renewal factors (Tanaka et al., 1998). The establishment of human TS cell lines would provide a valuable research tool for studying trophoblast development and differentiation, but efforts to derive human TS cell lines from human blastocysts have not been successful (Roberts & Fisher, 2011; Rossant, 2015). Therefore, genetic studies in mouse TS cells have, and will continue, to serve as the preeminent model for identifying genetic mechanisms that define TS cell identity.

While ES cells are pluripotent, and can give rise to any cell type in the body, it has long been appreciated that ES cells lack the intrinsic potential to efficiently produce cell types of the trophoblast lineage (Beddington & Robertson, 1989). Thus, even though ES and TS cells can be derived from the embryo at the same stage, these two cell lines are complementary in

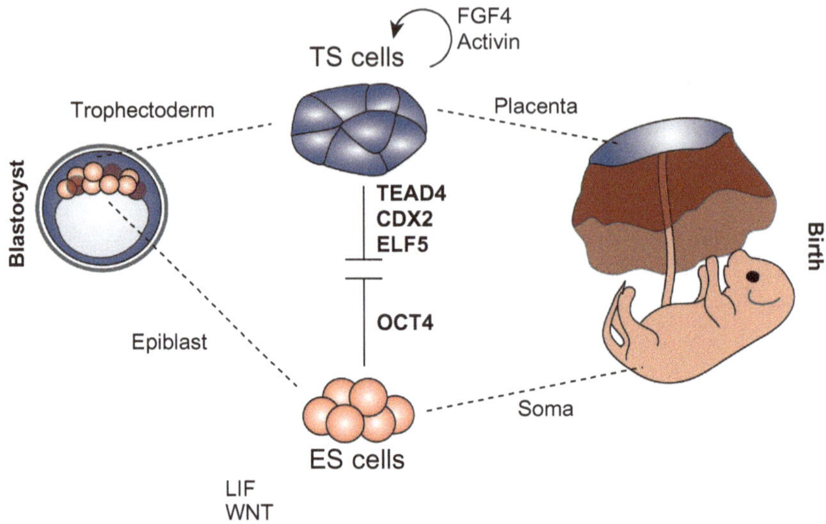

Fig. 1 Origins and regulation of ES and TS cell lines. The first lineage decision in development occurs prior to blastocyst stage and results in trophectoderm and inner cell mass (ICM). In mice, trophoblast stem (TS) cells can be derived from the trophoblast lineage as early as the blastocyst stage. The trophoblast is destined to produce extraembryonic tissues including the placenta. TS cells can self-renew in the presence of cues such as Fibroblast Growth Factor 4 (FGF4) and Activin. TS cells retain the capacity to differentiate in a lineage-appropriate manner on withdrawal of FGF4 or when introduced into host blastocysts. Embryonic stem (ES) cells are derived from a subset of cells present in the ICM, which are destined to produce the fetus, the umbilical cord, and portions of the surrounding fetal membranes. ES cells can self-renew in the presence of factors such as leukemia inhibitory factor (LIF) and WNT (Nichols et al., 1998; Willert et al., 2003; Williams et al., 1988; Ying et al., 2008), or to differentiate on withdrawal of these factors or when introduced into host blastocysts. Genes and pathways that reinforce the lineage-specific developmental potential of TS and ES cell lines include TEAD4, CDX2, ELF5, and OCT4.

terms of developmental potential. This observation is consistent with the hypothesis that progenitors of ES and TS cell lines are specified in parallel during development, as opposed to one lineage deriving from the other. Additionally, this observation is consistent with the hypothesis that a genetic barrier must exist within ES and TS cells, which prevents these two distinct cell fates from interconverting. Exciting studies in the last two decades have tested these hypotheses by identifying genes that are involved in establishing and maintaining the ES/TS lineage barrier.

One of the first studies to identify factors limiting conversion of ES cells to trophoblast was the pluripotency-promoting transcription factor OCT4 (encoded by *Pou5f1*) (Niwa, Miyazaki, & Smith, 2000; Niwa et al., 2005). Conditional inactivation of *Oct4* in ES cells led to the expression of TS cell genes such as *Cdx2* and *Eomesodermin* (*Eomes*) and to acquisition of trophoblast cell morphologies. Supplementation of *Oct4*-inactivated ES cells with FGF4 and fibroblast-conditioned medium, which support TS cell self-renewal (Tanaka et al., 1998), enabled *Oct4*-depleted ES cell lines to proliferate as TS-like cells (Niwa et al., 2000). These observations suggested that the pluripotency pathway overrides trophoblast fate in ES cells. Consistent with this interpretation, OCT4 can override trophoblast fate and induce pluripotency in TS cells on its own (Wu et al., 2011), or in the presence of other pluripotency-promoting factors (Kuckenberg et al., 2011). Therefore, OCT4 is both necessary and sufficient to define pluripotency in stem cell lines.

The requirement for other pluripotency factors in repressing TS cell fate in ES cells has also been investigated. The pluripotency gene *Sox2* has also been shown to be essential for repressing acquisition of TS cell fate in ES cells (Ivanova et al., 2006; Li et al., 2007; Masui et al., 2007). However, SOX2 has been shown to be essential for maintaining expression of *Oct4* in ES cells, arguing that SOX2 represses TS cell fate indirectly, via OCT4. By contrast, the pluripotency gene *Nanog* plays no role in repressing TS cell gene expression or cell fate (Chambers et al., 2007). Therefore, it is likely that only a subset of pluripotency genes represses trophoblast cell fate, especially those that act to stabilize expression of OCT4. For example, several *Cyclin* genes have been shown to limit acquisition of TS cell fate in ES cells, likely acting through OCT4 (Liu et al., 2017). Interestingly, this study also showed that Cyclins could phosphorylate OCT4, limiting OCT4 ubiquitination, and stabilizing OCT4 expression in ES cells.

The observation that *Cdx2* and *Eomes* are both upregulated in *Oct4*-inactivated ES cells led to the investigation of the effects of overexpression of each of these genes in ES cell lines. While overexpression of either *Cdx2* or *Eomes* induced formation of TS-like cells, only *Cdx2*-expressing cells were able to contribute to placenta development in chimeras (Niwa et al., 2005). Therefore, CDX2 is more instructive than EOMES in the conversion of ES to TS-like cells. The mechanism by which CDX2 drives trophoblast cell fate in ES cells is proposed to be by binding to OCT4 and preventing OCT4 from reinforcing its own expression (Niwa et al., 2005). These observations support the idea that CDX2 is a master regulator of trophoblast fate, which is consistent with the requirement for *Cdx2* in the embryo during establishment of TS cell progenitors (Strumpf et al., 2005). Initially, it was unclear whether the only role for CDX2 in TS cells is to repress OCT4, or if CDX2 could have additional essential functions. To test this, *Cdx2* and *Oct4* were simultaneously deleted in ES cell lines, with the expectation that, if the only role of CDX2 is to repress OCT4, then *Cdx2* should not be needed for cells to acquire or maintain TS cell fate in *Oct4* null ES cells. Interestingly, *Cdx2*; *Oct4* double knockout ES cell lines adopted trophoblast fate, but could not self-renew in the presence of FGF4 (Niwa et al., 2005). Therefore, CDX2 is thought to play two main roles in TS cells: repressing OCT4 and maintaining stem cell proliferation downstream of FGF4.

Given the central importance of OCT4/CDX2 in maintaining the ES/TS cell lineage barrier, much interest has focused on identifying factors that initially activate expression of *Cdx2* when the trophoblast lineage is first specified in the embryo. This led to the identification of TEAD4, which is essential for the initial expression of *Cdx2* in early embryos (Nishioka et al., 2008; Yagi et al., 2007). In addition, TEAD4 is sufficient to drive conversion of ES cell lines to TS-like cells, when fused to the transactivation domain of the viral protein VP16 (Nishioka et al., 2009). Therefore, genetic networks at play in the early embryo are also functional within ES cell lines. However, these studies were unable to provide insight into how similar converted ES cells are to embryo-derived TS cell lines. There are many ways to compare embryo- and ES cell-derived TS cell lines, including morphology, functional assays, transcriptome, and chromatin modifications. However, it is not necessarily obvious which of these readouts would provide the best measure of the authentic TS cell properties, nor what degree of difference would indicate biologically meaningful variation. Nevertheless, differences in all these readouts have been noted among ES cell-derived TS cell lines (Cambuli et al., 2014).

Whether there are other factors capable of inducing TS cell fate in ES cells downstream of TEAD4 is supported by analysis of the role of TEAD4 in vivo. For example, the embryonic *Tead4* null phenotype is more severe than is the embryonic *Cdx2* null phenotype, indicating that TEAD4 promotes expression of multiple trophectoderm regulators in parallel, and pointing to the existence of additional genes important for defining the ES/TS cell lineage barrier. One example is *Gata3*, which is regulated by TEAD4 and not CDX2, and is sufficient to induce trophoblast gene expression in ES cell lines (Ralston et al., 2010). While GATA3 was able to induce many trophoblast genes independently of CDX2, GATA3 was not sufficient to induce formation of self-renewing TS cell lines. Therefore, TEAD4 is likely to regulate expression of multiple downstream genes, each with differing roles in promoting trophoblast gene expression, proliferation, and establishment of stem cell progenitors.

Beyond transcription factors, other regulators of TS cell fate have been identified. For example, Wnt and Collagen have also been shown to influence TS-like differentiation of ES cells (He, Pant, Schiffmacher, Meece, &

Keefer, 2008; Schenke-Layland et al., 2007). In addition, the evidence that transcription factors help define the ES/TS cell lineage barrier led researchers to investigate how lineage-specific expression of transcription factors is achieved, and whether chromatin modification plays a role. For instance, ES cells deficient in DNA methylation pathway members such as *Dnmt1* acquired TS cell-like properties, indicating that DNA methylation normally represses expression of one or more trophoblast genes in ES cell lines (Ng et al., 2008). Interestingly, DNA methylation does not regulate expression of *Cdx2* or *Eomes* in ES cells. Rather, DNA methylation regulates expression of the trophoblast transcription factor *Elf5*, which is itself capable of inducing expression of *Cdx2* and *Eomes* when overexpressed in ES cell lines (Ng et al., 2008). During development, *Elf5* plays an essential role in maintaining the trophoblast lineage after implantation (Donnison et al., 2005) and is therefore later-acting than CDX2 and EOMES. Based on these observations, ELF5 is proposed to act as a gatekeeper of trophoblast fate, acting downstream of DNA methylation, which represses expression of trophoblast genes in ES cell lines.

Besides DNA methylation, additional epigenetic mechanisms have been identified that reinforce the ES/TS cell lineage barrier. For instance, TS cell-expressed microRNAs (miRNAs) were identified that induced formation of TS cell-like colonies, when transiently overexpressed in ES cell lines in the presence of the histone deacetylase inhibitor valproic acid (VPA) (Nosi, Lanner, Huang, & Cox, 2017). Curiously, however, miRNA-induced TS-like cells were not able to differentiate in vitro. When introduced into embryos, miRNA-induced TS-like cell lines contributed mainly to the mural trophectoderm, which is located distal to the ICM in the blastocyst and destined to produce Reichert's Membrane. Taken together, these observations suggest that the transient overexpression of trophoblast miRNAs leads ES cells to acquire a blastocyst-like trophoblast phenotype. The targets of the TS-inducing miRNAs are predicted to include known ES cell maintenance genes *Sall1*, *Sall4*, *Ccnd1*, *Ccdn2*, and *Lin28* (Nosi et al., 2017). However, it is unclear whether loss of one or more of these targets from ES cell lines would also produce TS-like cells with Reichert's Membrane-oriented developmental potential.

3. REPROGRAMMING AND THE TS CELL LINEAGE BARRIER

The discovery that factors such as CDX2 are sufficient to convert ES cells to TS-like cells did not address whether ES cells are uniquely responsive to TS cell-inducing factors, or whether TS cell-inducing factors are more broadly capable of inducing TS cell fate in other cell types. The ability of CDX2 to induce TS cell fate in other cell types has been investigated. One study focused on the ability of CDX2 to override pluripotency and induce trophoblast fate in pluripotent cells such as EC cells and iPS cells. Notably, CDX2 was able to induce TS-like cells in EC and iPS cells, arguing that EC and iPS cells are equivalent in their ability to respond to ectopic CDX2 (Blij, Parenti, Tabatabai-Yazdi, & Ralston, 2015). However, CDX2 was unable to induce TS cell fate in EpiSCs (Blij et al., 2015). Rather, CDX2 is thought to induce mesoderm formation in EpiSCs (Bernardo et al., 2011). These observations indicate that multiple pluripotent states exist, which can be defined functionally in terms of the response to overexpressed *Cdx2*.

Two subsequent studies investigated the abilities of trophoblast factors, including CDX2, to induce TS cell fate in differentiated somatic cells. The idea of identifying transcription factors capable of converting somatic cells to stem cells was first realized by screening a library of transcription factors to identify factors capable of reversing differentiation in somatic cells to produce induced pluripotent stem (iPS) cells (Takahashi & Yamanaka, 2006). Similarly, by screening a library of candidate trophoblast genes, four factors, *Eomes*, *Gata3*, *Tfap2c*, and *Ets2*, were identified as being sufficient to induce formation of TS-like cells (iTS cells) in mouse embryonic and adult tail tip fibroblasts (Benchetrit et al., 2015; Kubaczka et al., 2015). Interestingly, these same four factors were not able to convert ES cells to iTS cells (Kubaczka et al., 2015), indicating that the cellular context in which factors are overexpressed matters and can influence transcription factor activity. Similarly, *Cdx2* and *Elf5* were not able to induce TS cell fate in somatic cells (Benchetrit et al., 2015). One way to make sense of this observation is to consider the mechanisms by which ELF5 and CDX2 induce TS cell fate in ES cell lines. For example, CDX2 is thought mainly to induce TS cell fate in ES cells by repressing expression of OCT4 (Niwa et al., 2005). Similarly, ELF5 promotes expression of CDX2 in ES cells (Ng et al., 2008), which in turn represses OCT4. Since OCT4 is not present in fibroblasts, CDX2 and ELF5 lack TS cell-inducing ability in fibroblasts.

4. MECHANISMS REPRESSING XEN CELL FATE IN ES CELLS

During embryo development, extraembryonic endoderm is the second lineage to be specified and follows specification of the trophoblast lineage (Fig. 2A). Mature cells of the extraembryonic endoderm lineage play

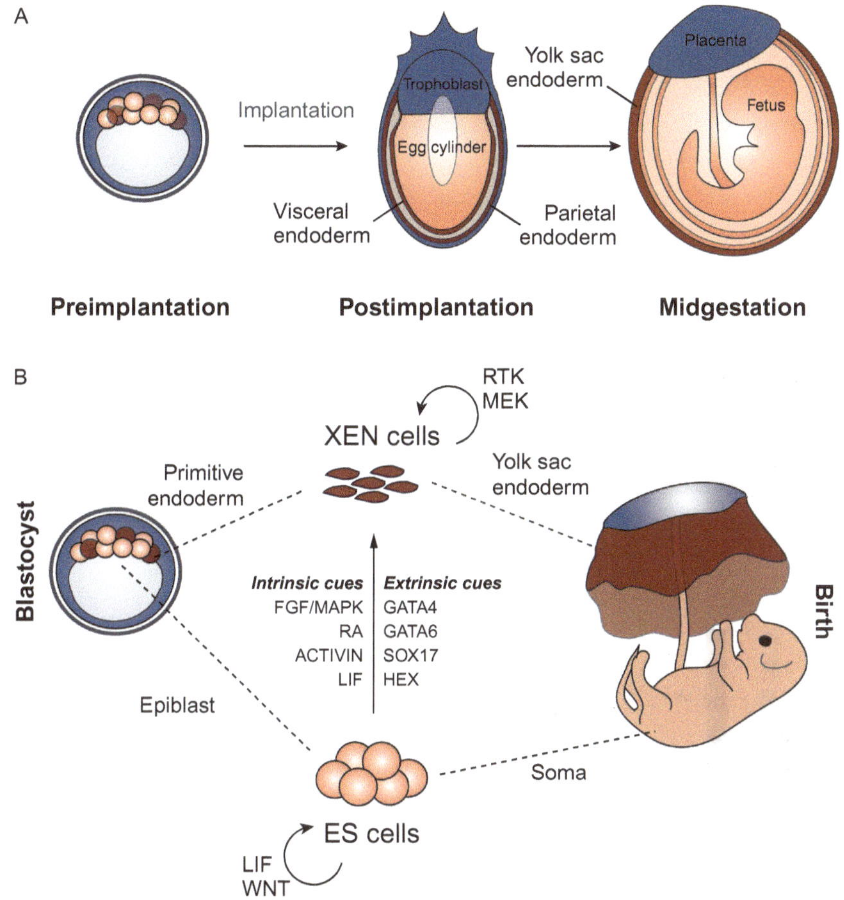

Fig. 2 Origins and regulation of ES and XEN cell lines. (A) Overview of extraembryonic endoderm development. Extraembryonic endoderm is first specified prior to implantation, in the blastocyst ICM. Following embryo implantation into the uterus, the primitive endoderm differentiates into parietal endoderm, which comprises Reichert's membrane, and visceral endoderm, which contacts and patterns the epiblast of the egg cylinder. The visceral endoderm goes on to comprise yolk sac endoderm midgestation. (B) The second lineage decision in development occurs during the blastocyst stage and produces epiblast, the progenitors of ES cells, and primitive endoderm, progenitors of XEN cells. XEN cells can either self-renew, in the presence of receptor tyrosine kinase (RTK) signaling, or give rise to differentiated cell types of the extraembryonic lineage. While ES cells give rise to extraembryonic endoderm-like cells at low efficiency, ES cells can be coaxed to a XEN-like state by factors that function intrinsically or extrinsically.

myriad roles in development, including nourishing the developing embryo, patterning the developing epiblast as visceral endoderm, inducing primitive hematopoiesis, contributing to the yolk sac, and even the definitive endoderm (Moerkamp et al., 2013). XEN cells are derived from the primitive endoderm in the blastocyst stage (Kunath et al., 2005) (Fig. 2B) or from the postimplantation extraembryonic endoderm (Lin et al., 2016), and are a useful in vitro tool for studying the differentiation of and inductive roles of the extraembryonic lineage (Artus et al., 2012; Brown et al., 2010; Paca et al., 2012). XEN cells self-renew when grown in the presence of serum, without need for additional growth factors. XEN cell self-renewal is thought to be dependent on PDGFRA-dependent activation of ERK (Artus, Panthier, & Hadjantonakis, 2010), while differentiation of XEN cells is guided by alternative pathways involving bone morphogenetic protein 2 (BMP2) (Artus et al., 2012; Paca et al., 2012).

While ES cells are extremely limited in their ability to produce TS cells in the absence of genetic manipulation, ES cells give rise more readily to primitive endoderm-like cells in response to appropriate extrinsic cues. The first evidence that pluripotent ES cells could give rise to extraembryonic endodermal cell types was through experiments with embryoid bodies, which result when ES cells are grown in suspension culture in the absence of the self-renewal factor leukemia inhibitory factor (LIF) (Brickman & Serup, 2017; Martin & Evans, 1975). As the embryoid body grows in size, the inner cells of the embryoid body express markers of the pluripotent epiblast lineage, whereas the outside cells epithelialize and express markers of the primitive endoderm lineage. While these outer cells express markers of extraembryonic endoderm cells, they fail to give rise to stable, self-renewing XEN cell lines (Coucouvanis & Martin, 1995).

Evidence that ES cells can spontaneously give rise to XEN-like cells has been provided by careful analyses of ES cell cultures, which showed that endodermal markers such as SOX17, GATA6, and HEX are expressed in subsets of ES cells (Canham, Sharov, Ko, & Brickman, 2010; Hamilton & Brickman, 2014; Niakan et al., 2010). Moreover, upon injection into blastocysts, these endoderm gene-expressing cells found in ES cell cultures contribute to the extraembryonic endoderm lineage (Canham et al., 2010; Niakan et al., 2010). Notably, LIF can enhance formation of XEN cell progenitors in vivo (Morgani et al., 2013). However, exogenous LIF is not needed for maintaining XEN cells in vitro, underscoring differences in pathways regulating the establishment of stem cell progenitors and the maintenance of bona fide stem cells.

Knowledge that XEN cells can spontaneously arise from ES cells, either in culture or through embryoid body formation points to the existence of specific signals that can direct ES cells to differentiate to XEN cells. Consistent with this proposal, expression of endodermal genes is repressed in ES cell cultures by interfering with FGF/ERK signaling (Canham et al., 2010; Hamilton & Brickman, 2014), consistent with the role of FGF/ERK signaling in promoting primitive endoderm development in the blastocyst (Chazaud, Yamanaka, Pawson, & Rossant, 2006; Kang, Garg, & Hadjantonakis, 2017; Kang, Piliszek, Artus, & Hadjantonakis, 2013; Molotkov, Mazot, Brewer, Cinalli, & Soriano, 2017; Nichols, Silva, Roode, & Smith, 2009; Yamanaka, Lanner, & Rossant, 2010).

While repressing acquisition of XEN cell fate in ES cell cultures can be advantageous, efforts have also focused on identifying ways to enrich XEN cell fate in ES cell cultures. Lessons from classical studies of EC differentiation pointed to a possible role for retinoic acid (RA) in extraembryonic endoderm differentiation. For example, F9 EC cells treated with RA in monolayer culture differentiate to primitive endoderm, whereas treatment with RA and dibutyryl cyclic AMP results in parietal endoderm differentiation (Strickland, Smith, & Marotti, 1980). Additionally, F9 EC cells grown as embryoid bodies in the presence of RA adopt to visceral endoderm cell fate (Hogan, Taylor, & Adamson, 1981). While F9 EC cells treated with RA exhibit morphological and molecular characteristics similar to embryo-derived XEN cells, it is important to note that as a result of their malignant origin, F9 EC cells often exhibit aneuploidy and do not reliably contribute to development in chimeric embryos (Cronmiller & Mintz, 1978; Papaioannou, Evans, Gardner, & Graham, 1979). Thus, the developmental potential of EC-derived XEN cells in vivo has not been tested.

Similar to its role in EC cells, RA can, in the presence of Activin, increase the proportion of ES cells that convert to XEN cells, a cell typed named cXEN (converted XEN) (Cho et al., 2012). These cXEN cells not only morphologically resemble embryo-derived XEN cells but also are molecularly indistinguishable from embryo-derived XEN cells and are able to differentiate to visceral endoderm upon BMP-induced differentiation, as well as contribute to the parietal endoderm when injected into blastocysts. These results suggested that ES cells, or a subpopulation of ES cells, can be prompted to differentiate to XEN cells upon receiving the right exogenous cues, in this case Activin/RA.

Just as FGF4/ERK signaling is essential for the spontaneous formation of XEN-like cells in ES cell cultures, FGF4/ERK signaling is also required for the Activin/RA-driven conversion of ES cells to XEN-like cells. Addition of Activin/RA was not sufficient to induce conversion of *FGF4* null ES cells to XEN cells nor were Activin/RA sufficient to induce conversion of ES cells to XEN cells in the presence of a MEK inhibitor (Cho et al., 2012). However, the role of FGF4/ERK signaling may be relatively late in the conversion process (Schröter, Rué, Mackenzie, & Martinez Arias, 2015). Thus, multiple extrinsic factors regulate the conversion of ES cells to XEN-like cells.

While Activin/RA induce XEN cell fate in ES cells, XEN cells cannot be derived from EpiSCs in the same way (Cho et al., 2012). This observation is consistent with differing developmental properties of ES cells and EpiSCs in vivo (Brons et al., 2007; Tesar et al., 2007), and in response trophoblast-inducing signals (Bernardo et al., 2011; Blij et al., 2015). Thus, neither TS nor XEN-like cells have been derived from EpiSCs. However, TS- and XEN-like cells have been derived from differentiated somatic cells (Benchetrit et al., 2015; Kubaczka et al., 2015; Parenti, Halbisen, Wang, Latham, & Ralston, 2016; Wamaitha et al., 2015), suggesting that the derivation of extraembryonic endoderm stem cell lines from EpiSCs should be possible, but may require identification of a unique set of extrinsic or intrinsic factors that are able to override primed pluripotency and induce extraembryonic pathways in EpiSCs. Identification of such factors could facilitate discovery of protocols for deriving human extraembryonic stem cell lines from human ES cells, since mouse EpiSCs are hypothesized to be more similar to human ES cells than are other mouse stem cell lines.

5. CELL-INTRINSIC REPRESSION OF XEN CELL FATE IN PLURIPOTENT CELLS

Evidence that XEN cells can be coaxed out of a population of otherwise pluripotent ES or EC cells by altering the signaling environment points to the existence of cell-autonomous regulators of XEN cell-specific gene expression. Similar to studies aimed at identifying transcription factors sufficient to convert ES cells to TS cells, studies have aimed at identifying transcription factors sufficient to convert ES cells to XEN cells. One of the first such studies showed that overexpression of GATA factors *Gata4* or *Gata6* could induce primitive endoderm gene expression (Fujikura et al., 2002; Shimosato, Shiki, & Niwa, 2007). When introduced into blastocysts, these XEN-like cells contributed to the parietal endoderm (Shimosato et al., 2007), indicating that not only do these ES cell-derived XEN-like cells express primitive endoderm markers, but they exhibit functional extraembryonic endoderm activity in vivo.

In addition to being sufficient to induce XEN fate in ES cells, *Gata6* is also necessary for the induction of XEN fate in ES cell lines. *Gata6* null ES cells cannot be differentiated to XEN cells in Activin/RA cXEN conditions (Cho et al., 2012). These cells fail to upregulate the primitive endoderm marker *Sox7*, and low levels of the pluripotency marker *Nanog* persist throughout several passages in XEN media, indicating that *Gata6* is necessary to override pluripotency and induce XEN gene expression (Cho et al., 2012). Notably, GATA factors are considered pioneer factors, capable of accessing and opening highly condensed chromatin to activate gene expression (Cirillo et al., 2002; Zaret & Carroll, 2011). Consistent with this, *Gata6* overexpression induced XEN (iXEN) cell fate in differentiated cells, whose chromatin is likely less open than ES cell chromatin (Wamaitha et al., 2015). These observations suggested that GATA6 regulates endodermal gene expression directly, and this was confirmed by genome-wide analysis of GATA6 binding sites in ES cells overexpressing *Gata6*. In parallel, GATA6 binds to regulatory regions of pluripotency genes, repressing their expression, and indicating that GATA6 induces XEN cell fate by two mechanisms: by direct activation of endodermal gene expression and by direct repression of pluripotent gene expression.

While studies have shown that GATA4/6 most potently and quickly induce XEN cell fate when overexpressed in ES cells, several other primitive endoderm transcription factors also force pluripotent cells to adopt a XEN cell fate. For example, overexpression of *Sox17* in ES cells results in downregulation of pluripotency markers and upregulation of primitive endoderm-specific genes (McDonald, Biechele, Rossant, & Stanford, 2014; Niakan et al., 2010), although SOX17 induces XEN cell fate much more slowly than do GATA4/6 (Wamaitha et al., 2015). SOX17 binding sites have been identified in EC cells treated with Activin/RA (Aksoy et al., 2013). Interestingly, SOX17 binding sites overlap with many OCT4 binding sites, suggesting that SOX17 and OCT4 promote endodermal fate together. While this idea is at odds with the role of OCT4 as an exclusive inducer of pluripotent cell fate, it is consistent with the discovery that OCT4 promotes primitive endoderm cell fate in parallel to epiblast fate in the blastocyst (Frum et al., 2013; Le Bin et al., 2014). By contrast, the ability of GATA6 to induce XEN cell fate in ES cells is independent of *Oct4* (Wamaitha et al., 2015). Therefore, SOX17 and OCT4 may induce XEN cell fate in parallel to GATA6 in ES cell lines.

6. REPROGRAMMING SOMATIC CELLS TO XEN-LIKE CELLS

Somatic cell reprogramming enables the derivation of colonies of iPS cells from mature, differentiated cells, such as fibroblasts following overexpression of the transcription factors *Oct4*, *Sox2*, *Klf4*, and *Myc* (OSKM)

(Takahashi & Yamanaka, 2006). While reprogramming is remarkable, it is inefficient, for reasons that are not yet resolved. In addition to low reprogramming efficiency, colonies with mixed or non-iPS cell phenotypes (Tonge et al., 2014), and cells that exist in a partially reprogrammed state (Meissner, Wernig, & Jaenisch, 2007; Mikkelsen et al., 2008; Silva et al., 2009; Sridharan et al., 2009) have been reported. Distinct from these, iXEN cell colonies have also been observed following OSKM reprogramming (Parenti et al., 2016) (Fig. 3A).

While OSKM are widely appreciated as pluripotency factors, OSKM have also been shown to promote primitive endoderm development in a variety of cellular contexts (Aksoy et al., 2013; Frum et al., 2013; Le Bin et al., 2014; Morgani & Brickman, 2015; Neri et al., 2012; Smith, Singh, & Dalton, 2010; Wicklow et al., 2014). In addition, expression of primitive endoderm genes has been observed during OSKM reprogramming (Serrano et al., 2013). These observations raised the possibility that OSKM could induce extraembryonic endoderm cell fate during reprogramming.

Consistent with this prediction, OSKM have been shown to induce formation of iXEN cells during reprogramming (Parenti et al., 2016). Colonies

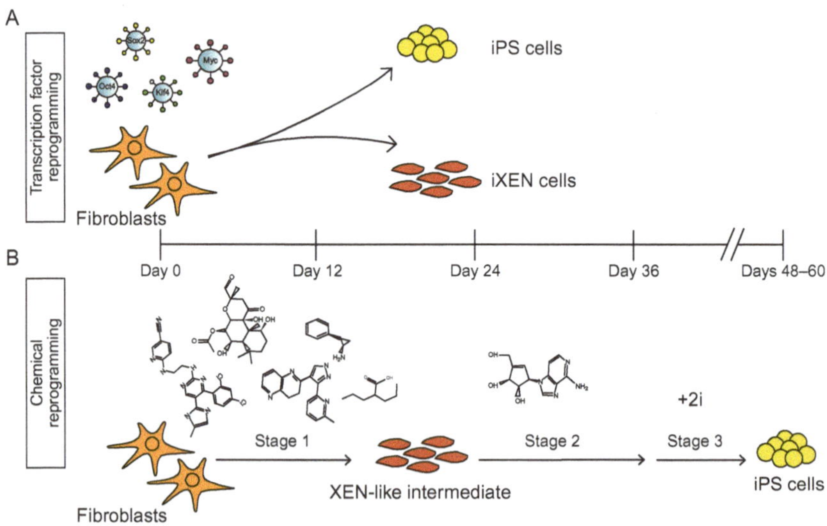

Fig. 3 XEN cell fate during somatic cell reprogramming. Either retroviral or transgenic overexpression of the reprogramming factors OSKM leads to parallel formation of iPS cells and iXEN cells. By contrast, chemical reprogramming is reported to produce a XEN-like state, which cannot be stabilized as a self-renewing population, but is proposed to serve as an intermediate between the fibroblast and iPS cell states.

of iXEN cells appeared larger and flatter than iPS cell colonies, and outnumbered iPS cell colonies by a factor of three. Individually, iXEN cells exhibited morphologies comparable to embryo-derived XEN cells, they homogeneously expressed endodermal markers such as GATA6, GATA4, SOX17, SOX7, and PDGFRA, and exhibited few significant differences in global transcriptional profile compared to embryo-derived XEN cell lines. Functionally, iXEN cells are multipotent, evidenced by their ability to differentiate into visceral endoderm in an in vitro differentiation assay, and to parietal endoderm in chimeric embryos. Finally, the majority of iXEN cell colonies isolated could be expanded and maintained their iXEN phenotype in culture, demonstrating self-renewal capacity, during continued passage.

In addition to murine iXEN cells, it was recently reported that canine embryonic fibroblasts are also capable of forming iXEN cells during OSKM reprogramming (Nishimura et al., 2017). Like murine iXEN cells, canine iXEN cells are multipotent, self-renewing cells that express XEN cell markers. Therefore, formation of both iXEN and iPS cell types during reprogramming is not limited to mice and indicates that OSKM promote primitive endoderm cell fate in multiple mammalian species.

In addition to transcription factors, small molecules have been shown to facilitate the conversion of somatic cells to pluripotency (Takahashi & Yamanaka, 2016). Interestingly, endodermal gene expression has also been observed during chemical reprogramming of somatic mouse cells (Hou et al., 2013) (Fig. 3). During chemical reprogramming, somatic cells are treated during three stages to produce iPS cell colonies. For the first 20 days, cells are treated with VPA, CHIR99021, 616452, tranylcypromine, and forskolin. During the next 20 days, cells are treated with DZNep. For the final 2 weeks, cells are treated with CHIR99021 and PD0325901, also known as 2i medium, which supports the self-renewal of naïve pluripotent stem cells (Marks et al., 2012; Ying et al., 2008). During chemical reprogramming, endodermal genes are thought to be expressed in the cells as they transition from somatic to pluripotent stem cell states (Zhao et al., 2015). This is at odds with evidence that iXEN and iPS cells arise via distinct lineages during OSKM reprogramming. For example, single-cell analysis of cells undergoing reprogramming revealed that markers of iXEN and iPS cell fate are detected in distinct populations throughout OSKM reprogramming (Parenti et al., 2016). Moreover, lineage tracing showed that iXEN cells are not derived from iPS cell colonies, nor are iPS cells derived from an endodermal population (Parenti et al., 2016). Therefore, intrinsic differences in the mechanisms of OSKM and chemical reprogramming may explain the differing uses of the XEN gene expression in reprogramming (Fig. 3).

In support of the view that chemical reprogramming and OSKM reprogramming utilize the XEN pathway in different ways, chemical reprogramming is much slower than OSKM reprogramming, taking 48–60 days to complete, whereas OSKM reprogramming is complete within 3 weeks or fewer. Moreover, stable, self-renewing iXEN cell lines could not be derived during chemical reprogramming (Zhao et al., 2015), in contrast with OSKM reprogramming, which gave rise to stable, self-renewing iXEN cell lines (Parenti et al., 2016). Understanding the mechanisms by which small molecules regulate cell fate could shed light on this issue. Components of the small molecule reprogramming cocktail could be replaced by overexpression of *SALL4* with either *GATA4* or *GATA6*, pointing to potential targets of some of the components of the chemical reprogramming cocktail. Further investigation is necessary to reveal the molecular targets and mechanisms of chemical reprogramming.

7. IN SEARCH OF A TOTIPOTENT CELL LINE

Thus far, we have discussed evidence that ES cells give rise relatively inefficiently to extraembryonic cell types. Thus, ES cell lines are considered pluripotent, but not totipotent. While methods exist for diverting ES cells to trophoblast or extraembryonic endoderm cell fates, it is also interesting to consider whether a totipotent cell line could be created from ES cells or derived from embryos. So far, experimental conditions that would capture cells of the embryo in the totipotent state have not been identified. However, several studies have attempted to expand the potential of ES cells, enabling them to contribute to both embryonic and extraembryonic cell fate simultaneously (Morgani, Nichols, & Hadjantonakis, 2017). For example, cells with properties of the two-cell (2C) embryo, including totipotency and gene expression profile, have been detected within ES cell populations (Macfarlan et al., 2012). Subsequent studies have identified pathways that repress the 2C state in ES cells. These studies pointed to a role for miRNAs in repressing the 2C state in ES cells by limiting activity of transposable elements of the MERVL family (Choi et al., 2017; Hendrickson et al., 2017; Morgani et al., 2017). While these studies have led to the identification of novel pathways limiting the potential of ES cell lines to differentiate to extraembryonic cell fates, 2C cells cannot be stably maintained as a homogenous population.

By contrast, small molecules that produce a self-renewing cell line with expanded developmental potential have recently been reported (Yang, Liu, et al., 2017; Yang, Ryan, et al., 2017). These studies, from two separate groups, identified two distinct chemical cocktails that reportedly enabled single cells to contribute to both fetal and extraembryonic lineages more efficiently than ES cells normally do. One group used inhibitors of JNK, p38, Src kinase, Axin1, in 2i + LIF medium, while the other group used only one of the 2i inhibitors (CHIR99021) + LIF, in the presence of a GPCR and a PARP1 inhibitor. It is not yet clear to what extent cells of expanded potential state resemble each other, and to what extent they resemble 2C cells.

The discovery of stem cells that, either as a population, or individually, exhibit expanded developmental potential provides several directions for future research. First, this discovery provides the opportunity to identify new pathways that define lineage barriers. Currently, the targets of small molecules are unknown in this context, so the mechanism by which these cocktails lead to expanded developmental potential are poorly defined. Second, the discovery of stem cells with expanded potential could be an important first step toward producing homogenous populations of human extraembryonic stem cell lines. As mentioned earlier, human blastocyst-derived extraembryonic stem cell lines would be a valuable resource, but do not currently exist. However, this goal is still in its infancy since protocols to produce mouse stem cells with expanded potential have failed to identify means for directing their differentiation down single, specific lineages. Finally, expanded potential stem cells provide an opportunity to discover what, if any, resemblance these cells bear to living embryos, and what these cells could teach us that we could not learn from studying embryos. The answers to these exciting unknowns await our continued investigation.

ACKNOWLEDGMENTS

The Ralston Lab is supported by funds from NIH R01 104009 to A.R. J.W. is supported by NIH T32 HD087166. We apologize to authors whose work was not discussed in this chapter.

REFERENCES

Aksoy, I., Jauch, R., Chen, J., Dyla, M., Divakar, U., Bogu, G. K., et al. (2013). Oct4 switches partnering from Sox2 to Sox17 to reinterpret the enhancer code and specify endoderm. *The EMBO Journal*, *32*, 938–953.

Artus, J., Douvaras, P., Piliszek, A., Isern, J., Baron, M. H., & Hadjantonakis, A. K. (2012). BMP4 signaling directs primitive endoderm-derived XEN cells to an extraembryonic visceral endoderm identity. *Developmental Biology*, *361*, 245–262.

Artus, J., Panthier, J. J., & Hadjantonakis, A. K. (2010). A role for PDGF signaling in expansion of the extra-embryonic endoderm lineage of the mouse blastocyst. *Development*, *137*, 3361–3372.

Beddington, R. S., & Robertson, E. J. (1989). An assessment of the developmental potential of embryonic stem cells in the midgestation mouse embryo. *Development*, *105*, 733–737.

Benchetrit, H., Herman, S., van Wietmarschen, N., Wu, T., Makedonski, K., Maoz, N., et al. (2015). Extensive nuclear reprogramming underlies lineage conversion into functional trophoblast stem-like cells. *Cell Stem Cell*, *17*, 543–556.

Bernardo, A. S., Faial, T., Gardner, L., Niakan, K. K., Ortmann, D., Senner, C. E., et al. (2011). BRACHYURY and CDX2 mediate BMP-induced differentiation of human and mouse pluripotent stem cells into embryonic and extraembryonic lineages. *Cell Stem Cell*, *9*, 144–155.

Blij, S., Parenti, A., Tabatabai-Yazdi, N., & Ralston, A. (2015). Cdx2 efficiently induces trophoblast stem-like cells in naïve, but not primed, pluripotent stem cells. *Stem Cells and Development*, *24*, 1352–1365.

Brickman, J. M., & Serup, P. (2017). Properties of embryoid bodies. *Wiley Interdisciplinary Reviews: Developmental Biology*, *6*(2). https://doi.org/10.1002/wdev.259.

Brons, I., Smithers, L., Trotter, M., Rugg-Gunn, P., Sun, B., Chuva de Sousa Lopes, S., et al. (2007). Derivation of pluripotent epiblast stem cells from mammalian embryos. *Nature*, *448*, 191–195.

Brown, K., Doss, M. X., Legros, S., Artus, J., Hadjantonakis, A. K., & Foley, A. C. (2010). eXtraembryonic ENdoderm (XEN) stem cells produce factors that activate heart formation. *PLoS One*, *5*, e13446.

Cambuli, F., Murray, A., Dean, W., Dudzinska, D., Krueger, F., Andrews, S., et al. (2014). Epigenetic memory of the first cell fate decision prevents complete ES cell reprogramming into trophoblast. *Nature Communications*, *5*, 5538.

Canham, M. A., Sharov, A. A., Ko, M. S., & Brickman, J. M. (2010). Functional heterogeneity of embryonic stem cells revealed through translational amplification of an early endodermal transcript. *PLoS Biology*, *8*, e1000379.

Chambers, I., Silva, J., Colby, D., Nichols, J., Nijmeijer, B., Robertson, M., et al. (2007). Nanog safeguards pluripotency and mediates germline development. *Nature*, *450*, 1230–1234.

Chazaud, C., Yamanaka, Y., Pawson, T., & Rossant, J. (2006). Early lineage segregation between epiblast and primitive endoderm in mouse blastocysts through the Grb2-MAPK pathway. *Developmental Cell*, *10*, 615–624.

Cho, L. T., Wamaitha, S. E., Tsai, I. J., Artus, J., Sherwood, R. I., Pedersen, R. A., et al. (2012). Conversion from mouse embryonic to extra-embryonic endoderm stem cells reveals distinct differentiation capacities of pluripotent stem cell states. *Development*, *139*, 2866–2877.

Choi, Y. J., Lin, C. P., Risso, D., Chen, S., Kim, T. A., Tan, M. H., et al. (2017). Deficiency of microRNA miR-34a expands cell fate potential in pluripotent stem cells. *Science*, *355*(6325), pii: eaag1927. https://doi.org/10.1126/science.aag1927.

Cirillo, L. A., Lin, F. R., Cuesta, I., Friedman, D., Jarnik, M., & Zaret, K. S. (2002). Opening of compacted chromatin by early developmental transcription factors HNF3 (FoxA) and GATA-4. *Molecular Cell*, *9*, 279–289.

Coucouvanis, E., & Martin, G. R. (1995). Signals for death and survival: A two-step mechanism for cavitation in the vertebrate embryo. *Cell*, *83*, 279–287.

Cronmiller, C., & Mintz, B. (1978). Karyotypic normalcy and quasi-normalcy of developmentally totipotent mouse teratocarcinoma cells. *Developmental Biology*, *67*, 465–477.

Donnison, M., Beaton, A., Davey, H. W., Broadhurst, R., L'Huillier, P., & Pfeffer, P. L. (2005). Loss of the extraembryonic ectoderm in Elf5 mutants leads to defects in embryonic patterning. *Development*, *132*, 2299–2308.

Erlebacher, A., Price, K. A., & Glimcher, L. H. (2004). Maintenance of mouse trophoblast stem cell proliferation by TGF-beta/activin. *Developmental Biology*, *275*, 158–169.

Evans, M., & Kaufman, M. (1981). Establishment in culture of pluripotential cells from mouse embryos. *Nature*, *292*, 154–156.

Frum, T., Halbisen, M. A., Wang, C., Amiri, H., Robson, P., & Ralston, A. (2013). Oct4 cell-autonomously promotes primitive endoderm development in the mouse blastocyst. *Developmental Cell*, *25*, 610–622.

Fujikura, J., Yamato, E., Yonemura, S., Hosoda, K., Masui, S., Nakao, K., et al. (2002). Differentiation of embryonic stem cells is induced by GATA factors. *Genes & Development*, *16*, 784–789.

Hamilton, W. B., & Brickman, J. M. (2014). Erk signaling suppresses embryonic stem cell self-renewal to specify endoderm. *Cell Reports*, *9*, 2056–2070.

He, S., Pant, D., Schiffmacher, A., Meece, A., & Keefer, C. L. (2008). Lymphoid enhancer factor 1-mediated Wnt signaling promotes the initiation of trophoblast lineage differentiation in mouse embryonic stem cells. *Stem Cells*, *26*, 842–849.

Hendrickson, P. G., Doráis, J. A., Grow, E. J., Whiddon, J. L., Lim, J. W., Wike, C. L., et al. (2017). Conserved roles of mouse DUX and human DUX4 in activating cleavage-stage genes and MERVL/HERVL retrotransposons. *Nature Genetics*, *49*(6), 925–934. https://doi.org/10.1038/ng.3844.

Hogan, B. L., Taylor, A., & Adamson, E. (1981). Cell interactions modulate embryonal carcinoma cell differentiation into parietal or visceral endoderm. *Nature*, *291*, 235–237.

Hou, P., Li, Y., Zhang, X., Liu, C., Guan, J., Li, H., et al. (2013). Pluripotent stem cells induced from mouse somatic cells by small-molecule compounds. *Science*, *341*, 651–654.

Ivanova, N., Dobrin, R., Lu, R., Kotenko, I., Levorse, J., DeCoste, C., et al. (2006). Dissecting self-renewal in stem cells with RNA interference. *Nature*, *442*, 533–538.

Kang, M., Garg, V., & Hadjantonakis, A. K. (2017). Lineage establishment and progression within the inner cell mass of the mouse blastocyst requires FGFR1 and FGFR2. *Developmental Cell*, *41*, 496–510, e495.

Kang, M., Piliszek, A., Artus, J., & Hadjantonakis, A. K. (2013). FGF4 is required for lineage restriction and salt-and-pepper distribution of primitive endoderm factors but not their initial expression in the mouse. *Development*, *140*, 267–279.

Kelly, G. M., & Gatie, M. I. (2017). Mechanisms regulating Stemness and differentiation in embryonal carcinoma cells. *Stem Cells International*, *2017*, 3684178.

Kubaczka, C., Senner, C., Araúzo-Bravo, M. J., Sharma, N., Kuckenberg, P., Becker, A., et al. (2014). Derivation and maintenance of murine trophoblast stem cells under defined conditions. *Stem Cell Reports*, *2*, 232–242.

Kubaczka, C., Senner, C. E., Cierlitza, M., Araúzo-Bravo, M. J., Kuckenberg, P., Peitz, M., et al. (2015). Direct induction of trophoblast stem cells from murine fibroblasts. *Cell Stem Cell*, *17*, 557–568.

Kuckenberg, P., Peitz, M., Kubaczka, C., Becker, A., Egert, A., Wardelmann, E., et al. (2011). Lineage conversion of murine extraembryonic trophoblast stem cells to pluripotent stem cells. *Molecular and Cellular Biology*, *31*, 1748–1756.

Kunath, T., Arnaud, D., Uy, G. D., Okamoto, I., Chureau, C., Yamanaka, Y., et al. (2005). Imprinted X-inactivation in extra-embryonic endoderm cell lines from mouse blastocysts. *Development*, *132*, 1649–1661.

Le Bin, G. C., Muñoz-Descalzo, S., Kurowski, A., Leitch, H., Lou, X., Mansfield, W., et al. (2014). Oct4 is required for lineage priming in the developing inner cell mass of the mouse blastocyst. *Development*, *141*, 1001–1010.

Li, J., Pan, G., Cui, K., Liu, Y., Xu, S., & Pei, D. (2007). A dominant-negative form of mouse SOX2 induces trophectoderm differentiation and progressive polyploidy in mouse embryonic stem cells. *The Journal of Biological Chemistry*, *282*, 19481–19492.

Lin, J., Khan, M., Zapiec, B., & Mombaerts, P. (2016). Efficient derivation of extraembryonic endoderm stem cell lines from mouse postimplantation embryos. *Scientific Reports*, *6*, 39457.

Liu, L., Michowski, W., Inuzuka, H., Shimizu, K., Nihira, N. T., Chick, J. M., et al. (2017). G1 cyclins link proliferation, pluripotency and differentiation of embryonic stem cells. *Nature Cell Biology*, *19*, 177–188.

Macfarlan, T. S., Gifford, W. D., Driscoll, S., Lettieri, K., Rowe, H. M., Bonanomi, D., et al. (2012). Embryonic stem cell potency fluctuates with endogenous retrovirus activity. *Nature*, *487*, 57–63.

Marks, H., Kalkan, T., Menafra, R., Denissov, S., Jones, K., Hofemeister, H., et al. (2012). The transcriptional and epigenomic foundations of ground state pluripotency. *Cell*, *149*, 590–604.

Martin, G. (1981). Isolation of a pluripotent cell line from early mouse embryos cultured in medium conditioned by teratocarcinoma stem cells. *Proceedings of the National Academy of Sciences of the United States of America*, *78*, 7634–7638.

Martin, G. R., & Evans, M. J. (1975). Differentiation of clonal lines of teratocarcinoma cells: Formation of embryoid bodies in vitro. *Proceedings of the National Academy of Sciences of the United States of America*, *72*, 1441–1445.

Masui, S., Nakatake, Y., Toyooka, Y., Shimosato, D., Yagi, R., Takahashi, K., et al. (2007). Pluripotency governed by Sox2 via regulation of Oct3/4 expression in mouse embryonic stem cells. *Nature Cell Biology*, *9*, 625–635.

McDonald, A. C., Biechele, S., Rossant, J., & Stanford, W. L. (2014). Sox17-mediated XEN cell conversion identifies dynamic networks controlling cell-fate decisions in embryo-derived stem cells. *Cell Reports*, *9*, 780–793.

Meissner, A., Wernig, M., & Jaenisch, R. (2007). Direct reprogramming of genetically unmodified fibroblasts into pluripotent stem cells. *Nature Biotechnology*, *25*, 1177–1181.

Mikkelsen, T. S., Hanna, J., Zhang, X., Ku, M., Wernig, M., Schorderet, P., et al. (2008). Dissecting direct reprogramming through integrative genomic analysis. *Nature*, *454*, 49–55.

Moerkamp, A. T., Paca, A., Goumans, M. J., Kunath, T., Kruithof, B. P., & Kruithof-de Julio, M. (2013). Extraembryonic endoderm cells as a model of endoderm development. *Development, Growth & Differentiation*, *55*, 301–308.

Molotkov, A., Mazot, P., Brewer, J. R., Cinalli, R. M., & Soriano, P. (2017). Distinct requirements for FGFR1 and FGFR2 in primitive endoderm development and exit from pluripotency. *Developmental Cell*, *41*, 511–526, e514.

Morgani, S. M., & Brickman, J. M. (2015). LIF supports primitive endoderm expansion during pre-implantation development. *Development*, *142*, 3488–3499.

Morgani, S. M., Canham, M. A., Nichols, J., Sharov, A. A., Migueles, R. P., Ko, M. S., et al. (2013). Totipotent embryonic stem cells arise in ground-state culture conditions. *Cell Reports*, *3*, 1945–1957.

Morgani, S., Nichols, J., & Hadjantonakis, A. K. (2017). The many faces of pluripotency: In vitro adaptations of a continuum of in vivo states. *BMC Developmental Biology*, *17*(1), 7. https://doi.org/10.1186/s12861-017-0150-4.

Neri, F., Zippo, A., Krepelova, A., Cherubini, A., Rocchigiani, M., & Oliviero, S. (2012). Myc regulates the transcription of the PRC2 gene to control the expression of developmental genes in embryonic stem cells. *Molecular and Cellular Biology*, *32*, 840–851.

Ng, R. K., Dean, W., Dawson, C., Lucifero, D., Madeja, Z., Reik, W., et al. (2008). Epigenetic restriction of embryonic cell lineage fate by methylation of Elf5. *Nature Cell Biology*, *10*, 1280–1290.

Niakan, K. K., Ji, H., Maehr, R., Vokes, S. A., Rodolfa, K. T., Sherwood, R. I., et al. (2010). Sox17 promotes differentiation in mouse embryonic stem cells by directly regulating extraembryonic gene expression and indirectly antagonizing self-renewal. *Genes & Development*, *24*, 312–326.

Nichols, J., Silva, J., Roode, M., & Smith, A. (2009). Suppression of Erk signalling promotes ground state pluripotency in the mouse embryo. *Development*, *136*, 3215–3222.

Nichols, J., Zevnik, B., Anastassiadis, K., Niwa, H., Klewe-Nebenius, D., Chambers, I., et al. (1998). Formation of pluripotent stem cells in the mammalian embryo depends on the POU transcription factor Oct4. *Cell*, *95*, 379–391.

Nishimura, T., Unezaki, N., Kanegi, R., Wijesekera, D. P. H., Hatoya, S., Sugiura, K., et al. (2017). Generation of canine induced extraembryonic endoderm-like cell line that forms both extraembryonic and embryonic endoderm derivatives. *Stem Cells and Development*, *26*, 1111–1120.

Nishioka, N., Inoue, K., Adachi, K., Kiyonari, H., Ota, M., Ralston, A., et al. (2009). The Hippo signaling pathway components Lats and Yap pattern Tead4 activity to distinguish mouse trophectoderm from inner cell mass. *Developmental Cell*, *16*, 398–410.

Nishioka, N., Yamamoto, S., Kiyonari, H., Sato, H., Sawada, A., Ota, M., et al. (2008). Tead4 is required for specification of trophectoderm in pre-implantation mouse embryos. *Mechanisms of Development*, *125*, 270–283.

Niwa, H., Miyazaki, J., & Smith, A. (2000). Quantitative expression of Oct-3/4 defines differentiation, dedifferentiation or self-renewal of ES cells. *Nature Genetics*, *24*, 372–376.

Niwa, H., Toyooka, T., Shimosato, D., Strumpf, D., Takahashi, K., Yagi, R., et al. (2005). Interaction between Oct3/4 and Cdx2 determines trophectoderm differentiation. *Cell*, *123*, 917–929.

Nosi, U., Lanner, F., Huang, T., & Cox, B. (2017). Overexpression of trophoblast stem cell-enriched MicroRNAs promotes trophoblast fate in embryonic stem cells. *Cell Reports*, *19*, 1101–1109.

Paca, A., Séguin, C. A., Clements, M., Ryczko, M., Rossant, J., Rodriguez, T. A., et al. (2012). BMP signaling induces visceral endoderm differentiation of XEN cells and parietal endoderm. *Developmental Biology*, *361*, 90–102.

Papaioannou, V. E., Evans, E. P., Gardner, R. L., & Graham, C. F. (1979). Growth and differentiation of an embryonal carcinoma cell line (C145b). *Journal of Embryology and Experimental Morphology*, *54*, 277–295.

Parenti, A., Halbisen, M., Wang, K., Latham, K., & Ralston, A. (2016). OSKM induce extraembryonic endoderm stem cells in parallel to iPS cells. *Stem Cell Reports*, *6*, 447–455.

Ralston, A., Cox, B., Nishioka, N., Sasaki, H., Chea, E., Rugg-Gunn, P., et al. (2010). Gata3 regulates trophoblast development downstream of Tead4 and in parallel to Cdx2. *Development*, *137*, 395–403.

Roberts, R. M., & Fisher, S. J. (2011). Trophoblast stem cells. *Biology of Reproduction*, *84*, 412–421.

Rossant, J. (2015). Mouse and human blastocyst-derived stem cells: Vive les differences. *Development*, *142*, 9–12.

Schenke-Layland, K., Angelis, E., Rhodes, K. E., Heydarkhan-Hagvall, S., Mikkola, H. K., & Maclellan, W. R. (2007). Collagen IV induces trophoectoderm differentiation of mouse embryonic stem cells. *Stem Cells*, *25*, 1529–1538.

Schröter, C., Rué, P., Mackenzie, J. P., & Martinez Arias, A. (2015). FGF/MAPK signaling sets the switching threshold of a bistable circuit controlling cell fate decisions in embryonic stem cells. *Development*, *142*, 4205–4216.

Serrano, F., Calatayud, C. F., Blazquez, M., Torres, J., Castell, J. V., & Bort, R. (2013). Gata4 blocks somatic cell reprogramming by directly repressing Nanog. *Stem Cells*, *31*, 71–82.

Shimosato, D., Shiki, M., & Niwa, H. (2007). Extra-embryonic endoderm cells derived from ES cells induced by GATA factors acquire the character of XEN cells. *BMC Developmental Biology*, *7*, 80.

Silva, J., Nichols, J., Theunissen, T., Guo, G., van Oosten, A., Barrandon, O., et al. (2009). Nanog is the gateway to the pluripotent ground state. *Cell*, *138*, 722–737.

Smith, K. N., Singh, A. M., & Dalton, S. (2010). Myc represses primitive endoderm differentiation in pluripotent stem cells. *Cell Stem Cell*, *7*, 343–354.

Sridharan, R., Tchieu, J., Mason, M. J., Yachechko, R., Kuoy, E., Horvath, S., et al. (2009). Role of the murine reprogramming factors in the induction of pluripotency. *Cell*, *136*, 364–377.

Strickland, S., Smith, K. K., & Marotti, K. R. (1980). Hormonal induction of differentiation in teratocarcinoma stem cells: Generation of parietal endoderm by retinoic acid and dibutyryl cAMP. *Cell*, *21*, 347–355.

Strumpf, D., Mao, C. A., Yamanaka, Y., Ralston, A., Chawengsaksophak, K., Beck, F., et al. (2005). Cdx2 is required for correct cell fate specification and differentiation of trophectoderm in the mouse blastocyst. *Development*, *132*, 2093–2102.

Takahashi, K., & Yamanaka, S. (2006). Induction of pluripotent stem cells from mouse embryonic and adult fibroblast cultures by defined factors. *Cell*, *126*, 663–676.

Takahashi, K., & Yamanaka, S. (2016). A decade of transcription factor-mediated reprogramming to pluripotency. *Nature Reviews. Molecular Cell Biology*, *17*, 183–193.

Tanaka, S., Kunath, T., Hadjantonakis, A. K., Nagy, A., & Rossant, J. (1998). Promotion of trophoblast stem cell proliferation by FGF4. *Science*, *282*, 2072–2075.

Tesar, P., Chenoweth, J., Brook, F., Davies, T., Evans, E., Mack, D., et al. (2007). New cell lines from mouse epiblast share defining features with human embryonic stem cells. *Nature*, *448*, 196–199.

Tonge, P. D., Corso, A. J., Monetti, C., Hussein, S. M., Puri, M. C., Michael, I. P., et al. (2014). Divergent reprogramming routes lead to alternative stem-cell states. *Nature*, *516*, 192–197.

Wamaitha, S. E., del Valle, I., Cho, L. T., Wei, Y., Fogarty, N. M., Blakeley, P., et al. (2015). Gata6 potently initiates reprograming of pluripotent and differentiated cells to extraembryonic endoderm stem cells. *Genes & Development*, *29*, 1239–1255.

Wicklow, E., Blij, S., Frum, T., Hirate, Y., Lang, R. A., Sasaki, H., et al. (2014). HIPPO pathway members restrict SOX2 to the inner cell mass where it promotes ICM fates in the mouse blastocyst. *PLoS Genetics*, *10*, e1004618.

Willert, K., Brown, J. D., Danenberg, E., Duncan, A. W., Weissman, I. L., Reya, T., et al. (2003). Wnt proteins are lipid-modified and can act as stem cell growth factors. *Nature*, *423*, 448–452.

Williams, R. L., Hilton, D. J., Pease, S., Willson, T. A., Stewart, C. L., Gearing, D. P., et al. (1988). Myeloid leukaemia inhibitory factor maintains the developmental potential of embryonic stem cells. *Nature*, *336*, 684–687.

Wu, T., Wang, H., He, J., Kang, L., Jiang, Y., Liu, J., et al. (2011). Reprogramming of trophoblast stem cells into pluripotent stem cells by Oct4. *Stem Cells*, *29*, 755–763.

Yagi, R., Kohn, M., Karavanova, I., Kaneko, K., Vullhorst, D., DePamphilis, M., et al. (2007). Transcription factor TEAD4 specifies the trophectoderm lineage at the beginning of mammalian development. *Development*, *134*, 3827–3836.

Yamanaka, Y., Lanner, F., & Rossant, J. (2010). FGF signal-dependent segregation of primitive endoderm and epiblast in the mouse blastocyst. *Development*, *137*, 715–724.

Yang, Y., Liu, B., Xu, J., Wang, J., Wu, J., Shi, C., et al. (2017). Derivation of pluripotent stem cells with in vivo embryonic and extraembryonic potency. *Cell*, *169*, 243–257, e225.

Yang, J., Ryan, D. J., Wang, W., Tsang, J. C., Lan, G., Masaki, H., et al. (2017). Establishment of mouse expanded potential stem cells. *Nature*, *550*, 393–397.

Ying, Q. L., Wray, J., Nichols, J., Batlle-Morera, L., Doble, B., Woodgett, J., et al. (2008). The ground state of embryonic stem cell self-renewal. *Nature*, *453*, 519–523.

Zaret, K. S., & Carroll, J. S. (2011). Pioneer transcription factors: Establishing competence for gene expression. *Genes & Development*, *25*, 2227–2241.

Zhao, Y., Zhao, T., Guan, J., Zhang, X., Fu, Y., Ye, J., et al. (2015). A XEN-like state bridges somatic cells to pluripotency during chemical reprogramming. *Cell*, *163*, 1678–1691.

CHAPTER NINE

From Germline to Soma: Epigenetic Dynamics in the Mouse Preimplantation Embryo

Michelle K.Y. Seah, Daniel M. Messerschmidt[1]
Developmental Epigenetics and Disease Group, Institute of Molecular and Cell Biology (IMCB), Agency for Science, Technology and Research (A*STAR), Singapore, Singapore
[1]Corresponding author: e-mail address: danielm@imcb.a-star.edu.sg

Contents

Abstract

When reflecting about cell fate commitment we think of differentiation. Be it during embryonic development or in an adult stem cell niche, where cells of a higher potency specialize and cell fate decisions are taken. Under normal circumstances this process is definitive and irreversible. Cell fate commitment is achieved by the establishment of cell-type-specific transcriptional programmes, which in turn are guided, reinforced, and ultimately locked-in by epigenetic mechanisms. Yet, this plunging drift in cellular potency linked to epigenetically restricted access to genomic information is problematic for reproduction. Particularly in mammals where germ cells are not set aside early on like in other species. Instead they are rederived from the embryonic ectoderm, a differentiating embryonic tissue with somatic epigenetic features. The epigenomes of germ cell precursors are efficiently reprogrammed against the differentiation trend, only to specialize once more into highly differentiated, sex-specific gametes: oocyte and sperm. Their differentiation state is reflected in their specialized epigenomes, and erasure of these features is required to enable the acquisition of the totipotent cell fate to kick start embryonic development of the next generation. Recent technological advances have enabled unprecedented insights into the epigenetic dynamics, first of DNA methylation and then of histone modifications, greatly expanding the historically technically limited understanding of this processes. In this chapter we will focus on the details of embryonic epigenetic reprogramming, a cell fate determination process against the tide to a higher potency.

Current Topics in Developmental Biology, Volume 128
ISSN 0070-2153
https://doi.org/10.1016/bs.ctdb.2017.10.011

1. OOCYTE-TO-EMBRYO TRANSITION

As far as cell fate changes go, fertilization—the fusion of male and female gametes—arguably induces the most profound change in the entirety of the developmental path of any mammalian organism. In a short period of time maternal and paternal epigenomes, now united in the zygote, undergo an switch from terminally differentiated and highly specialized to the only true totipotent state of the mammalian life cycle.

Next to the mere magnitude of this fate change a peculiarity is the reversed direction of this commitment. Classically development is viewed as a "downhill" process. Cells of a higher potency make lineage decisions and differentiate to adopt states of increasing specialization. This is concomitant with the loss of potency, a process, which is under normal circumstances irreversible. The changes to the epigenome (see Section 2), which directs these decisions, are referred to epigenetic programming, ever so restricting the genomic information available to the cell, while maintaining the total genetic information content.

How then is a new life cycle initiated? In mammals this is a particularly complex process. Mammalian germ cell precursors, the primordial germ cells, are not set aside very early in development, when cellular potency is high, but are instead derived from an embryonic cell lineage that already shows key characteristics of somatic cell differentiation (Hayashi, de Sousa Lopes, & Surani, 2007; Morgan, Dean, Coker, Reik, & Petersen-Mahrt, 2004; Saitou & Yamaji, 2012; Strome & Lehmann, 2007). Next to robust silencing of pluripotency and germ cell genes by DNA methylation, this also includes the presence of parental imprints and random X-chromosome inactivation in females. It requires an extensive effort to erase these and other features to bring cells close to a ground state with the potential to begin a new life cycle, a process coined reprogramming (Chuva de Sousa Lopes et al., 2008; Hajkova et al., 2002; Hemberger, Dean, & Reik, 2009; Messerschmidt, Knowles, & Solter, 2014; Morgan, Santos, Green, Dean, & Reik, 2005; Reik, 2007; Saitou & Yamaji, 2012; Sasaki & Matsui, 2008; Seisenberger et al., 2013). However, to functionally achieve this step, haploid gametes must be established. This renewed specialization, now in a sex-specific manner, produces sperm or oocytes with all their peculiarities and unique epigenetic states (McLaren & Lawson, 2005; Saitou & Yamaji, 2012).

The formation of male and female gametes serves the only purpose of sexual reproduction and is undone in a second, extensive reprogramming

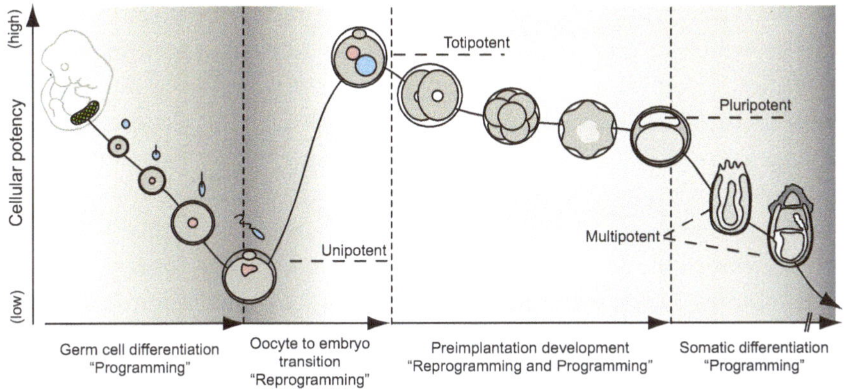

Fig. 1 Fate and potential during mouse reproduction and development. The developmental potential of cells during the mammalian life cycle varies drastically. Around birth the precursors of gametes have fully reprogrammed epigenomes and are (theoretically) totipotent. To achieve sexual reproduction, these cells specialize gender-specifically into oocyte and sperm. Highly differentiated and specialized these cells are very restricted in their potential, and they are unipotent. Fertilization of the oocyte by sperm triggers oocyte-to-embryo transition and a unique and dramatic epigenetic reprogramming process from uni- to totipotency. The one-cell and arguably the two-cell stage embryo are the only totipotent entities during mouse development. Later stages gradually lose developmental potential to ultimately establish the pluripotent epiblast of the blastocysts. Development ensues with cells gradually committing to germ layers and downstream differentiation, slowly losing potential and gaining specialization instead.

step in the newly conceived embryo. The embryo then enters a period of preimplantation development with the goal to on one hand allow for the first lineage segregation and the formation of the trophectoderm, quintessential for the implantation. On the other hand, it serves to establish the basic pluripotent state of the epiblast, the source of all embryonic tissues. These later processes are extensively studied and in part discussed within this issue in other chapters. In this chapter, we will focus our attention on the epigenetic reprogramming that takes place during this oocyte-to-embryo transition and the dynamic changes in both parental epigenomes when transitioning from germline to soma (Fig. 1).

2. EPIGENETICS

As in all fate decisions and differentiation events the genome of the cells undergoing these changes remains (with very few exceptions) stable. The changes are based in the accessibility and the usage of genetic information as regulated by epigenetic mechanisms. The epigenome is a heritable characteristic with the means to control the transcriptional readout of the genome which dictates, reinforces, and locks cell fate decisions in place. Most direct control in this respect is the methylation of DNA itself. DNA methylation imposes conformational changes and attracts specific binding partners. This can prevent the binding of activators, yet enhance recruitment of repressors. Naturally, the covalent posttranslational modification of proteins most intimately bound to DNA, particularly histones, can also control the accessibility to the genetic information, both promoting and inhibiting. Histone and DNA modifications are the predominant forms of epigenetic control, or at least the best studied to date, during oocyte to embryo transition and following cleavage stages. However, it should be mentioned that other epigenetic regulatory mechanism can play important roles. For instance, the composition of nucleosomes using permutations of histone variants, possibly cell-type or cell-cycle specific, can affect transcription (Ahmad & Henikoff, 2002; Bourc'his & Voinnet, 2010; Hake & Allis, 2006; Sarma & Reinberg, 2005; Smith, 2002). Last but not least, an interplay of chromatin and chromatin states with the overall nuclear architecture can influence and control the information flow from genome to function (Bernstein, Meissner, & Lander, 2007; Schneider & Grosschedl, 2007).

Although not changing genetic information itself epigenetic changes and modifications are, to some degree, heritable through mitosis and meiosis (Bonasio, Tu, & Reinberg, 2010). This is particularly well understood and shown for DNA methylation where semiconservative replication allows the inheritance of hemimethylated DNA strands to daughter cells (Holliday & Pugh, 1975; Riggs, 1975). A maintenance machinery is specifically dedicated to the restoration of full methylation at such sites during/after replication (see below) (Messerschmidt et al., 2014; Zhu & Reinberg, 2011). Also histone modifications and DNA-bound protein factors (epigenetic modulators and transcription factors) can be inherited by the daughter cell ensuring maintenance of the cell-type-specific transcriptional program (Estève et al., 2006; Zhu & Reinberg, 2011). Though, the mechanisms involved are less understood. Heritability of epigenomes is of major interest during reproduction, as it opens the possibility of the transgenerational transmission of nongenetic information, be it to the offspring's benefit or detriment (Heard & Martienssen, 2014). Numerous studies have indeed correlated offspring effects (particularly growth effects) to environmental exposure of parents, grandparents, or even great-grandparents. Mechanistically these transgenerational effects and inheritance of epigenetic information are little understood, yet are an exciting, timely field of research

(Heard & Martienssen, 2014; Zhu & Reinberg, 2011). In preparation for sexual reproduction the mammalian epigenome undergoes two rounds of reprogramming, essentially erasure of parental marks. To enable inheritance, such marks must escape these reprogramming waves. Paradigms exist for inheritance of germ cell features to the embryo and novel loci, genes, and regions are being described (Branco et al., 2016; Ferguson-Smith, 2011; Messerschmidt et al., 2012; Sampath Kumar et al., 2017). Inheritance of epigenomes across both waves, i.e., from parental soma through germ cells to the new embryo is reported yet less understood.

Toward a better understanding of the epigenetic reprogramming process, with all its implications, exciting new insights now reveal base-pair resolution analysis of not only DNA methylation dynamics (Borgel et al., 2010; Hackett et al., 2013; Kobayashi et al., 2013; Messerschmidt et al., 2014; Popp et al., 2010; Seisenberger et al., 2012; Smallwood et al., 2011; Smith et al., 2012; Wang et al., 2014) but also a detailed view of (some) histone modifications and their changes in oocytes and early embryos (Dahl et al., 2016; Liu et al., 2016; Wu et al., 2016; Zhang et al., 2016; Zheng et al., 2016). In the following we will be reviewing the progress made in researching the role of DNA methylation and histone modifications during oocyte-to-embryo transition and preimplantation development.

2.1 DNA Methylation

DNA methylation is experimentally more accessible than histone modifications and since the first observations of global DNA methylation dynamics in gametes and embryos much insights have been gained, both on the underlying mechanisms and on the functional impact during the reprogramming process (reviewed by Messerschmidt et al., 2014).

Covalent modification of DNA is the most direct way to alter its properties and possibly control informational content readout. Dozens of modifications affecting all four bases, predominantly by the addition of methyl groups, have been described (www.dnamod.hoffmanlab.org). While most of these variants are detectable in bacteria, the 5-methylcytosine (5mC) is the most common and significant modification observed in mammals (Bird, 2002). Because of its functional relevance often referred to as the "fifth base," 5mC is predominantly found in a CpG context in mammals, although, particularly in oocytes 5mC has also been found in other dinucleotide combinations (CpH) (Ramsahoye et al., 2000; Smallwood et al., 2011;

Ziller et al., 2011). However, whether these are functional, and what this function may be, remains unknown. CpG methylation on the other hand is well studied. Historically, CpG methylation was and remains predominantly associated with transcriptional repression (Deaton & Bird, 2011). The methyl group of the 5mC resides in the major groove of the DNA double helix where many DNA-interacting proteins bind. Methylation thus likely restricts binding, thereby modulating transcription. On the other hand, DNA methylation-specific "readers," methylation-binding proteins (MBPs), specifically recognize 5mC and serve as docking platforms for other interactors. During development DNA methylation also plays an essential role in silencing retroviral elements, regulating tissue-specific gene expression, genomic imprinting, and X-chromosome inactivation (Deaton & Bird, 2011; Ferguson-Smith, 2011; Messerschmidt et al., 2014; Smith & Meissner, 2013; Walsh, Chaillet, & Bestor, 1998; Weber et al., 2007).

2.1.1 DNA Methylation Dynamics: Easy Come–Easy Go?

DNA methylation is an active, enzyme-catalyzed process. DNA methyltransferases (DNMTs) transfer a methyl group from the donor *S*-adenosyl-L-methionine to the fifth position of cytosine and can be functionally categorized into two groups: de novo vs maintenance DNMTs (Goll & Bestor, 2005).

Two enzymes, DNMT3A and DNMT3B, were shown to have de novo DNA methylation activity (Okano, Xie, & Li, 1998). During development both enzymes play important roles in partially distinct expression domains, yet also have functional redundancy. DNMT3A is predominant in oocytes and preimplantation embryos (Kaneda et al., 2004; Kato et al., 2007), whereas DNMT3B is only transcribed after fertilization and later mostly in the epiblast of the developing embryo (Watanabe, Suetake, Tada, & Tajima, 2002). DNMT3B knockout is embryonic lethal, and a combined DNMT3A/B knockout causes even earlier lethality (Okano, Bell, Haber, & Li, 1999). Both de novo DNMTs interact with a third, yet catalytically inactive DNMT (DNMT3L), which enhances their activity. Though DNMT3L knockout mice are viable, they are infertile due to defects in sperm and oocyte epigenomes (Bourc'his & Bestor, 2004; Bourc'his, Xu, Lin, Bollman, & Bestor, 2001).

DNA methylation is well characterized as a heritable epigenetic mark. This is enabled through its predominant appearance in a palindromic CpG context, in combination with a semiconservative DNA replication mode in S-phase. DNMT1 shows a surprising affinity to hemimethylated

CpGs and is cell-cycle dependently regulated (Bestor, 1988; Kishikawa, Murata, Ugai, Yamazaki, & Yokoyama, 2003). It is recruited to sites of DNA synthesis through its PCNA-interacting binding partner NP95 where hemimethylated CpGs are then restored to full methylation (Arand et al., 2012; Bostick et al., 2007; Leonhardt, Page, Weier, & Bestor, 1992; Sharif et al., 2007). This function is essential during development. Mouse embryos lacking DNMT1 do not develop beyond gastrulation, stressing the importance of intact methylomes for development and differentiation (Arand et al., 2012; Hitt, Wu, Cohen, & Linn, 1988; Kurihara et al., 2008; Lei et al., 1996; Yoder, Soman, Verdine, & Bestor, 1997).

Global DNA demethylation is eminent during germ cell and embryonic epigenetic reprogramming. Yet, regional remodeling and removal of methyl marks may be required in somatic tissues or cell types, too. Enzymes with direct cytosine demethylation activity have not been identified in mammals, and likely do not exist due to the high energetic requirements to break the 5mC covalent bond. Yet, several alternative demethylation routes have been proposed (Messerschmidt et al., 2014; Wu & Zhang, 2010, 2014) (Fig. 2).

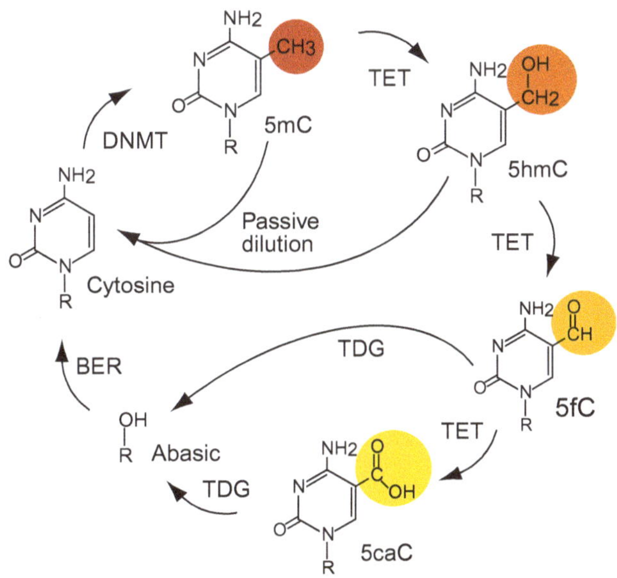

Fig. 2 DNA methylation and DNA demethylation pathways. Cytosine residues in a CpG context can be methylated at the fifth position by DNA methyltransferses forming 5mC. 5mC can be eliminated by passive dilution through DNA replication or be oxidized by TET enzymes to 5hmC. 5hmC is either removed by passive dilution or further oxidized by TET enzymes to 5fC and 5caC. Both oxidation products may be excised from DNA to reestablish unmodified cytosine.

First and easily conceivable is the prevention of DNA methylation maintenance, which through DNA replication in S-phase will result in passive DNA demethylation (see details below) (Ooi & Bestor, 2008). Downregulation of the DNA methylation maintenance machinery or even mere nuclear exclusion of essential factors during S-phase would result in demethylation. Passive demethylation however is proliferation-dependent and will be ineffective in nondividing cells. Locus-specific, targeted demethylation is not possible through passive demethylation, either. Active, enzymatically catalyzed DNA demethylation can on the other hand fulfill both requirements, and several mechanisms have been proposed and shown to be active during mammalian development and reproduction cycle (Messerschmidt et al., 2014; Wu & Zhang, 2010, 2014) (Fig. 2).

Ten-eleven translocation (TET) dioxygenases can catalyze the iterative oxidation of 5mC to 5hmC, 5fC, and finally 5caC (He et al., 2011; Inoue, Shen, Dai, He, & Zhang, 2011; Ito et al., 2010, 2011; Tahiliani et al., 2009) (Fig. 2). The latter could potentially be decarboxylated to unmodified 5mC, though evidence for this process is scarce (Ito et al., 2010). The derivatives may also be targeted by DNA repair mechanisms and be excised from DNA (Hayashi et al., 2007; He et al., 2011; Saitou & Yamaji, 2012; Shen et al., 2013; Strome & Lehmann, 2007). Lastly, oxidized derivatives of 5mC may not be further processed actively yet be cleared by a passive dilution through replication. DNMT1 was shown to have low affinity to 5hmC hemimethylation (Chuva de Sousa Lopes et al., 2008; Hajkova et al., 2002; Sasaki & Matsui, 2008; Valinluck & Sowers, 2007). Finally, TET-independent active DNA demethylation may involve deamination of 5mC to thymidine, e.g., by activation-induced deaminase, creating T:G mismatches. Thymidine-DNA glycosylase may catalyze the removal of this mismatch, reinstating an unmodified cytidine at this position (McLaren & Lawson, 2005; Morgan et al., 2004; Saitou & Yamaji, 2012) (Fig. 2).

2.1.2 DNA Methylation Dynamics in the Early Embryo

By the time oocyte and sperm meet for fertilization and the initiation of new life, they and their precursors have undergone extensive epigenetic changes, first erasure of epigenomes in primordial germ cells followed by sex-specific programming and differentiation during gametogenesis. Both processes are subject for reviews of their own and will not be discussed in detail here (Ahmad & Henikoff, 2002; Bourc'his & Voinnet, 2010; Hake & Allis, 2006; Hemberger et al., 2009; Messerschmidt et al., 2014; Morgan et al.,

2005; Reik, 2007; Saitou & Yamaji, 2012; Sarma & Reinberg, 2005; Sasaki & Matsui, 2008; Seisenberger et al., 2013; Smith, 2002).

Male and female mature gametes are grossly different far beyond their mere appearance. The sperm genome is close to fully methylated, packed densely with protamine's and only limited regions are histone associated. While the oocyte's genome is still substantially methylated, it is so to a much lower extent than the sperm's (Kobayashi et al., 2012; Popp et al., 2010). The oocyte genome also remains fully histone bound. As a result, in the newly formed embryo, parental genomes show stout epigenetic asymmetry (Reik, Santos, Mitsuya, Morgan, & Dean, 2003). Although the asymmetry is resolved in due course, the immediate changes make the discrepancy of maternal vs paternal epigenomes more pronounced. The paternal pronucleus becomes virtually fully demethylated after zygote formation, before the onset of DNA replication or pronuclear breakdown (~pronuclear stage (PN) 3) (Mayer, Niveleau, Walter, Fundele, & Haaf, 2000; Oswald et al., 2000; Popp et al., 2010) (Fig. 3). The immediate and replication-independent nature of the process suggests active demethylation. Indeed, shortly after fertilization, 5mC in the paternal genome undergoes extensive TET3-mediated conversion to 5hmC (Gu et al., 2011; Iqbal, Jin, Pfeifer, & Szabó, 2011; Wossidlo et al., 2011). Removal of TET3 from the zygote inhibits the accumulation of 5hmC and potentially delays paternal genome activation (Gu et al., 2011). This means that at the late pronuclear stages the paternal genome is 5mC low, yet 5hmC high. The fate of the latter is subject of intense debate. Originally only few sites in the paternal genome were shown to become "cytidine-unmodified" before replication. Higher oxidized forms (5fC and 5caC) were also detected in the zygote and paternal genomes of two- and four-cell embryos, suggesting continued oxidization by TET3. If and to what degree these modifications are removed by active means (i.e., DNA repair), or passive dilution is not clear (Gu et al., 2011; Hashimoto et al., 2012; He et al., 2011; Inoue et al., 2011; Inoue & Zhang, 2011) (Fig. 3).

Probing paternal pronucleus DNA methylation dynamics even deeper, a new study suggests that initial demethylation may be not only replication-, but also TET-independent (Amouroux et al., 2016). Detailed temporal resolution shows uncoupling of 5mC loss and 5hmC gain, occurring early and late in the zygote, respectively. While the former is believed to be dependent on excision/repair mechanisms, the latter was shown to actually counteract de novo DNA methylation mechanism active in the zygote (Amouroux et al., 2016; Messerschmidt, 2016) (Fig. 3).

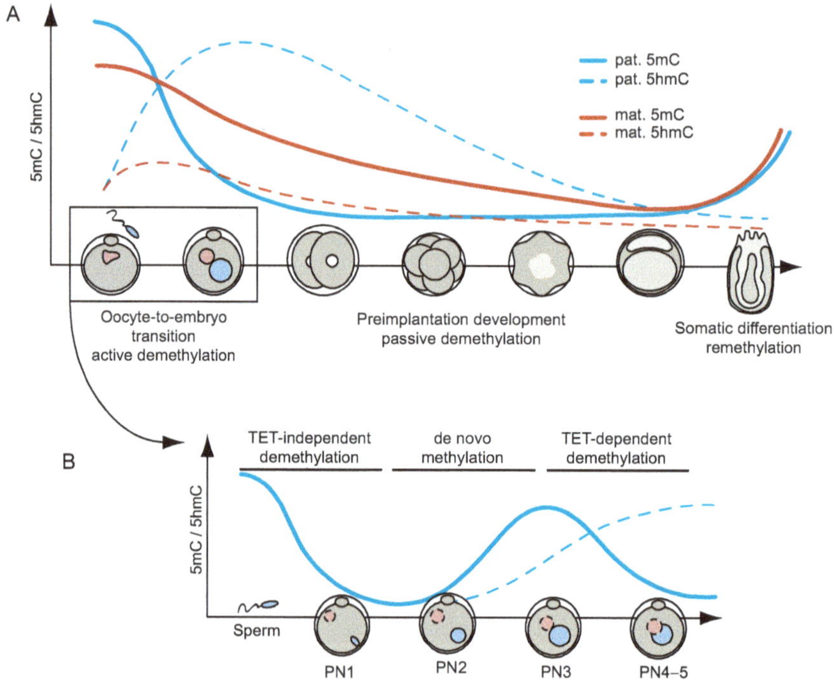

Fig. 3 DNA methylation dynamics during oocyte-to-embryo transition and preimplantation development. (A) Maternal and paternal genomes undergo complex, differential DNA demethylation dynamics postfertilization. The paternal 5mC is actively removed before zygotic S-phase by TET enzymes. This process results in a sudden increase of 5hmC. Most paternally acquired 5hmC is thought to be lost by passive dilution over subsequent cell divisions though other mechanisms (see main text) may contribute. Maternal 5mC levels are overall lower and predominantly removed by passive dilution through DNA replication. Maternal 5hmC is also detectable before first cleavage division attesting active DNA demethylation occurring in the maternal genome, too, albeit to a much lower degree. (B) Close scrutiny of 5mC/5hmC levels reveals that the paternal genome undergoes first TET-independent demethylation before TET-dependent demethylation counteracts de novo methylation activity in the paternal pronucleus.

In contrast to the sperm, only ~40% of CpGs in the mature oocyte genome are methylated (Kobayashi et al., 2012; Popp et al., 2010). Traditionally, and supported by early immunofluorescent data, the maternal genome is thought to be demethylated in a passive, replication-dependent manner postfertilization (Reik et al., 2003). 5mC signals persist and 5hmC is very low during zygotic progression (Fig. 3). Two criteria must be met to achieve such dynamics: (1) DNA methylation maintenance (DNMT1) must be inhibited and (2) TET3, which is active at the paternal genome, must be prevented from oxidizing maternal 5mC. Both prerequisites have been experimentally explored. DNMT1 is expressed in early preimplantation embryos, and two isoforms are detectable. *Dnmt1o* was described as an oocyte-specific variant, and *Dnmt1s* was shown to be activated later during preimplantation development. Expression at early stages is, however, very low, and existing DNMT1 protein appears to be largely excluded from the nucleus (Branco et al., 2016; Cardoso & Leonhardt, 1999; Hirasawa et al., 2008; Howell et al., 2001; Ratnam et al., 2002). The prerequisite for passive DNA demethylation is therefore fulfilled, yet raises new challenges and questions (see Section 2.1.4).

A highly expressed protein, STELLA, was shown to prevent active DNA demethylation in the maternal pronucleus (Nakamura et al., 2006, 2012). Deletion of STELLA causes a rapid loss of 5mC in the maternal pronucleus, mirroring the paternal pronucleus demethylation dynamics and resolving pronuclear asymmetry. It was later shown that loss of 5mC is indeed concurrent with 5hmC gain (Wossidlo et al., 2011). The underlying mechanisms to STELLA-mediated protection from TET3 oxidation of maternal 5mC are still not resolved. It appears however that H3K9me2 is required for a tight association of STELLA with chromatin and protective function. And although STELLA is also localized to the paternal genome, its chromatin interaction is weak and does not prevent TET3 activity globally (Nakamura et al., 2012). Interestingly, it appears that even on the paternal genome STELLA can mediate 5mC maintenance at sites of genomic imprinting, which are also rich in H3K9me2 (Nakamura et al., 2012). On the other extreme, when examined not by immunofluorescence, but direct sequencing, maternal DNA in zygotes and certainly two- and four-cell stage embryos, also shows clear evidence of active demethylation (5hmC) and TET activity (Guo et al., 2014; Wang et al., 2014) (Fig. 3).

2.1.3 What Is Going on?

When it comes to DNA methylation dynamics, we are currently faced with a deluge of information, based on multiple techniques, both traditional and novel, each having their own advantages and disadvantages. Although it appears at this time that not all of these findings are unifiable, each most likely represents a (somewhat biased) snapshot of the current state within the embryo at a given time point. Whereas there is a tendency to divide and categorize parental genomes and demethylation mechanisms, the reality most likely is a combination of active and passive mechanism affecting maternal and paternal genomes to different, possibly variable magnitude on genome-wide levels. It therefore seems that paternal and maternal pronuclei use at least four different routes to eradicate DNA methylation to variable extent and efficiency: (1) passive, replication-dependent, (2) TET- and replication-dependent, (3) TET-dependent and active, and last (4) TET-independent and active. Therefore, rather than struggling to comprehend DNA demethylation on a global level and segregating the process into paternal and maternal compartments, a functional analysis, defining the true requirements and the targets, would be the more prudent approach. Importantly, this notion goes either way: When and where reprogramming needs to be fast and active? Conversely, where is reprogramming undesired and even potentially harmful?

There is no question that epigenetic reprogramming of parental epigenomes in mammals is prerequisite to successful development. However, whether global demethylation is essential, or demethylation of a few crucial loci (e.g., *Nanog*) may suffice is unclear. More questionable in fact is if active removal of 5mC is indeed required. Round spermatids and somatic nuclei, when injected into zygotes, do not undergo (global) active demethylation, yet still have the capacity to develop normally (Kimura & Yanagimachi, 1995; Polanski, Motosugi, Tsurumi, Hiiragi, & Hoffmann, 2008). The very much reduced success rate however may suggest that active demethylation is a more efficient preparation of the epigenomes for development. Thus, perhaps the demethylation "cacophony" observed across the maternal and paternal epigenomes in the zygote and thereafter serve as a safety net, a redundant combination of demethylation processes ensuring that key loci are reprogrammed at high efficiency to establish totipotency when it counts.

2.1.4 Maintenance in the Light of Reprogramming: Imprinting and Beyond

In light of the redundant, reinforced DNA demethylation, reprogramming must not go too far. In mammals, a small subset of genes is monoallelically expressed in a parent-of-origin-specific manner during development and adult homeostasis. These genes are "imprinted" (Abramowitz & Bartolomei, 2012; Bartolomei, 2009; Ferguson-Smith, 2011). The expression patterns and whether paternal or maternal genes are expressed is defined by locus- and allele-specific DNA methylation established differentially in oocyte and sperm. Such methylation patterns at imprinting control regions (ICRs) must resist epigenetic reprogramming (Messerschmidt, 2012). Loss of imprinting causes severe developmental defects or abortion of development altogether (Amor & Halliday, 2008; Butler, 2009; Tomizawa &

Sasaki, 2012). Albeit DNMT1 being expressed at low levels and being excluded from preimplantation embryo nuclei, it is still required for the maintenance of DNA methylation at genomic imprints (Hirasawa et al., 2008; Howell et al., 2001). Hence, although not being detectable by conventional methods, DNMT1 is still present and required. Clearly, the carefully balanced low DNMT1 dosage is an attempt to coalesce genome-wide demethylation and locus-specific methylation maintenance. While canonically targeted to replicating DNA through NP95, DNMT1 is specifically targeted to ICRs through a KRAB-Zinc finger protein (ZFP) transcription factor (ZFP57) in a noncanonical fashion instead (Li et al., 2008; Messerschmidt et al., 2014; Quenneville et al., 2011). Interfering with this mechanism, i.e., by deleting ZFP57 causes loss of imprinting in mice and man (Li, 2013). The key to this mode of DNA methylation maintenance at ICRs is the selective binding of ZFP57 not only on a sequence-specific manner but also in DNA-methylation-dependent way (Quenneville et al., 2011). ZFP57 targets the methylated allele in ICRs, thus selectively maintaining methylation in an allele-specific fashion. DNMT1 does not bind to ZFP57 directly but is recruited by the KRAB-domain-interacting protein TRIM28 (Lorthongpanich et al., 2013; Messerschmidt et al., 2012; Quenneville et al., 2011). TRIM28 further interacts with de novo DNMTs which is thought to aid in the maintenance of DNA methylation in the light of 5mC removal, in addition essential to the maintenance function of DNMT1 (Quenneville et al., 2011). The maternal deletion of TRIM28 results in hypomethylation of ICRs, albeit a paternal rescue causes a mosaic imprinting defect pattern (Lorthongpanich et al., 2013; Messerschmidt et al., 2012).

TRIM28 interacts with many KRAB-domain proteins (Schultz, Friedman, & Rauscher, 2001). KRAB-ZFP proteins have evolved in tetrapods and mammals in particular, presumably in an arms race against endogenous retroviruses (ERVs) (Jacobs et al., 2014). Indeed, a key function of TRIM28 and various identified KRAB-ZFPs is to keep ERVs epigenetically silenced through DNA and histone methylation in embryos and embryonic stem cells (Iyengar & Farnham, 2011; Rowe et al., 2010; Schultz, Ayyanathan, Negorev, Maul, & Rauscher, 2002). Next to these two paradigmatic and long since acknowledged elements escaping epigenetic reprogramming (genomic imprints and ERVs), it has been postulated that other genomic regions or genes must maintain DNA methylation, or even inherit parental epigenetic marks for proper development. These marks may be of immediate importance or may modulate gene expression later in development through an epigenetic memory effect. The identification of such region is difficult, given ERV methylation is resisting reprogramming in part as well and potentially veils such sites from identification. It therefore requires functional genetic approaches to confirm the existence and identify such sites. The maternal TRIM28 knockout causes an extremely pleiotropic phenotype, in part based on the mosaic imprinting defects through ZFP57, but likely also on the disruption of other KRAB-ZFPs' functions in the early embryos (Lorthongpanich et al., 2013; Messerschmidt et al., 2012). It therefore represents a tractable study system for epigenetic reprogramming and inheritance in detail. One ZFP57-independent phenotype described in maternal TRIM28 mutants is early male-specific embryonic lethality (Sampath Kumar et al., 2017). Most male embryos conceived by maternal TRIM28 knockout mothers perish at implantation, with female embryos usually arresting only at later stages of gestation. This particular defect was mapped to a gene on the Y-chromosome, *Rbmy1a1*. Remarkably, *Rbmy1a1* is ectopically activated in males upon the loss of maternal TRIM28, yet is robustly silenced in controls. This is reflected in the promoter methylation of the *Rbmy1a1* gene, which is highly methylated in control sperm and throughout preimplantation, thus resisting reprogramming. Critically, this methylation is lost in maternal TRIM28 mutants. Close analysis showed that maternal TRIM28 is both essential and sufficient to maintain DNA methylation and *Rbmy1a1* repression, whereas zygotic TRIM28 is dispensable. *Rbmy1a1* therefore lies in a critical genomic region evading reprogramming demethylation, like genomic imprints. The silencing marks are inherited through the paternal germ cell and maintained by maternal factors present in the embryo (Sampath Kumar et al., 2017). Another recent report has shown the importance of maternally inherited DNA methylation instead (Branco et al., 2016). Oocyte-derived methylation, which persists through preimplantation development and resists demethylation, is essential for trophectoderm formation. Remarkably, this requirement goes beyond the previously reported importance of the imprinted gene *Ascl2*, but in addition multiple differentiation and physiological processes also rely on maternal methylation. The effects were most pronounced on a gene *Scml2*, which must remain suppressed to enable adequate trophoblast adhesion (Branco et al., 2016). It is unclear what mechanisms are involved in the maintenance of the required maternal DNA methylation patterns in this respect. Yet, it is only a matter of time for other such gene regions, with similar properties and importance, KRAB-ZFP/TRIM28-dependent or not, to be identified in the future.

2.2 Histone Modifications

The second prominent level of epigenetic regulation is the packaging of DNA. In eukaryotes DNA is packed around histone proteins and their covalent modification is used as epigenetic control mechanisms (Bannister & Kouzarides, 2011; Kouzarides, 2007; Sarma & Reinberg, 2005). Together, eight histones (2 each of H2A, H2B, H3, and H4) and the DNA wrapped around this complex make up a nucleosome. Using their highly abundant positively charged amino acid residues in their N-terminal "tails" histones closely interact with the negatively charged backbone of DNA, allowing the dense packaging required to store the close to 2-m-long DNA molecules in a single cell. These interactions open the possibility to intriguing regulatory mechanisms: (1) The use of histone variants to replace the canonical proteins can alter the nucleosome properties and ultimately influence the epigenome. Histone variants differ in amino acid sequence and are often cell-type specifically expressed. Multiple histone variants have been identified, in particular the histone variant H3.3 and its chaperone HIRA are of major interest in the oocyte-to-embryo transition context and discussed in detail elsewhere (Lim, Knowles, Solter, & Messerschmidt, 2016; Sarma & Reinberg, 2005). (2) On the other hand, by posttranslationally modifying histones, particularly, but not exclusively on their N-terminal "tails," properties such as charge can be altered to loosen or tighten the grip on DNA, making the underlying sequence and information more or less accessible to the transcriptional machinery (Bannister & Kouzarides, 2011; Kouzarides, 2007). Histone modifications may also act as docking platforms for chromatin "readers" attracted to specific modifications or combinations thereof. A countless number of modifications (methylation, acetylation, ubiquitination and phosphorylation to name a few) have been described in mammalian systems, yet most prominent and in the focus of this chapter are acetylation and methylation of lysine residues on the N-terminal tail of histone H3.

The addition of an acetyl group to a lysine neutralizes its positive charge and is therefore thought to favor an open chromatin conformation and transcriptional activity (Clayton, Hazzalin, & Mahadevan, 2006). Furthermore, acetylated lysine residues are recognized by bromo domain containing protein creating docking platforms for other regulative mechanisms (Zeng & Zhou, 2002). Prominent lysine methylations with a clear repressive function occur at histone 3, lysine 9 (H3K9me3) and histone 3, lysine 27 (H3K27me3). However, at other residues, e.g., K4 and K36, methylation is associated with active transcription. In line with this context-depending regulatory outcome, histone methylation does not alter the binding affinity of histone tails to DNA itself, yet serve as a docking platforms for regulatory proteins (Bannister & Kouzarides, 2011; Kouzarides, 2007).

2.2.1 Chromatin States and Changes in Germ Cells and Early Embryos

As for DNA methylation, the chromatin modification landscape in the early zygote is dominated by the asymmetry of paternal and maternal genomes (Burton & Torres-Padilla, 2010). This is hardly surprising, knowing that sperm DNA is loaded on basic protamine, and largely devoid of histones altogether to facilitate the functional needs (dense packaging) of these highly specialized cells. Only a small proportion of the genome (1%–10%), has been shown to, possibly by targeted mechanisms, retain histones in sperm chromatin. After fertilization, protamine–histone exchange takes place, loading the paternal DNA onto maternally provided and largely unmodified histones. Rapid paternal genome decondensation coincides with a strong hyperacetylation of various residues in the newly formed paternal pronucleus (Adenot, Mercier, Renard, & Thompson, 1997; Adenot, Szöllösi, Geze, Renard, & Debey, 1991; Santos, Hendrich, Reik, & Dean, 2002; van der Heijden et al., 2008).

In contrast the maternal genome is delivered as fully packaged and modified chromatin reminiscent, in its genome-wide composition, of somatic cells. The maternal pronucleus is thus rich in H3K9me3 and H3K27me3 both associated with transcriptional repression. Other marks such as H3K4me3 (active) and H3K36me3, H3K64me3, and H4K20me3 are also detectable in the maternal pronucleus (Bošković et al., 2012; Daujat et al., 2009; Kourmouli et al., 2004; Lepikhov & Walter, 2004; Santos, Peters, Otte, Reik, & Dean, 2005) (Fig. 4). Yet it is noteworthy at this point that both oocyte and early zygote are transcriptionally inert. The histone modifications of the maternal pronucleus remain largely stable during oocyte-to-embryo transition as the embryo prepares for first cleavage and transcriptional activation (Burton & Torres-Padilla, 2014) (Fig. 4).

Histone modifications are less accessible to study on a locus-specific level, yet global dynamics and peculiarities during and after oocyte-to-embryo transition have been observed. Only lately new technological advances have enabled a more detailed insight into the genome-wide distribution of some of these marks progress during this crucial time of development. In the following we will touch upon few prominent and best-studied histone modifications in the context of embryonic epigenetic reprogramming and oocyte-to-embryo transition.

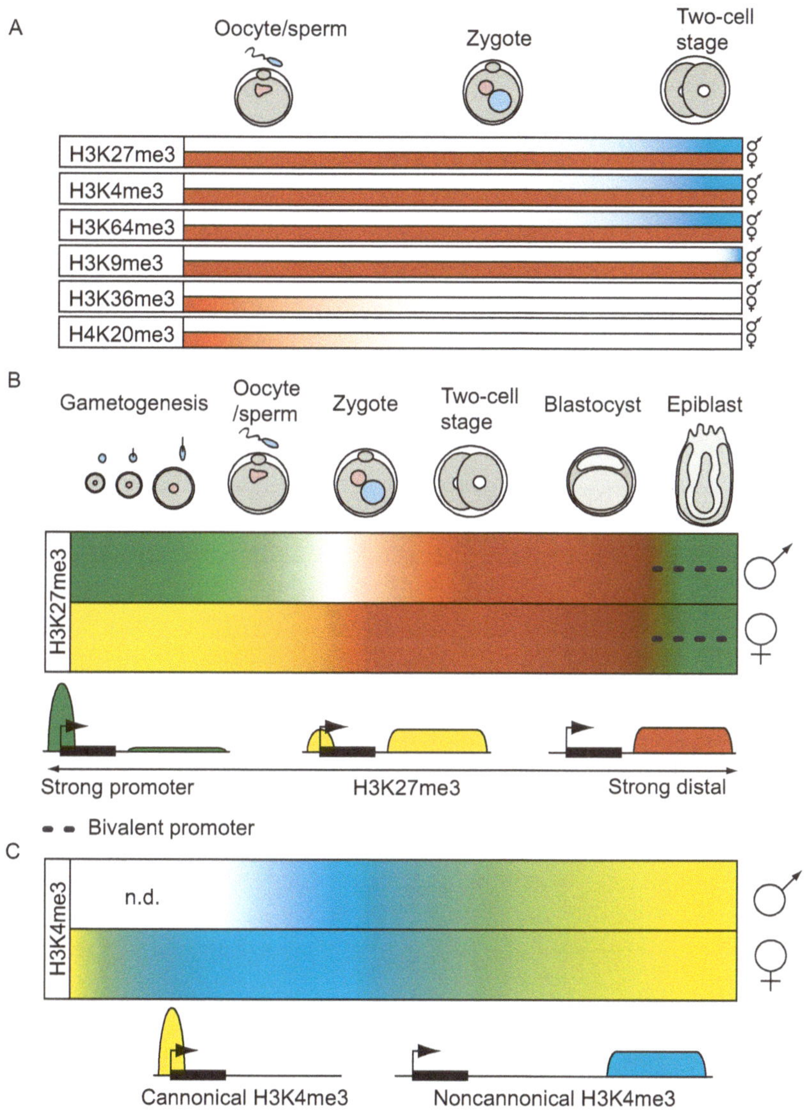

Fig. 4 Histone modification changes in OET and preimplantation development. (A) Selected histone modifications and their presence or absence and dynamic changes during oocyte-to-embryo transition in paternal and maternal genomes. All marks show substantial asymmetry in gametes and early zygote. Asymmetries appear mostly resolved at the two-cell stage. (B) H3K27me3 distribution in paternal and maternal genomes during gametogenesis, in mature gametes, zygote, preimplantation embryos, and postimplantation epiblast according to ChIP-seq data. In the paternal genome H3K27me3 comes in two flavors. Classical promoterH3K27me3 is observed during male

(Continued)

2.2.2 H3K27me3

The trimethylation of H3K27 (H3K27me3) is a prominent, repressive histone mark. It is established by the polycomb repressive complex 2 (PRC2), which is evolutionarily highly conserved and essential to development (Margueron & Reinberg, 2011). In mammals the PRC2 complex contains the catalytic enzymes EZH1 or EZH2 and their respective binding partners EED and SUZ12 are necessary for H3K27 methylation. In mice mutations of *Ezh2*, *Eed*, and *Suz12* all result in embryonic lethality around gastrulation. Canonically, H3K27me3 is found at the promoters of developmental genes, which is best studied in embryonic stem cells both of human and mouse origin. Here it was proposed to serve the temporal repression coined as "poising" of lineage-specific genes and is thus necessary for pluripotency.

H3K27me3 in the maternal pronucleus is inherited from the oocyte and highly abundant (Fig. 4). In contrast, it is barely detectable in the paternal pronucleus at first, yet increasing accumulation can be observed at the later stages, when replication ensues (Erhardt et al., 2003; Santenard et al., 2010; Santos et al., 2005; Wongtawan, Taylor, Lawson, Wilmut, & Pennings, 2011) (Fig. 4). In contrast the paternal pronucleus is rich in H3K27ac (Hayashi-Takanaka et al., 2011). Interestingly, on a global, staining-based level, the asymmetry of paternal and maternal H3K27 modification distribution seemingly rapidly resolves and is equalized at the two-cell stage (Santenard et al., 2010; Wongtawan et al., 2011) (Fig. 4).

In ES cells or somatic tissues H3K27me3 is enriched at developmental gene promoters. By extrapolation a similar distribution was expected in the preimplantation stages, which a recent effort of genome-wide H3K27me3 distribution analysis by ChIP-Seq in gametes and preimplantation embryos has shown to be only partially true (Liu et al., 2016; Zheng et al., 2016). In sperm H3K27me3 is indeed detectable at gene promoters,

Fig. 4—Cont'd gametogenesis and mature sperm, yet is quickly erased in the zygote. Preimplantation embryos display an unusual distal, not promoter enrichment of H3K27me3 on the paternal genome. Normal patterns are reestablished postimplantation. Female gametes display a combination of promoter and distal enrichment of H3K27me3, which is quickly reprogrammed to distal enrichment only in the zygote. Normal H3K27me3 patterns on the maternal genome are also reestablished postimplantation. Bivalent promoters on classical developmental genes are established in the epiblast. (C) Noncanonical H3K4me3 is present on paternal and maternal genomes in zygotes, yet is quickly erased and reprogrammed to canonical H3K4me3 with the initiation of embryonic transcriptional activation at the two-cell stage. H3K4me3 has not been determined (n.d.) in sperm precursors but has been found in growing oocytes, where is may be linked to transcriptional silencing (see main text).

particularly of developmental genes (Bernstein et al., 2006; Brykczynska et al., 2010; Hammoud et al., 2009). Previous imaging-based analysis had shown that H3K27me3 levels in paternal pronuclei are very low, however, could not conclusively answer if some transgenerational transmission might take place. The ChIP-Seq findings confirmed the canonical distribution in sperm and previous description of H3K27me3 erasure, thus excluding the possibility of, at least broad, inheritance of H3K27me3 through the paternal germ line (Zheng et al., 2016) (Fig. 4).

In contrast, H3K27me3 is abundant in the maternal genome and is largely inherited to the embryo. Yet again, the actual genomic distribution of the mark was unknown. The new studies reveal that in oocytes H3K27me3 is not solely found at the canonical sites, yet also at broad, non-promoter, CpG-poor intergenic regions and gene deserts (Zheng et al., 2016) (Fig. 4). As for the paternal genome, maternal canonical H3K27me3 is efficiently erased, particularly evident at classical polycomb target genes like at HOX genes, at *Pax6* and at *Sox1*. Thus, neither paternal nor maternal classical H3K27me3 information seems to be inherited to the embryo in larger scale. The classic marks remain absent throughout preimplantation development, and even in the inner cell mass of the blastocyst only weak promoter H3K27me3 enrichment can be found. Classical H3K27me3 is only established at developmental genes at periimplantation stages (Zheng et al., 2016) (Fig. 4).

Nonetheless, substantial increase in the paternal pronucleus and maintenance of H3K27me3 signal is observed in two-cell stage embryos. In the maternal genome the promiscuous enrichment at intergenic regions and gene deserts is maintained long after fertilization (up to the blastocysts). In the paternal genome an emergence of H3K27me3 at distal regions can be observed, too. Whether these sites are also marked in sperm but at much lower levels cannot be excluded but it appears that de novo H3K27 methylation takes place in the zygote and the two-cell stage (Zheng et al., 2016). Nonetheless, H3K27me3 distribution in maternal and paternal genomes remains strikingly asymmetric postfertilization. Thus, although seemingly assimilated at the two-cell stage as shown by immunofluorescence, the asymmetry of H3K27me3 persists throughout preimplantation development (Burton & Torres-Padilla, 2010; Zheng et al., 2016).

In part contrasting these observations, or perhaps rather emphasizing the importance of a detailed, locus-specific analysis, another recent study, has focussed on the description of parental (allele)-specific, DNase I hypersensitive sites (DHSs) in early embryos (Inoue, Jiang, Lu,

Suzuki, & Zhang, 2017). Unlike classical imprinted genes, for some paternal-specific DHSs, the corresponding maternal allele was not found DNA-methylated, but rather marked by H3K27me3. The maternal-specific silencing of these sites and linked paternal allele-specific expression was lost upon forced expression of KDM6B, an H3K27me3 demethylase. Thus, maternally inherited H3K27me3 is a previously unrecognized imprinting mechanism in the early mouse embryo (Inoue et al., 2017).

2.2.3 H3K4me3

The trimethylation of H3K4 is a very prominent and well-studied "active mark" (Bannister & Kouzarides, 2011; Kouzarides, 2007). It is established by the counteractor of the Polycomb group proteins, the homologs of *Drosophila* Trithorax group proteins. H3K4me3 enables gene activation and transcription via chromatin remodelers and further histone modifiers, which recognize and bind this modification. Genome-wide H3K4me3 distribution analyses, mainly performed on embryonic stem and mature cells, have shown that high levels of H3K4me3 are mostly associated with promoters of transcriptionally active genes. The enriched regions are commonly short, sharp, and coincide with other "active marks" (histone acetylation) and active polymerase II binding (Bernstein et al., 2006; Ng, Robert, Young, & Struhl, 2003; Pokholok et al., 2005; Santos-Rosa et al., 2002).

Historically only surveyed globally by immunofluorescence, H3K4me3 was shown to be very abundant in the maternal genome at the zygote stage, comparable with somatic cells (Fig. 4). Also, similar to the H3K27me3 mark, H3K4me3 is not or barely detectable in the paternal genome of the early zygote (Lepikhov & Walter, 2004; Santenard et al., 2010). However, signal emerges toward the later pronuclear stages, and the asymmetry is no longer apparent at the two-cell stage. Paternal and maternal H3K4me3 levels are thought to be assimilated then (Fig. 4).

Using highly sensitive ChIP-seq approaches several groups were recently able to generate a long-awaited genome-wide profile of H3K4me3 in gametes and preimplantation embryos (Dahl et al., 2016; Liu et al., 2016; Zhang et al., 2016). On the whole, these studies come to similar conclusions and confirm previous findings and assumptions. But, they also detail novel and unexpected insights. Reprogramming and erasure of sperm H3K4me3 patterns in the early zygote and thus paternal/maternal asymmetry was confirmed (Fig. 4). Furthermore, it was confirmed that strong H3K4me3 patterns reappear on the paternal genome by the late two-cell stage, though remains weaker in intensity compared to the maternal allele throughout preimplantation. Remarkably, weak two-cell-like paternal H3K4me3 patterns are already detectable at the late zygote stage, which may reflect a poising of these sites for transcription. Interestingly, these weak zygotic, paternal H3K4me3 domains are much expanded and atypical (Dahl et al., 2016; Liu et al., 2016; Zhang et al., 2016) (Fig. 4).

Turning to the maternal genome it becomes apparent that early oocyte expressed genes show H3K4me3 labeling in fully grown oocytes and zygotes, which are transcriptionally inert. This suggests inheritance of early oocyte-specific H3K4me3 marks to the embryo. Early embryo-specific genes showed strong enrichment at the two-cell stage, paralleling zygotic gene activation. However, low levels of H3K4me3 were also found at embryo-specific genes before activation of transcription, indicating (as in the paternal pronucleus) possible poising of promoters. After zygotic gene activation H3K4me3 marks correlate with active genes and gene transcription as expected (Dahl et al., 2016; Liu et al., 2016; Zhang et al., 2016) (Fig. 4).

More surprising is the finding of a noncanonical H3K4me3 (ncH3K4me3) distribution pattern very prominent in oocytes and persisting into early embryos (Dahl et al., 2016; Zhang et al., 2016). Low enrichment levels of H3K4me3 were found to spread over large regions (>10 kb), mostly distant from transcriptional start sites. The origin and possible function of maternal ncH3K4me3 are unexpected, too. It emerges in growing oocytes over regions of low DNA methylation, which are linked to low expression. It is not fully resolved yet, but it appears that DNA hypomethylation precedes ncH3K4me3 establishment. In a bizarre twist this could imply that in oocytes ncH3K4me3 may have a repressive function. This is further supported by aberrant transcriptional activity upon ncH3K4me3 removal in the usually transcriptionally inert oocyte (Zhang et al., 2016).

In mammals, unique H3K4me3 distribution has been observed in particular over developmental gene clusters. Here it coresides with H3K27me3, a repressing histone mark (see above). It has been proposed that this coexistence and interplay of activating and repressing marks, a "bivalent" condition, poise promoters for prompt activation upon differentiation stimuli (Bernstein et al., 2006; Boyer et al., 2006). The bivalency of developmental gene promoters is of particular importance at early differentiation commitments in the periimplantation embryo. It is yet unclear if the bivalent state is preexisting in the very early preimplantation embryo, or even inherited from parents. The ncH3K4me3 patterns transmitted by the oocyte do cover both, promoters and large gene-deprived regions. However, the promoters of developmental genes where poorly enriched for H3K4me3 in oocytes and zygotes and even further depleted in preimplantation embryos and ICMs. An increase of H3K4me3 at classical bivalent promoters is observed only in the epiblast (Dahl et al., 2016; Zhang et al., 2016). On the other hand, promoter H3K27me3 is only established at the periimplantation stages. Thus, bivalent promoters are, in large, established toward the end of preimplantation development and are not inherited through gametes to the embryo.

2.2.4 H3K9me3

H3K9me3 is a repressive histone mark, which is functionally distinctive from H3K27me3. Whereas H3K27me3 is more frequently found at gene promoters and involved in gene expression regulation, H3K9me3 is often found in gene-poor domains and at endogenous retrotransposons (Bannister & Kouzarides, 2011; Kouzarides, 2007; Mikkelsen et al., 2007).

Global ChIP-seq data for H3K9me3 is not available in gametes and embryos yet, and the observation of dynamic changes is based on classical staining methods. In the maternal pronucleus H3K9me3 is very abundant. Although H3K9 monomethylation is detectable in the paternal pronucleus, H3K9me3 is not (Lepikhov & Walter, 2004; Santos et al., 2005; Wongtawan et al., 2011; Yeo, Lee, Han, & Kang, 2005) (Fig. 4). This is intriguing as DNA methylation and H3K9me3 are intimately linked, with possibly the histone mark preceding the 5mC acquisition. Possibly, the lack of H3K9me3 in the paternal genome is a prerequisite to active DNA demethylation (see above). Of all the marks discussed in this chapter the asymmetry of H3K9me3 persists the longest and is still clearly detectable at the four-cell stage (Hayashi-Takanaka et al., 2011; Ribeiro-Mason et al., 2012). It appears to be properly resolved only after the eight-cell stage. However, as demonstrated by ChIP-seq for H3K27me3 and H3K4me3 in preimplantation embryos such global staining-derived conclusions are to be taken with caution as asymmetry may persist on a locus-specific level.

2.2.5 Other Histone Marks and Their Dynamic Changes

Next to the Histone modifications discussed in detail earlier, others also show similar dynamics and properties and may play crucial roles during the time of oocyte to embryo transition. For example, H3K36me3, which is usually spread within actively transcribed gene bodies, is enriched in the maternal, but not paternal pronucleus, so is H3K64me3 (often found in pericentromeric chromatin) and H4K20me3 (often found on repressed promoters). Yet, most of these asymmetries are resolved at the two-cell stage, at least on an immunofluorescent level (Bošković et al., 2012; Daujat et al., 2009; Probst, Santos, Reik, Almouzni, & Dean, 2007; van der Heijden et al., 2005; Wongtawan et al., 2011) (Fig. 4). Interestingly, unlike other marks where a gain of paternal marks is observed during assimilation, H4K20me3 and H3K36me3 are removed from the maternal genome and are undetected at the two-cell stage. H4K20me3 is only reestablished at the blastocyst stage (Wongtawan et al., 2011) (Fig. 4).

2.2.6 Functional Analysis of Histone Marks Through Effector Manipulations

As for DNA methylation, where good understanding of the global state and dynamics exists but detailed insights of reprogramming vs maintenance requirements on a target-specific level are lacking, we remain similarly oblivious in respect to histone modifications. Here the problem is even larger by magnitudes as a multitude of (1) mechanisms can affect a single modification and their dynamic and (2) combinations of multiple histone modifications may result in locus-specific outcomes and effects. Understanding the chromatin patterns and dynamics of individual marks during development is therefore an important resource to have, yet it is far from revealing the functional needs during reprogramming. The way forward here will be the manipulation of enzymes directly involved in the processes. This is either achievable by the overexpression of, e.g., methyltransferases or demethylases, or their maternal ablation—preferentially genetically—from oocytes to study specific effects and phenotypic outcomes.

For instance, H4K20me3 is found on heterochromatic regions, regions with a repressive, compact structure often at repeats, i.e., at telomeres, centromeres, and pericentromeric regions (Schotta et al., 2008). The absence of H4K20me3 during the very early developmental period was thought to be essential to achieve maximal chromatin plasticity and potency. Indeed, the enzymes driving trimethylation of H4K20me3 (*Kmt5b* and *Kmt5c*) are minimally expressed at the time. Ectopic overexpression of *Kmt5c*, which enforces maintenance of H4K20me3, causes developmental arrest at the two-cell stage (Eid, Rodriguez-Terrones, Burton, & Torres-Padilla, 2016). While this functionally shows the requirement of H4K20me3 removal for proper development, it still remains unclear which are the really relevant targets of this process. Another study addressed the function of KDM1A during oocyte-to-embryo transition. KDM1A is a H3K4 and

K9 demethylase and therefore likely to participate in the observed dynamics in maternal and paternal genomes described earlier. Maternal deletion of KDM1A has detrimental effects on development and causes a penetrant two-cell stage arrest (Ancelin et al., 2016). More such approaches in combination with genome-wide histone mark distribution readout by ChIP-seq will be required in the future. Through such experiments the relevant targets of the reprogramming process might come into reach and locus-specific manipulation approaches, for example, using CRISPR technology, may reveal the true mechanisms and needs for epigenetic reprogramming on the chromatin level.

3. CODA

It has been 35 years since, with the description of genomic imprinting in mice, the importance of epigenetic mechanisms in mammalian reproduction has been truly embraced (McGrath & Solter, 1983; Surani & Barton, 1983). Even long before the description of the "imprints" the initiation of new life through sexual reproduction was intensely studied. It is remarkable how much knowledge has been created since, and how far the field of epigenetics in general, yet also specifically during embryonic reprogramming, has come. Recently, the development of next-generation sequencing has hurled the field forward by a quantum leap and a provided wealth of information beyond imagination only few years back. Yet, our true understanding of the process, occurring in a simple biological system readily accessible in vitro, remains comparatively narrow. The imbalance of available data and a true, unquestionable comprehension of even specific events during, let alone, the whole reprogramming process is staggering. Do we not have the means to interpret the flood of data in a correct way? Or, are we simply asking the wrong questions? Are we missing a fundamental piece of the puzzle? A major chromatin-independent epigenetic mechanism perhaps? Or can we simply not see the wood for the trees? Whole genome analysis of chromatin states and its dynamics has undoubtedly its merits and will provide a valuable resource moving forward. It will now be important to integrate these data, identify cross talk of histone marks and DNA methylation, but most importantly of them all conduct the experiments that will reveal the functional aspects and true targets of embryonic epigenetic reprogramming.

REFERENCES

Abramowitz, L. K., & Bartolomei, M. S. (2012). Genomic imprinting: Recognition and marking of imprinted loci. *Current Opinion in Genetics & Development*, *22*, 72–78. https://doi.org/10.1016/j.gde.2011.12.001.

Adenot, P. G., Mercier, Y., Renard, J. P., & Thompson, E. M. (1997). Differential H4 acetylation of paternal and maternal chromatin precedes DNA replication and differential transcriptional activity in pronuclei of 1-cell mouse embryos. *Development*, *124*, 4615–4625.

Adenot, P. G., Szöllösi, M. S., Geze, M., Renard, J. P., & Debey, P. (1991). Dynamics of paternal chromatin changes in live one-cell mouse embryo after natural fertilization. *Molecular Reproduction and Development*, *28*, 23–34. https://doi.org/10.1002/mrd.1080280105.

Ahmad, K., & Henikoff, S. (2002). Epigenetic consequences of nucleosome dynamics. *Cell*, *111*, 281–284.

Amor, D. J., & Halliday, J. (2008). A review of known imprinting syndromes and their association with assisted reproduction technologies. *Human Reproduction*, *23*, 2826–2834. https://doi.org/10.1093/humrep/den310.

Amouroux, R., Nashun, B., Shirane, K., Nakagawa, S., Hill, P. W. S., D'Souza, Z., et al. (2016). De novo DNA methylation drives 5hmC accumulation in mouse zygotes. *Nature Cell Biology*, *18*, 225–233. https://doi.org/10.1038/ncb3296.

Ancelin, K., Syx, L., Borensztein, M., Ranisavljevic, N., Vassilev, I., Briseño-Roa, L., et al. (2016). Maternal LSD1/KDM1A is an essential regulator of chromatin and transcription landscapes during zygotic genome activation. *eLife*, *5*. https://doiorg/10.7554/eLife.08851.

Arand, J., Spieler, D., Karius, T., Branco, M. R., Meilinger, D., Meissner, A., et al. (2012). In vivo control of CpG and non-CpG DNA methylation by DNA methyltransferases. *PLoS Genetics*, *8*, e1002750. https://doi.org/10.1371/journal.pgen.1002750.

Bannister, A. J., & Kouzarides, T. (2011). Regulation of chromatin by histone modifications. *Cell Research*, *21*, 381–395. https://doi.org/10.1038/cr.2011.22.

Bartolomei, M. S. (2009). Genomic imprinting: Employing and avoiding epigenetic processes. *Genes & Development*, *23*, 2124–2133. https://doi.org/10.1101/gad.1841409.

Bernstein, B. E., Meissner, A., & Lander, E. S. (2007). The mammalian epigenome. *Cell*, *128*, 669–681. https://doi.org/10.1016/j.cell.2007.01.033.

Bernstein, B. E., Mikkelsen, T. S., Xie, X., Kamal, M., Huebert, D. J., Cuff, J., et al. (2006). A bivalent chromatin structure marks key developmental genes in embryonic stem cells. *Cell*, *125*, 315–326. https://doi.org/10.1016/j.cell.2006.02.041.

Bestor, T. H. (1988). Cloning of a mammalian DNA methyltransferase. *Gene*, *74*, 9–12.

Bird, A. (2002). DNA methylation patterns and epigenetic memory. *Genes & Development*, *16*, 6–21. https://doi.org/10.1101/gad.947102.

Bonasio, R., Tu, S., & Reinberg, D. (2010). Molecular signals of epigenetic states. *Science*, *330*, 612–616. https://doi.org/10.1126/science.1191078.

Borgel, J., Guibert, S., Li, Y., Chiba, H., Schübeler, D., Sasaki, H., et al. (2010). Targets and dynamics of promoter DNA methylation during early mouse development. *Nature Publishing Group*, *42*, 1093–1100. https://doi.org/10.1038/ng.708.

Bošković, A., Bender, A., Gall, L., Ziegler-Birling, C., Beaujean, N., & Torres-Padilla, M.-E. (2012). Analysis of active chromatin modifications in early mammalian embryos reveals uncoupling of H2A.Z acetylation and H3K36 trimethylation from embryonic genome activation. *Epigenetics*, *7*, 747–757. https://doi.org/10.4161/epi.20584.

Bostick, M., Kim, J. K., Estève, P.-O., Clark, A., Pradhan, S., & Jacobsen, S. E. (2007). UHRF1 plays a role in maintaining DNA methylation in mammalian cells. *Science*, *317*, 1760–1764. https://doi.org/10.1126/science.1147939.

Bourc'his, D., & Bestor, T. H. (2004). Meiotic catastrophe and retrotransposon reactivation in male germ cells lacking Dnmt3L. *Nature*, *431*, 96–99. https://doi.org/10.1038/nature02886.

Bourc'his, D., & Voinnet, O. (2010). A small-RNA perspective on gametogenesis, fertilization, and early zygotic development. *Science*, *330*, 617–622. https://doi.org/10.1126/science.1194776.

Bourc'his, D., Xu, G. L., Lin, C. S., Bollman, B., & Bestor, T. H. (2001). Dnmt3L and the establishment of maternal genomic imprints. *Science*, *294*, 2536–2539. https://doi.org/10.1126/science.1065848.

Boyer, L. A., Plath, K., Zeitlinger, J., Brambrink, T., Medeiros, L. A., Lee, T. I., et al. (2006). Polycomb complexes repress developmental regulators in murine embryonic stem cells. *Nature*, *441*, 349–353. https://doi.org/10.1038/nature04733.

Branco, M. R., King, M., Perez-Garcia, V., Bogutz, A. B., Caley, M., Fineberg, E., et al. (2016). Maternal DNA methylation regulates early trophoblast development. *Developmental Cell*, *36*, 152–163. https://doi.org/10.1016/j.devcel.2015.12.027.

Brykczynska, U., Hisano, M., Erkek, S., Ramos, L., Oakeley, E. J., Roloff, T. C., et al. (2010). Repressive and active histone methylation mark distinct promoters in human and mouse spermatozoa. *Nature Structural & Molecular Biology*, *17*, 679–687. https://doi.org/10.1038/nsmb.1821.

Burton, A., & Torres-Padilla, M.-E. (2010). Epigenetic reprogramming and development: A unique heterochromatin organization in the preimplantation mouse embryo. *Briefings in Functional Genomics*, *9*, 444–454. https://doi.org/10.1093/bfgp/elq027.

Burton, A., & Torres-Padilla, M.-E. (2014). Chromatin dynamics in the regulation of cell fate allocation during early embryogenesis. *Nature Reviews. Molecular Cell Biology*, *15*, 723–734. https://doi.org/10.1038/nrm3885.

Butler, M. G. (2009). Genomic imprinting disorders in humans: A mini-review. *Journal of Assisted Reproduction and Genetics*, *26*, 477–486. https://doi.org/10.1007/s10815-009-9353-3.

Cardoso, M. C., & Leonhardt, H. (1999). DNA methyltransferase is actively retained in the cytoplasm during early development. *The Journal of Cell Biology*, *147*, 25–32.

Chuva de Sousa Lopes, S. M., Hayashi, K., Shovlin, T. C., Mifsud, W., Surani, M. A., & McLaren, A. (2008). X chromosome activity in mouse XX primordial germ cells. *PLoS Genetics*, *4*, e30. https://doi.org/10.1371/journal.pgen.0040030.

Clayton, A. L., Hazzalin, C. A., & Mahadevan, L. C. (2006). Enhanced histone acetylation and transcription: A dynamic perspective. *Molecular Cell*, *23*, 289–296. https://doi.org/10.1016/j.molcel.2006.06.017.

Dahl, J. A., Jung, I., Aanes, H., Greggains, G. D., Manaf, A., Lerdrup, M., et al. (2016). Broad histone H3K4me3 domains in mouse oocytes modulate maternal-to-zygotic transition. *Nature*, *537*, 548–552. https://doi.org/10.1038/nature19360.

Daujat, S., Weiss, T., Mohn, F., Lange, U. C., Ziegler-Birling, C., Zeissler, U., et al. (2009). H3K64 trimethylation marks heterochromatin and is dynamically remodeled during developmental reprogramming. *Nature Structural & Molecular Biology*, *16*, 777–781. https://doi.org/10.1038/nsmb.1629.

Deaton, A. M., & Bird, A. (2011). CpG islands and the regulation of transcription. *Genes & Development*, *25*, 1010–1022. https://doi.org/10.1101/gad.2037511.

Eid, A., Rodriguez-Terrones, D., Burton, A., & Torres-Padilla, M.-E. (2016). SUV4-20 activity in the preimplantation mouse embryo controls timely replication. *Genes & Development*, *30*, 2513–2526. https://doi.org/10.1101/gad.288969.116.

Erhardt, S., Su, I.-H., Schneider, R., Barton, S., Bannister, A. J., Perez-Burgos, L., et al. (2003). Consequences of the depletion of zygotic and embryonic enhancer of zeste 2 during preimplantation mouse development. *Development*, *130*, 4235–4248.

Estève, P.-O., Chin, H. G., Smallwood, A., Feehery, G. R., Gangisetty, O., Karpf, A. R., et al. (2006). Direct interaction between DNMT1 and G9a coordinates DNA and

histone methylation during replication. *Genes & Development, 20*, 3089–3103. https://doi.org/10.1101/gad.1463706.
Ferguson-Smith, A. C. (2011). Genomic imprinting: The emergence of an epigenetic paradigm. *Nature Reviews. Genetics, 12*, 565–575. https://doi.org/10.1038/nrg3032.
Goll, M. G., & Bestor, T. H. (2005). Eukaryotic cytosine methyltransferases. *Annual Review of Biochemistry, 74*, 481–514. https://doi.org/10.1146/annurev.biochem.74.010904.153721.
Gu, T.-P., Guo, F., Yang, H., Wu, H.-P., Xu, G.-F., Liu, W., et al. (2011). The role of Tet3 DNA dioxygenase in epigenetic reprogramming by oocytes. *Nature, 477*, 606–610. https://doi.org/10.1038/nature10443.
Guo, F., Li, X., Liang, D., Li, T., Zhu, P., Guo, H., et al. (2014). Active and passive demethylation of male and female pronuclear DNA in the mammalian zygote. *Cell Stem Cell, 15*, 447–458. https://doi.org/10.1016/j.stem.2014.08.003.
Hackett, J. A., Sengupta, R., Zylicz, J. J., Murakami, K., Lee, C., Down, T. A., et al. (2013). Germline DNA demethylation dynamics and imprint erasure through 5-hydroxymethylcytosine. *Science, 339*, 448–452. https://doi.org/10.1126/science.1229277.
Hajkova, P., Erhardt, S., Lane, N., Haaf, T., El-Maarri, O., Reik, W., et al. (2002). Epigenetic reprogramming in mouse primordial germ cells. *Mechanisms of Development, 117*, 15–23.
Hake, S. B., & Allis, C. D. (2006). Histone H3 variants and their potential role in indexing mammalian genomes: The "H3 barcode hypothesis" *Proceedings of the National Academy of Sciences of the United States of America, 103*, 6428–6435. https://doi.org/10.1073/pnas.0600803103.
Hammoud, S. S., Nix, D. A., Zhang, H., Purwar, J., Carrell, D. T., & Cairns, B. R. (2009). Distinctive chromatin in human sperm packages genes for embryo development. *Nature, 460*, 473–478. https://doi.org/10.1038/nature08162.
Hashimoto, H., Liu, Y., Upadhyay, A. K., Chang, Y., Howerton, S. B., Vertino, P. M., et al. (2012). Recognition and potential mechanisms for replication and erasure of cytosine hydroxymethylation. *Nucleic Acids Research, 40*, 4841–4849. https://doi.org/10.1093/nar/gks155.
Hayashi, K., de Sousa Lopes, S. M. C., & Surani, M. A. (2007). Germ cell specification in mice. *Science, 316*, 394–396. https://doi.org/10.1126/science.1137545.
Hayashi-Takanaka, Y., Yamagata, K., Wakayama, T., Stasevich, T. J., Kainuma, T., Tsurimoto, T., et al. (2011). Tracking epigenetic histone modifications in single cells using fab-based live endogenous modification labeling. *Nucleic Acids Research, 39*, 6475–6488. https://doi.org/10.1093/nar/gkr343.
He, Y.-F., Li, B.-Z., Li, Z., Liu, P., Wang, Y., Tang, Q., et al. (2011). Tet-mediated formation of 5-carboxylcytosine and its excision by TDG in mammalian DNA. *Science, 333*, 1303–1307. https://doi.org/10.1126/science.1210944.
Heard, E., & Martienssen, R. A. (2014). Transgenerational epigenetic inheritance: Myths and mechanisms. *Cell, 157*, 95–109. https://doi.org/10.1016/j.cell.2014.02.045.
Hemberger, M., Dean, W., & Reik, W. (2009). Epigenetic dynamics of stem cells and cell lineage commitment: Digging Waddington's canal. *Nature Reviews. Molecular Cell Biology, 10*, 526–537. https://doi.org/10.1038/nrm2727.
Hirasawa, R., Chiba, H., Kaneda, M., Tajima, S., Li, E., Jaenisch, R., et al. (2008). Maternal and zygotic Dnmt1 are necessary and sufficient for the maintenance of DNA methylation imprints during preimplantation development. *Genes & Development, 22*, 1607–1616. https://doi.org/10.1101/gad.1667008.
Hitt, M. M., Wu, T. L., Cohen, G., & Linn, S. (1988). De novo and maintenance DNA methylation by a mouse plasmacytoma cell DNA methyltransferase. *The Journal of Biological Chemistry, 263*, 4392–4399.
Holliday, R., & Pugh, J. E. (1975). DNA modification mechanisms and gene activity during development. *Science, 187*, 226–232.

Howell, C. Y., Bestor, T. H., Ding, F., Latham, K. E., Mertineit, C., Trasler, J. M., et al. (2001). Genomic imprinting disrupted by a maternal effect mutation in the Dnmt1 gene. *Cell, 104*, 829–838. https://doi.org/10.1016/S0092-8674(01)00280-X.

Inoue, A., Jiang, L., Lu, F., Suzuki, T., & Zhang, Y. (2017). Maternal H3K27me3 controls DNA methylation-independent imprinting. *Nature, 547*, 419–424. https://doi.org/10.1038/nature23262.

Inoue, A., Shen, L., Dai, Q., He, C., & Zhang, Y. (2011). Generation and replication-dependent dilution of 5fC and 5caC during mouse preimplantation development. *Cell Research, 21*, 1670–1676. https://doi.org/10.1038/cr.2011.189.

Inoue, A., & Zhang, Y. (2011). Replication-dependent loss of 5-hydroxymethylcytosine in mouse preimplantation embryos. *Science, 334*, 194. https://doi.org/10.1126/science.1212483.

Iqbal, K., Jin, S.-G., Pfeifer, G. P., & Szabó, P. E. (2011). Reprogramming of the paternal genome upon fertilization involves genome-wide oxidation of 5-methylcytosine. *Proceedings of the National Academy of Sciences, 108*, 3642–3647. https://doi.org/10.1073/pnas.1014033108.

Ito, S., D'Alessio, A. C., Taranova, O. V., Hong, K., Sowers, L. C., & Zhang, Y. (2010). Role of Tet proteins in 5mC to 5hmC conversion, ES-cell self-renewal and inner cell mass specification. *Nature, 466*, 1129–1133. https://doi.org/10.1038/nature09303.

Ito, S., Shen, L., Dai, Q., Wu, S. C., Collins, L. B., Swenberg, J. A., et al. (2011). Tet proteins can convert 5-methylcytosine to 5-formylcytosine and 5-carboxylcytosine. *Science, 333*, 1300–1303. https://doi.org/10.1126/science.1210597.

Iyengar, S., & Farnham, P. J. (2011). KAP1 protein: An enigmatic master regulator of the genome. *The Journal of Biological Chemistry, 286*, 26267–26276. https://doi.org/10.1074/jbc.R111.252569.

Jacobs, F. M. J., Greenberg, D., Nguyen, N., Haeussler, M., Ewing, A. D., Katzman, S., et al. (2014). An evolutionary arms race between KRAB zinc-finger genes ZNF91/93 and SVA/L1 retrotransposons. *Nature, 516*, 242–245. https://doi.org/10.1038/nature13760.

Kaneda, M., Okano, M., Hata, K., Sado, T., Tsujimoto, N., Li, E., et al. (2004). Essential role for de novo DNA methyltransferase Dnmt3a in paternal and maternal imprinting. *Nature, 429*, 900–903. https://doi.org/10.1038/nature02633.

Kato, Y., Kaneda, M., Hata, K., Kumaki, K., Hisano, M., Kohara, Y., et al. (2007). Role of the Dnmt3 family in de novo methylation of imprinted and repetitive sequences during male germ cell development in the mouse. *Human Molecular Genetics, 16*, 2272–2280. https://doi.org/10.1093/hmg/ddm179.

Kimura, Y., & Yanagimachi, R. (1995). Development of normal mice from oocytes injected with secondary spermatocyte nuclei. *Biology of Reproduction, 53*, 855–862.

Kishikawa, S., Murata, T., Ugai, H., Yamazaki, T., & Yokoyama, K. K. (2003). Control elements of Dnmt1 gene are regulated in cell-cycle dependent manner. *Nucleic Acids Research. Supplement, 2001*, 307–308.

Kobayashi, H., Sakurai, T., Imai, M., Takahashi, N., Fukuda, A., Yayoi, O., et al. (2012). Contribution of intragenic DNA methylation in mouse gametic DNA methylomes to establish oocyte-specific heritable marks. *PLoS Genetics, 8*. e1002440. https://doi.org/10.1371/journal.pgen.1002440.s018.

Kobayashi, H., Sakurai, T., Miura, F., Imai, M., Mochiduki, K., Yanagisawa, E., et al. (2013). High-resolution DNA methylome analysis of primordial germ cells identifies gender-specific reprogramming in mice. *Genome Research, 23*, 616–627. https://doi.org/10.1101/gr.148023.112.

Kourmouli, N., Jeppesen, P., Mahadevhaiah, S., Burgoyne, P., Wu, R., Gilbert, D. M., et al. (2004). Heterochromatin and tri-methylated lysine 20 of histone H4 in animals. *Journal of Cell Science, 117*, 2491–2501. https://doi.org/10.1242/jcs.01238.

Kouzarides, T. (2007). Chromatin modifications and their function. *Cell, 128*, 693–705. https://doi.org/10.1016/j.cell.2007.02.005.
Kurihara, Y., Kawamura, Y., Uchijima, Y., Amamo, T., Kobayashi, H., Asano, T., et al. (2008). Maintenance of genomic methylation patterns during preimplantation development requires the somatic form of DNA methyltransferase 1. *Developmental Biology, 313*, 335–346. https://doi.org/10.1016/j.ydbio.2007.10.033.
Lei, H., Oh, S. P., Okano, M., Jüttermann, R., Goss, K. A., Jaenisch, R., et al. (1996). De novo DNA cytosine methyltransferase activities in mouse embryonic stem cells. *Development, 122*, 3195–3205.
Leonhardt, H., Page, A. W., Weier, H. U., & Bestor, T. H. (1992). A targeting sequence directs DNA methyltransferase to sites of DNA replication in mammalian nuclei. *Cell, 71*, 865–873.
Lepikhov, K., & Walter, J. (2004). Differential dynamics of histone H3 methylation at positions K4 and K9 in the mouse zygote. *BMC Developmental Biology, 4*, 12. https://doi.org/10.1186/1471-213X-4-12.
Li, X. (2013). Genomic imprinting is a parental effect established in mammalian germ cells. *Current Topics in Developmental Biology, 102*, 35–59. https://doi.org/10.1016/B978-0-12-416024-8.00002-7.
Li, X., Ito, M., Zhou, F., Youngson, N., Zuo, X., Leder, P., et al. (2008). A maternal-zygotic effect gene, Zfp57, maintains both maternal and paternal imprints. *Developmental Cell, 15*, 547–557. https://doi.org/10.1016/j.devcel.2008.08.014.
Lim, C. Y., Knowles, B. B., Solter, D., & Messerschmidt, D. M. (2016). Epigenetic control of early mouse development. *Current Topics in Developmental Biology, 120*, 311–360. https://doi.org/10.1016/bs.ctdb.2016.05.002.
Liu, X., Wang, C., Liu, W., Li, J., Li, C., Kou, X., et al. (2016). Distinct features of H3K4me3 and H3K27me3 chromatin domains in pre-implantation embryos. *Nature, 537*, 558–562. https://doi.org/10.1038/nature19362.
Lorthongpanich, C., Cheow, L. F., Balu, S., Quake, S. R., Knowles, B. B., Burkholder, W. F., et al. (2013). Single-cell DNA-methylation analysis reveals epigenetic chimerism in preimplantation embryos. *Science, 341*, 1110–1112. https://doi.org/10.1126/science.1240617.
Margueron, R., & Reinberg, D. (2011). The Polycomb complex PRC2 and its mark in life. *Nature, 469*, 343–349. https://doi.org/10.1038/nature09784.
Mayer, W., Niveleau, A., Walter, J., Fundele, R., & Haaf, T. (2000). Demethylation of the zygotic paternal genome. *Nature, 403*, 501–502. https://doi.org/10.1038/35000654.
McGrath, J., & Solter, D. (1983). Nuclear transplantation in the mouse embryo by microsurgery and cell fusion. *Science, 220*, 1300–1302.
McLaren, A., & Lawson, K. A. (2005). How is the mouse germ-cell lineage established? *Differentiation, 73*, 435–437. https://doi.org/10.1111/j.1432-0436.2005.00049.x.
Messerschmidt, D. M. (2012). Should I stay or should I go: Protection and maintenance of DNA methylation at imprinted genes. *Epigenetics*, 7, 969–975. https://doi.org/10.4161/epi.21337.
Messerschmidt, D. M. (2016). A twist in zygotic reprogramming. *Nature Cell Biology, 18*, 139–140. https://doi.org/10.1038/ncb3304.
Messerschmidt, D. M., de Vries, W., Ito, M., Solter, D., Ferguson-Smith, A., & Knowles, B. B. (2012). Trim28 is required for epigenetic stability during mouse oocyte to embryo transition. *Science, 335*, 1499–1502. https://doi.org/10.1126/science.1216154.
Messerschmidt, D. M., Knowles, B. B., & Solter, D. (2014). DNA methylation dynamics during epigenetic reprogramming in the germline and preimplantation embryos. *Genes & Development, 28*, 812–828. https://doi.org/10.1101/gad.234294.113.

Mikkelsen, T. S., Ku, M., Jaffe, D. B., Issac, B., Lieberman, E., Giannoukos, G., et al. (2007). Genome-wide maps of chromatin state in pluripotent and lineage-committed cells. *Nature, 448*, 553–560. https://doi.org/10.1038/nature06008.

Morgan, H. D., Dean, W., Coker, H. A., Reik, W., & Petersen-Mahrt, S. K. (2004). Activation-induced cytidine deaminase deaminates 5-methylcytosine in DNA and is expressed in pluripotent tissues: Implications for epigenetic reprogramming. *The Journal of Biological Chemistry, 279*, 52353–52360. https://doi.org/10.1074/jbc.M407695200.

Morgan, H. D., Santos, F., Green, K., Dean, W., & Reik, W. (2005). Epigenetic reprogramming in mammals. *Human Molecular Genetics, 14*(Spec No. 1), R47–R58. https://doi.org/10.1093/hmg/ddi114.

Nakamura, T., Arai, Y., Umehara, H., Masuhara, M., Kimura, T., Taniguchi, H., et al. (2006). PGC7/Stella protects against DNA demethylation in early embryogenesis. *Nature Cell Biology, 9*, 64–71. https://doi.org/10.1038/ncb1519.

Nakamura, T., Liu, Y.-J., Nakashima, H., Umehara, H., Inoue, K., Matoba, S., et al. (2012). PGC7 binds histone H3K9me2 to protect against conversion of 5mC to 5hmC in early embryos. *Nature, 486*, 415–419. https://doi.org/10.1038/nature11093.

Ng, H. H., Robert, F., Young, R. A., & Struhl, K. (2003). Targeted recruitment of Set1 histone methylase by elongating Pol II provides a localized mark and memory of recent transcriptional activity. *Molecular Cell, 11*, 709–719.

Okano, M., Bell, D. W., Haber, D. A., & Li, E. (1999). DNA methyltransferases Dnmt3a and Dnmt3b are essential for de novo methylation and mammalian development. *Cell, 99*, 247–257.

Okano, M., Xie, S., & Li, E. (1998). Cloning and characterization of a family of novel mammalian DNA (cytosine-5) methyltransferases. *Nature Genetics, 19*, 219–220. https://doi.org/10.1038/890.

Ooi, S. K. T., & Bestor, T. H. (2008). The colorful history of active DNA demethylation. *Cell, 133*, 1145–1148. https://doi.org/10.1016/j.cell.2008.06.009.

Oswald, J., Engemann, S., Lane, N., Mayer, W., Olek, A., Fundele, R., et al. (2000). Active demethylation of the paternal genome in the mouse zygote. *Current Biology, 10*, 475–478.

Pokholok, D. K., Harbison, C. T., Levine, S., Cole, M., Hannett, N. M., Lee, T. I., et al. (2005). Genome-wide map of nucleosome acetylation and methylation in yeast. *Cell, 122*, 517–527. https://doi.org/10.1016/j.cell.2005.06.026.

Polanski, Z., Motosugi, N., Tsurumi, C., Hiiragi, T., & Hoffmann, S. (2008). Hypomethylation of paternal DNA in the late mouse zygote is not essential for development. *The International Journal of Developmental Biology, 52*, 295–298. https://doi.org/10.1387/ijdb.072347zp.

Popp, C., Dean, W., Feng, S., Cokus, S. J., Andrews, S., Pellegrini, M., et al. (2010). Genome-wide erasure of DNA methylation in mouse primordial germ cells is affected by AID deficiency. *Nature, 463*, 1101–1105. https://doi.org/10.1038/nature08829.

Probst, A. V., Santos, F., Reik, W., Almouzni, G., & Dean, W. (2007). Structural differences in centromeric heterochromatin are spatially reconciled on fertilisation in the mouse zygote. *Chromosoma, 116*, 403–415. https://doi.org/10.1007/s00412-007-0106-8.

Quenneville, S., Verde, G., Corsinotti, A., Kapopoulou, A., Jakobsson, J., Offner, S., et al. (2011). In embryonic stem cells, ZFP57/KAP1 recognize a methylated hexanucleotide to affect chromatin and DNA methylation of imprinting control regions. *Molecular Cell, 44*, 361–372. https://doi.org/10.1016/j.molcel.2011.08.032.

Ramsahoye, B. H., Biniszkiewicz, D., Lyko, F., Clark, V., Bird, A. P., & Jaenisch, R. (2000). Non-CpG methylation is prevalent in embryonic stem cells and may be mediated by DNA methyltransferase 3a. *Proceedings of the National Academy of Sciences of the United States of America, 97*, 5237–5242.

Ratnam, S., Mertineit, C., Ding, F., Howell, C. Y., Clarke, H. J., Bestor, T. H., et al. (2002). Dynamics of Dnmt1 methyltransferase expression and intracellular localization during oogenesis and preimplantation development. *Developmental Biology*, *245*, 304–314. https://doi.org/10.1006/dbio.2002.0628.

Reik, W. (2007). Stability and flexibility of epigenetic gene regulation in mammalian development. *Nature*, *447*, 425–432. https://doi.org/10.1038/nature05918.

Reik, W., Santos, F., Mitsuya, K., Morgan, H., & Dean, W. (2003). Epigenetic asymmetry in the mammalian zygote and early embryo: Relationship to lineage commitment? *Philosophical Transactions of the Royal Society of London. Series B, Biological Sciences*, *358*, 1403–1409. discussion 1409. https://doi.org/10.1098/rstb.2003.1326.

Ribeiro-Mason, K., Boulesteix, C., Brochard, V., Aguirre-Lavin, T., Salvaing, J., Fleurot, R., et al. (2012). Nuclear dynamics of histone H3 trimethylated on lysine 9 and/or phosphorylated on serine 10 in mouse cloned embryos as new markers of reprogramming? *Cellular Reprogramming*, *14*, 283–294. https://doi.org/10.1089/cell.2011.0071.

Riggs, A. D. (1975). X inactivation, differentiation, and DNA methylation. *Cytogenetics and Cell Genetics*, *14*, 9–25.

Rowe, H. M., Jakobsson, J., Mesnard, D., Rougemont, J., Reynard, S., Aktas, T., et al. (2010). KAP1 controls endogenous retroviruses in embryonic stem cells. *Nature*, *463*, 237–240. https://doi.org/10.1038/nature08674.

Saitou, M., & Yamaji, M. (2012). Primordial germ cells in mice. *Cold Spring Harbor Perspectives in Biology*, *4*, 1–19. https://doi.org/10.1101/cshperspect.a008375.

Sampath Kumar, A., Seah, M. K. Y., Ling, K. Y., Wang, Y., Tan, J. H. L., Nitsch, S., et al. (2017). Loss of maternal Trim28 causes male-predominant early embryonic lethality. *Genes & Development*, *31*, 12–17. https://doi.org/10.1101/gad.291195.116.

Santenard, A., Ziegler-Birling, C., Koch, M., Tora, L., Bannister, A. J., & Torres-Padilla, M.-E. (2010). Heterochromatin formation in the mouse embryo requires critical residues of the histone variant H3.3. *Nature Cell Biology*, *12*, 853–862. https://doi.org/10.1038/ncb2089.

Santos, F., Hendrich, B., Reik, W., & Dean, W. (2002). Dynamic reprogramming of DNA methylation in the early mouse embryo. *Developmental Biology*, *241*, 172–182. https://doi.org/10.1006/dbio.2001.0501.

Santos, F., Peters, A. H., Otte, A. P., Reik, W., & Dean, W. (2005). Dynamic chromatin modifications characterise the first cell cycle in mouse embryos. *Developmental Biology*, *280*, 225–236. https://doi.org/10.1016/j.ydbio.2005.01.025.

Santos-Rosa, H., Schneider, R., Bannister, A. J., Sherriff, J., Bernstein, B. E., Emre, N. C. T., et al. (2002). Active genes are tri-methylated at K4 of histone H3. *Nature*, *419*, 407–411. https://doi.org/10.1038/nature01080.

Sarma, K., & Reinberg, D. (2005). Histone variants meet their match. *Nature Reviews. Molecular Cell Biology*, *6*, 139–149. https://doi.org/10.1038/nrm1567.

Sasaki, H., & Matsui, Y. (2008). Epigenetic events in mammalian germ-cell development: Reprogramming and beyond. *Nature Reviews. Genetics*, *2008*, 129–140. https://doi.org/10.1038/nrg2295.

Schneider, R., & Grosschedl, R. (2007). Dynamics and interplay of nuclear architecture, genome organization, and gene expression. *Genes & Development*, *21*, 3027–3043. https://doi.org/10.1101/gad.1604607.

Schotta, G., Sengupta, R., Kubicek, S., Malin, S., Kauer, M., Callén, E., et al. (2008). A chromatin-wide transition to H4K20 monomethylation impairs genome integrity and programmed DNA rearrangements in the mouse. *Genes & Development*, *22*, 2048–2061. https://doi.org/10.1101/gad.476008.

Schultz, D. C., Ayyanathan, K., Negorev, D., Maul, G. G., & Rauscher, F. J. (2002). SETDB1: A novel KAP-1-associated histone H3, lysine 9-specific methyltransferase that

contributes to HP1-mediated silencing of euchromatic genes by KRAB zinc-finger proteins. *Genes & Development*, *16*, 919–932. https://doi.org/10.1101/gad.973302.
Schultz, D. C., Friedman, J. R., & Rauscher, F. J. (2001). Targeting histone deacetylase complexes via KRAB-zinc finger proteins: The PHD and bromodomains of KAP-1 form a cooperative unit that recruits a novel isoform of the Mi-2alpha subunit of NuRD. *Genes & Development*, *15*, 428–443. https://doi.org/10.1101/gad.869501.
Seisenberger, S., Andrews, S., Krueger, F., Arand, J., Walter, J., Santos, F., et al. (2012). The dynamics of genome-wide DNA methylation reprogramming in mouse primordial germ cells. *Molecular Cell*, *48*, 849–862. https://doi.org/10.1016/j.molcel.2012.11.001.
Seisenberger, S., Peat, J. R., Hore, T. A., Santos, F., Dean, W., & Reik, W. (2013). Reprogramming DNA methylation in the mammalian life cycle: Building and breaking epigenetic barriers. *Philosophical Transactions of the Royal Society of London. Series B, Biological Sciences*, *368*, 20110330. https://doi.org/10.1098/rstb.2011.0330.
Sharif, J., Muto, M., Takebayashi, S.-I., Suetake, I., Iwamatsu, A., Endo, T. A., et al. (2007). The SRA protein Np95 mediates epigenetic inheritance by recruiting Dnmt1 to methylated DNA. *Nature*, *450*, 908–912. https://doi.org/10.1038/nature06397.
Shen, L., Wu, H., Diep, D., Yamaguchi, S., D'Alessio, A. C., Fung, H.-L., et al. (2013). Genome-wide analysis reveals TET- and TDG-dependent 5-methylcytosine oxidation dynamics. *Cell*, *153*, 692–706. https://doi.org/10.1016/j.cell.2013.04.002.
Smallwood, S. A., Tomizawa, S.-I., Krueger, F., Ruf, N., Carli, N., Segonds-Pichon, A., et al. (2011). Dynamic CpG island methylation landscape in oocytes and preimplantation embryos. *Nature Genetics*, *43*, 811–814. https://doi.org/10.1038/ng.864.
Smith, M. M. (2002). Histone variants and nucleosome deposition pathways. *Molecular Cell*, *9*, 1158–1160.
Smith, Z. D., Chan, M. M., Mikkelsen, T. S., Gu, H., Gnirke, A., Regev, A., et al. (2012). A unique regulatory phase of DNA methylation in the early mammalian embryo. *Nature*, *484*, 339–344. https://doi.org/10.1038/nature10960.
Smith, Z. D., & Meissner, A. (2013). DNA methylation: Roles in mammalian development. *Nature Reviews. Genetics*, *14*, 204–220. https://doi.org/10.1038/nrg3354.
Strome, S., & Lehmann, R. (2007). Germ versus soma decisions: Lessons from flies and worms. *Science*, *316*, 392–393. https://doi.org/10.1126/science.1140846.
Surani, M. A., & Barton, S. C. (1983). Development of gynogenetic eggs in the mouse: Implications for parthenogenetic embryos. *Science*, *222*, 1034–1036.
Tahiliani, M., Koh, K. P., Shen, Y., Pastor, W. A., Bandukwala, H., Brudno, Y., et al. (2009). Conversion of 5-methylcytosine to 5-hydroxymethylcytosine in mammalian DNA by MLL partner TET1. *Science*, *324*, 930–935. https://doi.org/10.1126/science.1170116.
Tomizawa, S.-I., & Sasaki, H. (2012). Genomic imprinting and its relevance to congenital disease, infertility, molar pregnancy and induced pluripotent stem cell. *Journal of Human Genetics*, *57*, 84–91. https://doi.org/10.1038/jhg.2011.151.
Valinluck, V., & Sowers, L. C. (2007). Endogenous cytosine damage products alter the site selectivity of human DNA maintenance methyltransferase DNMT1. *Cancer Research*, *67*, 946–950. https://doi.org/10.1158/0008-5472.CAN-06-3123.
van der Heijden, G. W., Dieker, J. W., Derijck, A. A. H. A., Muller, S., Berden, J. H. M., Braat, D. D. M., et al. (2005). Asymmetry in histone H3 variants and lysine methylation between paternal and maternal chromatin of the early mouse zygote. *Mechanisms of Development*, *122*, 1008–1022. https://doi.org/10.1016/j.mod.2005.04.009.
van der Heijden, G. W., Ramos, L., Baart, E. B., van den Berg, I. M., Derijck, A. A. H. A., van der Vlag, J., et al. (2008). Sperm-derived histones contribute to zygotic chromatin in humans. *BMC Developmental Biology*, *8*, 34. https://doi.org/10.1186/1471-213X-8-34.
Walsh, C. P., Chaillet, J. R., & Bestor, T. H. (1998). Transcription of IAP endogenous retroviruses is constrained by cytosine methylation. *Nature Genetics*, *20*, 116–117. https://doi.org/10.1038/2413.

Wang, L., Zhang, J., Duan, J., Gao, X., Zhu, W., Lu, X., et al. (2014). Programming and inheritance of parental DNA methylomes in mammals. *Cell*, *157*, 979–991. https://doi.org/10.1016/j.cell.2014.04.017.

Watanabe, D., Suetake, I., Tada, T., & Tajima, S. (2002). Stage- and cell-specific expression of Dnmt3a and Dnmt3b during embryogenesis. *Mechanisms of Development*, *118*, 187–190.

Weber, M., Hellmann, I., Stadler, M. B., Ramos, L., Pääbo, S., Rebhan, M., et al. (2007). Distribution, silencing potential and evolutionary impact of promoter DNA methylation in the human genome. *Nature Genetics*, *39*, 457–466. https://doi.org/10.1038/ng1990.

Wongtawan, T., Taylor, J. E., Lawson, K. A., Wilmut, I., & Pennings, S. (2011). Histone H4K20me3 and HP1α are late heterochromatin markers in development, but present in undifferentiated embryonic stem cells. *Journal of Cell Science*, *124*, 1878–1890. https://doi.org/10.1242/jcs.080721.

Wossidlo, M., Nakamura, T., Lepikhov, K., Marques, C. J., Zakhartchenko, V., Boiani, M., et al. (2011). 5-Hydroxymethylcytosine in the mammalian zygote is linked with epigenetic reprogramming. *Nature Communications*, *2*, 241. https://doi.org/10.1038/ncomms1240.

Wu, J., Huang, B., Chen, H., Yin, Q., Liu, Y., Xiang, Y., et al. (2016). The landscape of accessible chromatin in mammalian preimplantation embryos. *Nature*, *534*, 652–657. https://doi.org/10.1038/nature18606.

Wu, S. C., & Zhang, Y. (2010). Active DNA demethylation: Many roads lead to Rome. *Nature Reviews. Molecular Cell Biology*, *11*, 607–620. https://doi.org/10.1038/nrm2950.

Wu, H., & Zhang, Y. (2014). Reversing DNA methylation: Mechanisms, genomics, and biological functions. *Cell*, *156*, 45–68. https://doi.org/10.1016/j.cell.2013.12.019.

Yeo, S., Lee, K.-K., Han, Y.-M., & Kang, Y.-K. (2005). Methylation changes of lysine 9 of histone H3 during preimplantation mouse development. *Molecules and Cells*, *20*, 423–428.

Yoder, J. A., Soman, N. S., Verdine, G. L., & Bestor, T. H. (1997). DNA (cytosine-5)-methyltransferases in mouse cells and tissues. Studies with a mechanism-based probe. *Journal of Molecular Biology*, *270*, 385–395. https://doi.org/10.1006/jmbi.1997.1125.

Zeng, L., & Zhou, M.-M. (2002). Bromodomain: An acetyl-lysine binding domain. *FEBS Letters*, *513*, 124–128.

Zhang, B., Zheng, H., Huang, B., Li, W., Xiang, Y., Peng, X., et al. (2016). Allelic reprogramming of the histone modification H3K4me3 in early mammalian development. *Nature*, *537*, 553–557. https://doi.org/10.1038/nature19361.

Zheng, H., Huang, B., Zhang, B., Xiang, Y., Du, Z., Xu, Q., et al. (2016). Resetting epigenetic memory by reprogramming of histone modifications in mammals. *Molecular Cell*, *63*, 1066–1079. https://doi.org/10.1016/j.molcel.2016.08.032.

Zhu, B., & Reinberg, D. (2011). Epigenetic inheritance: Uncontested? *Cell Research*, *21*, 435–441. https://doi.org/10.1038/cr.2011.26.

Ziller, M. J., Müller, F., Liao, J., Zhang, Y., Gu, H., Bock, C., et al. (2011). Genomic distribution and inter-sample variation of non-CpG methylation across human cell types. *PLoS Genetics*, *7*, e1002389. https://doi.org/10.1371/journal.pgen.1002389.

CHAPTER TEN

Pre-gastrula Development of Non-eutherian Mammals

Stephen Frankenberg
School of BioSciences, University of Melbourne, Parkville, VIC, Australia

Contents

Abstract

Marsupials and monotremes differ from eutherian mammals in many features of their reproduction and development. Some features appear to be representative of transitional stages in evolution from therapsid reptiles to humans and mice, particularly with

Current Topics in Developmental Biology, Volume 128
ISSN 0070-2153
https://doi.org/10.1016/bs.ctdb.2017.10.013

respect to the extraembryonic tissues that have undergone remarkable modifications to accommodate reduced egg size and quantity of yolk/deutoplasm, and increasing emphasis on viviparity and placentation. Trophoblast and hypoblast contribute the epithelial layers in most of the extraembryonic membranes and are the first two lineages to differentiate from the embryonic lineage. How they are specified varies greatly among mammals, perhaps largely due to heterochrony in the stage at which they must function. Differences probably also exist in the stage at which lineages are specified relative to the stage at which they fully commit to differentiation. The dogma of sequential commitment to trophoblast and hypoblast with progressive loss of potency may not be a fundamental feature of early mammalian development, but merely a recently acquired developmental pattern in eutherians, or at least mice.

1. INTRODUCTION

1.1 The Evolution of Amniote Development

Gastrulation is a key stage of development at which the body plan begins to unfold. The three body axes (anteroposterior, dorsoventral, and left–right) become defined and the germ layers (ectoderm, mesoderm, and endoderm) segregate and rearrange themselves to form the gut—the most primitive of all organs. In most metazoans, gastrulation is the first step in the generation of morphological complexity. However, most vertebrates have an entire extra phase of development preceding gastrulation to produce the extraembryonic tissues, which support later development and are eventually discarded. These are especially elaborate in the amniotes, whose extant members include birds, reptiles, and mammals.

The early reptilian ancestors of mammals ("therapsid" reptiles) are likely to have laid large, yolky eggs similar to those of modern-day birds and reptiles. These "cleidoic" eggs had a thick shell to prevent desiccation of the embryo, a key adaptation in the evolution of fully terrestrial vertebrates from amphibians. Unlike amphibians, which are capable of feeding as an aquatic larval stage while still developing, terrestrial reptiles must attain a complete adult body plan by the time of hatching in order to feed in a terrestrial environment. Thus, large quantities of yolk to support development are also characteristic of the cleidoic egg.

The evolution of extraembryonic membranes (or "fetal membranes") in amniotes was coupled with the evolution of these large, yolky eggs: the *yolk sac* functioned as an "external gut" to access the stored yolk for development; the *amnion* provided a fluid-filled environment to support development of the embryo; the *allantois* compartmentalized the storage of waste products that could not diffuse passively into the nonaqueous environment; and the *chorion* provided a link with the external environment to facilitate gas exchange. Most of the fetal membranes are constructed from different arrangements of extraembryonic versions of the primary germ layers—ectoderm, mesoderm, and endoderm (Fig. 1). The basic organization of the fetal membranes is conserved in all amniotes (Ferner & Mess, 2011; Sheng & Foley, 2012), although there is variation in the extent of their fusion or separation. For example, in many amniotes the extracoelomic cavity extends to completely surround and internalize the yolk sac, whereas in marsupials the yolk sac maintains a partially superficial position. Thus the marsupial yolk sac wall is constructed from three different membrane types: somatopleure (mesoderm + endoderm), bilaminar omphalopleure (endoderm + ectoderm), and trilaminar omphalopleure (endoderm + mesoderm + ectoderm).

1.2 Oviparity to Viviparity in Mammals

Coupled with the evolution of viviparity and the placenta, most mammals have dispensed with yolk completely and reverted to very small eggs, but retained and modified the fetal membranes for other functions associated

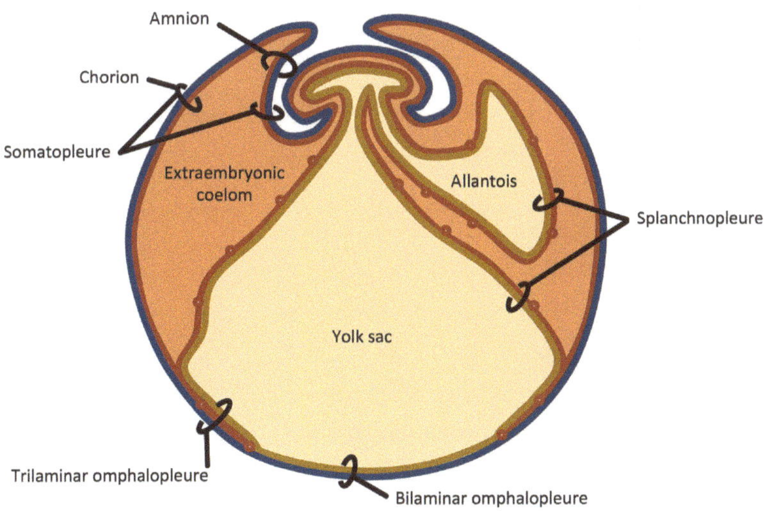

Fig. 1 Arrangement of fetal membranes and their germ-layer composition in a generalized amniote. In mammals, trophoblast is homologous to the extraembryonic ectoderm and contributes to the chorion and omphalopleure (but not the amnion, whose ectoderm layer is epiblast derived), while the hypoblast contributes to the yolk sac (but not the allantois, whose endodermal layer is epiblast derived).

with intrauterine development, such as formation of a placenta. This "backflip" in evolution complicates comparisons with the early development of birds and reptiles by distorting morphological homologies. Monotremes and marsupials, which together with eutherians constitute the three main branches of the class Mammalia (Fig. 2), have retained many features of early development that reflect stages in the transition from ancestral egg-laying reptiles to eutherians such as humans and mice. Thus, the study of monotreme and marsupial development can provide fascinating insights into the evolution of viviparity and the placenta, as well as the mechanisms by which mammals segregate their embryonic and extraembryonic tissues. This comparative approach also provides a strategy for identifying cellular mechanisms that are more ancient and thus fundamental, vs those that are more recently evolved and potentially more plastic.

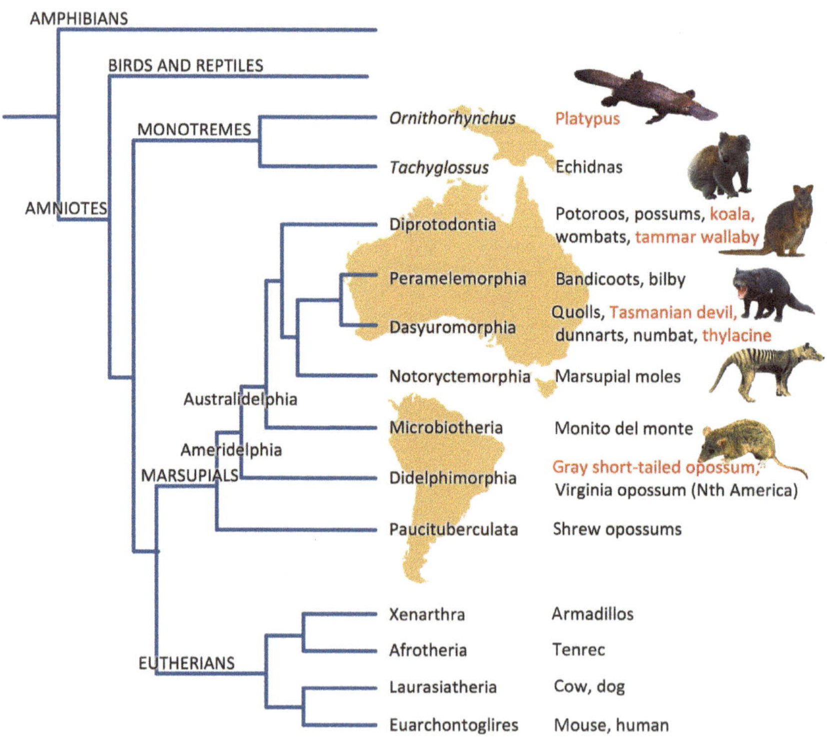

Fig. 2 Phylogeny of amniotes. The cladogram shows the phylogenetic relationship between major groups of mammals and with other amniotes. Common names of selected species are listed, some of which are referred to in this review. Marsupials and monotremes whose genomes have been sequenced are highlighted in *red*. (Branch lengths are not to scale.)

1.3 Mammalian Phylogeny and Model Species

Extant mammals share a common ancestor approximately 220 million years ago (Meredith et al., 2011), when the monotreme lineage diverged from other mammals. Today's monotremes comprise only the Australian platypus and the echidnas, of which the common short-beaked echidna inhabits Australia and several threatened species of long-beaked echidna inhabit New Guinea. Marsupials and eutherians diverged around 190 million years ago (Meredith et al., 2011). Extant marsupials are believed to have a South American common ancestor, from which three main lineages arose. Two lineages remain only in South America, with the exception of one species, the Virginia opossum, that dispersed to North America. The third lineage, the Australidelphia, apparently reached Australia via Gondwana and subsequently radiated to produce all of Australia's marsupials (Nilsson et al., 2010). Only one Australidelphid species—the monito del monte (*Dromiciops gliroides*)—is still found in South America and is thus more closely related to Australian marsupials than to other South American species (Mitchell et al., 2014) (Fig. 2).

Much of our knowledge of early monotreme and marsupial development comes from studies in the late 19th and early 20th centuries. W.H. Caldwell was the first to report egg-laying in the platypus in a famous telegram sent to the British Association for the Advancement of Science meeting in Canada in 1884. He subsequently published observations on a number of embryonic stages (Caldwell, 1887), but far more of our knowledge of monotreme embryology is based on stages collected by J.P. Hill and his colleagues J.T. Wilson and T.T. Flynn in the early 1900s (Flynn & Hill, 1939, 1942, 1947; Wilson & Hill, 1908). Later publications by other authors only reexamined their original material.

Although marsupial embryo specimens have been comparatively much more accessible, the impressively meticulous studies in the early 20th century by C.G. Hartman and E. McCrady (on the Virginia opossum), and by J.P. Hill (on the eastern quoll), have contributed most of our knowledge on early marsupial development. These were supplemented by more recent studies by others, especially L. Selwood with regard to cell-lineage specification. The best-studied marsupials (apart from those mentioned earlier) include dunnarts (*Sminthopsis* spp.), bandicoots (*Isoodon* and *Perameles*), the brown antechinus, the brushtail possum, the tammar wallaby, and the gray short-tailed opossum. Their phylogenetic relationships are shown in Fig. 2.

With the advent of molecular tools and genomics, monotreme and marsupial biology is progressing once again. Genomes have been sequenced and published for the platypus (Warren et al., 2008), the gray short-tailed opossum (Mikkelsen et al., 2007), the tammar wallaby (Renfree et al., 2011), and the Tasmanian devil (Murchison et al., 2012). A high-quality genome assembly for the koala is also now available online, while the next few years are set to see genome assemblies made available for a substantial number of additional marsupial species, including the fat-tailed dunnart.

2. OVERVIEW OF MONOTREME DEVELOPMENT

2.1 Early Monotreme Development

In reptiles and birds, early cleavage is meroblastic and localized at one pole of the large, yolky egg. Peripheral cells of the newly formed blastodisc must undergo a lengthy period of proliferation toward the abembryonic pole in order to form an epithelium that completely encloses the yolk. Monotreme eggs, although substantially smaller, undergo a similar phase of development. At ovulation, the monotreme egg has a diameter of about 4 mm, small compared with birds and reptiles but considerably larger than those of other mammals. Meroblastic cleavage results in a blastodisc at the embryonic pole of the conceptus, from which peripheral cells proliferate toward the abembryonic pole and eventually enclose the yolk (Fig. 3). A key difference with birds and reptiles, however, is that once the epithelium is complete, the conceptus expands in volume by a mechanism that is presumed homologous to blastocyst expansion in other mammals. This compensates for the smaller diameter of monotreme eggs (due to less yolk) compared to those of birds and reptiles.

Shortly after fertilization, the monotreme egg becomes enveloped by a thin shell that stretches as the conceptus expands in volume. The zona pellucida that surrounds the egg from the time of ovulation is lost during this expansion period. After conceptus expansion ceases, an additional thick, leathery shell is deposited and protects the conceptus after laying, similar to shells of oviparous reptiles and birds (Hill, 1933). The physical separation of the conceptus from maternal tissues may be a requirement to prevent its rejection by the maternal immune system, though it may also serve other important roles.

2.2 Fetal Membranes and Placentation

Although famously oviparous, monotremes develop for a significant period of time within the uterus and develop a functional placenta, as do marsupials. The eutherian crown group "Placentalia" is thus misleadingly named as it

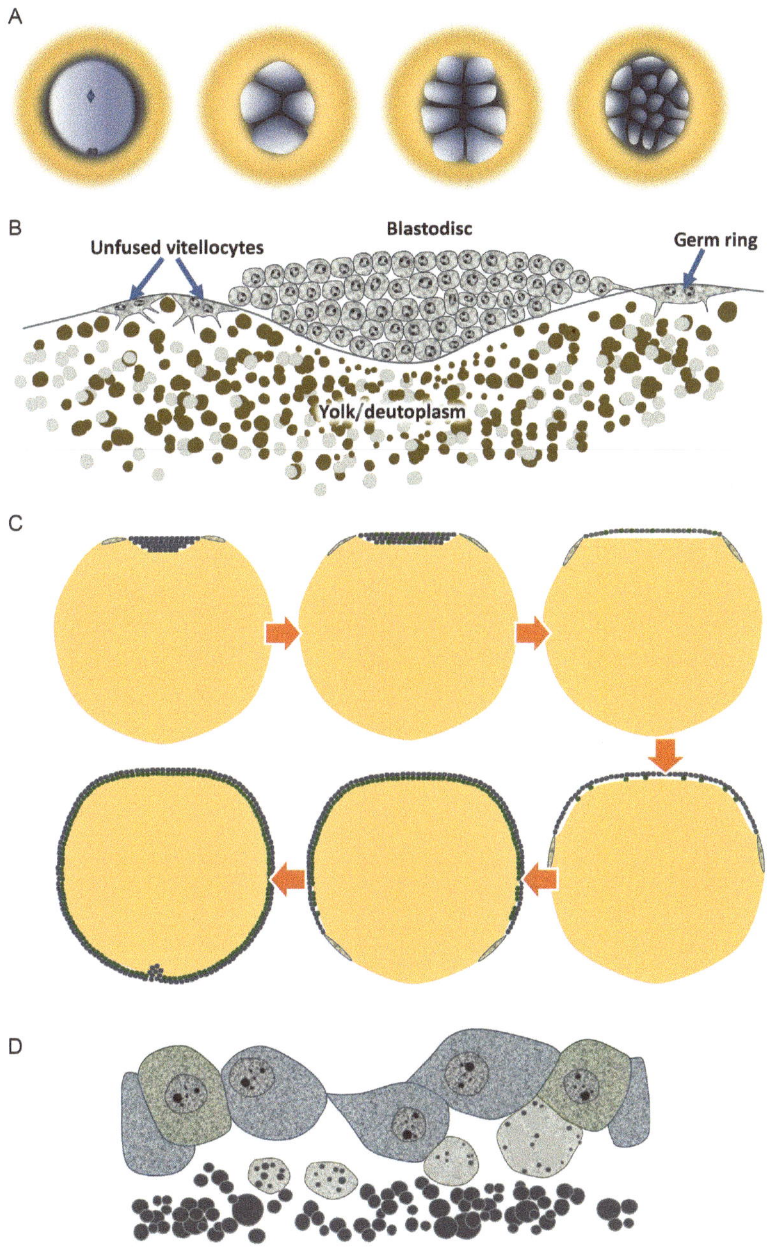

Fig. 3 Early monotreme development. (A) Early cleavage in platypus and echidna. At the zygote stage, asymmetry is seen in the position of the polar bodies. The first cleavage spindle is located toward the opposite pole and orientates so as to produce differently sized blastomeres at the two-cell stage. By the eight-cell stage, bilateral symmetry is overt in both the arrangement of blastomeres and the ovoid shape of the blastodisc. *(Continued)*

implies that they are unique in this respect. Although direct contact is not made between the monotreme placenta and the uterine epithelium, it is adapted for nutrient transfer and thus undergoes a period of expansion to maximize its surface area (Hughes, 1993; Luckett, 1977). The placental epithelium is not in direct contact with the maternal endometrium, as in other mammals, but is separated by a thin shell membrane which begins to form after fertilization and stretches to accommodate conceptus expansion. These additional nutrients are necessary to support development not only in utero but also after the egg is laid at around the 18–19-somite stage (Hill & Gatenby, 1926). At laying, the egg (including the leathery shell) measures about 15 mm in diameter and the fetal membranes consist of a rudimentary proamniotic head fold, no allantois, and a yolk sac with bilaminar and trilaminar omphalopleure (Luckett, 1977). After oviposition, the echidna carries its egg in a pouch, whereas the platypus incubates it within a nesting burrow. During incubation, the omphalopleure becomes completely trilaminar and the allantois expands outward and fuses with the chorion. The extremely altricial young hatches about 10 days later, after which nutrition depends on the sucking of milk from its mother's mammary glands (Griffiths, 2012).

The above mode of reproduction is likely to have been similar in the most recent common ancestor of monotremes and other therian mammals. Curiously, monotremes have an independently evolved mechanism of chromosomal sex determination, since their sex chromosomes are not homologous to those of marsupials and eutherians (Veyrunes et al., 2008). It is thus likely that the most recent common mammalian ancestor employed temperature-dependent sex determination, consistent with the fact that monotreme gonadal sex differentiation is initiated after laying.

Fig. 3—Cont'd By the 16-cell stage, 2-cell populations are distinguishable by their central or peripheral position. (B) The multilayered blastodisc. Multinucleate vitellocytes are present outside the blastodisc, some of which have fused to form the germ ring. (C) Blastocyst formation. As the germ ring extends toward the abembryonic pole, the blastodisc follows it and thins until it is unilaminar. Hypoblast precursors then delaminate to form a bilaminar epithelium by the time the blastocyst is complete. (D) Unilaminar blastoderm showing putative epiblast (*blue*) and hypoblast (*green*) precursors. *Panel (A): Redrawn from Flynn, T. T., & Hill, J. P. (1939). The development of the Monotremata. Part IV. Growth of the ovarian ovum, maturation, fertilisation, and early cleavage.* The Zoological Society of London, 24, *445–622. Panel (D): Redrawn after Flynn, T. T., & Hill, J. P. (1947). The development of the monotremata. Part VI. The later stages of cleavage and the formation of the primary germ layers.* The Transactions of the Zoological Society of London, *21, 1–151.*

3. OVERVIEW OF MARSUPIAL DEVELOPMENT

3.1 From Ovulation to Birth

Marsupials have short gestation periods compared with most eutherians and give birth to highly altricial young that are comparable to newly hatched monotremes in their stage of development, even in large species such as kangaroos.

The early marsupial conceptus is surrounded by three extracellular coats: a zona pellucida, which surrounds the oocyte in all mammals; a thick mucoid coat; and a thin outer shell coat. The mucoid and shell coats are deposited sequentially by secretions from the reproductive tract (Arnold & Shorey, 1985a, 1985b; Casey, Martinus, & Selwood, 2002; Roberts, Breed, & Mayrhofer, 1994). As the blastocyst expands, the mucoid coat becomes compressed until it and the zona pellucida eventually disappear. By contrast, the shell coat stretches to accommodate the increasing volume of the blastocyst while continuing to accumulate material (Shaw, 1996), just as the early monotreme shell coat does. Unlike in monotremes, the marsupial shell coat breaks down about two-thirds of the way through pregnancy, allowing direct contact between the maternal endometrium and the placenta. Thus the expanding shell was probably a feature of the most recent mammalian ancestor that was lost in eutherians, while shell breakdown followed by direct endometrial-placental contact was probably a feature of the most recent therian ancestor, and would have been coupled with the evolution of mechanisms for maternal immune tolerance of the conceptus.

All marsupials develop a choriovitelline placenta, with the yolk sac comprising splanchnopleure, bilaminar omphalopleure, and trilaminar omphalopleure parts. In most species, the placenta makes only superficial contact with the endometrium. Some species also develop a chorioallantoic placenta, which in bandicoots is highly invasive (Renfree, 1982).

3.2 Cleavage and Deutoplasmolysis

Marsupial ova are around 200–250 μm in diameter, considerably smaller than those of monotremes, but about twice the diameter of eutherian ova. Although they are often described as "yolky," much of the ooplasm consists of translucent vesicles that are expelled into the extracellular space during cleavage, with little evidence for a nutritional role. Thus "deutoplasm" is a preferable term that conveys no assumption as to its role.

Accordingly, monotreme yolk can be considered a type of deutoplasm. Although some yolk-like platelets are present in marsupial deutoplasm, the genes encoding the three vitellogenin proteins (VIT1, 2, and 3) were found to be either absent or disabled in sequenced marsupial genomes (Babin, 2008; Brawand, Wahli, & Kaessmann, 2008). Curiously, *VIT2* transcripts are expressed in marsupials (S. Frankenberg and M. B. Renfree, unpublished data), but they are predicted to encode severely truncated proteins lacking any of the lipid-binding domains of canonical vitellogenin proteins. However, it remains plausible that they have retained some minimal role.

The zygotes of most or possibly all marsupials are polarized, with the pronuclei localized at the embryonic pole. Cleavage is described as holoblastic, but associated with progressive elimination of deutoplasm into the cleavage cavity toward the abembryonic pole (Fig. 4). In a sense, marsupial cleavage has similarities to the meroblastic cleavage of other birds and reptiles, which is followed by later complete separation from the yolk. The difference in marsupials is that both steps occur almost simultaneously. Much of the eliminated deutoplasm remains within large, membrane-bounded bodies, previously termed "yolk masses" but more cautiously referred to as "deutoplasts." Depending on the species, they can be multiple in number, or single, such as in the dasyurid marsupials.

During early cleavage in marsupials, there is a rapid reduction in total cell volume due to expulsion of deutoplasm and numerous small vesicles into the extracellular space (Frankenberg & Selwood, 1998; Sathananthan, Selwood, Douglas, & Nanayakkara, 1997; Selwood & Smith, 1990) blastomeres are localized toward the embryonic pole. Unlike in eutherians, contact between blastomeres is initially minimal and they instead adhere to the inner surface of the zona pellucida. With subsequent establishment of cell–cell adhesion, a rudimentary epithelium begins to form, and with further cell divisions, it expands toward the abembryonic pole. Thus, the establishment of cell-zona adhesion during very early cleavage ensures that cells never occupy an inner position and the nascent blastocyst consists of a unilaminar epithelium with no inner cell mass (Selwood, 1992).

4. PLURIBLAST–TROPHOBLAST SEGREGATION

4.1 Terms and Definitions

While the term "trophoblast" is mammal specific, in terms of its developmental origins it is clearly homologous to the extraembryonic ectoderm layer of the chorion in other amniotes. In the mouse, trophoblast is specified

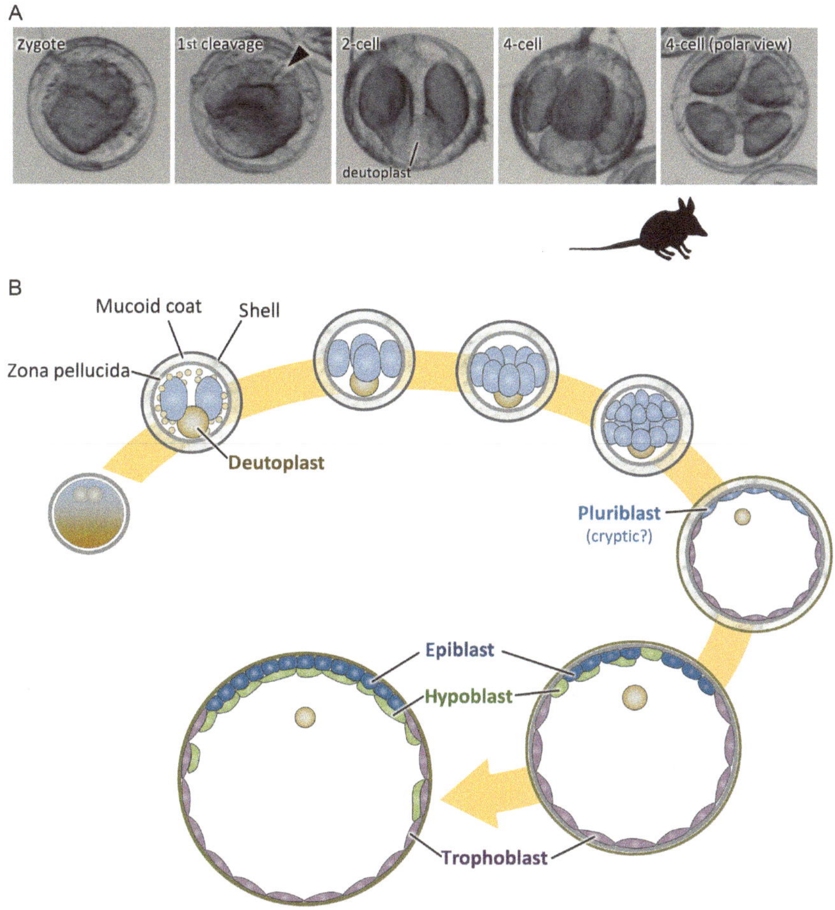

Fig. 4 Early marsupial development. (A) Cleavage in the fat-tailed dunnart. As the first cleavage furrow (*arrowhead*) progresses, the deutoplasm is simultaneously extruded toward the abembryonic pole where it forms a single, large deutoplast. (B) Generalized scheme for marsupial blastocyst formation and early lineage segregation, basely largely on the dasyurid mode (dunnart, quoll, Tasmanian devil).

by positional cues that drive the establishment of cell polarity. Inner cells of the morula, which lack an apical–basal axis, activate Hippo pathway signaling which suppresses trophoblast-specific genes (see Chapters "Cell polarity-dependent regulation of cell allocation and the first lineage specification in the preimplantation mouse embryo" by Saini and Yamanaka and "Our first choice: Cellular and genetic underpinnings of trophectoderm identity and differentiation in the mammalian embryo" by Menchero et al., in this issue).

The mechanism of trophoblast specification by segregating inner and outer cells cannot be the ancestral mammalian mechanism, since neither monotremes nor marsupials form an inner cell mass during development. Even Afrotherian and perhaps Xenarthran eutherians form a unilaminar blastocyst but nevertheless generate an inner cell mass by subsequent asymmetric divisions of epithelial cells (Frankenberg, de Barros, Rossant, & Renfree, 2016). This may be simply a case of heterochrony with respect to the relative onset of full epithelial function and the initiation of asymmetric divisions. It is likely that in many eutherians trophoblast-fated epithelial cells retain totipotency for some time after blastocyst formation. A distinguishing feature of all eutherians, however, is having a blastocyst stage in which an inner cell mass is *fully* enclosed by a trophoblast epithelium.

In both monotremes and marsupials, the embryonic lineage remains exposed on the surface of the conceptus until the amnion envelopes it much later in development. The term *pluriblast* refers to the early population of cells that is fated to form all lineages except the trophoblast, and is applicable to all mammals (Johnson & Selwood, 1996; Selwood & Johnson, 2006). In eutherians, the inner cell mass is the pluriblast.

4.2 Pluriblast–Trophoblast Segregation in Monotremes

Essentially all that is known of early development in monotremes is from the studies by Flynn and Hill (1939, 1947) and Wilson and Hill (1908), with later interpretations by other authors. In monotremes, two populations of cells with putatively distinguishable fates emerge by the time the blastodisc has 16 cells: central cells and marginal cells (Fig. 3A). Marginal cells were suggested to give rise to both surface blastodisc cells and subsurface vitellocytes, the latter of which fuse to form a syncytial germ ring that expands toward the abembryonic pole. The vitellocytes do not appear to have any homologue in other mammals and appear more akin to the periblast cells of birds and the merocytes of reptiles (Flynn & Hill, 1947). While the germ ring is forming, the surface cells proliferate to form a multilayer blastodisc (Fig. 3B). As the germ ring expands toward the abembryonic pole, the blastodisc appears to follow it and in doing so thins to a unilaminar blastoderm. By the time the blastoderm reaches the abembryonic pole, hypoblast cells have delaminated and it has become bilaminar (Fig. 3C).

No information exists on when a difference between trophoblast and pluriblast cells first emerges, although eventually this must occur. Although the central cells of the 16-cell stage must be more likely to give rise to pluriblast cells, there is no evidence to suggest whether they actually become specified at this stage, or whether positional signals associated with the expanding blastoderm provide the necessary cues.

4.3 Pluriblast–Trophoblast Segregation in Marsupials

In marsupials, early blastomeres are arranged in what can be viewed as a rudimentary two-dimensional blastodisc, analogous to that of monotremes (as well as birds and reptiles). This is because, very early in cleavage, blastomeres adhere to the inner surface of the zona pellucida and are localized to the embryonic pole. In the dunnart, "marginal cells" of this disc proliferate and spread toward the abembryonic pole, thus completing the unilaminar blastocyst. This pattern is highly stereotypical in the cleavage stages of dasyurid marsupials (Fig. 4). In the dunnart, the eight-cell stage comprises a single ring of cells encircling the equator of the conceptus. At the next division, each cell divides in a plane that results in an upper tier of eight cells in the embryonic hemisphere and a lower tier of eight cells in the abembryonic hemisphere. These tiers are hypothesized to give rise to pluriblast and trophoblast, respectively (Selwood, 1992). The two subpopulations are still morphologically distinguishable in the nascent blastocyst of most marsupial species but become indistinguishable during early blastocyst expansion. Thus, while it is highly likely that the upper tier of eight cells is predisposed to forming pluriblast, it is unclear if this is to the complete exclusion of trophoblast fate. The early blastocyst cells remain truly totipotent, and other mechanisms that regulate relative numbers of committed pluriblast and trophoblast might influence the proportions of each lineage contributed by each tier of cells at the 16-cell stage.

There have been very few molecular studies examining pluriblast–trophoblast segregation in marsupials. An attractive hypothesis was that cell-density cues involving Hippo signaling, homologous to the mechanism in mouse, could drive lineage segregation in a two-dimensional rather than a three-dimensional context. However, immunolocalization of the Hippo effectors YAP and WWTR1 revealed no differences in nuclear localization at any stage prior to overt lineage segregation (Frankenberg, Shaw, Freyer, Pask, & Renfree, 2013). WWTR1 was only localized to the nuclei of trophoblast cells after they had already differentiated, indicating a likely role in trophoblast proliferation during blastocyst expansion. By contrast, the earliest differential expression of YAP was specific to the embryonic disc, and this appears to be regulated at the transcriptional level (Frankenberg & Renfree, unpublished data).

The stage at which molecular differences in pluriblast and trophoblast emerge in the wallaby corresponds almost precisely with the onset of hypoblast differentiation (see Section 5). The failure to detect any differences among cells of unilaminar blastocysts using a large number of key, evolutionarily conserved lineage markers suggests that cells are not yet fully determined to one lineage or the other, although they may have been specified by inheritance of asymmetrically distributed cytoplasmic factors. Hill (1910) first noted that a sharp "sutural line" only appears between the trophoblast and pluriblast at the time of hypoblast formation, indicating an abrupt transition in cell identity.

4.4 The Evolution of Pluriblast–Trophoblast Segregation in Mammals

The above observations suggest that Hippo signaling does indeed have an ancient role in the trophoblast but not in its specification, at least in the therian common ancestor. It seems more plausible that trophoblast specification in this ancestor and in extant marsupials depends on conceptus polarity derived from asymmetric distribution of maternal determinants associated with the deutoplasm. An attractive model is that in the eutherian lineage, Hippo signaling has been coopted for an additional role in pluriblast–trophoblast specification to compensate for loss of zygotic asymmetry (Fig. 5). In the early stem eutherian lineage, this evolutionary process may have been driven by a need for much earlier trophoblast differentiation to facilitate precocious implantation. This selection pressure resulted in increasingly precocious Hippo pathway-driven trophoblast proliferation, until it occurred even prior to trophoblast specification, which was still dependent on conceptus polarity. Differential epithelialization along the embryonic–abembryonic axis also resulted in early internalization of pluriblast-fated cells, allowing cell-density mechanisms to additionally interact with the Hippo pathway (a conserved interaction that occurs in many other tissues (Pan, 2010)). Once these two prerequisites were in place, the Hippo pathway was able to evolve novel cellular mechanisms that fed back into the gene regulatory network specifying trophoblast. With this new mechanism in place, zygote asymmetry became entirely redundant for lineage specification. For its other role in specifying a polarized embryonic–abembryonic axis, it also became redundant due to stochastic

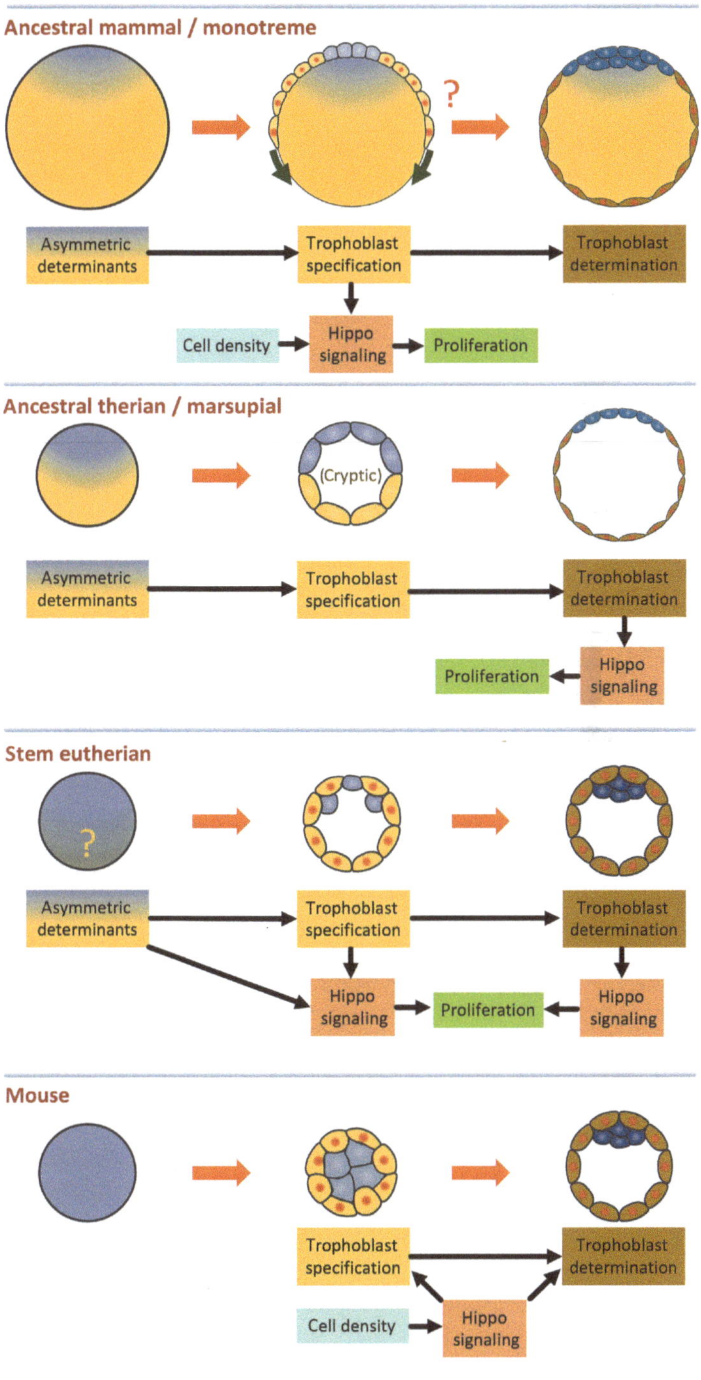

Fig. 5 See legend on next page.

processes establishing asymmetry during blastocyst cavitation. It would be very informative to examine YAP/WWTR1 localization during blastocyst formation in monotremes, as well as in Afrotherian and Xenarthran mammals, at least some of which form a unilaminar blastocyst prior to generating an inner cell mass (Frankenberg et al., 2016).

5. EPIBLAST–HYPOBLAST SEGREGATION

5.1 Epiblast–Hypoblast Segregation in Eutherians

The pluriblast in all mammals segregates into the epiblast (which gives rise to the embryo proper) and the hypoblast (which contributes to the yolk sac). In eutherians, the epiblast and hypoblast segregate from the inner cell mass, with hypoblast forming an epithelium separating the epiblast from the blastocyst cavity. For many years, it was thought that positional signals specify early pluripotent ICM cells adjacent to the cavity to become hypoblast. It is now known that epiblast and hypoblast precursors appear much earlier and are distributed in a "salt-and-pepper" pattern throughout the ICM (Chazaud, Yamanaka, Pawson, & Rossant, 2006; Rossant, Chazaud, & Yamanaka, 2003) and then sort into their respective layers by a combination of cell movements and apoptosis (Meilhac et al., 2009; Plusa, Piliszek, Frankenberg, Artus, & Hadjantonakis, 2008). NANOG and GATA6 are the two earliest known markers of epiblast and hypoblast precursors, respectively. In the early ICM, they exhibit broadly overlapping expression but become progressively restricted by mutually inhibitory mechanisms at the

Fig. 5 Hypothetical scheme for the evolution of trophoblast-specification mechanisms in mammals. Cell density-driven Hippo signaling (indicated in cells with red nuclei) may have constituted a mechanism for proliferation of trophoblast-fated cells in an ancestral mammal, but without an instructive role in specifying trophoblast identity, which instead relied on robust cues related to conceptus asymmetry. With precocious blastocyst formation in the therian ancestor, the pathway's role in proliferation was delayed until after trophoblast specification, which still depended on conceptus asymmetry. Because of an absence in differences in cell density, activation of Hippo signaling switched to a cell autonomous mechanism dependent on trophoblast identity. During evolution of the stem eutherian lineage, Hippo-induced proliferation of weakly specified trophoblast precursors occurred in progressively earlier stages of development, coupled with increasing envelopment of the pluriblast population. With greatly diminished conceptus asymmetry in later-evolved eutherians, internalization of cells allowed cell-density mechanisms to once again drive differential Hippo signaling, which evolved novel links with the gene regulatory network specifying trophoblast and causing conceptus asymmetry mechanisms to become completely redundant.

transcriptional level. For a more detailed description of epiblast–hypoblast segregation in eutherians, see Chapter "Primitive endoderm differentiation: From specification to epithelialization" by Bassalert et al. (in this issue).

5.2 Epiblast–Hypoblast Segregation in Monotremes

In monotremes, Flynn and Hill (1947) described two distinct populations of cells based on staining pattern even before the multilayered blastoderm has thinned to a unilaminar epithelium. The authors concluded that these represented progenitors of epiblast and hypoblast, respectively, on the basis of their distinct histological staining patterns. They remain distinguishable in the unilaminar blastoderm (Fig. 3C and D), before delaminating to produce the bilaminar blastoderm. Although molecular analyses of monotreme embryos would be greatly informative, these early observations are quite consistent with the apparently stochastic allocation of epiblast and hypoblast precursors in the eutherian inner cell mass.

5.3 Epiblast–Hypoblast Segregation in Marsupials

Three types of hypoblast formation have been recognized in marsupials based on morphology (Selwood, 1986, 1992):

Type 1: Hypoblast formation occurs precociously in the early non-expanded blastocyst. A subset of cells delaminate and form a multilayered mass beneath the nascent epiblast. This type is seen in the Virginia opossum.

Type 2: The pluriblast is distinguishable from the trophoblast before hypoblast cells begin to emerge.

Type 3: The pluriblast is indistinguishable from the trophoblast until hypoblast cells begin to emerge.

Regardless of the type of formation, there is no evidence in any species that hypoblast cells form by asymmetric division. They are initially identifiable within the unilaminar epithelium and subsequently delaminate.

In the Virginia opossum (Type 1), hypoblast cells delaminate first at the margins of the embryonic disc and then subsequently from more centrally (Hartman, 1916, 1919). In the dasyurids (eastern quoll and stripe-faced dunnart; Type 3), the first hypoblast cells also arise in small clusters at the margins of the nascent disc but localized to one side of it (Hill, 1910; Kress & Selwood, 2006; Selwood & Woolley, 1991), possibly marking the future anteroposterior (A-P) axis (see Section 6.2).

Only one study, of the tammar wallaby (Type 3 hypoblast formation), has examined marsupial epiblast–hypoblast segregation at the molecular level (Frankenberg & Renfree, 2013). In the tammar, all cells appear identical in the diapausing unilaminar blastocyst, even when stained for several conserved lineage-specific transcription factors such as POU5F1, POU5F3, SOX2, NANOG, and CDX2. In the early expanding blastocyst (after reactivation from diapause), strong staining of NANOG and GATA6 proteins was restricted to the nascent embryonic disc and was largely mutually exclusive. Most GATA6-positive hypoblast cells had already delaminated, although some nondelaminated cells could be found at the perimeter of the disc. Within the superficial disc epithelium, nuclei positive for both NANOG and GATA6 were never observed; however in mitotic cells the cytoplasm was either NANOG positive and GATA6 negative, or positive for both proteins, suggesting that downregulation of NANOG might precede that of GATA6 at the transcriptional level. Thus, at least some of the molecular mechanisms establishing epiblast–hypoblast segregation appear to be conserved between mouse and marsupial.

6. AXES AND ASYMMETRY

6.1 Embryonic–Abembryonic (Dorsoventral) Axis

In reptiles and birds, the asymmetric position of the pronuclei relative to the yolk defines the embryonic–abembryonic axis, which corresponds to the later dorsoventral axis. The same is true of monotremes, having yolk-laden eggs, as well as marsupials, in which the asymmetric position of the deutoplasm marks the future abembryonic pole of the conceptus. The origin of the dorsoventral axis in marsupials and monotremes thus ultimately lies in oogenesis and the asymmetric positioning of the germinal vesicle relative to the deutoplasm.

In eutherians, the embryonic–abembryonic axis is defined by the site of formation of the blastocyst cavity relative to the inner cell mass. No definitive determinant has been identified in eutherians that specifies the cite of cavity formation, and it is likely that it is essentially random and any subtle cues have little biological importance (Motosugi, Dietrich, Polanski, Solter, & Hiiragi, 2006). It is curious to note, however, that in the unilaminar blastocyst of the tenrec (an Afrotherian), the inner cell mass is generated by asymmetric divisions in a localized region of the blastocyst epithelium (Bluntschli, 1938; Goetz, 1939). Thus, in the tenrec, the embryonic–abembryonic axis arises from a conceptus asymmetry in inner cell generation.

6.2 Anteroposterior Axis

The A-P axis is the second body axis specified during vertebrate development. Establishing bilateral symmetry can either be a preliminary step in the specification of the A-P axis or can arise as an outcome of directly specifying A-P polarity within a radial blastodisc.

Monotremes exhibit morphological bilateral symmetry immediately after fertilization, when the germinal disc becomes elliptical in shape (Fig. 3A). Flynn and Hill (1939) observed that the polar bodies were positioned toward one pole of the long axis of the disc, while the pronuclei were located toward the opposite pole. From the latter pole, there was also an infiltration of yolky cytoplasm overlying the germ plasm. It seems likely therefore that the position of sperm entry provides a cue for establishing this polarity.

Caldwell (1887) as well as Flynn and Hill (1939) concluded that the first cleavage division is asymmetrical, resulting in one smaller and one larger blastomere. However, this seems to be based solely on the arrangement of blastomeres after the second division, so should be regarded with caution. Nevertheless, bilateral symmetry is overt at the four- and eight-cell stages in both the arrangement of blastomeres and the shape of the disc. At the four-cell stage, blastomeres are arranged in a rhomboidal pattern that is somewhat at odds with Flynn and Hill's interpretation that the first and second cleavage planes are, respectively, perpendicular and parallel to the long axis of the disc (see Fig. 3A), but by the eight-cell stage, blastomeres are regularly arranged and with differences in size between opposite ends of the disc's long axis. After the fourth round of divisions leading to the 32-cell stage, blastomeres become less regularly arranged and overt bilateral symmetry disappears. It is therefore not certain whether a cryptic polarity persists that links the initial postfertilization polarity with the later A-P axis; however it seems highly plausible.

In marsupials, there is little-to-no evidence of bilateral symmetry until close to gastrulation. Four-cell stages of the brushtail possum show a stereotypical rhomboidal arrangement of blastomeres at the embryonic pole of the conceptus reminiscent of that seen in monotremes (Frankenberg & Selwood, 1998). A role for sperm entry point has also been suggested, but it seems unlikely that a subtle positional cue acquired so early would be robustly inherited until the onset of gastrulation in the absence of any observable cytoplasmic rearrangements. In the mouse, a role was suggested for sperm entry point in patterning the much earlier-established embryonic–abembryonic axis (Piotrowska & Zernicka-Goetz, 2001), but follow-up studies failed to find robust support for such a mechanism and the hypothesis appears to have been abandoned. If the sperm entry point has any influence on embryonic patterning in either marsupials or eutherians, it is likely to be minimal and redundant. Cell-division asynchrony has also been proposed as a mechanism in marsupials (Selwood, 1992; Selwood & Smith, 1990), which cannot be ruled out.

Whatever the initial cue, it is probable that reactivation-diffusion mechanisms involving NODAL signaling are used to amplify weak asymmetries into a robust asymmetry that determines the A-P axis in all vertebrates (Schier, 2003). In the small conceptuses of marsupials and eutherians, small, stochastic asymmetries are probably sufficient to "seed" reaction–diffusion, whereas in the much larger conceptuses of birds, reptiles, and monotremes, reaction–diffusion might face additional challenges over greater distances. Therefore, nonstochastic cues (such as gravity in the chick (Kochav & Eyal-Giladi, 1971)) can provide an initial symmetry-breaking "kick-start."

Although the A-P axis can be specified earlier, it is in the endodermal layer that it first manifests. In the chick, the first extraembryonic endodermal germ layer—the primary hypoblast—is positioned centrally within the area pellucida (blastodisc). The A-P axis emerges with the appearance of Koller's sickle on the posterior side of the disc, marking where the primitive streak develops at the onset of gastrulation, after which the hypoblast migrates anteriorly (Chuai & Weijer, 2008). Similarly, the mouse early egg cylinder is essentially morphologically radially symmetrical until the hypoblast-derived distal visceral endoderm migrates anteriorly, heralding the onset of gastrulation (see Chapter "The head's tale: anterior–posterior axis formation in the mouse embryo" by Stower and Srinivas, in this issue).

In a morphological study of hypoblast formation in the stripe-faced dunnart, Kress and Selwood (2006) observed the first appearance of hypoblast cells on only one side of the pluriblast at its edge. Consistent with this, in the tammar wallaby, an asymmetry was observed in the embryonic disc at the very earliest stage of hypoblast formation, well before gastrulation. Thus, GATA6-expressing nascent hypoblast cells were present only on one side of the disc, while YAP-expressing pluriblast/epiblast cells were similarly localized to this same side (Frankenberg et al., 2013).

It would be unexpected for marsupials to initiate morphological A-P axis asymmetry at the onset of hypoblast formation rather than just before gastrulation. In the mouse, there is little evidence for asymmetry during hypoblast formation in the blastocyst that could define a future A-P axis. It is possible that the observed early asymmetry in marsupial hypoblast formation is not instructive for later events, and that it is caused by biologically irrelevant stochastic asymmetries that are either the same as or independent from those that are amplified into A-P axis asymmetry during gastrulation. An open question is therefore whether marsupials differ fundamentally from eutherians by directly coupling A-P axis specification with hypoblast formation instead of with later movements of hypoblast derivatives that specify the position of the primitive streak at the onset of gastrulation.

7. DISCUSSION: HOMOLOGIES AMONG VERTEBRATES IN LINEAGE SPECIFICATION AND REGULATION OF POTENCY

At least two distinct phases of endoderm formation can be identified in amniotes: the first generates the hypoblast, while the second generates the definitive endoderm of the gut. In the mouse, they are separated by approximately 2 days, with no evidence of additional endoderm formation in the intervening period. Although mouse hypoblast derivatives were shown to contribute a small proportion of cells to the fetal gut (Kwon, Viotti, & Hadjantonakis, 2008), a clear dichotomy can be made between hypoblast-derived (predominantly extraembryonic) and gastrulation-derived (solely embryonic) endoderm. The mechanisms of hypoblast formation can therefore be considered as a temporal duplication of the much more ancient gastrulation module, which was modified to suppress mesoderm formation.

Other fundamental properties also characterize each endoderm type. In female mice, inactivation of the X chromosome is paternally inherited in hypoblast derivatives, whereas gastrulation-derived endoderm inherits a random pattern of X-inactivation that is established in the postimplantation epiblast (Barakat, Jonkers, Monkhorst, & Gribnau, 2010; Monkhorst, Jonkers, Rentmeester, Grosveld, & Gribnau, 2008; Sugawara, Takagi, & Sasaki, 1985; Takagi & Sasaki, 1975; West, Frels, Chapman, & Papaioannou, 1977). The process of X-reactivation and subsequent random inactivation is coupled to other mechanisms associated with the regulation of pluripotency (Minkovsky, Patel, & Plath, 2012; Silva et al., 2009). Female marsupials do not exhibit random X-inactivation and instead all embryonic and extraembryonic lineages inherit a paternally imprinted inactive X (Cooper, VandeBerg, Sharman, & Poole, 1971; Johnston & Robinson, 1987; Sharman, 1971; Wang, Douglas, Vandeberg, Clark, & Samollow, 2014). The long noncoding RNA *XIST* that regulates X-inactivation in eutherians is absent in other mammals. Paternally imprinted X-activation in marsupials is instead regulated by another lncRNA, *RSX*, which has no discernible homology with *XIST* (Grant et al., 2012).

In the mouse, expression of the core pluripotency factor NANOG occurs in two distinct phases. The first phase in the blastocyst is associated with the acquisition of naive pluripotency and is coupled with hypoblast specification; subsequently NANOG is acutely downregulated. The second phase is associated with the postimplantation egg cylinder and is associated with the onset of gastrulation and primed pluripotency (Hart, Hartley, Ibrahim, & Robb, 2004; Silva et al., 2009). During the first phase, NANOG itself appears to be directly involved in the process of reactivation of the paternally inherited inactive X (Silva et al., 2009; Williams, Kalantry, Starmer, & Magnuson, 2011). Transcriptome analysis of conceptus stages of the tammar wallaby (Frankenberg and Renfree, unpublished data) suggests that NANOG expression is downregulated for a brief period after reactivation of the diapausing unilaminar blastocyst and then upregulated again at around the time that hypoblast cells appear. However, a period of decreased NANOG after hypoblast formation was not identified; thus it is still unclear whether equivalent separate phases of endoderm formation or naive vs primed pluripotency exist in marsupials. The possibility cannot be ruled out that endoderm formation in marsupials is a continuous process with no precise temporal boundary between extraembryonic and embryonic specifications. Endodermal fates (and properties) might instead be determined solely and directly by the environment in which cells ultimately find themselves.

In marsupials, the available evidence suggests that trophoblast might not be fully determined until epiblast–hypoblast segregation has already initiated. If so, it would challenge the doctrine of a stepwise differentiation of each lineage as fundamental to development. Under certain culture conditions, mouse embryonic stem cells appear capable of interconverting between any of trophoblast, hypoblast, and epiblast (Morgani et al., 2013), while in bovine blastocysts, trophoblast cells maintain *POU5F1* expression until comparatively late and can contribute to ICM when reaggregated with early blastomeres (Berg et al., 2011). Monotreme blastodiscs appear to contain hypoblast precursors well before it thins to a unilaminar epithelium and proliferates toward the abembryonic pole, suggesting that hypoblast specification might even occur before trophoblast specification. Consistent with this, Flynn and Hill did not report any regional differences in sites of hypoblast delamination, which suggests that spatial separation of pluriblast and trophoblast is not a prerequisite in monotremes. In marsupials, full determination of epiblast, hypoblast, and trophoblast appears to be almost simultaneous. Thus, it is possible that pluriblast–trophoblast segregation is not a fundamentally primary developmental step, but merely an outcome of precocious trophoblast differentiation in eutherians, or at least mice. Lineage tracing or cell transplantation experiments will be crucial for demonstrating totipotency in marsupial unilaminar blastocysts.

In the mouse, naive and primed pluripotency are also distinguished by how they regulate expression of POU5F1 (OCT4), a transcription factor with a central role in the regulation of pluripotency. In naive pluripotency, *Pou5f1* expression is driven by a distal enhancer, while in primed pluripotency it is driven by a proximal enhancer. The distal enhancer is also necessary for *Pou5f1* expression in germ cells (Yeom et al., 1996). The *POU5F1* gene family has a curious evolutionary history in other vertebrates. *POU5F3* (previously called *POU2* (Frankenberg et al., 2014)) is a paralogue of *POU5F1* present in the genomes of many other vertebrate lineages, but not in eutherian mammals. The two genes arose by duplication of an ancestral gene in a common ancestor of all jawed vertebrates. Both genes are conserved in some lineages, such as monotremes, marsupials, turtles, salamanders, and coelacanths, while one or the other has become extinct in other lineages (Fig. 6A) (Frankenberg & Renfree, 2013). While the reasons for this are unclear, current evidence suggests that they may relate to mechanisms of germ-cell specification. Vertebrates that lack POU5F1 tend to specify their germ cells via inheritance of germ plasm from the oocyte. Vertebrates that have retained POU5F1 (regardless of whether they also retained POU5F3) tend to specify their germ cells via inductive mechanisms later in development (Frankenberg, Pask, & Renfree, 2010; Johnson et al., 2003). POU5F1 is also the paralogue that more consistently shows germ-cell expression. In the wallaby (which has both genes), POU5F1 is expressed strongly in primordial germ cells and during cleavage stages, but in the later epiblast POU5F3 is a more specific marker of pluripotent cells (Frankenberg et al., 2010, 2013). It is noteworthy that mechanisms for inducing primordial germ-cell identity have strong similarities to those inducing naive pluripotency (Magnusdottir & Surani, 2014). Collectively the data suggest that POU5F1 is distinguished from POU5F3 by its role in a germ-cell/totipotent state. The evolution of germ plasm may cause partial redundancy of POU5F1 since germ plasm factors (e.g., DDX4, DAZL, and NANOS) may be sufficient to maintain germ-cell identity.

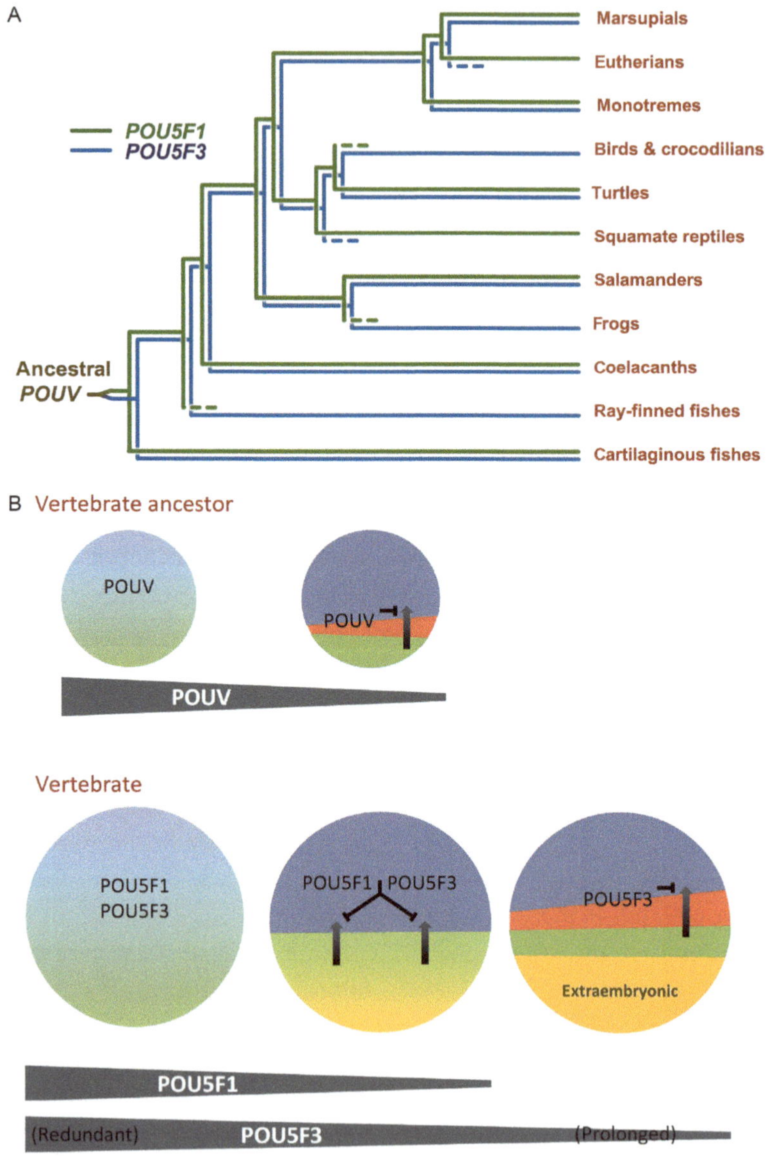

Fig. 6 Model for the evolution of POUV transcription factors in vertebrates. (A) *POU5F1* and *POU5F3* arose by gene duplication in a common ancestor of jawed vertebrate. Subsequent extinctions of one or the other gene occurred in some vertebrate lineages. (B) Divergence in the roles of POU5F1 and POU5F3 may have contributed to the evolution of extraembryonic tissues in an early vertebrate ancestor. Before the evolution of extraembryonic endoderm (*upper panel*), a single POUV gene was sufficient to protect prospective ectoderm from short-range mesendoderm-inducing signals (*black arrows*). The evolution of larger eggs and more yolk favored precocious specification of

If POU5F1 became redundant in vertebrates with germ plasm, what made POU5F3 redundant in other vertebrates? The distinction between POU5F1 and POU5F3 may be largely (but not entirely) due to differences in their expression pattern. Indeed, this could provide an explanation for the conservation of both genes after the original duplication event. Even before the acquisition of meroblastic cleavage, extraembryonic endoderm is thought to have been a feature of primitive vertebrates that was lacking in more primitive deuterostomes (Takeuchi, Takahashi, Okabe, & Aizawa, 2009). A "double-dose" of POUV genes might have allowed the conceptus to achieve an extended period of endoderm production before the onset of gastrulation with concomitant mesoderm induction. With both genes expressed, totipotency would be preserved in the epiblastic hemisphere despite mesendoderm-inducing signals emanating from cells in the yolky hemisphere, which are able to induce themselves to form endoderm. Later in development, downregulation of just one gene (*POU5F1*) is sufficient to allow mesoderm induction in epiblast nearest to the signal source (Fig. 6B).

POU5F3 redundancy (as has occurred in eutherians and squamate reptiles) suggests that *POU5F1* might have undergone convergent evolution with *POU5F3* to annex its role. Regulation of expression imposes fewer evolutionary constraints than does protein sequence. Nevertheless, there do appear to be some differences between POU5F1 and POU5F3 protein functions that are dependent on their sequence. In the wallaby, POU5F1 appears to undergo differential nuclear localization in the embryo, whereas POU5F3 does not (Frankenberg et al., 2013). Also, *POU5F1* and *POU5F3* from various species differ in their capacity for reprogramming or maintaining pluripotency in mouse ES cells (Morrison & Brickman, 2006; Tapia et al., 2012). One of the few conserved differences between the two paralogues (except in cartilaginous fishes) is the deletion of a single arginine residue within the nuclear localization signal of POU5F1 (Frankenberg & Renfree, 2013; Pan, Qin, Liu, Scholer, & Pei, 2004). With

extraembryonic endoderm (but not mesoderm) to facilitate yolk utilization. With a "double-dose" of POUV genes, a simple temporal difference in their expression allowed enhanced totipotency in the epiblast to prevent mesoderm induction. Downregulation of POU5F1 later in development allowed mesoderm induction during gastrulation. *Panel (A): Adapted from Frankenberg, S. R., Frank, D., Harland, R., Johnson, A. D., Nichols, J., Niwa, et al. (2014). The POU-er of gene nomenclature.* Development, 141, *2921–2923.*

the limited data available, a viable hypothesis is that POU5F3 redundancy is achievable because POU5F1 protein has a broader functionality (i.e., only the expression pattern of POU5F1 needs to change), whereas POU5F1 redundancy is achievable by altering the context of its role (i.e., by evolving germ plasm). Clearly, much more needs to be understood about the mechanisms of early development in different vertebrate species before we can test these ideas. Marsupials will be key models for future research since they have both POU5F1 and POU5F3 and are the most closely related to eutherian mammals.

GLOSSARY

Deutoplasm Secondary cytoplasm that eventually becomes separated from the cellular part of the conceptus during cleavage, sometimes in the form of a membrane-bounded *deutoplast*.

Determination Full commitment of a cell to a particular lineage, which cannot be redirected to another lineage even if placed in different environment. Determination implies the imposition of robust epigenetic changes.

Omphalopleure A part of the yolk sac wall that includes both trophoblast-derived and hypoblast-derived layers. It may include both trilaminar (including mesoderm) and bilaminar (excluding mesoderm) parts. The term was originally used by Hill (1898), who credited its coining to J.T. Wilson.

Pluriblast A population of cells that are committed to and have the potential to form both epiblast and hypoblast, but not trophoblast.

Somatopleure Any embryonic or extraembryonic tissue comprising ectodermal and mesodermal layers, but not endodermal.

Specification Direction of a cell to a particular lineage without full commitment. Specification may be achieved through inheritance of cytoplasmic factors.

Splanchnopleure Any embryonic or extraembryonic tissue comprising endodermal and mesodermal layers, but not ectodermal.

Yolk deutoplasm that has a chiefly nutritive role in the support of development. Collectively, it may contain some components that are not necessarily nutritive.

REFERENCES

Arnold, R., & Shorey, C. D. (1985a). Structure of the oviducal epithelium of the brush-tailed possum (Trichosurus vulpecula). *Journal of Reproduction and Fertility, 73*, 9–19.

Arnold, R., & Shorey, C. D. (1985b). Structure of the uterine luminal epithelium of the brush-tailed possum (Trichosurus vulpecula). *Journal of Reproduction and Fertility, 74*, 565–573.

Babin, P. J. (2008). Conservation of a vitellogenin gene cluster in oviparous vertebrates and identification of its traces in the platypus genome. *Gene, 413*, 76–82.

Barakat, T. S., Jonkers, I., Monkhorst, K., & Gribnau, J. (2010). X-changing information on X inactivation. *Experimental Cell Research, 316*, 679–687.

Berg, D. K., Smith, C. S., Pearton, D. J., Wells, D. N., Broadhurst, R., Donnison, M., et al. (2011). Trophectoderm lineage determination in cattle. *Developmental Cell, 20*, 244–255.

Bluntschli, H. R. (1938). Le developpement primaire et l'implantation chez un centetine (Hemicentetes). *Comptes Rendus de l'Association des Anatomistes*, *44*, 39–46.

Brawand, D., Wahli, W., & Kaessmann, H. (2008). Loss of egg yolk genes in mammals and the origin of lactation and placentation. *PLoS Biology*, *6*. e63.

Caldwell, W. H. (1887). The embryology of Monotremata and Marsupialia. Part I. *Proceedings of the Royal Society of London*, *42*, 177–180.

Casey, N. P., Martinus, R., & Selwood, L. (2002). Outer egg coats of the marsupial conceptus: Secretion and protein composition. *Molecular Reproduction and Development*, *62*, 181–194.

Chazaud, C., Yamanaka, Y., Pawson, T., & Rossant, J. (2006). Early lineage segregation between epiblast and primitive endoderm in mouse blastocysts through the Grb2-MAPK pathway. *Developmental Cell*, *10*, 615–624.

Chuai, M., & Weijer, C. J. (2008). The mechanisms underlying primitive streak formation in the chick embryo. *Current Topics in Developmental Biology*, *81*, 135–156.

Cooper, D. W., VandeBerg, J. L., Sharman, G. B., & Poole, W. E. (1971). Phosphoglycerate kinase polymorphism in kangaroos provides further evidence for paternal X inactivation. *Nature: New Biology*, *230*, 155–157.

Ferner, K., & Mess, A. (2011). Evolution and development of fetal membranes and placentation in amniote vertebrates. *Respiratory Physiology & Neurobiology*, *178*, 39–50.

Flynn, T. T., & Hill, J. P. (1939). The development of the Monotremata. Part IV. Growth of the ovarian ovum, maturation, fertilisation, and early cleavage. *Transactions of the Zoological Society of London*, *24*, 445–622.

Flynn, J. J., & Hill, J. P. (1942). The later stages of cleavage and the formation of the primary germ-layers in the Monotremata. (Preliminary communication.). *Proceedings of the Zoological Society of London*, *A111*, 233–253.

Flynn, T. T., & Hill, J. P. (1947). The development of the monotremata. Part VI. The later stages of cleavage and the formation of the primary germ layers. *Transactions of the Zoological Society of London*, *21*, 1–151.

Frankenberg, S. R., de Barros, F. R., Rossant, J., & Renfree, M. B. (2016). The mammalian blastocyst. *Wiley Interdisciplinary Reviews: Developmental Biology*, *5*, 210–232.

Frankenberg, S. R., Frank, D., Harland, R., Johnson, A. D., Nichols, J., Niwa, H., et al. (2014). The POU-er of gene nomenclature. *Development*, *141*, 2921–2923.

Frankenberg, S., Pask, A., & Renfree, M. B. (2010). The evolution of class V POU domain transcription factors in vertebrates and their characterisation in a marsupial. *Developmental Biology*, *337*, 162–170.

Frankenberg, S., & Renfree, M. B. (2013). On the origin of POU5F1. *BMC Biology*, *11*, 56.

Frankenberg, S., & Selwood, L. (1998). An ultrastructural study of the role of an extracellular matrix during normal cleavage in a marsupial, the brushtail possum. *Molecular Reproduction and Development*, *50*, 420–433.

Frankenberg, S., Shaw, G., Freyer, C., Pask, A. J., & Renfree, M. B. (2013). Early cell lineage specification in a marsupial: A case for diverse mechanisms among mammals. *Development*, *140*, 965–975.

Goetz, R. H. (1939). On the early development of the Tenrecoidea (*Hemicentetes semispinosus*). *Biomorphosis*, *1*, 67–79.

Grant, J., Mahadevaiah, S. K., Khil, P., Sangrithi, M. N., Royo, H., Duckworth, J., et al. (2012). Rsx is a metatherian RNA with Xist-like properties in X-chromosome inactivation. *Nature*, *487*, 254–258.

Griffiths, M. (2012). *The biology of the monotremes*. Elsevier.

Hart, A. H., Hartley, L., Ibrahim, M., & Robb, L. (2004). Identification, cloning and expression analysis of the pluripotency promoting Nanog genes in mouse and human. *Developmental Dynamics*, *230*, 187–198.

Hartman, C. G. (1916). Studies in the development of the opossum Didelphys virginiana. I. History of the early cleavage. II. Formation of the blastocyst. *Journal of Morphology*, *27*, 1–83.

Hartman, C. G. (1919). Studies in the development of the opossum Didelphys virginiana. III. Description of new material on maturation, cleavage and endoderm formation. IV. The bilaminar blastocyst. *Journal of Morphology*, *32*(1), 139.

Hill, J. P. (1898). The placentation of Perameles. *Quarterly Journal of Microscopical Science*, *40*, 385–446.

Hill, J. P. (1910). The early development of the Marsupialia, with special reference to the native cat (Dasyurus viverrinus). Contributions to the embryology of the Marsupialia. IV. *Quarterly Journal of Microscopical Science*, *56*, 1–134.

Hill, J. P. (1933). V. The development of the monotremata. Part II. The structure of the egg-shell. *Journal of Zoology*, *21*, 443–476.

Hill, J., & Gatenby, J. B. (1926). 35. The corpus luteum of the monotremata. *Journal of Zoology*, *96*, 715–763.

Hughes, R. L. (1993). Monotreme development with particular reference to the extraembryonic membranes. *The Journal of Experimental Zoology*, *266*, 480–494.

Johnson, A. D., Crother, B., White, M. E., Patient, R., Bachvarova, R. F., Drum, M., et al. (2003). Regulative germ cell specification in axolotl embryos: A primitive trait conserved in the mammalian lineage. *Philosophical Transactions of the Royal Society of London*, *358*, 1371–1379.

Johnson, M. H., & Selwood, L. (1996). Nomenclature of early development in mammals. *Reproduction, Fertility, and Development*, *8*, 759–764.

Johnston, P. G., & Robinson, E. S. (1987). X chromosome inactivation in female embryos of a marsupial mouse (Antechinus stuartii). *Chromosoma*, *95*, 419–423.

Kochav, S., & Eyal-Giladi, H. (1971). Bilateral symmetry in chick embryo determination by gravity. *Science (New York, N.Y.)*, *171*, 1027–1029.

Kress, A., & Selwood, L. (2006). Marsupial hypoblast: Formation and differentiation of the bilaminar blastocyst in Sminthopsis macroura. *Cells, Tissues, Organs*, *182*, 155–170.

Kwon, G. S., Viotti, M., & Hadjantonakis, A.-K. (2008). The endoderm of the mouse embryo arises by dynamic widespread intercalation of embryonic and extraembryonic lineages. *Developmental Cell*, *15*, 509–520.

Luckett, W. P. (1977). Ontogeny of amniote fetal membranes and their application to phylogeny. In W. P. Luckett (Ed.), *Major patterns in vertebrate evolution* (pp. 439–516): Springer.

Magnusdottir, E., & Surani, M. A. (2014). How to make a primordial germ cell. *Development*, *141*, 245–252.

Meilhac, S. M., Adams, R. J., Morris, S. A., Danckaert, A., Le Garrec, J. F., & Zernicka-Goetz, M. (2009). Active cell movements coupled to positional induction are involved in lineage segregation in the mouse blastocyst. *Developmental Biology*, *331*, 210–221.

Meredith, R. W., Janecka, J. E., Gatesy, J., Ryder, O. A., Fisher, C. A., Teeling, E. C., et al. (2011). Impacts of the cretaceous terrestrial revolution and KPg extinction on mammal diversification. *Science (New York, N.Y.)*, *334*, 521–524.

Mikkelsen, T. S., Wakefield, M. J., Aken, B., Amemiya, C. T., Chang, J. L., Duke, S., et al. (2007). Genome of the marsupial Monodelphis domestica reveals innovation in non-coding sequences. *Nature*, *447*, 167–177.

Minkovsky, A., Patel, S., & Plath, K. (2012). Concise review: Pluripotency and the transcriptional inactivation of the female mammalian X chromosome. *Stem Cells*, *30*, 48–54.

Mitchell, K. J., Pratt, R. C., Watson, L. N., Gibb, G. C., Llamas, B., Kasper, M., et al. (2014). Molecular phylogeny, biogeography, and habitat preference evolution of marsupials. *Molecular Biology and Evolution*, *31*, 2322–2330.

Monkhorst, K., Jonkers, I., Rentmeester, E., Grosveld, F., & Gribnau, J. (2008). X inactivation counting and choice is a stochastic process: Evidence for involvement of an X-linked activator. *Cell*, *132*, 410–421.

Morgani, S. M., Canham, M. A., Nichols, J., Sharov, A. A., Migueles, R. P., Ko, M. S., et al. (2013). Totipotent embryonic stem cells arise in ground-state culture conditions. *Cell Reports*, *3*, 1945–1957.

Morrison, G. M., & Brickman, J. M. (2006). Conserved roles for Oct4 homologues in maintaining multipotency during early vertebrate development. *Development*, *133*, 2011–2022.

Motosugi, N., Dietrich, J. E., Polanski, Z., Solter, D., & Hiiragi, T. (2006). Space asymmetry directs preferential sperm entry in the absence of polarity in the mouse oocyte. *PLoS Biology*, *4*, e135.

Murchison, E. P., Schulz-Trieglaff, O. B., Ning, Z., Alexandrov, L. B., Bauer, M. J., Fu, B., et al. (2012). Genome sequencing and analysis of the Tasmanian devil and its transmissible cancer. *Cell*, *148*, 780–791.

Nilsson, M. A., Churakov, G., Sommer, M., Tran, N. V., Zemann, A., Brosius, J., et al. (2010). Tracking marsupial evolution using archaic genomic retroposon insertions. *PLoS Biology*, *8*, e1000436.

Pan, D. (2010). The hippo signaling pathway in development and cancer. *Developmental Cell*, *19*, 491–505.

Pan, G., Qin, B., Liu, N., Scholer, H. R., & Pei, D. (2004). Identification of a nuclear localization signal in OCT4 and generation of a dominant negative mutant by its ablation. *The Journal of Biological Chemistry*, *279*, 37013–37020.

Piotrowska, K., & Zernicka-Goetz, M. (2001). Role for sperm in spatial patterning of the early mouse embryo. *Nature*, *409*, 517–521.

Plusa, B., Piliszek, A., Frankenberg, S., Artus, J., & Hadjantonakis, A. K. (2008). Distinct sequential cell behaviours direct primitive endoderm formation in the mouse blastocyst. *Development*, *135*, 3081–3091.

Renfree, M. B. (1982). Implantation and placentation. In C. R. Austin & R. V. Short (Eds.), *Reproduction in mammals* (pp. 26–69). Cambridge: Cambridge University Press.

Renfree, M. B., Papenfuss, A. T., Deakin, J. E., Lindsay, J., Heider, T., Belov, K., et al. (2011). Genome sequence of an Australian kangaroo, Macropus eugenii, provides insight into the evolution of mammalian reproduction and development. *Genome Biology*, *12*, R81.

Roberts, C. T., Breed, W. G., & Mayrhofer, G. (1994). Origin of the oocyte shell membrane of a dasyurid marsupial: An immunohistochemical study. *The Journal of Experimental Zoology*, *270*, 321–331.

Rossant, J., Chazaud, C., & Yamanaka, Y. (2003). Lineage allocation and asymmetries in the early mouse embryo. *Philosophical Transactions of the Royal Society of London*, *358*, 1341–1348. discussion 1349.

Sathananthan, H., Selwood, L., Douglas, I., & Nanayakkara, K. (1997). Early cleavage to formation of the unilaminar blastocyst in the marsupial Antechinus stuartii: Ultrastructure. *Reproduction, Fertility, and Development*, *9*, 201–212.

Schier, A. F. (2003). Nodal signaling in vertebrate development. *Annual Review of Cell and Developmental Biology*, *19*, 589–621.

Selwood, L. (1986). The marsupial blastocyst-a study of the blastocysts in the Hill Collection. *Australian Journal of Zoology*, *34*, 177–187.

Selwood, L. (1992). Mechanisms underlying the development of pattern in marsupial embryos. *Current Topics in Developmental Biology*, *27*, 175–233.

Selwood, L., & Johnson, M. H. (2006). Trophoblast and hypoblast in the monotreme, marsupial and eutherian mammal: Evolution and origins. *BioEssays*, *28*, 128–145.

Selwood, L., & Smith, D. (1990). Time-lapse analysis and normal stages of development of cleavage and blastocyst formation in the marsupials the brown antechinus and the stripe-faced dunnart. *Molecular Reproduction and Development*, *26*, 53–62.

Selwood, L., & Woolley, P. A. (1991). A timetable of embryonic development, and ovarian and uterine changes during pregnancy, in the stripe-faced dunnart, Sminthopsis macroura (Marsupialia: Dasyuridae). *Journal of Reproduction and Fertility*, *91*, 213–227.

Sharman, G. B. (1971). Late DNA replication in the paternally derived X chromosome of female kangaroos. *Nature*, *230*, 231–232.

Shaw, G. (1996). The uterine environment in early pregnancy in the tammar wallaby. *Reproduction, Fertility, and Development*, *8*, 811–818.

Sheng, G., & Foley, A. C. (2012). Diversification and conservation of the extraembryonic tissues in mediating nutrient uptake during amniote development. *Annals of the New York Academy of Sciences*, *1271*, 97–103.

Silva, J., Nichols, J., Theunissen, T. W., Guo, G., van Oosten, A. L., Barrandon, O., et al. (2009). Nanog is the gateway to the pluripotent ground state. *Cell*, *138*, 722–737.

Sugawara, O., Takagi, N., & Sasaki, M. (1985). Correlation between X-chromosome inactivation and cell differentiation in female preimplantation mouse embryos. *Cytogenetics and Cell Genetics*, *39*, 210–219.

Takagi, N., & Sasaki, M. (1975). Preferential inactivation of the paternally derived X chromosome in the extraembryonic membranes of the mouse. *Nature*, *256*, 640–642.

Takeuchi, M., Takahashi, M., Okabe, M., & Aizawa, S. (2009). Germ layer patterning in bichir and lamprey; an insight into its evolution in vertebrates. *Developmental Biology*, *332*, 90–102.

Tapia, N., Reinhardt, P., Duemmler, A., Wu, G., Arauzo-Bravo, M. J., Esch, D., et al. (2012). Reprogramming to pluripotency is an ancient trait of vertebrate Oct4 and Pou2 proteins. *Nature Communications*, *3*, 1279.

Veyrunes, F., Waters, P. D., Miethke, P., Rens, W., McMillan, D., Alsop, A. E., et al. (2008). Bird-like sex chromosomes of platypus imply recent origin of mammal sex chromosomes. *Genome Research*, *18*, 965–973.

Wang, X., Douglas, K. C., Vandeberg, J. L., Clark, A. G., & Samollow, P. B. (2014). Chromosome-wide profiling of X-chromosome inactivation and epigenetic states in fetal brain and placenta of the opossum, Monodelphis domestica. *Genome Research*, *24*, 70–83.

Warren, W. C., Hillier, L. W., Marshall Graves, J. A., Birney, E., Ponting, C. P., Grutzner, F., et al. (2008). Genome analysis of the platypus reveals unique signatures of evolution. *Nature*, *453*, 175–183.

West, J. D., Frels, W. I., Chapman, V. M., & Papaioannou, V. E. (1977). Preferential expression of the maternally derived X chromosome in the mouse yolk sac. *Cell*, *12*, 873–882.

Williams, L. H., Kalantry, S., Starmer, J., & Magnuson, T. (2011). Transcription precedes loss of Xist coating and depletion of H3K27me3 during X-chromosome reprogramming in the mouse inner cell mass. *Development*, *138*, 2049–2057.

Wilson, J. T., & Hill, J. P. (1908). Observations on the development of Ornithorhynchus. *Philosophical Transactions of the Royal Society B: Biological Sciences*, *199*, 31–168.

Yeom, Y. I., Fuhrmann, G., Ovitt, C. E., Brehm, A., Ohbo, K., Gross, M., et al. (1996). Germline regulatory element of Oct-4 specific for the totipotent cycle of embryonal cells. *Development*, *122*, 881–894.

CHAPTER ELEVEN

Pre-implantation Development of Domestic Animals

Anna Piliszek*[,1], **Zofia E. Madeja**[†]
*Institute of Genetics and Animal Breeding, Polish Academy of Sciences, Jastrzebiec, Poland
[†]Faculty of Veterinary Medicine and Animal Sciences, Poznan University of Life Sciences, Poznan, Poland
[1]Corresponding author: e-mail address: a.piliszek@ighz.pl

Contents

Abstract

During the first days following fertilization, cells of mammalian embryo gradually lose totipotency, acquiring distinct identity. The first three lineages specified in the mammalian embryo are pluripotent epiblast, which later gives rise to the embryo proper, and two extraembryonic lineages, hypoblast (also known as primitive endoderm) and trophectoderm, which form tissues supporting development of the fetus in utero. Most of our knowledge regarding the mechanisms of early lineage specification in mammals comes from studies in the mouse. However, the growing body of evidence points to both similarities and species-specific differences. Understanding molecular and cellular mechanisms of early embryonic development in nonrodent mammals expands our understanding of basic mechanisms of differentiation and is essential for the development of effective protocols for assisted reproduction in agriculture, veterinary medicine, and for biomedical research. This review summarizes the current state of knowledge on key events in epiblast, hypoblast, and trophoblast differentiation in domestic mammals.

Current Topics in Developmental Biology, Volume 128
ISSN 0070-2153
https://doi.org/10.1016/bs.ctdb.2017.11.005

1. INTRODUCTION

The earliest stages of embryonic development in mammals are devoted to the specification of three distinct cell lineages—epiblast, hypoblast, and trophectoderm. Pluripotent epiblast (EPI) gives rise to the embryo proper and subsequently forms all of the tissues of the adult organism, while the other two lineages form the extraembryonic structures. Trophectoderm (TE) is necessary for embryo implantation in the uterus and provides grounds for the development of the embryonic part of the placenta. Hypoblast (called primitive endoderm, PrE in the mouse) is involved in epiblast patterning and yolk sac development.

Our understanding of mechanisms governing these processes is based primarily on the data obtained from the mouse model. Studies of other mammalian species, including human, nonhuman primates, and domestic animals, prove that while many aspects of early development are conserved across species, the noted differences in embryogenesis, gene expression patterns, and lineage specification create questions that still need to be answered. Comparisons between species can be difficult due to significant variation in the events surrounding lineage specification such as the timing of embryonic genome activation (EGA), the length of the cell cycle, timing and mode of implantation, and size of the embryo. In this review, we focus on the events that lead to the differentiation of the first three cell lineages in domestic mammals.

With increasing application of assisted reproduction techniques in farm animal breeding and growing prospects for the application of nonrodent stem cell research in medicine and animal biotechnology, cell lineage formation in species of economic significance such as cattle and sheep is of increasing interest. Most importantly, sharing many similarities with human embryos, domestic mammals' embryos present an attractive model for investigating the fundamental mechanisms of early development.

2. TIMING OF PREIMPLANTATION DEVELOPMENT

The end of the preimplantation period is marked by implantation of the embryo in the uterus. However, the length of this period and the extent of differentiation preceding implantation vary significantly between the species. While in the mouse and human implanting blastocyst comprises only three lineages (TE, EPI, and PrE) (reviewed by Piliszek, Grabarek, Frankenberg, & Plusa, 2016), in most of the domestic mammals differentiation is significantly more advanced before implantation, including hypoblast migration, further differentiation of the trophoblast, and gastrulation of the EPI compartment (Bazer, Spencer, Johnson, Burghardt, & Wu, 2009; Kuijk, Geijsen, & Cuppen, 2015; Pfeffer, Smith, Maclean, & Berg, 2017; van Leeuwen, Berg, & Pfeffer, 2015).

In cattle developing embryo reaches morula stage at 5 days postinsemination (dpi), cavitation takes place at 7–8 dpi and hatching around 8–9 dpi. Hatched blastocyst undergoes elongation, and finally the implantation is initiated around 20–21 dpi (Peippo, Machaty, & Peter, 2011; van Soom, Ysebaert, & de Kruif, 1997). Similar pattern of elongation is characteristic for some other ungulates (specifically Artiodactyla)—in sheep morula is formed around 4–5 days postcoitum (dpc), cavitation and formation of blastocyst—around 7–8 dpc, and implantation takes place at 14–16 dpc, when embryo reaches around 70 mm (Bindon, 1971). In pig morula is formed around 4–5 dpc, blastocyst around 6–7 dpc, attachment to the uterus occurs around 13 dpc, and implantation—at 18 dpc and around 100 mm in length (Perry & Rowlands, 1962). In horse cavitation occurs around 6–7 dpc, embryo attachment to the uterus around 17 dpc, and final implantation—at 35 dpc when embryo reaches diameter of 20 mm (Ginther, 1992). In the rabbit embryo cavitation takes places at 3 dpc, and implantation—around 6.5 dpc, when embryo measures around 5 mm. Timing of preimplantation development in different mammalian species is summarized in Fig. 1.

3. EMBRYONIC GENOME ACTIVATION

The earliest stages of mammalian embryonic development rely on maternal mRNAs and proteins that have been synthesized and stored in oocyte's cytoplasm during its maturation and growth, and are able to drive early development when the newly formed embryo is transcriptionally inactive. The underlying mechanisms of embryogenesis highly depend on a coordinated cascade of genetically controlled events, including demethylation of maternal and paternal genomes and embryonic genome activation (EGA, also referred to as zygotic genome activation, ZGA). Differences in timing and mechanism of lineage differentiation might result from divergence in EGA between species.

During the transition phase between maternal and embryonic genome activation (maternal-to-zygotic transition, MZT), early development relies on posttranscriptional gene regulation. The key events of MZT involve the elimination of maternal mRNA, which initially is accompanied by

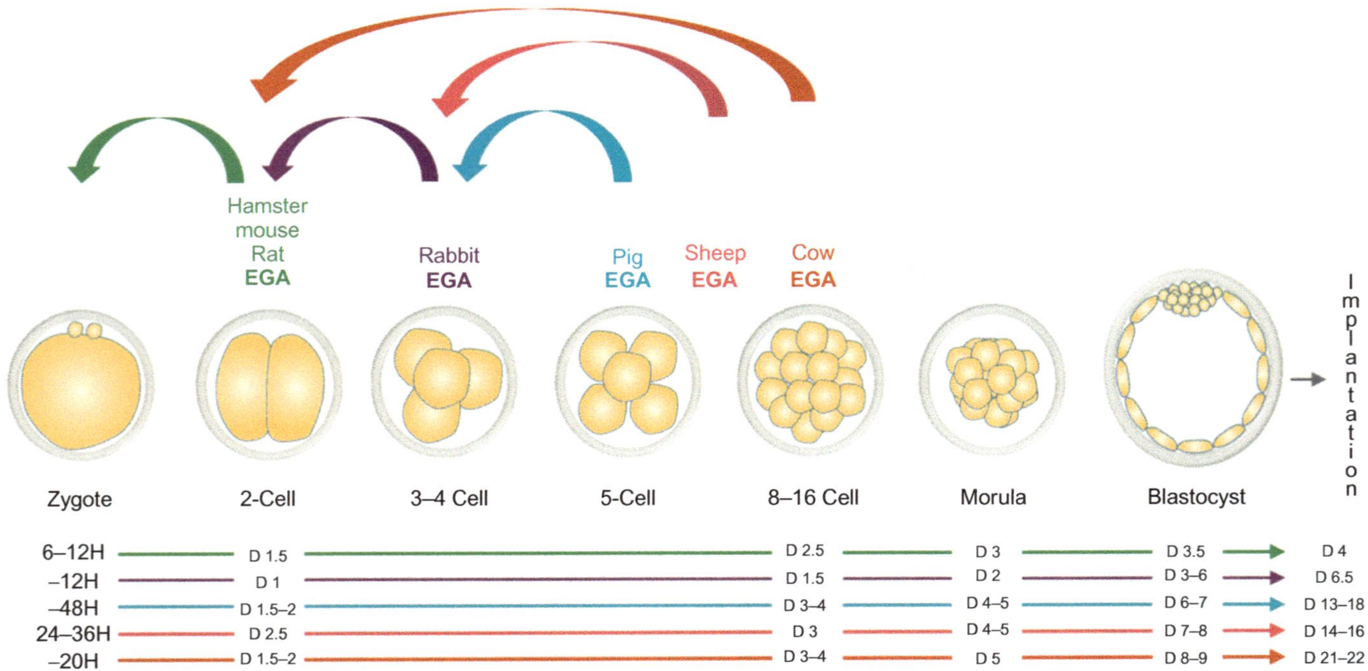

Fig. 1 From zygote to implantation—timing of development. A comparison between rodents and nonrodent mammals. *Arrows* point to the time of detection of the first embryonic transcripts, which occur prior to the major embryonic genome activation (EGA); D, days postoocyte insemination; H, hours postinsemination.

maternally encoded factors. However, as zygotic/embryonic transcription intensifies, the production of new proteins and microRNAs (miRNAs) enhances the efficiency of maternal mRNA degradation. The process in which the mammalian embryo begins to rely on its innate molecular signaling is strictly species specific. In case of rodents ZGA/EGA occurs during the initial stages of development. In the mouse the minor burst of EGA takes place toward the end of the 1-cell stage and is instantly followed by a major burst during the 2-cell stage (Aoki, Worrad, & Schultz, 1997; Latham, Solter, & Schultz, 1992). Similarly in rat (Zernicka-Goetz, 1994) and hamster (Seshagiri, McKenzie, Bavister, Williamson, & Aiken, 1992) the initiation of the EGA occurs late at the 2-cell stage. In other mammals this process may be much extended in time, spans several cell cycles, and occurs closer to the first cell differentiation events (Fig. 1). In nonrodent mammals, this process is divided into two steps: the initial (minor) EGA and the main (major) EGA. In bovine and sheep embryos the major EGA occurs at the 8–16-cell stage (Crosby, Gandolfi, & Moor, 1988; Meirelles et al., 2004); however, studies show that even zygotes are to a small degree transcriptionally active (Memili & First, 1999). The de novo RNA synthesis was observed as early as the 2–5-cell stage in cattle (Natale, Kidder, Westhusin, & Watson, 2000; Viuff, Avery, Greve, King, & Hyttel, 1996), 2-cell stage in rabbit, and 4-cell stage in sheep; however, the major EGA in these species takes place at 4- and 8-cell stages (reviewed by Kanka, 2003). In pig embryos obtained in vivo the major burst of transcription occurs during the third cell cycle after fertilization, at the 4–5-cell stage (Maddox-Hyttel et al., 2001), similar to human embryos (Braude, Bolton, & Moore, 1988).

Studies suggest that processes underlying EGA in mammals may be universal; however, in case of the rabbit more features were observed to be common with Xenopus than with the mouse. Although rabbit embryonic mRNAs begin to be transcribed in a steadily increasing manner during the first cleavage divisions, maternal mRNAs remain stable up to the 8–16-cell stage (Henrion, Brunet, Renard, & Duranthon, 1997). In contrast to other mammals (such as mice or cattle) the total RNA content in rabbit embryo remains high. Embryonic transcripts are synthesized at high levels even when their maternal counterparts have not undergone marked degradation (Brunet-Simon, Henrion, Renard, & Duranthon, 2001). Adequately the protein content remains stable up to the late blastocyst stage, which distinguishes rabbit embryos from mice, where the protein content deteriorates from the 2-cell stage.

The key regulators of lineage specification belong to the large group of transcription factors (TFs). Studies in mouse showed that MZT is triggered by DNA replication during the first cleavage division. However, in non-rodent species this process is delayed to another two or three rounds of cell division (Fig. 1), suggesting another level of regulation. In species such as cow, sheep, and pig, transcriptional silence is probably due to unfavorable chromatin configuration and inefficient transcriptional apparatus. After maternal mRNA degradation, transcriptionally silent chromatin must undergo remodeling in order to acquire a transcriptionally permissive state. Study of chromatin arrangements during major EGA in bovine embryos revealed its close correlation with chromosome territory (CT) localization—after EGA gene poor chromosomes tend to locate at the nuclear periphery (Koehler et al., 2009). Gene locus position within the CT may be related to stage-specific process of gene activation and lineage differentiation. A close relationship was observed between locus position and transcript abundance of OCT4, NANOG, and CDX2 genes during bovine preimplantation development (Orsztynowicz et al., 2017).

Lineage-related transcription factors such as OCT4, NANOG, TEAD, and CDX2 follow the general trend of embryonic genes, with a visible trough in the expression level around the time of EGA (summarized in Fig. 2). The mRNA for the OCT4 gene, which is also important for the acquisition of oocyte's developmental potential, degrades after fertilization and is de novo synthesized around the time of EGA (Cao et al., 2014; Fukuda et al., 2016; Khan et al., 2012; Mamo, Gal, Polgar, & Dinnyes, 2008). On the contrary, transcripts for NANOG, TEAD4, and CDX2 genes, which do not have a reported role in oogenesis, are either maintained at low level before the EGA—mouse, rabbit, pig (Cao et al., 2014; Henderson et al., 2014; Wang, Wang, Xie, & Wang, 2016; Wu et al., 2010), or are not expressed—bovine embryos (Madeja et al., 2013; Orsztynowicz et al., 2017).

4. METABOLIC REQUIREMENTS OF PREIMPLANTATION MAMMALIAN EMBRYOS

Mammalian embryos prior to compaction have a relatively low requirement for external energy sources, which are mostly processed by oxidative phosphorylation. At the blastocyst stage, growing embryos exhibit increased requirement for oxygen and energy substrates such as pyruvate, amino acids, and glucose (Partridge & Leese, 1996; Thompson, Partridge, Houghton, Cox, & Leese, 1996). Glucose concentration in the uterus is relatively high compared to the oviduct (where precompaction development takes place) which may favor glycolytic activity. Glycolysis is responsible for

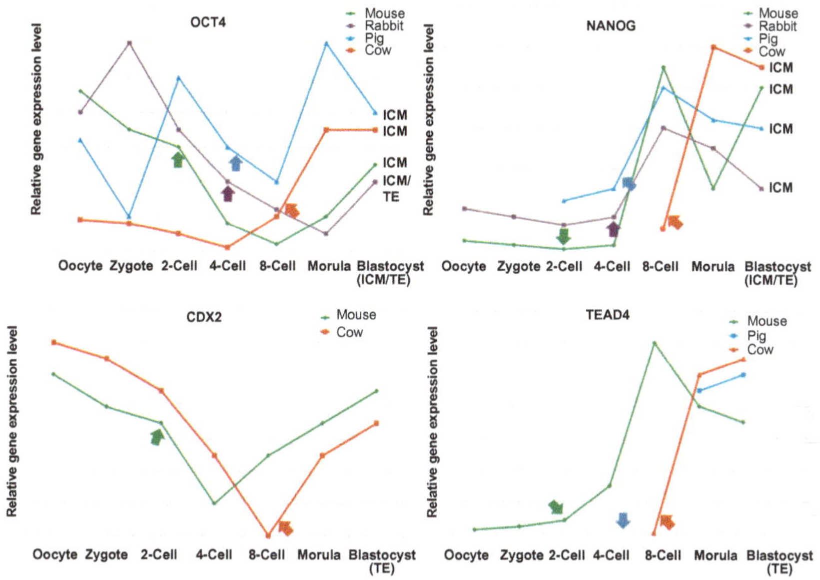

	OCT4	NANOG	CDX2	TEAD4
Mouse	Fukuda et al. (2016)	Cao et al. (2014)	Wu et al. (2010)	Yagi et al. (2007)
Rabbit	Mamo et al. (2008)	Henderson et al. (2014)		
Pig	Cao et al. (2014)	Cao et al. (2014) Wang et al. (2016)		Emura et al. (2016)
Cow	Khan et al. (2012) Orsztynowicz et al. (2017)	Khan et al. (2012) Orsztynowicz et al. (2017) Madeja et al. (2013)	Orsztynowicz et al. (2017)	Sakurai et al. (2017)

Fig. 2 A schematic representation of spatiotemporal changes of lineage-specific transcripts abundance in preimplantation embryos. *Arrows* indicate species-specific moment of embryonic genome activation (EGA). *Table* indicates appropriate references.

the majority of glucose uptake at the blastocyst stage in human, rat, sheep, pig, and bovine embryos (Brison & Leese, 1991; Flood & Wiebold, 1988; Gardner, Lane, & Batt, 1993; Gott, Hardy, Winston, & Leese, 1990; Thompson et al., 1996).

Recent findings suggest the role played by metabolic pathways in cell fate specification in mammalian embryo, particularly in acquisition of naïve vs primed pluripotency (characteristic of murine ICM vs postimplantation epiblast, respectively). Oxidative phosphorylation and glycolysis are significantly reduced in human and mouse ES cells (Kalkan et al., 2017; Sperber et al., 2015; Takashima et al., 2014; Zhou et al., 2012). In the mouse development this metabolic transition is coincident with implantation, and therefore with significant changes in embryo environment and availability of oxygen and energy substrates. Additionally, oxygen levels are reduced in the uterine fluid at the time of implantation, as shown in hamsters and rabbits (Fischer & Bavister, 1993). However, in many domestic animals these two events are not coincident, as the implantation often occurs long after segregation of epiblast and hypoblast lineage, and possibly after the transition from naïve to primed pluripotency. It remains to be uncovered whether states of pluripotency rely in the same manner on cellular metabolism in domestic animals as in the mouse.

5. FIRST CELL FATE DECISION—SPECIFICATION OF INNER CELL MASS AND TROPHECTODERM

At the earliest stages of development mouse embryo blastomeres are totipotent and equal in their developmental potential (Mintz, 1965; Tam & Rossant, 2003; Tarkowski, 1959; Tarkowski & Wróblewska, 1967) (also reviewed in the chapter by Klimczewska, Kasperczuk, & Suwińska, n.d.). It has been confirmed that blastomeres remain totipotent at least up to the 8-cell stage in the rabbits (Moore, Adams, & Rowson, 1968), sheep (Willadsen, 1981), pig (Saito & Niemann, 1991), horse (Allen & Pashen, 1984), and cattle (Johnson, Loskutoff, Plante, & Betteridge, 1995; Willadsen & Polge, 1981). The first differentiation process is initiated by the polarization of outside blastomeres, coincident with compaction (Johnson & Ziomek, 1981), when inside (apolar) and outside (polarized) cells acquire different identity and fate for the first time (Tarkowski & Wróblewska, 1967). Establishment of cell polarity is marked by the localization of microvilli on the apical surface of outside (polar) blastomeres (Reeve & Ziomek, 1981). The timing of compaction and blastomere polarization differs between species—in the mouse it takes place as early as the 8-cell stage (Reeve & Ziomek, 1981), in the rabbit it is not evident until the 32–64-cell stage morula (Ziomek, Chatot, & Manes, 1990), and porcine embryos undergo gradual waves of compaction and decompaction, culminating shortly before cavitation (Reima, Lehtonen, Virtanen, & Fléchon, 1993).

Establishment of cell polarity, marked by polarized localization of the microvilli in the outside cells, is observed after 16-cell stage in bovine embryos (Koyama, Suzuki, Yang, Jiang, & Foote, 1994) and at 32–64-cell stage in the rabbit (Koyama et al., 1994; Ziomek et al., 1990), confirming that the timing of cell polarization is closely linked to compaction and may differ between species.

In the mouse morula outside and inside cells are also distinguished by the localization of junction-associated proteins angiomotin (Amot) and Amotl2, resulting in differential activity of Hippo pathway in these compartments (Hirate et al., 2013; Nishioka et al., 2008; Sasaki, 2010). Active Hippo pathway leads to phosphorylation and degradation of YAP (Yes-associated protein) in the inside cells, while in the outside cells inactive Hippo signaling allows binding of YAP to TEAD4 (TEA-domain protein 4). Subsequently, this first differentiation event sets aside two distinct compartments—TE and ICM (reviewed in the chapter by Saini & Yamanaka, n.d.). The differentiation of TE may be regarded as the hallmark event in mammalian preimplantation development, as it is the first tissue that becomes differentiated during embryogenesis. At the molecular level the initiation of the TE lineage formation is promoted by the action of TEAD4, which was shown to regulate multiple transcription factors important for trophoblast development, such as *Cdx2* and *Gata3* (Nishioka et al., 2008, 2009; Ralston et al., 2010; Yagi et al., 2007). Hippo pathway components have been also found in preimplantation embryos in domestic animals. *TEAD4* and *YAP* are expressed in the horse blastocyst (Iqbal, Chitwood, Meyers-Brown, Roser, & Ross, 2014), and *TEAD4* in pig morulae (Emura, Sakurai, Takahashi, Hashizume, & Sawai, 2016) and bovine 8-cell embryos, morulae, and blastocysts (Fujii, Moriyasu, Hirayama, Hashizume, & Sawai, 2010; Ozawa et al., 2012; Sakurai, Takahashi, Emura, Hashizume, & Sawai, 2017). The specific role of TEAD4 in lineage differentiation in these species has yet to be uncovered, as in bovine embryos no difference in the gene expression level of *TEAD4* and *YAP* was found between ICM and TE (Fujii et al., 2010; Ozawa et al., 2012), and TEAD4 downregulation did not affect expression of TE and EPI markers or development until blastocyst stage (Sakurai et al., 2017).

Specification of the TE lineage from outside cells of the morula is regulated by expression of several transcription factors acting downstream of Hippo pathway, including Cdx2 (Strumpf et al., 2005) and Gata3 (Home et al., 2009; Ralston et al., 2010).

CDX2 (Caudal-related homeodomain transcription factor) has been shown to be the main TF driving TE specification in the mouse embryo (Strumpf et al., 2005). *Cdx2* is expressed in mouse morula and early blastocyst and becomes restricted to TE as the blastocyst expands (Strumpf et al., 2005). Contrary to the mouse, morula stage localization of CDX2 has not been confirmed in most of the studies concerning domestic mammals. At blastocyst stage CDX2 is expressed in the TE of various mammalian species including human, pig, bovine, rabbit (Berg et al., 2011; Bou et al., 2017; Goissis & Cibelli, 2014a; Kirchhof et al., 2000; Kuijk et al., 2008; Madeja et al., 2013; Sakurai et al., 2016), dog (Wilcox et al., 2009), and sheep (Moradi et al., 2015), and upregulation of *CDX2* in porcine embryos results in preferential differentiation to TE lineage (Bou et al., 2017). Detailed analysis of CDX2 localization shows that cavitation and initial TE specification are executed before initiation of *CDX2* expression in rabbit and pig (Bou et al., 2017; Liu et al., 2015; A. Piliszek, unpublished data). In bovine and porcine embryos, downregulation of CDX2 has no effect on pluripotency gene expression or initial specification of the TE (Berg et al., 2011; Goissis & Cibelli, 2014a; Sakurai et al., 2016), but results in loss of epithelial integrity and failure of TE maintenance at later blastocyst stages (Bou et al., 2017; Goissis & Cibelli, 2014a). *CDX2* might be also required for TE cell proliferation in posthatching bovine blastocyst (Berg et al., 2011), and for acquisition of classical protein kinase C (PKC)-mediated TE polarity in porcine blastocyst (Bou et al., 2017). It should be noted that also in the mouse, *Cdx2* knockout or knockdown does not disrupt initial stages of cavitation and blastocyst formation, but affects maintenance of TE epithelial integrity and results in ectopic expression of *Oct4* and *Nanog* (Strumpf et al., 2005; Wu et al., 2010).

CDX2 is initially coexpressed with core pluripotency factor OCT4/POU5F1 (octamer-binding protein 4/POU domain, class 5 transcription factor 1) (Schöler, Ruppert, Suzuki, Chowdhury, & Gruss, 1990) in morula and in TE of the early mouse blastocyst (Dietrich & Hiiragi, 2007). At stages of blastocyst development their expression becomes mutually exclusive, with CDX2 localizing to TE and OCT4—to ICM, and reciprocal inhibition of these two factors is suggested to be the driving mechanism of TE differentiation (Niwa et al., 2005) (Fig. 2).

In nonrodent species OCT4 expression is retained in TE until the late blastocyst stage, colocalizing with CDX2 in human (Kirchhof et al., 2000), bovine (Berg et al., 2011; Kuijk et al., 2008; Madeja et al., 2013), horse (Choi et al., 2009; Iqbal et al., 2014), rabbit (Kobolak et al., 2009), and dog embryos (Wilcox et al., 2009). This suggests that downregulation of *OCT4* is not necessary for the initial differentiation of TE in domestic mammals. Downregulation of *CDX2* does not increase *OCT4* expression levels (Berg et al., 2011), while *OCT4* knockdown results in decreased levels of *CDX2* in bovine embryos (Sakurai et al., 2016). This divergence in CDX2/OCT4 dynamics might result from differences in OCT4 regulatory regions between mice and other species. The study by Berg et al. (2011)

demonstrated that bovine *OCT4* locus does not contain the *cis*-acting regulatory region which is present in the mouse, and required for suppression of TE-specific transcription. Continuous expression of *OCT4* until later stages of blastocyst expansion possibly affects TE developmental potency, as the TE cells have been shown to contribute to the ICM lineage in bovine (Berg et al., 2011) and human embryonic chimeras (De Paepe et al., 2013). In later stage blastocyst this ability is lost, probably concomitant with diminished *OCT4* expression in the TE.

GATA3 (GATA-binding protein 3) is another transcription factor acting downstream of the TEAD4, which has been linked to the TE cell fate specification in the mouse (Home et al., 2009; Ralston et al., 2010). GATA3 is coexpressed with CDX2 in the mouse TE and is able to induce TE differentiation of ES cells, yet mouse embryos depleted of GATA3 show no defects in TE specification or in implantation (Ralston et al., 2010). *GATA3* is specifically expressed in TE of horse (Iqbal et al., 2014), cattle (Ozawa et al., 2012; Smith, Berg, Berg, & Pfeffer, 2010), and rabbit blastocyst (A. Piliszek, unpublished data). However, a recent report indicates that the lack of TE-specific defects in mice results from genetic redundancy with GATA2 (GATA-binding protein 2) (Home et al., 2017), as blastocyst cavity and TE layer are formed in the GATA2/GATA3 double knockout embryos with lower frequency. Indeed, both of these factors are expressed in bovine embryos and overexpression of GATA2 and/or GATA3 is able to induce expression of other TE-specific genes in fibroblasts upon transfection (Bai et al., 2011). GATA3 is likely to play a role in later stages of TE differentiation, rather than initial specification—in the mouse GATA2/GATA3 double knockout predominantly affects embryo implantation (Home et al., 2017), and in the rabbit GATA3 appears only in expanded blastocyst at 4 dpc (A. Piliszek, unpublished data).

T-box transcription factor Eomes (Eomesodermin) is also associated with TE fate in the mouse (Ciruna & Rossant, 1999; Hancock, Agulnik, Silver, & Papaioannou, 1999; McConnell, Petrie, Stennard, Ryan, & Nichols, 2005; Russ et al., 2000; Strumpf et al., 2005). It is expressed in mouse morula and TE lineage (McConnell et al., 2005) and is required for proper implantation (Russ et al., 2000; Strumpf et al., 2005). At later stages Eomes is involved in gastrulation and organogenesis (Ciruna & Rossant, 1999; Hancock et al., 1999). This later TE-independent role is probably more conserved across the species, since *EOMES* transcripts are absent in TE of bovine (Berg et al., 2011; Hall, Ruddock, & French, 2005; Ozawa et al., 2012; Smith et al., 2010), sheep (Guillomot, Turbe, Hue, & Renard, 2004), and rabbit embryos (A. Piliszek, unpublished data) but found at later stages in embryonic disc in those species (Guillomot et al., 2004; van Leeuwen et al., 2015).

Taken together, these data suggest a considerable delay in TE lineage commitment compared to initial specification in domestic animal blastocyst. In the mouse embryo outside blastomeres of 32-cell morula lose totipotency and adopt TE fate even before the appearance of blastocoel cavity and morphologically distinguished TE (Suwińska, Czołowska, Ozdzeński, & Tarkowski, 2008). This early specification might be an exception rather than the rule in mammalian development, allowing for TE differentiation and implantation in a relatively short time of development and a small size of a murine preimplantation embryo. The relatively late onset of TE-specific TF expression together with maintained expression of pluripotency factor OCT4 appears to be a common phenomenon in preimplantation development of many domestic mammalian species. What is more, even after morphological separation of TE and ICM lineages, TE cells of bovine embryo are able to contribute to pluripotent ICM in chimera (Berg et al., 2011), raising possibility that they retain the high level of plasticity. It should be noted that in many mammalian species preimplantation embryo significantly increases its size, and even after the specification of EPI and hypoblast lineages, TE further differentiates in preparation for implantation. One notable event is the fate of Rauber's layer—the part of TE overlying epiblast (polar trophoblast). In the mouse embryo polar TE participates in implantation, and after implantation expands to form extraembryonic ectoderm (ExE). In nonrodent species, such as rabbit (Williams & Biggers, 1990), sheep, pig, and cattle (Maddox-Hyttel et al., 2003), Rauber's layer disappears in the late blastocyst, exposing the epiblast, and therefore only mural trophectoderm is involved in the implantation process.

Unlike mouse and human, ungulate preimplantation development extends beyond blastocyst hatching, and includes period of posthatching trophoblast elongation (reviewed by Blomberg, Hashizume, & Viebahn, 2008). During this process embryo progresses from spherical through ovoid and tubular blastocyst stages, eventually forming a highly elongated, filamentous blastocyst with a relatively small embryonic disc. In nonungulate species, such as the horse and rabbit, spherical blastocyst increases in size without hatching, but considerable growth (up to 5 mm in rabbit and 20 mm in horse) is facilitated by remodeling of embryonic coats enveloping the embryo (reviewed by Betteridge, 1989; Denker, 2000). Increased size of the trophoblast implantation (up to 150 mm in pig and cattle) (Degrelle et al., 2005; Maddox-Hyttel et al., 2003) allows for a greater area of attachment and subsequently of nutrient exchange (Stroband & Van der Lende, 1990). Differences in trophoblast development and morphology across the species coincide with variation in mode of implantation and placentation in eutherian mammals (Bazer et al., 2009; McGowen, Erez, Romero, & Wildman, 2014).

Studies in the mouse show a requirement for FGF signaling in maintenance of TE lineage (Goldin & Papaioannou, 2003). FGF/MEK pathway components are expressed in horse blastocyst in both ICM and TE (Iqbal et al., 2014). In porcine embryos expression of *bFGF* and *FGFR2* was confirmed in both EPI and TE at similar levels (Hall, Christensen, Gao, Schmidt, & Hyttel, 2009). In ovine embryos (Moradi et al., 2015) FGF2 treatment or FGFR inhibition did not affect TE cell number, blastocyst formation, or expression of TE markers (CDX2) significantly. Similarly in bovine embryos FGF4 treatment did not significantly affect CDX2-positive cell number (Kuijk et al., 2012); however, in a different study addition of FGF2 increased the rate of development to blastocyst stage in vitro (Fields, Hansen, & Ealy, 2011).

6. SPECIFICATION OF EPIBLAST AND HYPOBLAST IN MAMMALIAN BLASTOCYST

After the formation of a blastocyst, ICM further differentiates into two distinct lineages—pluripotent EPI and extraembryonic hypoblast/PrE. In mouse embryos, PrE- and EPI-specific transcription factors are initially coexpressed at the morula and early blastocyst stages (Plusa, Piliszek, Frankenberg, Artus, & Hadjantonakis, 2008). At the mid-blastocyst stage their expression becomes mutually exclusive, and nascent PrE and EPI cells become distributed in an apparently random, so-called salt-and-pepper manner (Chazaud, Yamanaka, Pawson, & Rossant, 2006). Eventually, these two populations become sorted into two distinct layers—PrE facing the blastocyst cavity, and EPI encapsulated by PrE and polar TE (Chazaud et al., 2006; Meilhac et al., 2009; Plusa et al., 2008; also reviewed in the chapter by Bessalert, Valverde, & Chazaud, n.d.).

In mouse embryos, EPI lineage is defined by the expression of core pluripotency transcription factors including POU5F1/OCT4 (Nichols et al., 1998; Schöler et al., 1990), NANOG (Chambers et al., 2003), and SOX2 (Avilion et al., 2003). Although the information about detailed expression pattern is not available for all of the domestic mammals, none of these factors were confirmed to be missing in the ICM of blastocysts of these species.

POU5F1/OCT4 is expressed in cattle (Berg et al., 2011; Kuijk et al., 2008; Madeja et al., 2013), horse (Choi et al., 2009; Iqbal et al., 2014), pig (Blomberg, Schreier, & Talbot, 2008), sheep (Sanna et al., 2009), goat (HosseinNia et al., 2016), rabbit (Kobolak et al., 2009; Táncos, Bock, Nemes, Kobolák, & Dinnyés, 2015), and dog preimplantation embryos (Wilcox et al., 2009). As mentioned earlier, POU5F1/OCT4 is already expressed at the morula stage, found in both ICM and TE lineage of several species at the blastocyst stage, and only later specifically restricted to the ICM. The expression of OCT4 is diminished in goat embryos by the blastocyst stage (HosseinNia et al., 2016) and becomes restricted to the EPI lineage in 8 dpc blastocyst in porcine embryos (Blomberg, Schreier, et al., 2008).

SOX2 (SRY-related HMG box-containing transcription factor 2) appears to be a more specific EPI marker in mammals. It is expressed specifically in the ICM of porcine (Hall et al., 2009; Liu et al., 2015), equine (Iqbal et al., 2014), and rabbit embryos (Piliszek, Madeja, & Plusa, 2017; Táncos et al., 2015) not coexpressed with TE markers at any of the stages. In goat in vitro-derived and parthenogenetic embryos, *SOX2* is found at the morula and blastocyst stages, although its expression is not limited to ICM (Yu, Zhao, Wang, & Ma, 2015). In bovine embryos, *SOX2* is already expressed at the 8-cell stage, with a significant increase in blastocysts at 7 dpc, and restricted to the nuclei of ICM cells at 7–9 dpc (Goissis & Cibelli, 2014b; Khan et al., 2012; Ozawa et al., 2012). SOX2 silencing by siRNA in bovine embryos decreases the rate of development to the blastocyst stage, yet some SOX2-negative cells are able to contribute to the bovine ICM (Goissis & Cibelli, 2014b), although it cannot be excluded that this population represents the SOX2-negative hypoblast lineage.

NANOG is one of the key factors associated with EPI fate and pluripotency in the mouse (Chambers et al., 2003). It is found in the ICM of most mammals, although its specific expression pattern varies among species (Fig. 2). In bovine embryos, *NANOG* is expressed already at 8-cell stage, and in 7–9 dpc blastocyst is found exclusively in the ICM (Khan et al., 2012; Madeja et al., 2013). As shown by another study, NANOG protein is not found at the morula stage and is distributed in a salt-and-pepper pattern in the bovine ICM (Kuijk et al., 2008, 2012) suggestive of restriction to EPI precursors. In rabbit embryos, it is also expressed from 2-cell stage (Táncos et al., 2015) up to 4 dpc blastocyst, when it is specifically localized to the ICM (Piliszek et al., 2017), and in the goat from oocyte up to later stages, with increased levels at morula and blastocyst stages (Yu et al., 2015). In porcine embryos, NANOG expression is even more specific, as it is not found until 8 dpc (Cao et al., 2014; Kuijk et al., 2008; Wolf, Serup, & Hyttel, 2011), when it is coexpressed with SOX2 exclusively in the EPI (Hall et al., 2009). *NANOG* has been also confirmed to be specifically expressed in the ICM of equine blastocyst (Iqbal et al., 2014).

In mouse embryos, emerging PrE cell population is marked by sequential activation of GATA6, SOX17, GATA4, and SOX7 (Artus, Piliszek, & Hadjantonakis, 2011; Koutsourakis, Langeveld, Patient, Beddington, & Grosveld, 1999; Niakan et al., 2010; Schrode, Saiz, Di Talia, & Hadjantonakis, 2014; Soudais et al., 1995). GATA6, together with EPI marker NANOG, is initially expressed in all cells of the mouse embryo (Plusa et al., 2008). Its subsequent restriction to nascent PrE is accompanied by the initiation of GATA4 and SOX17 expression in those cells (Artus et al., 2011; Niakan et al., 2010; Plusa et al., 2008).

In pig embryos, GATA6 expression is increased at blastocyst stage and found at significantly higher levels in some of the ICM cells (Kuijk et al., 2008). Interestingly, GATA6 was not found in in vitro-produced porcine embryos, but authors speculate that this might result from the differences in the developmental age (Kuijk et al., 2008). GATA6 protein distribution was also found in a pattern resembling salt-and-pepper distribution in bovine ICM, and sometimes lining the blastocyst cavity, suggestive of an already sorted hypoblast layer (Kuijk et al., 2008).

On the other hand, GATA4 appears to be a less-specific hypoblast marker in some ungulates. In bovine embryos, GATA4 marks both ICM and TE (Kuijk et al., 2008, 2012). In porcine embryos, *GATA4* expression was diminished by the blastocyst stage and GATA4 protein not detected in one study (Kuijk et al., 2008), while it showed a hypoblast-specific localization in a study from another group using in vitro-derived embryos (Rodríguez, Allegrucci, & Alberio, 2012).

In horse embryos, *GATA4* and *GATA6* were found to be expressed in both ICM and TE samples, with notably higher levels in ICM; however, it is possible that due to the method of sample collection, these groups might include hypoblast underlying ICM and hypoblast (parietal endoderm) underlying TE (Iqbal et al., 2014). In another study, GATA6 protein was found in both TE and hypoblast of parthenogenetic blastocysts at 7 dpc, but was specifically restricted to hypoblast in in vivo-derived equine embryos at the same time (Desmarais et al., 2011). In the rabbit embryos *GATA6* is expressed throughout preimplantation development, becoming progressively restricted to nascent PrE lineage, and SOX17 expression emerges in the GATA6-positive ICM cells at 3.5 dpc (Piliszek et al., 2017).

Mutual repression of *Nanog* and *Gata6* has been suggested as a main mechanism of PrE vs EPI specification in the mouse, as the salt-and-pepper pattern of lineage precursors emerges by the restricted expression of NANOG and GATA6 in distinct but intermingled cell populations of the ICM (Bessonnard et al., 2014; Frankenberg et al., 2011; Schröter, Rué, Mackenzie, & Martinez Arias, 2015; Singh, Hamazaki, Hankowski, & Terada, 2007; Xenopoulos, Kang, Puliafito, Di Talia, & Hadjantonakis, 2015). During the lineage specification of domestic mammals, increasing restriction of TF expression and emergence of the seemingly salt-and-pepper pattern has been observed. In bovine embryos, GATA6 expression becomes mutually exclusive from NANOG, and resulting hypoblast and epiblast precursor cells segregate into two distinct layers (Denicol et al., 2014). In porcine blastocysts, GATA6 is found in a subset of ICM cells (Kuijk et al., 2008). In rabbit embryos, although GATA6 follows a similar pattern of progressive restriction, NANOG does not exhibit the salt-and-pepper pattern and is found in all ICM cells until the segregation of EPI and PrE layers (Piliszek et al., 2017). Initial absence of NANOG in porcine embryos, and its persistent expression in rabbit embryos suggests that mutual inhibition of GATA6 and NANOG might not be a strict requirement for epiblast vs hypoblast specification in all mammals.

7. FGF/MEK PATHWAY IN SPECIFICATION OF EPIBLAST AND HYPOBLAST

Primitive endoderm vs epiblast specification is dependent on FGF/MEK signaling in the mouse embryos (reviewed in the chapter by Bessalert et al., n.d.). Knockout of *Fgf4* (Kang, Piliszek, Artus, & Hadjantonakis, 2013) and inhibition of FGF receptor or its downstream pathway component, MAPK/ERK kinase (MEK), using small molecule inhibitors (Nichols, Silva, Roode, & Smith, 2009; Yamanaka, Lanner, & Rossant, 2010) drive ICM cells toward EPI fate, while culture in excess amounts of FGF4 or FGF2 results in specification of ICM cells into PrE (Arman, Haffner-Krausz, Chen, Heath, & Lonai, 1998; Frankenberg et al., 2011; Kang et al., 2013; Nichols et al., 2009; Yamanaka et al., 2010). Contrary to the mouse, inhibition of FGF receptor or ERK signaling in human embryos does not ablate hypoblast, confirmed by the sustained presence of GATA6 and GATA4 factors (Kuijk et al., 2012; Roode et al., 2012). In domestic mammals, modulation of FGF signaling affects the second cell fate specification event to different extents, and the specific mechanisms may vary greatly among different species.

Culturing bovine embryos in the presence of MEK inhibitors increases the proportion of NANOG-positive (EPI) cells within the ICM, but does not fully ablate GATA6-positive (hypoblast) population, while culturing them in the presence of upstream FGF receptor inhibitors does not alter the ICM composition. On the other hand, culture in the presence of exogenous FGF increases the proportion of hypoblast cells and ablates the EPI lineage of bovine embryos (Kuijk et al., 2012). FGF2 treatment of preimplantation sheep embryos also results in increased expression of *GATA4* and decreased expression of *NANOG* at blastocyst stage, but their expression levels remain unaffected by FGF receptor inhibition (Moradi et al., 2015). Similarly, MEK inhibition in porcine embryos does not completely ablate hypoblast, but decreases the proportion of GATA4-positive cells and increases NANOG proportion, while FGFR inhibition does not affect epiblast vs hypoblast proportion, but diminishes the overall cell number of the ICM. FGF4 treatment results in loss of EPI, and a significant increase in numbers of GATA4-positive hypoblast cells of porcine embryos (Rodríguez et al., 2012). Another study using parthenogenetic porcine embryos and a different FGFR inhibitor reported an increased expression of pluripotency and trophoblast markers *SOX2*, *OCT4*, and *KLF4*, but no change in the expression of *NANOG* or *GATA4* upon the inhibition of FGFR (Li et al., 2016). In the rabbit embryos, providing a saturating level of FGF4 also results in the ablation of EPI lineage, and MEK inhibition leads to the loss of SOX17-positive PrE population (Piliszek et al., 2017). Despite that, the expression of early PrE marker *GATA6* remains unaffected, and remaining SOX2-positive EPI compartment does not exhibit an elevated level of *NANOG* expression, a characteristic phenotype of naive EPI (Piliszek et al., 2017) as it does in the mouse upon MEK inhibition (Nichols et al., 2009).

Several components of FGF/MEK pathway are expressed in the preimplantation embryos of domestic mammals. In porcine embryos *bFGF*, *FGFR1*, and *FGFR2* were found to be expressed at the blastocyst stage. However, only *FGFR1* was detected exclusively in the EPI (Hall et al., 2009). Expression of *FGF4* and FGF receptors (*FGFR1*, *FGFR2*) was confirmed in the preimplantation rabbit embryos (Piliszek et al., 2017). In the IVF ovine embryos, expression of FGF receptors 1–4 was detected during the preimplantation period, with a peak at the morula stage (Moradi et al., 2015).

In mouse preimplantation embryos it has been hypothesized that EPI cells secrete FGF4 ligand, which acts through FGFR2 to specify the PrE lineage (Bessonnard et al., 2014; Guo et al., 2010; Kang et al., 2013). However, a recent work has shown critical requirements for both FGFR1 and FGFR2 for the lineage specification in murine ICM (Kang, Garg, & Hadjantonakis, 2017; Molotkov, Mazot, Brewer, Cinalli, & Soriano, 2017). This is in agreement with the previous study using human embryos where *FGFR2* is not present in 6 dpc blastocyst (Kunath et al., 2014), as well as studies with porcine (Fujii et al., 2013; Hall et al., 2009) and rabbit (Piliszek et al., 2017) embryos, where *FGFR1*, but not *FGFR2*, is expressed during the period of EPI vs hypoblast specification. In sheep blastocysts, *FGFR1* and *FGFR3* are expressed at relatively highest levels (Moradi et al., 2015), while *FGFR1* and *FGFR4* are significantly upregulated in the horse ICM (Iqbal et al., 2014).

Detailed comparisons among domestic mammals are hampered by differences in the timing of the lineage specification as well as discrepancies in specific treatments and markers used in different studies. However, these results collectively show a conserved role for FGF signaling in certain aspects of hypoblast development in mammals and suggest that the epiblast specification may require additional signals.

8. CONCLUSIONS

The key players of TE, EPI, and PrE specification, such as transcription factors and growth factors, might be conserved among mammals, but their precise relationship differs across the species. To uncover the fundamental mechanism of early lineage specification in mammals, and distinguish between the core circuit and species-specific variants, comparative studies of different systems are necessary. Precise comparisons are greatly impeded by scarcity of data available in nonmurine species, obvious differences in the developmental timing, lack of consistent staging systems, and differences among specific strains and breeds of the same species. Additionally, the available information in large animals often comes from studies on in vitro-obtained embryos. Although techniques of assisted reproduction can produce developmentally competent, or at least morphologically normal embryos, studies have shown that parthenogenetic activation, or different composition of in vitro culture media, can affect lineage specification (Choi et al., 2015; Saenz-de-Juano, Naturil-Alfonso, Vicente, & Marco-Jiménez, 2013). Despite these obstacles, the growing interest in nonrodent species for both scientific and biomedical research increases the number of studies performed on domesticated mammals.

ACKNOWLEDGMENTS

This work was funded by Narodowe Centrum Nauki (NCN) Grant Nos. 2011/03/D/NZ3/03992 (A.P.), 4429/B/P01/2010/39 (Z.E.M.), and 2012/05/B/NZ9/03349 (Z.E.M.).

REFERENCES

Allen, W. R., & Pashen, R. L. (1984). Production of monozygotic (identical) horse twins by embryo micromanipulation. *Journal of Reproduction and Fertility*, *71*, 607–613.

Aoki, F., Worrad, D. M., & Schultz, R. M. (1997). Regulation of transcriptional activity during the first and second cell cycles in the preimplantation mouse embryo. *Developmental Biology*, *181*, 296–307. https://doi.org/10.1006/dbio.1996.8466.

Arman, E., Haffner-Krausz, R., Chen, Y., Heath, J. K., & Lonai, P. (1998). Targeted disruption of fibroblast growth factor (FGF) receptor 2 suggests a role for FGF signaling in pregastrulation mammalian development. *Proceedings of the National Academy of Sciences of the United States of America*, *95*, 5082–5087.

Artus, J., Piliszek, A., & Hadjantonakis, A.-K. (2011). The primitive endoderm lineage of the mouse blastocyst: Sequential transcription factor activation and regulation of differentiation by Sox17. *Developmental Biology*, *350*, 393–404. https://doi.org/10.1016/j.ydbio.2010.12.007.

Avilion, A. A., Nicolis, S. K., Pevny, L. H., Perez, L., Vivian, N., & Lovell-Badge, R. (2003). Multipotent cell lineages in early mouse development depend on SOX2 function. *Genes & Development*, *17*, 126–140. https://doi.org/10.1101/gad.224503.

Bai, H., Sakurai, T., Someya, Y., Konno, T., Ideta, A., Aoyagi, Y., et al. (2011). Regulation of trophoblast-specific factors by GATA2 and GATA3 in bovine trophoblast CT-1 cells. *The Journal of Reproduction and Development*, *57*, 518–525.

Bazer, F. W., Spencer, T. E., Johnson, G. A., Burghardt, R. C., & Wu, G. (2009). Comparative aspects of implantation. *Reproduction (Cambridge, England)*, *138*, 195–209. https://doi.org/10.1530/REP-09-0158.

Berg, D. K., Smith, C. S., Pearton, D. J., Wells, D. N., Broadhurst, R., Donnison, M., et al. (2011). Trophectoderm lineage determination in cattle. *Developmental Cell*, *20*, 244–255. https://doi.org/10.1016/j.devcel.2011.01.003.

Bessalert, C., Valverde, E.L., Chazaud, C., n.d. Specification of extra-embryonic lineages during mouse pre-implantation development. *Current Topics in Developmental Biology*, 128.

Bessonnard, S., De Mot, L., Gonze, D., Barriol, M., Dennis, C., Goldbeter, A., et al. (2014). Gata6, Nanog and Erk signaling control cell fate in the inner cell mass through a tristable regulatory network. *Development (Cambridge, England)*, *141*, 3637–3648. https://doi.org/10.1242/dev.109678.

Betteridge, K. J. (1989). The structure and function of the equine capsule in relation to embryo manipulation and transfer. *Equine Veterinary Journal*, *21*, 92–100. https://doi.org/10.1111/j.2042-3306.1989.tb04690.x.

Bindon, B. M. (1971). Systematic study of preimplantation stages of pregnancy in the sheep. *Australian Journal of Biological Sciences*, *24*, 131–147.

Blomberg, L., Hashizume, K., & Viebahn, C. (2008). Blastocyst elongation, trophoblastic differentiation, and embryonic pattern formation. *Reproduction (Cambridge, England)*, *135*, 181–195. https://doi.org/10.1530/REP-07-0355.

Blomberg, L. A., Schreier, L. L., & Talbot, N. C. (2008). Expression analysis of pluripotency factors in the undifferentiated porcine inner cell mass and epiblast during in vitro culture. *Molecular Reproduction and Development*, *75*, 450–463. https://doi.org/10.1002/mrd.20780.

Bou, G., Liu, S., Sun, M., Zhu, J., Xue, B., Guo, J., et al. (2017). CDX2 is essential for cell proliferation and polarity in porcine blastocysts. *Development (Cambridge, England)*, *144*, 1296–1306. https://doi.org/10.1242/dev.141085.

Braude, P., Bolton, V., & Moore, S. (1988). Human gene expression first occurs between the four- and eight-cell stages of preimplantation development. *Nature*, *332*, 459–461. https://doi.org/10.1038/332459a0.

Brison, D. R., & Leese, H. J. (1991). Energy metabolism in late preimplantation rat embryos. *Journal of Reproduction and Fertility*, *93*, 245–251.

Brunet-Simon, A., Henrion, G., Renard, J. P., & Duranthon, V. (2001). Onset of zygotic transcription and maternal transcript legacy in the rabbit embryo. *Molecular Reproduction and Development*, *58*, 127–136. https://doi.org/10.1002/1098-2795(200102)58:2<127::AID-MRD1>3.0.CO;2-A.

Cao, S., Han, J., Wu, J., Li, Q., Liu, S., Zhang, W., et al. (2014). Specific gene-regulation networks during the pre-implantation development of the pig embryo as revealed by deep sequencing. *BMC Genomics*, *15*, 4. https://doi.org/10.1186/1471-2164-15-4.

Chambers, I., Colby, D., Robertson, M., Nichols, J., Lee, S., Tweedie, S., et al. (2003). Functional expression cloning of Nanog, a pluripotency sustaining factor in embryonic stem cells. *Cell*, *113*, 643–655.

Chazaud, C., Yamanaka, Y., Pawson, T., & Rossant, J. (2006). Early lineage segregation between epiblast and primitive endoderm in mouse blastocysts through the Grb2-MAPK pathway. *Developmental Cell*, *10*, 615–624. https://doi.org/10.1016/j.devcel.2006.02.020.

Choi, Y. H., Harding, H. D., Hartman, D. L., Obermiller, A. D., Kurosaka, S., McLaughlin, K. J., et al. (2009). The uterine environment modulates trophectodermal POU5F1 levels in equine blastocysts. *Reproduction (Cambridge, England)*, *138*, 589–599. https://doi.org/10.1530/REP-08-0394.

Choi, Y.-H., Ross, P., Velez, I. C., Macías-García, B., Riera, F. L., & Hinrichs, K. (2015). Cell lineage allocation in equine blastocysts produced in vitro under varying glucose concentrations. *Reproduction (Cambridge, England)*, *150*, 31–41. https://doi.org/10.1530/REP-14-0662.

Ciruna, B. G., & Rossant, J. (1999). Expression of the T-box gene Eomesodermin during early mouse development. *Mechanisms of Development*, *81*, 199–203.

Crosby, I. M., Gandolfi, F., & Moor, R. M. (1988). Control of protein synthesis during early cleavage of sheep embryos. *Journal of Reproduction and Fertility*, *82*, 769–775.

Degrelle, S. A., Campion, E., Cabau, C., Piumi, F., Reinaud, P., Richard, C., et al. (2005). Molecular evidence for a critical period in mural trophoblast development in bovine blastocysts. *Developmental Biology*, *288*, 448–460. https://doi.org/10.1016/j.ydbio.2005.09.043.

Denicol, A.C., Block, J., Kelley, D.E., Pohler, K.G., Dobbs, K.B., Mortensen, C.J., Ortega, M.S., Hansen, P.J., 2014. The WNT signaling antagonist Dickkopf-1 directs lineage commitment and promotes survival of the preimplantation embryo. The FASEB Journal: Official Publication of the Federation of American Societies for Experimental Biology 28, 3975–3986. doi:https://doi.org/10.1096/fj.14-253112

Denker, H. W. (2000). Structural dynamics and function of early embryonic coats. *Cells, Tissues, Organs*, *166*, 180–207. https://doi.org/10.1159/000016732.

De Paepe, C., Cauffman, G., Verloes, A., Sterckx, J., Devroey, P., Tournaye, H., et al. (2013). Human trophectoderm cells are not yet committed. *Human Reproduction (Oxford, England)*, *28*, 740–749. https://doi.org/10.1093/humrep/des432.

Desmarais, J. A., Demers, S.-P., Suzuki, J., Laflamme, S., Vincent, P., Laverty, S., et al. (2011). Trophoblast stem cell marker gene expression in inner cell mass-derived cells from parthenogenetic equine embryos. *Reproduction (Cambridge, England)*, *141*, 321–332. https://doi.org/10.1530/REP-09-0536.

Dietrich, J.-E., & Hiiragi, T. (2007). Stochastic patterning in the mouse pre-implantation embryo. *Development (Cambridge, England)*, *134*, 4219–4231. https://doi.org/10.1242/dev.003798.

Emura, N., Sakurai, N., Takahashi, K., Hashizume, T., & Sawai, K. (2016). OCT-4 expression is essential for the segregation of trophectoderm lineages in porcine preimplantation embryos. *The Journal of Reproduction and Development, 62*, 401–408. https://doi.org/10.1262/jrd.2016-040.

Fields, S. D., Hansen, P. J., & Ealy, A. D. (2011). Fibroblast growth factor requirements for in vitro development of bovine embryos. *Theriogenology, 75*, 1466–1475. https://doi.org/10.1016/j.theriogenology.2010.12.007.

Fischer, B., & Bavister, B. D. (1993). Oxygen tension in the oviduct and uterus of rhesus monkeys, hamsters and rabbits. *Journal of Reproduction and Fertility, 99*, 673–679. https://doi.org/10.1530/jrf.0.0990673.

Flood, M. R., & Wiebold, J. L. (1988). Glucose metabolism by preimplantation pig embryos. *Journal of Reproduction and Fertility, 84*, 7–12.

Frankenberg, S., Gerbe, F., Bessonnard, S., Belville, C., Pouchin, P., Bardot, O., et al. (2011). Primitive endoderm differentiates via a three-step mechanism involving Nanog and RTK signaling. *Developmental Cell, 21*, 1005–1013. https://doi.org/10.1016/j.devcel.2011.10.019.

Fujii, T., Moriyasu, S., Hirayama, H., Hashizume, T., & Sawai, K. (2010). Aberrant expression patterns of genes involved in segregation of inner cell mass and trophectoderm lineages in bovine embryos derived from somatic cell nuclear transfer. *Cellular Reprogramming, 12*, 617–625. https://doi.org/10.1089/cell.2010.0017.

Fujii, T., Sakurai, N., Osaki, T., Iwagami, G., Hirayama, H., Minamihashi, A., et al. (2013). Changes in the expression patterns of the genes involved in the segregation and function of inner cell mass and trophectoderm lineages during porcine preimplantation development. *The Journal of Reproduction and Development, 59*, 151–158.

Fukuda, A., Mitani, A., Miyashita, T., Kobayashi, H., Umezawa, A., & Akutsu, H. (2016). Spatiotemporal dynamics of OCT4 protein localization during preimplantation development in mice. *Reproduction (Cambridge, England), 152*, 417–430. https://doi.org/10.1530/REP-16-0277.

Gardner, D. K., Lane, M., & Batt, P. (1993). Uptake and metabolism of pyruvate and glucose by individual sheep preattachment embryos developed in vivo. *Molecular Reproduction and Development, 36*, 313–319. https://doi.org/10.1002/mrd.1080360305.

Ginther, O. J. (1992). *Reproductive biology of the mare: Basic and applied aspects*. Equiservices.

Goissis, M. D., & Cibelli, J. B. (2014a). Functional characterization of CDX2 during bovine preimplantation development in vitro. *Molecular Reproduction and Development, 81*, 962–970. https://doi.org/10.1002/mrd.22415.

Goissis, M. D., & Cibelli, J. B. (2014b). Functional characterization of SOX2 in bovine preimplantation embryos. *Biology of Reproduction, 90*, 30. https://doi.org/10.1095/biolreprod.113.111526.

Goldin, S. N., & Papaioannou, V. E. (2003). Paracrine action of FGF4 during periimplantation development maintains trophectoderm and primitive endoderm. *Genes, 36*, 40–47. https://doi.org/10.1002/gene.10192.

Gott, A. L., Hardy, K., Winston, R. M., & Leese, H. J. (1990). Non-invasive measurement of pyruvate and glucose uptake and lactate production by single human preimplantation embryos. *Human Reproduction (Oxford, England), 5*, 104–108.

Guillomot, M., Turbe, A., Hue, I., & Renard, J.-P. (2004). Staging of ovine embryos and expression of the T-box genes Brachyury and Eomesodermin around gastrulation. *Reproduction, 127*, 491–501. https://doi.org/10.1530/rep.1.00057.

Guo, G., Huss, M., Tong, G. Q., Wang, C., Li Sun, L., Clarke, N. D., et al. (2010). Resolution of cell fate decisions revealed by single-cell gene expression analysis from zygote to blastocyst. *Developmental Cell, 18*, 675–685. https://doi.org/10.1016/j.devcel.2010.02.012.

Hall, V. J., Christensen, J., Gao, Y., Schmidt, M. H., & Hyttel, P. (2009). Porcine pluripotency cell signaling develops from the inner cell mass to the epiblast during early development. *Developmental Dynamics: An Official Publication of the American Association of Anatomists*, *238*, 2014–2024. https://doi.org/10.1002/dvdy.22027.

Hall, V. J., Ruddock, N. T., & French, A. J. (2005). Expression profiling of genes crucial for placental and preimplantation development in bovine in vivo, in vitro, and nuclear transfer blastocysts. *Molecular Reproduction and Development*, *72*, 16–24. https://doi.org/10.1002/mrd.20337.

Hancock, S. N., Agulnik, S. I., Silver, L. M., & Papaioannou, V. E. (1999). Mapping and expression analysis of the mouse ortholog of Xenopus Eomesodermin. *Mechanisms of Development*, *81*, 205–208.

Henderson, G. R. W., Brahmasani, S. R., Yelisetti, U. M., Konijeti, S., Katari, V. C., & Sisinthy, S. (2014). Candidate gene expression patterns in rabbit preimplantation embryos developed in vivo and in vitro. *Journal of Assisted Reproduction and Genetics*, *31*, 899–911. https://doi.org/10.1007/s10815-014-0233-0.

Henrion, G., Brunet, A., Renard, J. P., & Duranthon, V. (1997). Identification of maternal transcripts that progressively disappear during the cleavage period of rabbit embryos. *Molecular Reproduction and Development*, *47*, 353–362. https://doi.org/10.1002/(SICI)1098-2795(199708)47:4<353::AID-MRD1>3.0.CO;2-J.

Hirate, Y., Hirahara, S., Inoue, K.-I., Suzuki, A., Alarcon, V. B., Akimoto, K., et al. (2013). Polarity-dependent distribution of angiomotin localizes Hippo signaling in preimplantation embryos. *Current Biology*, *23*, 1181–1194. https://doi.org/10.1016/j.cub.2013.05.014.

Home, P., Kumar, R. P., Ganguly, A., Saha, B., Milano-Foster, J., Bhattacharya, B., et al. (2017). Genetic redundancy of GATA factors in the extraembryonic trophoblast lineage ensures the progression of preimplantation and postimplantation mammalian development. *Development (Cambridge, England)*, *144*, 876–888. https://doi.org/10.1242/dev.145318.

Home, P., Ray, S., Dutta, D., Bronshteyn, I., Larson, M., & Paul, S. (2009). GATA3 is selectively expressed in the trophectoderm of peri-implantation embryo and directly regulates Cdx2 gene expression. *The Journal of Biological Chemistry*, *284*, 28729–28737. https://doi.org/10.1074/jbc.M109.016840.

HosseinNia, P., Hajian, M., Tahmoorespur, M., Hosseini, S. M., Ostadhosseini, S., Nasiri, M. R., et al. (2016). Expression profile of developmentally important genes in pre and peri-implantation goat embryos produced in vitro. *International Journal of Fertility and Sterility*, *10*, 310–319.

Iqbal, K., Chitwood, J. L., Meyers-Brown, G. A., Roser, J. F., & Ross, P. J. (2014). RNA-seq transcriptome profiling of equine inner cell mass and trophectoderm. *Biology of Reproduction*, *90*, 61. https://doi.org/10.1095/biolreprod.113.113928.

Johnson, W. H., Loskutoff, N. M., Plante, Y., & Betteridge, K. J. (1995). Production of four identical calves by the separation of blastomeres from an in vitro derived four-cell embryo. *The Veterinary Record*, *137*, 15–16.

Johnson, M. H., & Ziomek, C. A. (1981). The foundation of two distinct cell lineages within the mouse morula. *Cell*, *24*, 71–80.

Kalkan, T., Olova, N., Roode, M., Mulas, C., Lee, H. J., Nett, I., et al. (2017). Tracking the embryonic stem cell transition from ground state pluripotency. *Development (Cambridge, England)*, *144*, 1221–1234. https://doi.org/10.1242/dev.142711.

Kang, M., Garg, V., & Hadjantonakis, A.-K. (2017). Lineage establishment and progression within the inner cell mass of the mouse blastocyst requires FGFR1 and FGFR2. *Developmental Cell*, *41*, 496.e5–510.e5. https://doi.org/10.1016/j.devcel.2017.05.003.

Kang, M., Piliszek, A., Artus, J., & Hadjantonakis, A.-K. (2013). FGF4 is required for lineage restriction and salt-and-pepper distribution of primitive endoderm factors but not their initial expression in the mouse. *Development (Cambridge, England)*, *140*, 267–279. https://doi.org/10.1242/dev.084996.

Kanka, J. (2003). Gene expression and chromatin structure in the pre-implantation embryo. *Theriogenology, 59*, 3–19.
Khan, D. R., Dubé, D., Gall, L., Peynot, N., Ruffini, S., Laffont, L., et al. (2012). Expression of pluripotency master regulators during two key developmental transitions: EGA and early lineage specification in the bovine embryo. *PLoS One,* 7. e34110. https://doi.org/10.1371/journal.pone.0034110.
Kirchhof, N., Carnwath, J. W., Lemme, E., Anastassiadis, K., Schöler, H., & Niemann, H. (2000). Expression pattern of Oct-4 in preimplantation embryos of different species. *Biology of Reproduction, 63*, 1698–1705.
Klimczewska, K., Kasperczuk, A., Suwińska, A., n.d.. The regulative nature of mammalian embryos. *Current Topics in Developmental Biology,* 128.
Kobolak, J., Kiss, K., Polgar, Z., Mamo, S., Rogel-Gaillard, C., Tancos, Z., et al. (2009). Promoter analysis of the rabbit POU5F1 gene and its expression in preimplantation stage embryos. *BMC Molecular Biology, 10*, 88. https://doi.org/10.1186/1471-2199-10-88.
Koehler, D., Zakhartchenko, V., Froenicke, L., Stone, G., Stanyon, R., Wolf, E., et al. (2009). Changes of higher order chromatin arrangements during major genome activation in bovine preimplantation embryos. *Experimental Cell Research, 315*, 2053–2063. https://doi.org/10.1016/j.yexcr.2009.02.016.
Koutsourakis, M., Langeveld, A., Patient, R., Beddington, R., & Grosveld, F. (1999). The transcription factor GATA6 is essential for early extraembryonic development. *Development (Cambridge, England), 126*, 723–732.
Koyama, H., Suzuki, H., Yang, X., Jiang, S., & Foote, R. H. (1994). Analysis of polarity of bovine and rabbit embryos by scanning electron microscopy. *Biology of Reproduction, 50*, 163–170.
Kuijk, E. W., Du Puy, L., Van Tol, H. T. A., Oei, C. H. Y., Haagsman, H. P., Colenbrander, B., et al. (2008). Differences in early lineage segregation between mammals. *Developmental Dynamics: An Official Publication of the American Association of Anatomists, 237*, 918–927. https://doi.org/10.1002/dvdy.21480.
Kuijk, E., Geijsen, N., & Cuppen, E. (2015). Pluripotency in the light of the developmental hourglass. *Biological Reviews of the Cambridge Philosophical Society, 90*, 428–443. https://doi.org/10.1111/brv.12117.
Kuijk, E. W., van Tol, L. T. A., Van de Velde, H., Wubbolts, R., Welling, M., Geijsen, N., et al. (2012). The roles of FGF and MAP kinase signaling in the segregation of the epiblast and hypoblast cell lineages in bovine and human embryos. *Development (Cambridge, England), 139*, 871–882. https://doi.org/10.1242/dev.071688.
Kunath, T., Yamanaka, Y., Detmar, J., MacPhee, D., Caniggia, I., Rossant, J., et al. (2014). Developmental differences in the expression of FGF receptors between human and mouse embryos. *Placenta, 35*, 1079–1088. https://doi.org/10.1016/j.placenta.2014.09.008.
Latham, K. E., Solter, D., & Schultz, R. M. (1992). Acquisition of a transcriptionally permissive state during the 1-cell stage of mouse embryogenesis. *Developmental Biology, 149*, 457–462.
Li, L. Y., Li, M. M., Yang, S. F., Zhang, J., Li, Z., Zhang, H., et al. (2016). Inhibition of FGF signalling pathway augments the expression of pluripotency and trophoblast lineage marker genes in porcine parthenogenetic blastocyst. *Reproduction in Domestic Animals = Zuchthygiene, 51*, 649–656. https://doi.org/10.1111/rda.12725.
Liu, S., Bou, G., Sun, R., Guo, S., Xue, B., Wei, R., et al. (2015). Sox2 is the faithful marker for pluripotency in pig: Evidence from embryonic studies. *Developmental Dynamics, 244*, 619–627. https://doi.org/10.1002/dvdy.24248.
Maddox-Hyttel, P., Alexopoulos, N. I., Vajta, G., Lewis, I., Rogers, P., Cann, L., et al. (2003). Immunohistochemical and ultrastructural characterization of the initial post-hatching development of bovine embryos. *Reproduction (Cambridge, England), 125*, 607–623.

Maddox-Hyttel, P., Dinnyés, A., Laurincik, J., Rath, D., Niemann, H., Rosenkranz, C., et al. (2001). Gene expression during pre- and peri-implantation embryonic development in pigs. *Reproduction. Supplement*, *58*, 175–189.

Madeja, Z. E., Sosnowski, J., Hryniewicz, K., Warzych, E., Pawlak, P., Rozwadowska, N., et al. (2013). Changes in sub-cellular localisation of trophoblast and inner cell mass specific transcription factors during bovine preimplantation development. *BMC Developmental Biology*, *13*, 32. https://doi.org/10.1186/1471-213X-13-32.

Mamo, S., Gal, A. B., Polgar, Z., & Dinnyes, A. (2008). Expression profiles of the pluripotency marker gene POU5F1 and validation of reference genes in rabbit oocytes and preimplantation stage embryos. *BMC Molecular Biology*, *9*, 67. https://doi.org/10.1186/1471-2199-9-67.

McConnell, J., Petrie, L., Stennard, F., Ryan, K., & Nichols, J. (2005). Eomesodermin is expressed in mouse oocytes and pre-implantation embryos. *Molecular Reproduction and Development*, *71*, 399–404. https://doi.org/10.1002/mrd.20318.

McGowen, M. R., Erez, O., Romero, R., & Wildman, D. E. (2014). The evolution of embryo implantation. *The International Journal of Developmental Biology*, *58*, 155–161. https://doi.org/10.1387/ijdb.140020dw.

Meilhac, S. M., Adams, R. J., Morris, S. A., Danckaert, A., Le Garrec, J.-F., & Zernicka-Goetz, M. (2009). Active cell movements coupled to positional induction are involved in lineage segregation in the mouse blastocyst. *Developmental Biology*, *331*, 210–221. https://doi.org/10.1016/j.ydbio.2009.04.036.

Meirelles, F. V., Caetano, A. R., Watanabe, Y. F., Ripamonte, P., Carambula, S. F., Merighe, G. K., et al. (2004). Genome activation and developmental block in bovine embryos. *Animal Reproduction Science*, *82–83*, 13–20. https://doi.org/10.1016/j.anireprosci.2004.05.012.

Memili, E., & First, N. L. (1999). Control of gene expression at the onset of bovine embryonic development. *Biology of Reproduction*, *61*, 1198–1207.

Mintz, B. (1965). Genetic mosaicism in adult mice of quadriparental lineage. *Science*, *148*, 1232–1233.

Molotkov, A., Mazot, P., Brewer, J. R., Cinalli, R. M., & Soriano, P. (2017). Distinct requirements for FGFR1 and FGFR2 in primitive endoderm development and exit from pluripotency. *Developmental Cell*, *41*, 511–526.e4. https://doi.org/10.1016/j.devcel.2017.05.004.

Moore, N. W., Adams, C. E., & Rowson, L. E. (1968). Developmental potential of single blastomeres of the rabbit egg. *Journal of Reproduction and Fertility*, *17*, 527–531.

Moradi, M., Riasi, A., Ostadhosseini, S., Hajian, M., Hosseini, M., Hosseinnia, P., et al. (2015). Expression profile of FGF receptors in preimplantation ovine embryos and the effect of FGF2 and PD173074. *Growth Factors*, *33*, 393–400. https://doi.org/10.3109/08977194.2015.1102138.

Natale, D. R., Kidder, G. M., Westhusin, M. E., & Watson, A. J. (2000). Assessment by differential display-RT-PCR of mRNA transcript transitions and alpha-amanitin sensitivity during bovine preattachment development. *Molecular Reproduction and Development*, *55*, 152–163. https://doi.org/10.1002/(SICI)1098-2795(200002)55:2<152::AID-MRD4>3.0.CO;2-N.

Niakan, K. K., Ji, H., Maehr, R., Vokes, S. A., Rodolfa, K. T., Sherwood, R. I., et al. (2010). Sox17 promotes differentiation in mouse embryonic stem cells by directly regulating extraembryonic gene expression and indirectly antagonizing self-renewal. *Genes & Development*, *24*, 312–326. https://doi.org/10.1101/gad.1833510.

Nichols, J., Silva, J., Roode, M., & Smith, A. (2009). Suppression of Erk signalling promotes ground state pluripotency in the mouse embryo. *Development (Cambridge, England)*, *136*, 3215–3222. https://doi.org/10.1242/dev.038893.

Nichols, J., Zevnik, B., Anastassiadis, K., Niwa, H., Klewe-Nebenius, D., Chambers, I., et al. (1998). Formation of pluripotent stem cells in the mammalian embryo depends on the POU transcription factor Oct4. *Cell*, *95*, 379–391.

Nishioka, N., Inoue, K., Adachi, K., Kiyonari, H., Ota, M., Ralston, A., et al. (2009). The Hippo signaling pathway components Lats and Yap pattern Tead4 activity to distinguish mouse trophectoderm from inner cell mass. *Developmental Cell*, *16*, 398–410. https://doi.org/10.1016/j.devcel.2009.02.003.

Nishioka, N., Yamamoto, S., Kiyonari, H., Sato, H., Sawada, A., Ota, M., et al. (2008). Tead4 is required for specification of trophectoderm in pre-implantation mouse embryos. *Mechanisms of Development*, *125*, 270–283. https://doi.org/10.1016/j.mod.2007.11.002.

Niwa, H., Toyooka, Y., Shimosato, D., Strumpf, D., Takahashi, K., Yagi, R., et al. (2005). Interaction between Oct3/4 and Cdx2 determines trophectoderm differentiation. *Cell*, *123*, 917–929. https://doi.org/10.1016/j.cell.2005.08.040.

Orsztynowicz, M., Lechniak, D., Pawlak, P., Kociucka, B., Kubickova, S., Cernohorska, H., et al. (2017). Changes in chromosome territory position within the nucleus reflect alternations in gene expression related to embryonic lineage specification. *PLoS One*, *12*, e0182398. https://doi.org/10.1371/journal.pone.0182398.

Ozawa, M., Sakatani, M., Yao, J., Shanker, S., Yu, F., Yamashita, R., et al. (2012). Global gene expression of the inner cell mass and trophectoderm of the bovine blastocyst. *BMC Developmental Biology*, *12*, 33. https://doi.org/10.1186/1471-213X-12-33.

Partridge, R. J., & Leese, H. J. (1996). Consumption of amino acids by bovine preimplantation embryos. *Reproduction, Fertility, and Development*, *8*, 945–950.

Peippo, J., Machaty, Z., & Peter, A. (2011). Terminologies for the pre-attachment bovine embryo. *Theriogenology*, *76*, 1373–1379. https://doi.org/10.1016/j.theriogenology.2011.06.018.

Perry, J. S., & Rowlands, I. W. (1962). Early pregnancy in the pig. *Journal of Reproduction and Fertility*, *4*, 175–188. https://doi.org/10.1530/jrf.0.0040175.

Pfeffer, P. L., Smith, C. S., Maclean, P., & Berg, D. K. (2017). Gene expression analysis of bovine embryonic disc, trophoblast and parietal hypoblast at the start of gastrulation. *Zygote (Cambridge, England)*, *25*, 265–278. https://doi.org/10.1017/S0967199417000090.

Piliszek, A., Grabarek, J. B., Frankenberg, S. R., & Plusa, B. (2016). Cell fate in animal and human blastocysts and the determination of viability. *Molecular Human Reproduction*, *22*, 681–690. https://doi.org/10.1093/molehr/gaw002.

Piliszek, A., Madeja, Z. E., & Plusa, B. (2017). Suppression of ERK signalling abolishes primitive endoderm formation but does not promote pluripotency in rabbit embryo. *Development*, *144*, 3719–3730. https://doi.org/10.1242/dev.156406.

Plusa, B., Piliszek, A., Frankenberg, S., Artus, J., & Hadjantonakis, A.-K. (2008). Distinct sequential cell behaviours direct primitive endoderm formation in the mouse blastocyst. *Development (Cambridge, England)*, *135*, 3081–3091. https://doi.org/10.1242/dev.021519.

Ralston, A., Cox, B. J., Nishioka, N., Sasaki, H., Chea, E., Rugg-Gunn, P., et al. (2010). Gata3 regulates trophoblast development downstream of Tead4 and in parallel to Cdx2. *Development (Cambridge, England)*, *137*, 395–403. https://doi.org/10.1242/dev.038828.

Reeve, W. J., & Ziomek, C. A. (1981). Distribution of microvilli on dissociated blastomeres from mouse embryos: Evidence for surface polarization at compaction. *Journal of Embryology and Experimental Morphology*, *62*, 339–350.

Reima, I., Lehtonen, E., Virtanen, I., & Fléchon, J.-E. (1993). The cytoskeleton and associated proteins during cleavage, compaction and blastocyst differentiation in the pig. *Differentiation*, *54*, 35–45. https://doi.org/10.1111/j.1432-0436.1993.tb00657.x.

Rodríguez, A., Allegrucci, C., & Alberio, R. (2012). Modulation of pluripotency in the porcine embryo and iPS cells. *PLoS One* 7, e49079. https://doi.org/10.1371/journal.pone.0049079.

Roode, M., Blair, K., Snell, P., Elder, K., Marchant, S., Smith, A., et al. (2012). Human hypoblast formation is not dependent on FGF signalling. *Developmental Biology, 361*, 358–363. https://doi.org/10.1016/j.ydbio.2011.10.030.

Russ, A. P., Wattler, S., Colledge, W. H., Aparicio, S. A., Carlton, M. B., Pearce, J. J., et al. (2000). Eomesodermin is required for mouse trophoblast development and mesoderm formation. *Nature, 404*, 95–99. https://doi.org/10.1038/35003601.

Saenz-de-Juano, M. D., Naturil-Alfonso, C., Vicente, J. S., & Marco-Jiménez, F. (2013). Effect of different culture systems on mRNA expression in developing rabbit embryos. *Zygote (Cambridge, England), 21*, 103–109. https://doi.org/10.1017/S0967199411000414.

Saini, D., Yamanaka, Y., n.d. Cell polarity dependent regulation of cell allocation and the first lineage specification in the preimplantation mouse embryo. *Current Topics in Developmental Biology*, 128.

Saito, S., & Niemann, H. (1991). Effects of extracellular matrices and growth factors on the development of isolated porcine blastomeres. *Biology of Reproduction, 44*, 927–936.

Sakurai, N., Takahashi, K., Emura, N., Fujii, T., Hirayama, H., Kageyama, S., et al. (2016). The necessity of OCT-4 and CDX2 for early development and gene expression involved in differentiation of inner cell mass and trophectoderm lineages in bovine embryos. *Cellular Reprogramming, 18*, 309–318. https://doi.org/10.1089/cell.2015.0081.

Sakurai, N., Takahashi, K., Emura, N., Hashizume, T., & Sawai, K. (2017). Effects of downregulating TEAD4 transcripts by RNA interference on early development of bovine embryos. *The Journal of Reproduction and Development, 63*, 135–142. https://doi.org/10.1262/jrd.2016-130.

Sanna, D., Sanna, A., Mara, L., Pilichi, S., Mastinu, A., Chessa, F., et al. (2009). Oct4 expression in in-vitro-produced sheep blastocysts and embryonic-stem-like cells. *Cell Biology International, 34*, 53–60. https://doi.org/10.1042/CBI20090008.

Sasaki, H. (2010). Mechanisms of trophectoderm fate specification in preimplantation mouse development. *Development, Growth & Differentiation, 52*, 263–273. https://doi.org/10.1111/j.1440-169X.2009.01158.x.

Schöler, H. R., Ruppert, S., Suzuki, N., Chowdhury, K., & Gruss, P. (1990). New type of POU domain in germ line-specific protein Oct-4. *Nature, 344*, 435–439. https://doi.org/10.1038/344435a0.

Schrode, N., Saiz, N., Di Talia, S., & Hadjantonakis, A.-K. (2014). GATA6 levels modulate primitive endoderm cell fate choice and timing in the mouse blastocyst. *Developmental Cell, 29*, 454–467. https://doi.org/10.1016/j.devcel.2014.04.011.

Schröter, C., Rué, P., Mackenzie, J. P., & Martinez Arias, A. (2015). FGF/MAPK signaling sets the switching threshold of a bistable circuit controlling cell fate decisions in embryonic stem cells. *Development (Cambridge, England), 142*, 4205–4216. https://doi.org/10.1242/dev.127530.

Seshagiri, P. B., McKenzie, D. I., Bavister, B. D., Williamson, J. L., & Aiken, J. M. (1992). Golden hamster embryonic genome activation occurs at the two-cell stage: Correlation with major developmental changes. *Molecular Reproduction and Development, 32*, 229–235. https://doi.org/10.1002/mrd.1080320307.

Singh, A. M., Hamazaki, T., Hankowski, K. E., & Terada, N. (2007). A heterogeneous expression pattern for Nanog in embryonic stem cells. *Stem Cells (Dayton, Ohio), 25*, 2534–2542. https://doi.org/10.1634/stemcells.2007-0126.

Smith, C. S., Berg, D. K., Berg, M., & Pfeffer, P. L. (2010). Nuclear transfer-specific defects are not apparent during the second week of embryogenesis in cattle. *Cellular Reprogramming, 12*, 699–707. https://doi.org/10.1089/cell.2010.0040.

Soudais, C., Bielinska, M., Heikinheimo, M., MacArthur, C. A., Narita, N., Saffitz, J. E., et al. (1995). Targeted mutagenesis of the transcription factor GATA-4 gene in mouse embryonic stem cells disrupts visceral endoderm differentiation in vitro. *Development (Cambridge, England)*, *121*, 3877–3888.

Sperber, H., Mathieu, J., Wang, Y., Ferreccio, A., Hesson, J., Xu, Z., et al. (2015). The metabolome regulates the epigenetic landscape during naive-to-primed human embryonic stem cell transition. *Nature Cell Biology*, *17*, 1523–1535. https://doi.org/10.1038/ncb3264.

Stroband, H. W., & Van der Lende, T. (1990). Embryonic and uterine development during early pregnancy in pigs. *Journal of Reproduction and Fertility. Supplement*, *40*, 261–277.

Strumpf, D., Mao, C.-A., Yamanaka, Y., Ralston, A., Chawengsaksophak, K., Beck, F., et al. (2005). Cdx2 is required for correct cell fate specification and differentiation of trophectoderm in the mouse blastocyst. *Development (Cambridge, England)*, *132*, 2093–2102. https://doi.org/10.1242/dev.01801.

Suwińska, A., Czołowska, R., Ozdzeński, W., & Tarkowski, A. K. (2008). Blastomeres of the mouse embryo lose totipotency after the fifth cleavage division: Expression of Cdx2 and Oct4 and developmental potential of inner and outer blastomeres of 16- and 32-cell embryos. *Developmental Biology*, *322*, 133–144. https://doi.org/10.1016/j.ydbio.2008.07.019.

Takashima, Y., Guo, G., Loos, R., Nichols, J., Ficz, G., Krueger, F., et al. (2014). Resetting transcription factor control circuitry toward ground-state pluripotency in human. *Cell*, *158*, 1254–1269. https://doi.org/10.1016/j.cell.2014.08.029.

Tam, P. P. L., & Rossant, J. (2003). Mouse embryonic chimeras: Tools for studying mammalian development. *Development (Cambridge, England)*, *130*, 6155–6163. https://doi.org/10.1242/dev.00893.

Táncos, Z., Bock, I., Nemes, C., Kobolák, J., & Dinnyés, A. (2015). Cloning and characterization of rabbit POU5F1, SOX2, KLF4, C-MYC and NANOG pluripotency-associated genes. *Gene*, *566*, 148–157. https://doi.org/10.1016/j.gene.2015.04.034.

Tarkowski, A. K. (1959). Experiments on the development of isolated blastomers of mouse eggs. *Nature*, *184*, 1286–1287.

Tarkowski, A. K., & Wróblewska, J. (1967). Development of blastomeres of mouse eggs isolated at the 4- and 8-cell stage. *Journal of Embryology and Experimental Morphology*, *18*, 155–180.

Thompson, J. G., Partridge, R. J., Houghton, F. D., Cox, C. I., & Leese, H. J. (1996). Oxygen uptake and carbohydrate metabolism by in vitro derived bovine embryos. *Journal of Reproduction and Fertility*, *106*, 299–306.

van Leeuwen, J., Berg, D. K., & Pfeffer, P. L. (2015). Morphological and gene expression changes in cattle embryos from hatched blastocyst to early gastrulation stages after transfer of in vitro produced embryos. *PLoS One*, *10*, e0129787. https://doi.org/10.1371/journal.pone.0129787.

van Soom, A., Ysebaert, M. T., & de Kruif, A. (1997). Relationship between timing of development, morula morphology, and cell allocation to inner cell mass and trophectoderm in in vitro-produced bovine embryos. *Molecular Reproduction and Development*, *47*, 47–56. https://doi.org/10.1002/(SICI)1098-2795(199705)47:1<47::AID-MRD7>3.0.CO;2-Q.

Viuff, D., Avery, B., Greve, T., King, W. A., & Hyttel, P. (1996). Transcriptional activity in in vitro produced bovine two- and four-cell embryos. *Molecular Reproduction and Development*, *43*, 171–179. https://doi.org/10.1002/(SICI)1098-2795(199602)43:2<171::AID-MRD6>3.0.CO;2-O.

Wang, N., Wang, Y., Xie, Y., & Wang, H. (2016). OTX2 impedes self-renewal of porcine iPS cells through downregulation of NANOG expression. *Cell Death Discovery*, *2*, 16090. https://doi.org/10.1038/cddiscovery.2016.90.

Wilcox, J. T., Semple, E., Gartley, C., Brisson, B. A., Perrault, S. D., Villagómez, D. A. F., et al. (2009). Characterization of canine embryonic stem cell lines derived from different niche microenvironments. *Stem Cells and Development, 18*, 1167–1178. https://doi.org/10.1089/scd.2008.0336.

Willadsen, S. M. (1981). The development capacity of blastomeres from 4- and 8-cell sheep embryos. *Journal of Embryology and Experimental Morphology, 65*, 165–172.

Willadsen, S. M., & Polge, C. (1981). Attempts to produce monozygotic quadruplets in cattle by blastomere separation. *The Veterinary Record, 108*, 211–213.

Williams, B. S., & Biggers, J. D. (1990). Polar trophoblast (Rauber's layer) of the rabbit blastocyst. *The Anatomical Record, 227*, 211–222. https://doi.org/10.1002/ar.1092270210.

Wolf, X. A., Serup, P., & Hyttel, P. (2011). Three-dimensional localisation of NANOG, OCT4, and E-cadherin in porcine pre- and peri-implantation embryos. *Developmental Dynamics, 240*, 204–210. https://doi.org/10.1002/dvdy.22491.

Wu, G., Gentile, L., Fuchikami, T., Sutter, J., Psathaki, K., Esteves, T. C., et al. (2010). Initiation of trophectoderm lineage specification in mouse embryos is independent of Cdx2. *Development (Cambridge, England), 137*, 4159–4169. https://doi.org/10.1242/dev.056630.

Xenopoulos, P., Kang, M., Puliafito, A., Di Talia, S., & Hadjantonakis, A.-K. (2015). Heterogeneities in Nanog expression drive stable commitment to pluripotency in the mouse blastocyst. *Cell Reports, 10*(9), 1508–1520. https://doi.org/10.1016/j.celrep.2015.02.010.

Yagi, R., Kohn, M. J., Karavanova, I., Kaneko, K. J., Vullhorst, D., DePamphilis, M. L., et al. (2007). Transcription factor TEAD4 specifies the trophectoderm lineage at the beginning of mammalian development. *Development (Cambridge, England), 134*, 3827–3836. https://doi.org/10.1242/dev.010223.

Yamanaka, Y., Lanner, F., & Rossant, J. (2010). FGF signal-dependent segregation of primitive endoderm and epiblast in the mouse blastocyst. *Development (Cambridge, England), 137*, 715–724. https://doi.org/10.1242/dev.043471.

Yu, X., Zhao, X., Wang, H., & Ma, B. (2015). Expression patterns of OCT4, NANOG, and SOX2 in goat preimplantation embryos from in vivo and in vitro. *Journal of Integrative Agriculture, 14*, 1398–1406. https://doi.org/10.1016/S2095-3119(14)60923-0.

Zernicka-Goetz, M. (1994). Activation of embryonic genes during preimplantation rat development. *Molecular Reproduction and Development, 38*, 30–35. https://doi.org/10.1002/mrd.1080380106.

Zhou, W., Choi, M., Margineantu, D., Margaretha, L., Hesson, J., Cavanaugh, C., et al. (2012). HIF1α induced switch from bivalent to exclusively glycolytic metabolism during ESC-to-EpiSC/hESC transition. *The EMBO Journal, 31*, 2103–2116. https://doi.org/10.1038/emboj.2012.71.

Ziomek, C. A., Chatot, C. L., & Manes, C. (1990). Polarization of blastomeres in the cleaving rabbit embryo. *The Journal of Experimental Zoology, 256*, 84–91. https://doi.org/10.1002/jez.1402560111.

CHAPTER TWELVE

Human Pre-gastrulation Development

Sissy E. Wamaitha, Kathy K. Niakan[1]
Human Embryo and Stem Cell Laboratory, The Francis Crick Institute, London, United Kingdom
[1]Corresponding author: e-mail address: kathy.niakan@crick.ac.uk

Contents

Abstract

Understanding the progression of early human embryonic development prior to implantation is of fundamental biological importance. Greater insights into early developmental events may lead to clinical improvements, not only via the establishment of novel stem cell models with increased potential or more physiological relevance, but also by uncovering some underlying causes of infertility, miscarriages, and developmental disorders. The majority of human embryos available for study are those donated to research once they are surplus to family building following in vitro fertilization, though in some countries it is also possible to create embryos using donated gametes. As human embryo development is surprisingly inefficient, with only 40% reaching the blastocyst stage in vitro (French, Sabanegh, Goldfarb, & Desai, 2010; Gardner, Lane, Stevens, Schlenker, & Schoolcraft, 2000), many embryos may not develop to a stage suitable for study. Where legally permitted, the oversight of human embryo research is subject to either ethics approval from a local institutional review board (i.e., China and the United States) or both a national regulator as well as a regional research ethics committee (i.e., the United Kingdom). The study of human development has historically been by necessity comparative, relying on model organisms and stem cell lines to inform analyses.

Current Topics in Developmental Biology, Volume 128
ISSN 0070-2153
https://doi.org/10.1016/bs.ctdb.2017.11.004

Preimplantation mouse and human embryos in particular exhibit remarkably similar gross morphologies at these early stages of development, although key differences have been identified in gene expression patterns and developmental timing. While recent advances in high-resolution transcriptomic analyses at the single cell level have improved our capability to interrogate expression patterns directly in the human embryo, we still lack an understanding of basic molecular events in the human embryo, including how the first cell lineages become specified. Here, we present a current overview of the major developmental events during human preimplantation development, from fertilization to delineation of the embryonic and extraembryonic lineages prior to implantation. Comparisons to both the mouse and alternative models are included where these have formed the basis for similar investigations in a human context.

1. STAGES OF PREIMPLANTATION DEVELOPMENT

1.1 Fertilization and Cleavage

In humans, as in other mammals, embryonic development begins with the fusion of the oocyte and sperm to form the diploid zygote. Fertilization occurs in the fallopian tube, where the sperm penetrate the cumulus cells surrounding the oocyte and bind to the zona pellucida, a protective glycoprotein membrane formed during oogenesis. This induces the release of cortical granules via the acrosome (Okabe, 2013) and alters the composition of the zona, which hardens to prevent the binding of multiple sperm (polyspermy). The sperm then fuses with the oocyte plasma membrane, with the resulting "plasma membrane block" also preventing polyspermy. Fusion of the sperm and oocyte plasma membranes is sufficient to generate a polyspermy block when the zona pellucida is removed (Sengoku et al., 1995, 1999).

Much of the understanding of zona physiology following fertilization is based on studies in the mouse, where the zona is composed of three glycoproteins, ZP1, ZP2, and ZP3 (Bleil & Wassarman, 1980). ZP3 reportedly acts as the primary sperm receptor, binding the acrosome-intact sperm and inducing the acrosome reaction; ZP2 is the secondary receptor for the acrosome-reacted sperm; and ZP1 forms cross-links between the ZP2 and ZP3 proteins (Wassarman, Jovine, & Litscher, 2004). In addition, cleavage of ZP2 by ovastacin, a protease contained in cortical granules and released via exocytosis following sperm–oocyte fusion, alters the architecture of the zona and prevents polyspermy (Burkart, Xiong, Baibakov, Jimenez-Movilla, & Dean, 2012; Quesada, Sanchez, Alvarez, & Lopez-Otin, 2004). However, the human zona pellucida comprises an additional glycoprotein, ZP4 (Lefievre et al., 2004), which is present as a pseudogene in the mouse genome, and the human glycoproteins may have divergent roles to their mouse homologues (Gupta et al., 2012). Consequently, though the morphological consequences appear similar, the mechanisms following sperm binding to the zona may be distinct in the human.

Fertilization initiates the process of egg activation, whereby the gamete transitions into an embryo, beginning with reinitiating of the cell cycle (Clift & Schuh, 2013). At the time of ovulation, the oocyte is arrested at the metaphase stage of the second meiotic division, following extrusion of first polar body during egg maturation. When the sperm binds, it introduces the phospholipase c zeta isoform (PLCζ) into the oocyte, which initiates a cascade culminating in binding of inositol 1,4,5-triphosphate (IP3) to its receptor IP3R on the endoplasmic reticulum, and release of calcium ions from reserves (Wakai, Vanderheyden, & Fissore, 2011). The resulting calcium oscillations reactivate the meiotic cycle, leading to chromosome segregation for production and extrusion of the second polar body, and formation of the haploid maternal pronucleus (Clift & Schuh, 2013; Miyazaki & Ito, 2006). The fertilized oocyte is now called the zygote.

Calcium uptake is followed by release of multiple cortical granules, a proportion of which facilitate the exocytosis of zinc ions accumulated during oocyte maturation back into the extracellular environment (Kim, Vogt, O'Halloran, & Woodruff, 2010; Que et al., 2015). This "zinc spark" has been observed following parthenogenetic activation of human and primate oocytes, and in both mouse parthenotes and fertilized oocytes (Duncan et al., 2016; Kim et al., 2011; Zhang, Duncan, Que, O'Halloran, & Woodruff, 2016). Zinc expulsion may be crucial for zona hardening, as exposing mouse oocytes to exogenous zinc also prevents sperm binding (Que et al., 2017). The rise in intracellular calcium also prompts the transition from meiosis to mitosis for subsequent cell divisions in embryogenesis.

After the mitotic spindle assembles following breakdown of the pronuclear envelopes, the zygote undergoes a series of mitotic divisions (Fig. 1). Cytoplasmic volume does not increase and is instead segregated into increasingly smaller cells, resulting in a high nuclear to cytoplasmic ratio (Aiken, Swoboda, Skepper, & Johnson, 2004). These cell divisions form the 2-cell, and subsequently the 4-cell stage embryo, at 1 and 2 days postfertilization (dpf), respectively. At this stage the embryo is transcriptionally silent (Braude, Bolton, & Moore, 1988), depending instead on maternal proteins and the translation of maternal mRNAs provided in the oocyte cytoplasm, which become activated after the surge in calcium. Finally,

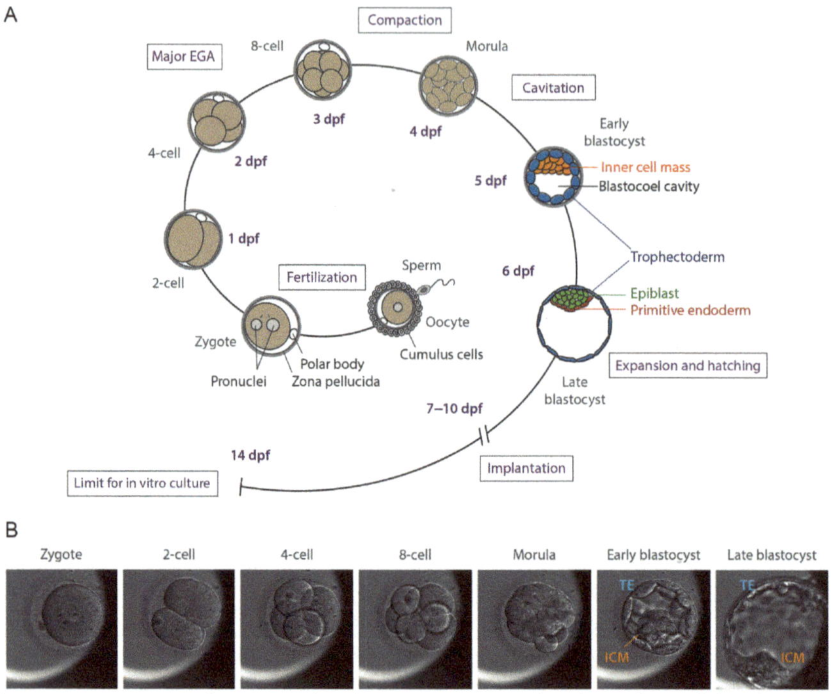

Fig. 1 Human preimplantation development. (A) A stylized timeline of development. Following fusion of the oocyte and sperm at fertilization, the one-cell zygote undergoes a series of mitotic divisions, forming the 2-cell and 4-cell embryo in relative transcriptional silence. Activation of the embryonic transcriptional program (embryonic genome activation, EGA) occurs between the 4- and 8-cell stage, although a minor wave may occur as early as the 2-cell stage. Between the 8- and 16-cell stage, the blastomeres undergo compaction, and likely begin to exhibit polarity, though this remains poorly understood in a human context. Cavitation marks the formation of a blastocyst comprising an inner cell mass (ICM) and trophectoderm between 5 and 6 days postfertilization (dpf). In the late blastocyst, the ICM is further segregated into a pluripotent epiblast that will give rise to all three germ layers of the embryo proper, and a primitive endoderm layer that gives rise to extraembryonic endoderm cells that will form the yolk sac. The blastocyst then expands and eventually hatches from the zona pelucida, and implants into the uterine cell wall between 7 and 10 dpf. The limit for in vitro culture of human embryos is set at 14 dpf, or the equivalent time of the appearance of the primitive streak. (B) Images of an embryo progressing from one-cell zygote to late blastocyst, acquired using an Embryo-Scope™ time-lapse system.

embryo genome activation (EGA) and the onset of active transcription mark the transition from oocyte to zygote gene programs and occur between the 4- and 8-cell stage at 2–2.5 dpf in the human (Blakeley et al., 2015; Braude et al., 1988). A discussion surrounding the mechanisms responsible for the onset of human EGA is presented later in this chapter.

1.2 Blastulation and Implantation

The embryo enters the uterine cavity at the 8-cell stage (3 dpf). Between the 8- and 16-cell stage (3–4 dpf) embryos undergo compaction, whereby individual blastomeres become flattened and begin to adhere more strongly to each other, and cell boundaries become nearly indistinguishable, forming a tightly packed ball of cells known as the morula (Fig. 1) (Edwards, Purdy, Steptoe, & Walters, 1981; Nikas, Ao, Winston, & Handyside, 1996; Steptoe, Edwards, & Purdy, 1971). Although compaction has been observed prior to the 8-cell stage, embryos where this occurred were less likely to continue their developmental progression toward the blastocyst (Iwata et al., 2014).

In the mouse, compaction occurs at the 8-cell stage and is also characterized by flattened blastomeres, alongside formation of tight and gap junction structures, and expression of associated proteins such as tight junction protein ZO-1 (Ducibella, Albertini, Andersen, & Biggers, 1975; Fleming, McConnel, Johnson, & Stevenson, 1989; Magnuson, Demsey, & Stackpole, 1977). It is unclear precisely what regulates the timing of compaction, which occurs independently of cell number, though initiation is thought to involve PKC-alpha and β-catenin (Cockburn & Rossant, 2010; White, Bissiere, Alvarez, & Plachta, 2016). Blastomeres develop a polarized distribution of surface microvilli at one pole of the cell (the apical pole), in contrast to the uniform distribution observed at the 4-cell stage (Calarco & Epstein, 1973; Ducibella, Ukena, Karnovsky, & Andersen, 1977; Reeve & Ziomek, 1981). There is also an increase in calcium-dependent cell–cell adhesion mediated by E-cadherin, which is required for formation of adherens junctions, and is also localized in filopodia extending onto neighboring cells that regulate cell shape during compaction (Fierro-Gonzalez, White, Silva, & Plachta, 2013; Pauken & Capco, 1999; Shirayoshi, Okada, & Takeichi, 1983; Vestweber, Gossler, Boller, & Kemler, 1987; White et al., 2016; White & Plachta, 2015).

In human embryos, scanning electron microscopy identified early signs of polarization of surface microvilli at the 8- to 12-cell stage around 3 dpf, with more pronounced effects at the 10- and 18-cell stage (Nikas et al., 1996). Gap junction structures were also observed in the compacted morula (Gualtieri, Santella, & Dale, 1992), and the connexin gap junction protein GJA1 (CX43) was detected at the cell membrane as early as the 8-cell stage (Hardy, Warner, Winston, & Becker, 1996). Membrane-localized E-cadherin was observed from 4 dpf at regions of cell–cell contact (Alikani, 2005). Altogether, this suggests that some mechanisms of compaction may potentially be conserved in the human context, but this has not been extensively studied.

In the mouse, compaction occurs concomitantly with cell polarization, whereby blastomeres at the 8-cell stage change their morphology and become polarized along the apico-basal axis (Johnson & Ziomek, 1981b). Cytoskeletal components accumulate at the apical region, while the nuclei reposition at the basal part. Many proteins have been shown to be involved in the establishment of cell polarization, including, among others, the partitioning defective (PAR) complex (Ahringer, 2003; Etemad-Moghadam, Guo, & Kemphues, 1995; Guo & Kemphues, 1995; Joberty, Petersen, Gao, & Macara, 2000; Lin et al., 2000). In *Drosophila* and mouse, the PAR3/PAR6/aPKC complex is located at the apical region, whereas the protein PAR1 is located at the basolateral side (Hurov, Watkins, & Piwnica-Worms, 2004; Suzuki et al., 2004). During the 8- to 16-cell stage, cell divisions are asymmetrical, giving rise to inner cells and outer cells, which exhibit distinct apical and basolateral domains. Only the outer cells conserve the apical cortical domain, which allows these cells to reestablish polarity and to become polar, while the inner cells remain apolar (Johnson & Ziomek, 1981a).

Polarization differences have been suggested to determine subsequent cell fate through differential activation of the Hippo signaling pathway in inner and outer cells in mouse embryos (Korotkevich et al., 2017; Watanabe, Biggins, Tannan, & Srinivas, 2014), which is discussed later in this chapter. Apicobasal polarity affects contractility and surface tension of individual blastomeres, which in turn determines their propensity to segregate as inner or outer cells (Maitre et al., 2016). Inhibiting contractility affects Yap localization and causes blastomeres to adopt inner cell characteristics, but cell position remains unchanged. Supporting this, following asymmetric division apolar daughter cells express increased phosphorylated Yap, and if these cells are positioned in the outer layer, they are likely to be internalized (Anani, Bhat, Honma-Yamanaka, Krawchuk, & Yamanaka, 2014). Consequently, Yap differential expression is likely established soon after asymmetric division of polar and apolar cells, but prior to segregation into inner or outer positions. E-cadherin is also involved in cell polarization via catenin-mediated connections with the actin cytoskeleton (Stephenson, Yamanaka, & Rossant, 2010; White & Plachta, 2015).

These mechanisms have not yet been elucidated in the context of human development, and none of the components of the PAR complex or the Hippo pathway have been extensively studied thus far. It has been hypothesized that appropriate cleavage patterns are important for human embryo progression (Ajduk & Zernicka-Goetz, 2016), based on an observation that human embryos exhibit a planar morphology with fewer intercellular contacts and have reduced rates of blastocyst formation (Ebner et al., 2012). Defining the expression pattern of cell polarization proteins that are known to regulate mouse lineage specification in the context of human development would be informative, as would the analysis of possible alternative pathways that may regulate this critical stage of human development.

Embryos subsequently undergo cavitation to form a blastocyst consisting of a fluid-filled cavity (the blastocoel) and an inner cell mass (ICM), surrounded by an outer layer of TE cells at 5 dpf (Fig. 1) (Hertig, Rock, & Adams, 1956; Hertig, Rock, Adams, & Mulligan, 1954; Steptoe et al., 1971). The segregation of the ICM and the TE marks the first lineage specification event in the developing embryo, with the TE eventually contributing to the fetal portion of the placenta, but also playing a key role in blastocoel formation. In the mouse, cavitation is dependent on fluid exchange by Na^+/K^+-ATPase and aquaporins in the TE, with junctional complexes between the cells creating a tightly sealed layer to allow retention of fluid pumped into the cavity and subsequent expansion of the blastocyst (Cockburn & Rossant, 2010). Tight and gap junction structures are present in the human morula (Gualtieri et al., 1992), and blastocysts express transcripts associated with tight junction and desmosome formation, such as ZO-1 and claudins, as well as connexin gap junction proteins (Bloor et al., 2004; Ghassemifar, 2003). This suggests that some aspects of this process are likely conserved in the human.

Between 5 and 6 dpf the ICM is thought to further segregate into pluripotent epiblast (Epi) progenitor cells, which form the embryo proper, and to primitive endoderm (PE) cells, which contribute predominantly to the yolk sac, or amnion. These lineage contributions are largely based on analysis of mouse postimplantation phenotypes for the equivalent lineages (Gardner, 1985; Gardner, Papaioannou, & Barton, 1973; Gardner & Rossant, 1979), which for ethical reasons cannot be carried out in the human, but it is presumed likely that these assignations hold true in a human context. After further expansion, the human blastocyst hatches from the zona pellucida and begins to implant into the uterine wall at around 7–10 dpf.

In preparation for implantation, the uterine endometrium undergoes a process of decidualization in response to progesterone, and possibly oestrogen, increasing stromal cell proliferation in the uterine lining in order to accommodate the blastocyst (Gellersen & Brosens, 2014). There is a narrow window of endometrial receptivity during which the uterus can support implantation, occurring in the mid-secretory phase of the menstrual cycle, 7–10 days after ovulation, and coincident with a surge of luteinising hormone (Acosta et al., 1999; Aplin & Ruane, 2017). Ultimately, implantation is likely a two-way interaction, requiring both a competent blastocyst and a receptive uterus, and several factors have been implicated in mediating this process in the human (Aplin & Ruane, 2017; Norwitz, Schust, & Fisher, 2001b; Wang & Dey, 2006). These include adhesion molecules such as integrins, mucins, selectin, and trophinin (Fukuda et al., 1995; Genbacev et al., 2011; Lessey et al., 1992; Meseguer et al., 2001; Sugihara et al., 2007), the integrin-binding protein osteopontin (Apparao et al., 2001; Johnson, Burghardt, Bazer, & Spencer, 2003), the implantation-associated factor COX-2 (Brosens et al., 2014), and the heparin-binding EFG-like (HB-EFG) growth factor (Chobotova et al., 2002; Leach et al., 1999).

As human implantation cannot be observed directly in vivo, much of current morphological knowledge is based on the Carnegie histological sample series, which spans the first 60 days of development (Hertig et al., 1956; O'Rahilly & Muller, 1987), and on extrapolations from primates and other model organisms (Enders, Schlafke, & Hendrickx, 1986; Lee & DeMayo, 2004). Implantation first involves apposition of the blastocyst, whereby it correctly orients itself to the uterine epithelium (Hertig et al., 1956). This is followed by attachment and adhesion of the blastocyst to the epithelium, then invasion through the epithelium into the uterine lining (Fig. 2). Around the timing of implantation, the TE proliferates and differentiates into cytotrophoblast cells, whose subsequent differentiation and fusion results in the formation of the multinucleated syncytiotrophoblast, which is thought to be the initial invading interface (Aplin & Ruane, 2017). The blastocyst eventually embeds itself into the stromal vasculature of the uterine lining (Norwitz, Schust, & Fisher, 2001a). The Epi and PE cavitate to form the amniotic cavity and yolk sac, respectively, with the two layers of Epi and PE between these cavities forming the bilaminar disc (Enders et al., 1986; Luckett, 1975). Following this, the embryo proceeds toward gastrulation and further development.

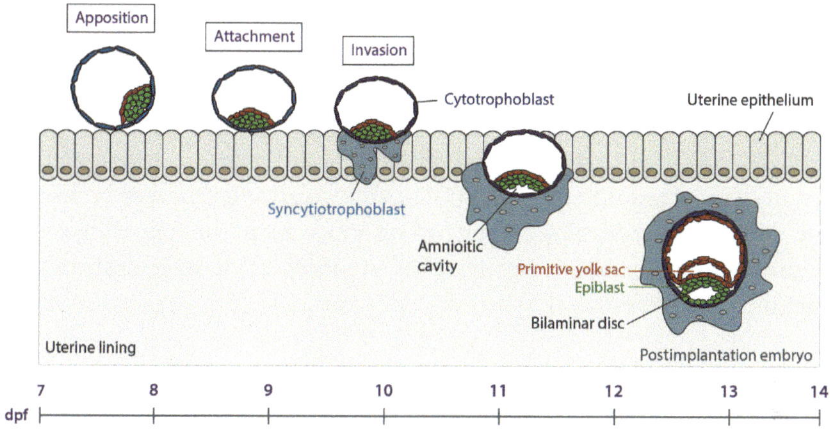

Fig. 2 Human embryo implantation. The blastocyst implants into the uterine cell wall between 7 and 10 dpf. The embryo first correctly orients itself to the uterine epithelium (apposition), generally attaching via the polar TE, closest to the ICM, before attaching more securely to the epithelium. During implantation, the trophectoderm gives rise to syncytiotrophoblast cells which form a multinucleated syncytium that invades the uterine epithelium, and cytotrophoblast cells. Both these derivatives contribute to the fetal portion of the placenta. The blastocyst eventually embeds itself into the stromal vasculature of the uterine lining, and the Epi and PE cavitate to form the amniotic cavity and yolk sac, respectively.

2. REGULATING GENE EXPRESSION DURING HUMAN PREIMPLANTATION DEVELOPMENT

Many of the factors involved in regulating the morphological events of development (compaction, polarization, blastocyst formation) have been well characterized in the mouse, and a number of these have homologues in the human. However, in general, human embryos exhibit a protracted developmental timeline compared to the mouse (Niakan, Han, Pedersen, Simon, & Pera, 2012). Compaction of blastomeres after the 8-cell stage occurs between 3 and 4 dpf in the human vs 2 and 3 dpf in the mouse. Morphological segregation of the ICM and TE at the blastocyst stage occurs between 5 and 6 dpf in the human (64–128 cells), but 3 and 4 dpf in the mouse (32–64 cells). Implantation in the mouse occurs between 4 and 4.5 dpf (>128 cells), and between 7 and 10 dpf (>200 cells) in the human. Human embryonic development is also more susceptible to genetic instability and aneuploidies, perhaps due to suboptimal culture conditions that have historically been developed to be optimal for mouse preimplantation development and/or differences in cell cycle checkpoints in early cleavage stage embryos compared to somatic cells (Delhanty & Handyside, 1995; Harrison, Kuo, Scriven, Handyside, & Ogilvie, 2000; Vanneste et al., 2009; Vassena et al., 2011). By contrast, spontaneous aneuploidy rates in mouse embryos are quite low (Hassold & Hunt, 2001). Molecular differences may underlie these differences in developmental timing and as such, the following section investigates the regulatory mechanisms involved at these preimplantation stages of development.

2.1 Epigenetic Regulation of Gene Expression

The epigenetic landscape of a given cell or lineage is established by a number of mechanisms, including DNA methylation, histone modifications, and the formation of histone variants. While the methylation pattern of differentiated tissues is relatively static, human preimplantation development involves vast and dynamic changes associated with reprogramming of the genome. Early human development first involves decondensation of the sperm genome, followed by chromatin remodeling, protamine-to-histone exchange, and global demethylation of both paternal and maternal DNA, with more rapid active demethylation of the paternal genome (Clift & Schuh, 2013; Guo et al., 2014; Smith et al., 2014). DNA methylation occurs predominantly at CpG dinucleotides, where DNA methyltransferases (DNMTs) mediate the transfer of a methyl group to cytosines, generating 5-methylcytosine (5mC).

The major wave of demethylation in humans occurs between fertilization and the 2-cell stage (Guo et al., 2014; Smith et al., 2014). Demethylation levels then reach a low point in the ICM of the pluripotent blastocyst, and erasure of epigenetic memory is thought to be a requirement for acquisition of pluripotency (Lee, Hore, & Reik, 2014). However, DNA methylation associated with promoter regions still represses the expression of corresponding genes. Postimplantation, the hypomethylated state of early mammalian embryos is reversed, and methylation patterns take on a genome-wide profile broadly resembling adult tissues (Guo et al., 2014). It remains unclear how these methylation dynamics are regulated in the human embryo, though ten-eleven-translocation (TET) dioxygenases have been implicated in regulating the conversion of 5mC to 5hmC (Perera et al., 2015; Tahiliani et al., 2009), and are expressed at these early stages of development (Guo et al., 2014).

Another mechanism of early epigenetic regulation involves the inactivation of one X chromosome to ensure dosage compensation between male and female mammals (Lyon, 1961). X chromosome inactivation (XCI) is linked to the expression of the noncoding RNA *Xist/XIST* from the future inactive X (Brockdorff et al., 1992; Brown et al., 1992; Penny, Kay, Sheardown, Rastan, & Brockdorff, 1996). Xist/XIST coats the X-chromosome in *cis*, and this is followed by a series of chromosome-wide epigenetic changes that prevent the expression of some, but not all genes on the inactive X (Lu, Carter, & Chang, 2017; Tukiainen et al., 2017; Yang, Babak, Shendure, & Disteche, 2010). Although it was suggested that initiation of gene silencing on the paternal X chromosome occurred despite the absence of paternal *Xist* (Kalantry, Purushothaman, Bowen, Starmer, & Magnuson, 2009), more recent genome-wide expression analysis shows *Xist*-dependent paternal X-chromosome gene silencing (Borensztein et al., 2017), consistent with previous studies (Namekawa, Payer, Huynh, Jaenisch, & Lee, 2010). Paternal XCI is maintained in extraembryonic tissues, however, in the ICM the paternal X is reactivated for a brief period, followed by random X inactivation in the epiblast (Mak et al., 2004; Okamoto, Otte, Allis, Reinberg, & Heard, 2004; Takagi & Sasaki, 1975).

In human embryos, *XIST* is upregulated between the 4- and 8-cell stage (Briggs, Dominguez, Chavez, & Reijo Pera, 2015; Daniels, Zuccotti, Kinis, Serhal, & Monk, 1997; Ray, Winston, & Handyside, 1997). In contrast to the mouse, XIST has been detected in both female and male human embryos (Daniels et al., 1997; Okamoto et al., 2011; Ray et al., 1997) and accumulation observed on both X-chromosomes in the female embryo at the late blastocyst stage (Okamoto et al., 2011). In addition, X-linked transcripts were detected up to the blastocyst stage despite XIST accumulation (Okamoto et al., 2011). This would suggest that the presence of XIST at these stages does not necessarily indicate XCI is occurring, and thus XCI may only begin after the blastocyst is formed or around implantation. However, an alternative study observed distinct XIST expression patterns in male and female embryos, and inactive X-specific chromatin modifications, and concluded that XCI was in fact functioning (van den Berg et al., 2009). Additionally, although a recent study also observed accumulation of XIST on both female X chromosomes, they suggested that this instead resulted in dampening of gene expression, with biallelic expression of X-linked dosage-compensated genes whose expression was gradually downregulated over time (Petropoulos et al., 2016). However, reanalysis of this data and additional data sets by an independent group found no evidence of dampening of X chromosome gene expression in human preimplantation embryos, instead their data are consistent with the suggestion of random XCI (Moreira de Mello, Fernandes, Vibranovski, & Pereira, 2017).

Interestingly, an alternative X-linked long noncoding RNA, XACT, coaccumulates with XIST on X chromosomes in both male and female embryos during preimplantation development, though with differential distribution (Vallot et al., 2017). It will be interesting to determine how this relates to the mechanisms of human X-inactivation. There also does not appear to be strict paternal imprinting in human extraembryonic lineages either in the blastocyst or later in the placenta (Hamada et al., 2016; Moreira de Mello et al., 2010; Okamoto et al., 2011; Petropoulos et al., 2016), further suggesting divergent regulation of XCI in humans compared to the mouse. Altogether, further investigation should provide additional insights into this fascinating aspect of gene regulation.

2.2 Activating Embryonic Gene Expression

In human embryos, EGA occurs between the 4- and 8-cell stage (Braude et al., 1988), although minor human EGA may occur as early as the 2-cell stage (Taylor, Ray, Ao, Winston, & Handyside, 1997; Vassena et al., 2011). These dynamics are likely linked to developmental timing, rather than cell number, as transcription is activated even in arrested embryos, as long as they have undergone the first mitotic division (Dobson et al., 2004). By contrast, EGA in the mouse occurs between the 1- and 2-cell stage (Flach, Johnson, Braude, Taylor, & Bolton, 1982; Hamatani, Carter, Sharov, & Ko, 2004), although in some other mammalian species, such as the cow and sheep, EGA occurs at later stages, similar to the human (Jukam, Shariati, & Skotheim, 2017). EGA has been suggested to involve the degradation of residual maternal mRNA in two distinct waves (Vassena et al., 2011) alongside upregulation of embryonic gene expression (Blakeley et al., 2015; Dobson et al., 2004; Galan et al., 2010; Tohonen et al., 2015; Vassena et al., 2011; Xue et al., 2013; Yan et al., 2013).

It remains unclear what drives the onset of human EGA. Gene sets upregulated early in EGA vary greatly between species, with only 40% of EGA-activated transcripts conserved between human and mouse, and 18.5% between human and bovine (Xie et al., 2010). A number of genes uniquely expressed during EGA have been identified in mice, such as *Zscan4*

(Falco et al., 2007), though several are not conserved in the human (Ko, 2016). However, *ZSCAN4* is also upregulated early in human EGA, along with additional novel PRD-like homeobox genes, such as *ARGFX* and *LEUTX* (Jouhilahti et al., 2016; Madissoon et al., 2016; Töhönen et al., 2017). A number of transposable elements are also upregulated, including the endogenous retrovirus HERV-K (Goke et al., 2015; Grow et al., 2015; Tohonen et al., 2015), as well as *Alu* repeats, short interspersed nuclear elements (SINE) (Töhönen et al., 2017). Notably, several of these transposable elements do not appear to be activated in somatic tissues (Goke et al., 2015), suggesting they may have a specific role during early embryonic development or that there is an absence of their repression in this context. Transposable elements have also been linked to the regulation of pluripotency genes in human embryonic stem (ES) cells, and in some cases also contain pluripotency factor binding motifs (Fort et al., 2014; Friedli & Trono, 2015; Grow et al., 2015; Kunarso et al., 2010). A number of pluripotency factors are expressed early during EGA (Vassena et al., 2011), and it is possible they may also regulate, or be regulated by, aspects of this process.

Recent studies have also identified a role for the DUX transcription factor family during EGA (De Iaco et al., 2017; Hendrickson et al., 2017; Töhönen et al., 2017). DUX4 mRNA and protein expression is restricted to the nucleus of 4-cell human embryos and activates EGA-associated genes including *ZSCAN4* when overexpressed in human-induced pluripotent stem (iPS) cells (Hendrickson et al., 2017) or human ES cells (De Iaco et al., 2017). The mouse orthologue, Dux, is restricted to the 2-cell stage in mouse embryos (coincident with the earlier onset of EGA in the mouse) and when expressed in mouse ES cells generates an open chromatin state similar to that in 2-cell mouse embryos (Hendrickson et al., 2017). Utilizing the CRISPR (clustered regularly interspaced, short palindromic repeat)–Cas9 (CRISPR-associated) gene editing system (Jinek et al., 2012) to inactivate *Dux* in mouse zygotes resulted in failure to progress to morula or blastocyst stages and affected transcriptional activation of genes associated with EGA (De Iaco et al., 2017; Hendrickson et al., 2017). However, it remains unclear what triggers *Dux* expression in the mouse, and as Dux only activates a subset of EGA-associated genes, it is unlikely to be the sole driver of this transition. Nevertheless, DUX proteins are interesting candidates for further interrogation during this critical cell stage in human development.

2.3 Lineage-Specific Gene Expression Patterns

As highlighted earlier, preimplantation development first involves segregation of the ICM and the TE, followed by separation of the cells within the ICM into either pluripotent Epi cells, or PE cells. The mechanisms by which these lineages are specified in the human embryo have only recently begun to be characterized, and were initially largely defined by studies in the mouse. The mouse TE, Epi, and PE lineages are each characterized by the expression of distinct genes, whose role was largely identified through mouse mutant phenotypes affecting lineage emergence or maintenance. These distinctions were also confirmed by genome-wide expression analysis (Guo et al., 2010; Kurimoto et al., 2006) and cell-surface marker expression (Rugg-Gunn et al., 2012), which provide a useful means of tracking lineage specification during mouse preimplantation development.

In the mouse TE, the transcription factor Tead4 regulates the expression of *Cdx2* and *Gata3*, which are expressed from the morula stage and themselves regulate downstream factors such as *Eomes* and *Elf5* that are required for TE maintenance (Latos et al., 2015; Ng et al., 2008; Nishioka et al., 2008; Ralston et al., 2010; Ralston & Rossant, 2008; Russ et al., 2000; Strumpf et al., 2005). Genetic manipulations of these transcription factors have revealed the hierarchy in which they act in TE specification. *Tead4*$^{-/-}$ embryos express the ICM/Epi-associated transcription factor Oct4 in all blastomeres, and fail to form a blastocoel cavity, remaining compacted (Yagi et al., 2007). *Cdx2*$^{-/-}$ embryos form a blastocoel cavity, but this then collapses and cannot be maintained, likely due to a failure to maintain the integrity of tight junctions within the TE (Strumpf et al., 2005). Other markers of the TE are also absent from *Cdx2*$^{-/-}$ mouse embryos (Ralston & Rossant, 2008; Strumpf et al., 2005). In contrast, while *Eomes*$^{-/-}$ embryos do initiate a decidual response and are able to implant, they fail to undergo differentiation and proliferation (Russ et al., 2000; Strumpf et al., 2005). Ap-2γ (*Tcfap2c*) is also required for the maintenance of the mouse TE lineage, regulating genes involved in tight junction assembly and fluid accumulation, as well as *Cdx2* (Cao et al., 2015; Choi, Carey, Wilson, & Knott, 2012; Kuckenberg et al., 2010).

A number of these TE lineage-associated factors are also expressed in the human embryo, though expression patterns are not necessarily always conserved with the mouse. *GATA3* is expressed in the human TE, but homologues of key mouse TE factors such as *ELF5* and *EOMES* are absent during preimplantation development (Blakeley et al., 2015). Following implantation, ELF5 expression has been detected in human villous cytotrophoblast cells (Hemberger, Udayashankar, Tesar, Moore, & Burton, 2010), suggesting important differences in the timing of key placental-associated genes. Additional genes such as the tight junction-associated *CLDN10* and placenta-enriched *PLAC8* have been identified as TE-specific in the human, but are also absent in mouse preimplantation embryos (Blakeley et al., 2015). Curiously, AP-2γ (*TFAP2C*) is expressed in both the Epi and the TE in the human blastocyst (Blakeley et al., 2015), though it is expressed in all trophoblast lineages in the human placenta (Biadasiewicz et al., 2011). Furthermore, CDX2 is only detected in the human blastocyst, after cavitation (Chen et al., 2009; Niakan & Eggan, 2013) similar to cow and pig embryos (Berg et al., 2011; Bou et al., 2017), suggesting distinct roles for these TE factors compared to the mouse.

Within the mouse ICM, the Epi gene expression network includes the pluripotency-associated transcription factors Oct4 (*Pou5f1*), Sox2, and Nanog, while the PE instead expresses factors including Gata6, Gata4, and Sox17, and the cell-surface receptor Pdgfra. Oct4 is initially expressed in all cells within the ICM, and only becomes restricted to the Epi in the late blastocyst (Grabarek et al., 2012). The earliest markers of the presumptive Epi and PE are Nanog and Gata6, respectively, which are initially coexpressed before resolving into a mosaic salt-and-pepper pattern in the ICM (Plusa, Piliszek, Frankenberg, Artus, & Hadjantonakis, 2008). Experiments using $Pdgfra^{H2B\text{-}GFP}$ reporter mice to mark presumptive PE cells showed that ICM cells initially randomly segregate, with Epi or PE commitment correlated with Nanog or Gata4 upregulation, respectively, and cell migration and subsequent apoptosis ensuring that cells are correctly located (Plusa et al., 2008). Both $Pou5f1^{-/-}$ and $Nanog^{-/-}$ mouse embryos fail to form a pluripotent epiblast (Chambers et al., 2007; Mitsui et al., 2003; Nichols et al., 1998), and all cells in the $Nanog^{-/-}$ ICM instead express Gata6 (Frankenberg et al., 2011). Conversely, $Gata6^{-/-}$ mouse embryos fail to form the PE, and Oct4, Nanog, and Sox2 are instead expressed across the ICM (Bessonnard et al., 2014; Schrode, Saiz, Di Talia, & Hadjantonakis, 2014). Although $Sox17^{-/-}$ embryos form a PE layer, PE cells are progressively lost if implantation is delayed in these embryos (Artus, Piliszek, & Hadjantonakis, 2011), suggesting Sox17 is required for PE maintenance.

NANOG and SOX2 are expressed in the human Epi at the early blastocyst stage from 5 dpf (Cauffman, De Rycke, Sermon, Liebaers, & Van de Velde, 2009; Hyslop et al., 2005). OCT4 is initially expressed from the 8-cell stage and is then detectable in all cells in the embryo (including the TE), and OCT4-high expression is only restricted to the human Epi in the late blastocyst at 6 dpf, which corresponds to the optimal timing for successful derivation of human embryonic stem cells (Chen et al., 2009; Niakan & Eggan, 2013). GATA6 exhibits a broader expression pattern in the human late-blastocyst and is present in both the PE and the TE (Deglincerti et al., 2016; Roode et al., 2012), while SOX17 is localized to the PE, where it overlaps with GATA6 (Niakan & Eggan, 2013). Additional novel lineage markers have been identified, such as KLF17 in the Epi (Blakeley et al., 2015), which are not detected in the equivalent mouse lineage. Interestingly, the gene expression patterns observed during human development have occasionally been observed in nonrodent model organisms. GATA6 is similarly broadly expressed in the Epi and TE of primate (rhesus macaque and cynomolgus monkey) and bovine embryos (Boroviak et al., 2015; Kuijk et al., 2012; Nakamura et al., 2016), while *POU5F1* is expressed in both the rabbit ICM and TE (Cauffman et al., 2009). Cynomolgus monkey embryos also exhibit remarkably similar OCT4 and CDX2 expression dynamics to those in the human embryo (Nakamura et al., 2016).

Investigating mRNA expression in single human blastomeres from the 5- to 8-cell stages did not determine any distinction between cell fates based on lineage-specific gene expression (Galan et al., 2010). Intriguingly, recent single cell transcriptomic analysis of cynomolgus monkey (Nakamura et al., 2016) and human embryos (Petropoulos et al., 2016) seemingly identifies the TE, Epi, and PE emerging concurrently at the blastocyst stage, rather than in two sequential steps as in the mouse. This may reflect greater plasticity during early development, and both inner and outer cells of human blastocysts disaggregated at 5 dpf are capable of forming a blastocyst with both an ICM and TE (De Paepe et al., 2013). Cells in the early human embryo may thus retain developmental plasticity for longer than the mouse, though it is not clear why this is the case—one hypothesis is that the shorter developmental timeline in the mouse necessitates earlier lineage segregation (Rossant, 2014), ensuring a fully committed TE is present in time for appropriate implantation to occur.

Another possibility is that the methods that have been used to assess whether later markers of the Epi, TE, and PE are differentially expressed prior to or during compaction may not capture as yet uncharacterized upstream regulators that drive ICM vs TE specification. Alternatively, specification may be driven via posttranscriptional modifications rather than at the transcriptional level. Further transcriptional analysis of these early stages, while ideally maintaining positional information, would be informative. Moreover, as methods to investigate the proteins enriched in a small number of cells are being developed (Budnik, Levy, & Slavov, 2017; Heath, Ribas, & Mischel, 2015), these technologies could be a useful complement to transcriptional analyses.

Given the distinctions between gene expression patterns in the human compared to the mouse, further studies would benefit from interrogating gene function directly in the human preimplantation embryo. This is especially true for novel human-specific factors, whose mouse homologues are not expressed at an equivalent stage, or in the required lineage, but also where distinctions in timing of expression suggest an alternative function (i.e., KLF17 or CLDN10). Methods such as CRISPR/Cas9-mediated genome editing system could provide a precise and efficient means to address the functional significance of these putative developmental regulators in human embryos and to determine whether there is a conserved role for factors known to regulate mouse preimplantation development (i.e., OCT4 or GATA6). An efficient and precise method to inactivate these genes is particularly important given the relative scarcity of human embryos available for research, especially at earlier stages of development. Coupled with current advances enabling high-resolution genomic and transcriptomic analyses, which can then be validated by protein expression or morphological analyses, this will further our understanding of the gene regulatory mechanisms involved in early human lineage specification.

Indeed, using CRISPR/Cas9 as a proof-of-principle to examine the role of OCT4 in lineage specification in the human embryo has not only demonstrated the feasibility of this method but also revealed a surprisingly distinct role for this transcription factor in human embryogenesis compared to the mouse. *Oct4*$^{-/-}$ mice form a blastocyst but cannot be maintained due to defects in the ICM, including the establishment of the PE (Frum et al., 2013; Le Bin et al., 2014). However, in the human, OCT4-targeted embryos are compromised in their ability to form a blastocyst and surprisingly exhibit downregulation of genes associated with Epi, TE, and PE (Fogarty et al., 2017). For example, while *Oct4*$^{-/-}$ mouse blastocysts retain Nanog expression in the epiblast, NANOG is undetectable in OCT4-targeted human embryos. This suggests that in the mouse Nanog expression is regulated independently of Oct4, perhaps via STAT3 signaling, and that this pathway or others that might compensate for the absence of *Oct4* are not conserved in the human. Alternatively, the loss of OCT4 in the human may indirectly lead to the absence of NANOG as the ICM fails to form. Again, in striking contrast to the mouse, OCT4-targeted human embryos down-regulate a number of TE markers including *CDX2*, *GATA2*, *GATA3*, and *HAND1*. While Gata2 expression has not yet been reported in *Oct4*$^{-/-}$ mouse embryos, other markers such as *Cdx2*, *Gata3*, and *Hand1* are not only expressed, but importantly, are transcriptionally upregulated or ectopically expressed in the ICM. This suggests that in the mouse Oct4 negatively regulates the expression of trophectoderm genes. By contrast in the human embryo, OCT4 possibly functions to positively regulate key trophectoderm genes, consistent with its perdurance in the TE (Niakan & Eggan, 2013).

2.4 The Role of Extracellular Signaling Networks

Extracellular signaling has been linked to both ICM and TE segregation, and subsequent Epi and PE specification in the mouse, but it is unclear whether this also holds true in a human context. In the mouse embryo, differential activation of components of the Hippo signaling pathway after the 8-cell stage delineates inner and outer cells, which then resolve into the ICM and TE. In outer cells, Hippo signaling is curtailed as Lats1/2 protein kinases are sequestered at the apical surface by PAR complex proteins and are thus unable to phosphorylate the transcriptional coactivator Yap (Cockburn, Biechele, Garner, & Rossant, 2013; Hirate et al., 2013; Kono, Tamashiro, & Alarcon, 2014; Nishioka et al., 2009). Yap subsequently translocates to the nucleus and facilitates Tead4 induction of a TE-specific gene expression program as detailed earlier (Kono et al., 2014; Nishioka et al., 2009; Yagi et al., 2007). Notch signaling is also thought to be involved in this lineage segregation, as it has been shown to regulate Cdx2 in outer cells (Rayon et al., 2014). Conversely, in inner cells, Lats1/2 are free in the cytosol to phosphorylate Yap and exclude it from the nucleus, allowing an ICM gene expression profile to take hold. Indeed, knockdown of the PAR component *Pard6b* in mouse embryos suppresses formation of the blastocyst cavity and reduces Cdx2 expression, while Nanog is ectopically expressed (Alarcon, 2010).

Subsequently, fibroblast growth factor (FGF) signaling, and consequent activation of the mitogen-activated protein kinase (MAPK) pathway, is required to facilitate the segregation of Epi and PE lineages within the ICM (Lanner & Rossant, 2010). A reciprocal receptor ligand relationship exists between the Epi and PE, whereby FGF ligands secreted by the Epi binds to FGF receptors that are enriched on the PE (Guo et al., 2010). Mutating genes coding for FGF signaling pathway components detrimentally affects PE formation and disrupts PE gene expression patterns, phenocopying the effects seen with mutations in PE lineage specifying genes (Arman, Haffner-Krausz, Chen, Heath, & Lonai, 1998; Chazaud, Yamanaka, Pawson, & Rossant, 2006; Cheng et al., 1998; Feldman, Poueymirou, Papaioannou, DeChiara, & Goldfarb, 1995). Inhibiting FGF receptors or blocking Erk phosphorylation by the MAPK Mek also increases the proportion of Nanog-expressing cells within the ICM (Nichols, Silva, Roode, & Smith, 2009; Yamanaka, Lanner, & Rossant, 2010), suggesting that FGF signaling primarily functions via the Raf-Mek-Erk pathway. Conversely, treatment with FGF ligands alone or coupled with the FGF receptor binding facilitator heparin resulted in down-regulation of Nanog expression and conversion to Gata6-positive PE progenitors (Yamanaka et al., 2010).

Despite conservation of some lineage-specific genes between mouse and human embryos, it is likely that distinct signaling pathways are required at these early stages. Although human embryos seemingly undergo a similar process of compaction and cavitation as that observed in the mouse, the later expression of CDX2 (blastocyst, 5 dpf), and absence of CDX2-regulated factors such as ELF5 and EOMES (Blakeley et al., 2015; Niakan & Eggan, 2013) suggests that Hippo signaling may not necessarily drive ICM-TE segregation or that Hippo signaling in humans regulates alternative TE factors upstream of CDX2. Nuclear YAP expression has been detected in the ICM at the early blastocyst stage (5 dpf) in human embryos, and is only restricted to the TE in the late blastocyst (6 dpf) (Noli, Capalbo, Ogilvie, Khalaf, & Ilic, 2015), though it is unclear if it is expressed earlier in human development. In the mouse, Yap begins to be specifically localized in outer cells after the 8-cell stage, coincident with Cdx2 expression, and importantly it is not detected in the ICM (Hirate, Cockburn, Rossant, & Sasaki, 2012; Nishioka et al., 2009). Furthermore, inhibiting FGF/Erk signaling in human embryos has no effect on either PE or Epi formation in the human blastocyst, and gene expression patterns are unchanged (Kuijk et al., 2012; Roode et al., 2012). However, the FGF receptor was not targeted independently of inhibiting Erk, so it remains possible that FGF might function via an alternative downstream pathway to regulate lineage specification. Altogether, this suggests that the Epi cells in vivo may not require FGF signaling, which is distinct to the requirement for this signaling pathway in existing human ES cells described in the chapter further later. Moreover, it suggests that the mechanisms regulating the PE are fundamentally distinct in humans compared to mice.

Interestingly, in bovine embryos, although FGF stimulation resulted in an ICM made up entirely of GATA6-expressing cells, FGF/Erk inhibition drives a switch to all NANOG-expressing cells as is the case in the mouse (Kuijk et al., 2012). Erk inhibition increased the proportion of NANOG-expressing cells but only partially blocked GATA6 expression, while FGF receptor inhibition had no effect (Kuijk et al., 2012). This suggests that although modulating FGF signaling does influence the Epi–PE lineages in bovine embryos, unlike in the mouse it is likely not the major pathway involved. Similarly, in marmoset embryos, Erk or Wnt inhibition increased the proportion of NANOG expressing cells within the ICM and diminished, but did not abolish, GATA6-only PE cells (Boroviak et al., 2015). Intriguingly, a proportion of cells in control marmoset blastocysts were allocated as unstained (did not express either NANOG, GATA6, or CDX2) (Boroviak et al., 2015), suggesting that as-yet-undetermined alternative lineage markers may also be required for optimal lineage assignation. Inhibiting Wnt or Erk signaling decreased the proportion of unstained cells, but increased the occurrence of cells with ectopic coexpression of two or all of the three factors (Boroviak et al., 2015). Examining the identity of these unstained cells and gene expression patterns prior to the blastocyst stage may elucidate how lineage overlap following signal modulation is related to developmental progression.

Alternative signaling pathways have been implicated in regulating gene expression in the human embryo, although how they are related to lineage specification per se remains to be elucidated. Both insulin-like growth factor 1 (IGF1) and granulocyte-macrophage colony-stimulating factor (GM-CSF or CSF2) have been implicated in improving ICM cell survival, though it is unclear if this also affects lineage allocation (Robertson, Sjoblom, Jasper, Norman, & Seamark, 2001; Sjoblom, Wikland, & Robertson, 2002; Spanos, Becker, Winston, & Hardy, 2000). Components of the Nodal signaling pathway are enriched in the human Epi, and inhibiting TGFβ/Nodal in the human embryo was recently shown to result in loss of NANOG expression (Blakeley et al., 2015). Embryos were treated from 3 to 6 dpf with the small molecule inhibitor SB-431542, which targets the TGFβ/Activin/Nodal type 1 receptors ALK5, ALK4, and ALK7 (Inman et al., 2002). Earlier experiments inhibiting TGFβ/Nodal signaling instead observed a positive effect on NANOG expression (Van der Jeught et al., 2013), though as a lower concentration of SB-431542 was used (10 μM compared to 40 μM), it is possible TGFβ/Nodal signaling may not have been fully suppressed. Inhibiting TGFβ/Nodal signaling in mouse embryos, or in marmosets did not disrupt lineage segregation (Blakeley et al., 2015; Boroviak et al., 2015; Granier et al., 2011), suggesting this pathway may potentially be specific to human lineage specification. Altogether, this implies that distinct signaling pathways may be relevant in each model organism, with occasional, but not absolute, conservation of the resulting effect on the expression of lineage-associated genes. How this tallies with conservation of gene expression patterns across species is unclear, but nevertheless highlights the importance of studying signaling patterns directly in context and across a number of species.

3. MODELING HUMAN PREGASTRULATION DEVELOPMENT IN VITRO

3.1 Stem Cell Lines From Preimplantation Embryos

Although much can be gleaned from studies involving the culture of preimplantation human embryos in vitro, donated embryos are a finite resource and the relatively small number of cells within each embryo further restricts the types of analyses that can be carried out. Given the links between lineage-specific gene networks and signaling pathways, various studies have modulated signaling in an attempt to derive cell lines in vitro that retain the characteristics of their embryonic cell type of origin. Thus far, only human ES cells have been successfully established.

The first human ES-like cells were isolated from human embryos that were plated and allowed to hatch onto a human oviduct epithelial feeder layer, after which the ICM clumps that attached to the monolayer were disaggregated and subcultured in an attempt to maintain a stable self-renewing line (Bongso, Fong, Ng, & Ratnam, 1994). Although cells retained stem cell-like morphology and normal karyotype, they could not be maintained for more than two passages (Bongso et al., 1994). Similar experiments performed in nonhuman primates were more successful and refined the techniques that enabled successful derivation of ES cells from human embryos, namely, removing the TE using immunosurgery and plating the intact ICM onto a mouse embryonic fibroblast (MEF) layer (Thomson et al., 1998, 1995, 1996). This gave rise to the first stable ES cell lines (H1, H7, H9, H13, H14) that could be propagated for multiple passages and had a normal karyotype. A number of human ES cell lines have subsequently been derived (Aflatoonian et al., 2010; Cowan et al., 2004; Mitalipova et al., 2003; Suemori et al., 2006), though the H1 and H9 lines are predominantly used in human ES cell studies (Kobold, Guhr, Kurtz, & Loser, 2015; Loser, Schirm, Guhr, Wobus, & Kurtz, 2010).

ES cells are both self-renewing and pluripotent, can be maintained indefinitely in culture, and have the potential to differentiate into the three lineages that comprise the embryo proper (ectoderm, endoderm, and mesoderm). Human ES cells require OCT4, NANOG, and SOX2 to maintain pluripotency (Wang, Oron, Nelson, Razis, & Ivanova, 2012), and induction of OCT4 and SOX2 along with KLF4 and c-MYC in ES cell culture conditions is sufficient to reprogram human fibroblasts and other somatic cells to iPS cells (Park, Lerou, Zhao, Huo, & Daley, 2008; Takahashi & Yamanaka, 2006; Yu et al., 2007). Similar to the human Epi, Nodal- or Activin-driven TGFβ signaling has been shown to have a role in maintaining human ES and iPS cells, regulating pluripotency gene expression by binding directly to the *NANOG* promoter, as well as reinforcing expression of Nodal signaling components (Besser, 2004; Brown et al., 2011; James, Levine, Besser, & Brivanlou, 2005; Vallier, Alexander, & Pedersen, 2005; Vallier et al., 2009).

Additionally, FGF is present in the majority of human ES cell culture media, if not overtly via addition of exogenous ligand, then by culture on MEF layers or in MEF-conditioned media (Amit et al., 2000; Chen et al., 2011; Cowan et al., 2004; Levenstein et al., 2006; Ludwig et al., 2006; Reubinoff, Pera, Fong, Trounson, & Bongso, 2000; Thomson et al., 1998; Xu et al., 2001). FGF/Erk inhibition in human ES cells affects Nanog expression and promotes neural differentiation (Greber et al., 2011, 2010), implying that FGF signaling is required to maintain human ES cell pluripotency, which is at odds with the FGF inhibition studies in the human embryo (Kuijk et al., 2012; Roode et al., 2012). However, it has also been suggested that FGF only indirectly promotes human ES cell pluripotency, instead stimulating either the supportive MEF layer (Greber, Lehrach, & Adjaye, 2007) or fibroblast-like cells differentiated from human ES cells themselves (Bendall et al., 2007), to secrete factors, such as IGF2, that subsequently promote pluripotency.

Recent analyses suggest that existing human ES cells do not fully recapitulate human Epi gene expression patterns and signaling requirements (Blakeley et al., 2015; Yan et al., 2013), which presents challenges in using these as a model to explore the basic biology of this developmental stage. This is in contrast to the mouse system where mouse pluripotent cells show similar gene expression profiles to, and cluster with, their embryonic cell types of origin (Boroviak, Loos, Bertone, Smith, & Nichols, 2014;

Boroviak et al., 2015). Several studies have developed culture systems utilizing a combination of inhibitors and growth factors in an attempt to generate human ES cells that are more similar to the pluripotent cells of the human Epi (Chan et al., 2013; Gafni et al., 2013; Guo et al., 2016; Takashima et al., 2014; Theunissen et al., 2014; Valamehr et al., 2014; Ware et al., 2014). Indeed, the resulting cell lines exhibit some embryo-associated characteristics, such as global demethylation, which correlates with the hypomethylated status observed in the human ICM (Gafni et al., 2013; Guo et al., 2016; Smith et al., 2014; Theunissen et al., 2016), and upregulate novel human-specific factors such as KLF17 in some cases (Blakeley et al., 2015; Collier et al., 2017; Guo et al., 2016). Nevertheless, these cells remain somewhat transcriptionally distinct from the human Epi, and lose imprinting marks that are established during blastocyst development (Pastor et al., 2016), suggesting that current in vitro culture conditions do not fully reflect the conditions required for pluripotency maintenance in vivo. However, given that they share a core transcriptional network despite their different culture systems (Huang, Maruyama, & Fan, 2014), it would be interesting to further analyze these genes to determine if they lend insight into the fundamental underpinnings of the human pluripotent state. Moreover, it would be informative to determine whether the growth factors or inhibitors used to establish these alternative human ES cells reflect the signaling requirements for the establishment or maintenance of the in vivo Epi.

Stem cell lines representing the human TE and PE lineages are being refined or developed. By contrast, detailed molecular genetic analysis has enabled the establishment of mouse trophoblast stem (TS) cells have been derived both from the embryo and from directed reprogramming of fibroblasts. Plating extraembryonic ectoderm from E6.5 mouse conceptuses cultured on MEFs in FGF4 and heparin allows for the emergence of TS cell colonies (Tanaka, Kunath, Hadjantonakis, Nagy, & Rossant, 1998). Transient expression of the combination of transcription factors *Tcfap2c/Eomes/Gata3/Ets2* or *Tcfap2c/Eomes/Gata3/Myc* in mouse fibroblasts (Benchetrit et al., 2015; Kubaczka et al., 2015) leads to the emergence of mouse TS cells resembling those derived from the blastocyst. However, similar attempts to derive human TS cells have not yet been successful, although, modulating BMP signaling in human ES cells results in a loss of pluripotency and upregulation of some TE-associated genes (Das et al., 2007; Li et al., 2013; Xu et al., 2002). These cells cannot be propagated as self-renewing trophoblast stem cell lines and it has been suggested that the resulting cells are in fact extraembryonic mesoderm derivatives, rather than human TS cells (Bernardo et al., 2011), this likely reflects differences arising from experimental culture conditions, and BMP4 likely promotes emergence of a mixture of distinct trophoblast and mesoderm progenitors (Amita et al., 2013; Drukker et al., 2012). TS-like cells have also been isolated from EB outgrowths generated from human ES cell lines (Zdravkovic et al., 2015), using conditions previously used to generate cells from human chorionic tissue in an attempt to identify an alternative TS cell niche (Genbacev et al., 2011). However, though these cells express TE-associated factors such as GATA3 and CDX2 (Zdravkovic et al., 2015), they also express ELF5 and EOMES, which as discussed earlier, are absent in human blastocysts. Therefore, these cells may not necessarily reflect a preimplantation TE identity and may represent a later population of trophoblast cells.

TS cell derivation from whole human blastocysts has been attempted using FGF-supplemented media on a MEF supportive layer (Kunath et al., 2014), based on conditions developed in the mouse (Tanaka et al., 1998; Uy, Downs, & Gardner, 2002). However, only one of 60 embryos gave rise to cells that could be propagated, which did not resemble mouse TS cells or human ES cells, but could not be maintained beyond three passages (Kunath et al., 2014). Given that mouse TS cells also rely on TGFβ/Activin for maintenance (Erlebacher, Price, & Glimcher, 2004; Kubaczka et al., 2014), which is instead relevant for maintaining pluripotency in the human, it is likely that human TS cells instead require a distinct signaling environment. More recently self-renewing TS cells have been established from both human blastocysts and villous cytotrophoblast cells by blocking TGF-β signaling, histone deacetylase, and Rho-associated kinase and activating EGF and WNT signaling (Okae et al., 2018). The resulting human TS cells have a gene expression pattern that is similar to primary trophoblast cells. Further functional analysis of these cells, including their in vivo potential (e.g., in chimera contribution), would inform which in vivo trophoblast cell type they most closely resemble as well as their clinical utility.

Extraembryonic endoderm (XEN) cells are yet to be derived from the human PE, though they have been successfully isolated from mouse embryos (Kunath et al., 2005). Mouse XEN cells can also be generated using growth factor modulation (Cho et al., 2012) or ectopic expression of GATA factors in mouse ES cells (Fujikura et al., 2002; Shimosato, Shiki, & Niwa, 2007; Wamaitha et al., 2015), generating stable XEN cell lines that contribute to PE lineages in vitro. Addition of MEF-conditioned media to human

ES cells upregulates PE-associated factors, but with minimal pluripotency factor downregulation (Drukker et al., 2012), indicating MEFs secrete both pluripotency- and differentiation-promoting factors. Overexpression of PE-associated transcription factors SOX7 and SOX17 has been shown to upregulate PE-associated factors in human ES cells, though the resulting cultures retained expression of pluripotency factors NANOG and OCT4 (Seguin, Draper, Nagy, & Rossant, 2008). However, ectopic expression of GATA6 in human ES cell cultures upregulated a number of PE-associated factors while also downregulating pluripotency gene expression, and generated cells that were morphologically distinct from pluripotent human ES cells (Wamaitha et al., 2015). Although these cells could be expanded and maintained their morphology for more than three passages, they could not be maintained indefinitely, again suggesting alternative conditions or factors may be required to derive stable human XEN cell lines.

3.2 In Vitro Implantation Models

Studying human embryo development postimplantation in vivo is practically and ethically unfeasible, and thus investigating this time period has been challenging. Following implantation, the human epiblast forms a pseudostratified columnar epithelium with an amniotic cavity (Hertig et al., 1956). These transformations have been suggested to be coincident with changes in pluripotency states in both humans and mice (Shahbazi et al., 2017). Human and primate embryos exhibit some distinct implantation mechanisms compared to the mouse. For example, in contrast to the invasive interstitial implantation of the human and primate trophoblast derivatives, in mice and rats the uterine epithelium invaginates to surround the trophoblast, while in bovine and porcine embryos the blastocyst expands to fuse with the epithelium without penetration (Carter, Enders, & Pijnenborg, 2015; Lee & DeMayo, 2004). Additionally, despite overall similarities, implantation in humans and other primates is not identical (Carter et al., 2015), and consequently it is also challenging to extrapolate mechanisms of human placentation from alternative model organisms. In vitro research on human embryos is currently subject to a 14-day limit on embryo experimentation postfertilization (or the equivalent time of the emergence of the primitive streak at gastrulation), either by law or in scientific guidelines (Hyun, Wilkerson, & Johnston, 2017; Pera, 2017). However, once the blastocyst hatches from the zona at 7 dpf, it needs to undergo implantation in order to carry on development, and thus without appropriate implantation models, most studies fall short of the 14-day limit.

Foundational studies using mouse ES cells (Coucouvanis & Martin, 1995) have been further developed more recently into models of early postimplantation human development. These include: geometrically organized human ES cells that resemble early stages of gastrulation (Warmflash, Sorre, Etoc, Siggia, & Brivanlou, 2014), human pluripotent stem cell-derived cysts that form a proamniotic cavity (Shahbazi et al., 2017; Taniguchi et al., 2015), human ES cell-derived postimplantation amniotic sac-like structures (Shao et al., 2017) and embryonic organoids or gastruloids (Turner et al., 2016, 2017). In the case of human ES cell-derived in vitro models, they reconstruct some aspects of early human embryogenesis including the generation of what resembles a proamniotic cavity and asymmetric squamous epithelial structures, suggesting that they will be complementary to the continued direct study of human embryos. Further refinement of these models provides an exciting opportunity to accurately recreate the shape and cellular organization of human embryos shortly after implantation and to study their critical interaction with extraembryonic cells. Moreover, complementary detailed characterization in organisms that more closely resemble the human, such as marmosets and cynomolgus monkeys (Boroviak et al., 2015; Nakamura et al., 2016), will further inform this critical stage of embryogenesis. In addition to human ES cell-derived models, trophoblast spheroids generated from primary human trophoblast cells or cytotrophoblast cells have been differentiated from human ES cells (Weimar, Post Uiterweer, Teklenburg, Heijnen, & Macklon, 2013). These are then cocultured with uterine endometrial explants, monocultures of endometrial epithelial cell lines, or on multiple layers of endometrial and stromal cells seeded in an extracellular matrix. Although some of these implantation substrates can exhibit characteristics of the uterine implantation niche following stimulation with ovarian hormones, such as decidualization, it is unclear to what extent they accurately mimic the complex in vivo uterine environment. Consequently, they may only provide insights into selected parts of the implantation process, rather than a complete model. In addition, these conditions have not thus far been shown to enable embryo culture beyond 9 dpf (Carver et al., 2003; Teklenburg & Macklon, 2009).

Intriguingly, two recent studies cultured human embryos to the equivalent of 13 dpf in vitro in the absence of any maternal tissues

(Deglincerti et al., 2016; Shahbazi et al., 2016), using culture conditions refined using mouse embryos (Bedzhov, Leung, Bialecka, & Zernicka-Goetz, 2014; Bedzhov & Zernicka-Goetz, 2014). When plated, human embryos preferentially attached via the polar TE (Deglincerti et al., 2016) and, notably, were capable of self-organization even in the absence of maternal tissue. They showed a clear distinction between OCT4-expressing Epi and GATA6-expressing PE, and epiblast cells acquired apico-basal polarity by 8–9 dpf (Deglincerti et al., 2016; Shahbazi et al., 2016). PAR6B expression was also observed, suggesting that polarization of TE had also occurred (Shahbazi et al., 2016). Attached embryos also formed putative amniotic and yolk sac cavities, though these could not be maintained. Both studies noted that the morphological characteristics they observed in stages of in vitro attachment up to 12 dpf were reminiscent of those described in the Carnegie series (Hertig et al., 1956; O'Rahilly & Muller, 1987). This suggests that this culture system may provide a useful means of interrogating this window of development.

Though both studies concluded their cultures prior to 14 dpf, upholding current ethical conventions, this demonstration of prolonged in vitro culture has opened up discussion around a potential review of the 14-day limit (Hyun et al., 2017; Pera, 2017). However, in vitro attached embryos at 14 dpf were less comparable to equivalent Carnegie stages Deglincerti et al., 2016). It would be useful to further refine the existing system (e.g., by coaxing the trophectoderm-derived cells to grow away from the embryo to mimic trophoblast invasion and encouraging these cells or other in vitro cultured cells to facilitate nutrient and gas exchange) and to more broadly characterize the mechanisms behind the morphogenetic events.

4. CONCLUSIONS

The human embryo represents an invaluable resource for study, despite the limited availability of embryos donated to research and a relatively short time window permitted for study. Although the mouse has long been used as a model organism, the emerging differences between mouse and human in early development highlight the importance of further investigations in a human context. Future comparisons with a wider variety of model organisms will no doubt vastly increase our understanding of the foundations of early development including conserved and divergent mechanisms, especially in tandem with the growing body of research from a human embryonic context.

A number of fundamental questions remain to be addressed. What drives the key morphological events and initiates gene expression during human embryo development? How is lineage actually specified in the human embryo? What are the relevant signaling pathways, and how do they influence these events? In addition, there remains a crucial requirement for available cell populations in vitro that are most reflective of their associated Epi, TE, or PE lineage counterpart in vivo. The demonstrable consistency in signaling requirements and gene expression between mouse ES cells and the preimplantation Epi (Boroviak et al., 2014) allows mouse ES cells to be used as a suitable model to test regulatory mechanisms. However, as existing human ES cells are somewhat distinct from the Epi, and representative human TE or PE cell lines are still being developed or refined, this presents difficulties in exploring the basic biology of this developmental stage on a large scale. Exploiting gene expression datasets compiled during early human development may identify lineage specification factors or signaling pathways that could inform further attempts to isolate these cell types. Similarly, further interrogating both preimplantation embryos and postimplantation in vitro culture systems using emerging technologies will shed light on this important stage of early development.

ACKNOWELDGEMENTS

S.E.W and K.K.N are supported by the Francis Crick Institute which receives its core funding from Cancer Research UK (FC001120), the UK Medical Research Council (FC001120), and the Wellcome Trust (FC001120).

REFERENCES

Acosta, A., Elberger, L., Borghi, M., Calamera, J. C., Chemes, H., Doncel, G. F., et al. (1999). Endometrial dating and determination of the window of implantation in healthy fertile women. *Fertility and Sterility*, *73*, 788–798.

Aflatoonian, B., Ruban, L., Shamsuddin, S., Baker, D., Andrews, P., & Moore, H. (2010). Generation of Sheffield (Shef) human embryonic stem cell lines using a microdrop culture system. *In Vitro Cellular & Developmental Biology. Animal*, *46*, 236–241.

Ahringer, J. (2003). Control of cell polarity and mitotic spindle positioning in animal cells. *Current Opinion in Cell Biology*, *15*, 73–81.

Aiken, C. E., Swoboda, P. P., Skepper, J. N., & Johnson, M. H. (2004). The direct measurement of embryogenic volume and nucleo-cytoplasmic ratio during mouse pre-implantation development. *Reproduction*, *128*, 527–535.

Ajduk, A., & Zernicka-Goetz, M. (2016). Polarity and cell division orientation in the cleavage embryo: From worm to human. *Molecular Human Reproduction*, *22*, 691–703.

Alarcon, V. B. (2010). Cell polarity regulator PARD6B is essential for trophectoderm formation in the preimplantation mouse embryo. *Biology of Reproduction*, *83*, 347–358.

Alikani, M. (2005). Epithelial cadherin distribution in abnormal human pre-implantation embryos. *Human Reproduction*, *20*, 3369–3375.

Amit, M., Carpenter, M. K., Inokuma, M. S., Chiu, C.-P., Harris, C. P., Waknitz, M. A., et al. (2000). Clonally derived human embryonic stem cell lines maintain pluripotency and proliferative potential for prolonged periods of culture. *Developmental Biology*, *227*, 271–278.

Amita, M., Adachi, K., Alexenko, A. P., Sinha, S., Schust, D. J., Schulz, L. C., et al. (2013). Complete and unidirectional conversion of human embryonic stem cells to trophoblast by BMP4. *Proceedings of the National Academy of Sciences of the United States of America*, *110*, E1212–E1221.

Anani, S., Bhat, S., Honma-Yamanaka, N., Krawchuk, D., & Yamanaka, Y. (2014). Initiation of Hippo signaling is linked to polarity rather than to cell position in the pre-implantation mouse embryo. *Development*, *141*, 2813–2824.

Aplin, J. D., & Ruane, P. T. (2017). Embryo-epithelium interactions during implantation at a glance. *Journal of Cell Science*, *130*, 15–22.

Apparao, K. B. C., Murray, M. J., Fritz, M. A., Meyer, W. R., Chambers, A. F., Truong, P. R., et al. (2001). Osteopontin and its receptor avb3 integrin are coexpressed in the human endometrium during the menstrual cycle but regulated differentially. *Journal of Clinical Endocrinology and Metabolism*, *86*, 4991–5000.

Arman, E., Haffner-Krausz, R., Chen, Y., Heath, J. K., & Lonai, P. (1998). Targeted disruption of fibroblast growth factor (FGF) receptor 2 suggests a role for FGF signaling in pregastrulation mammalian development. *Proceedings of the National Academy of Sciences of the United States of America*, *95*, 5082–5087.

Artus, J., Piliszek, A., & Hadjantonakis, A. K. (2011). The primitive endoderm lineage of the mouse blastocyst: Sequential transcription factor activation and regulation of differentiation by Sox17. *Developmental Biology*, *350*, 393–404.

Bedzhov, I., Leung, C. Y., Bialecka, M., & Zernicka-Goetz, M. (2014). In vitro culture of mouse blastocysts beyond the implantation stages. *Nature Protocols*, *9*, 2732–2739.

Bedzhov, I., & Zernicka-Goetz, M. (2014). Self-organizing properties of mouse pluripotent cells initiate morphogenesis upon implantation. *Cell*, *156*, 1032–1044.

Benchetrit, H., Herman, S., van Wietmarschen, N., Wu, T., Makedonski, K., Maoz, N., et al. (2015). Extensive nuclear reprogramming underlies lineage conversion into functional trophoblast stem-like cells. *Cell Stem Cell*, *17*, 543–556.

Bendall, S. C., Stewart, M. H., Menendez, P., George, D., Vijayaragavan, K., Werbowetski-Ogilvie, T., et al. (2007). IGF and FGF cooperatively establish the regulatory stem cell niche of pluripotent human cells in vitro. *Nature*, *448*, 1015–1021.

Berg, D. K., Smith, C. S., Pearton, D. J., Wells, D. N., Broadhurst, R., Donnison, M., et al. (2011). Trophectoderm lineage determination in cattle. *Developmental Cell*, *20*(2), 244–255. https://doi.org/10.1016/j.devcel.2011.01.003.

Bernardo, A. S., Faial, T., Gardner, L., Niakan, K. K., Ortmann, D., Senner, C. E., et al. (2011). BRACHYURY and CDX2 mediate BMP-induced differentiation of human and mouse pluripotent stem cells into embryonic and extraembryonic lineages. *Cell Stem Cell*, *9*, 144–155.

Besser, D. (2004). Expression of nodal, lefty-A, and lefty-B in undifferentiated human embryonic stem cells requires activation of Smad2/3*. *Journal of Biological Chemistry*, *279*, 45076–45084.

Bessonnard, S., De Mot, L., Gonze, D., Barriol, M., Dennis, C., Goldbeter, A., et al. (2014). Gata6, Nanog and Erk signaling control cell fate in the inner cell mass through a tristable regulatory network. *Development*, *141*, 3637–3648.

Biadasiewicz, K., Sonderegger, S., Haslinger, P., Haider, S., Saleh, L., Fiala, C., et al. (2011). Transcription factor AP-2alpha promotes EGF-dependent invasion of human trophoblast. *Endocrinology*, *152*, 1458–1469.

Blakeley, P., Fogarty, N. M., Del Valle, I., Wamaitha, S. E., Hu, T. X., Elder, K., et al. (2015). Defining the three cell lineages of the human blastocyst by single-cell RNA-seq. *Development*, *142*, 3613.

Bleil, J. D., & Wassarman, P. M. (1980). Structure and function of the zona pellucida: Identification and characterization of the proteins of the mouse oocyte's zona pellucida. *Developmental Biology*, *76*, 185–202.

Bloor, D., Wilson, Y., Kibschull, M., Traub, O., Leese, H. J., Winterhager, E., et al. (2004). Expression of connexins in human preimplantation embryos in vitro. *Reproductive Biology and Endocrinology*, *2*, 25.

Bongso, A., Fong, C.-Y., Ng, S.-C., & Ratnam, S. (1994). Isolation and culture of inner cell mass cells from human blastocysts. *Human Reproduction*, *9*, 2110–2117.

Borensztein, M., Syx, L., Ancelin, K., Diabangouaya, P., Picard, C., Liu, T., et al. (2017). Xist-dependent imprinted X inactivation and the early developmental consequences of its failure. *Nature Structural & Molecular Biology*, *24*(3), 226–233. https://doi.org/10.1038/nsmb.3365.

Boroviak, T., Loos, R., Bertone, P., Smith, A., & Nichols, J. (2014). The ability of inner-cell-mass cells to self-renew as embryonic stem cells is acquired following epiblast specification. *Nature Cell Biology*, *16*, 516–528.

Boroviak, T., Loos, R., Lombard, P., Okahara, J., Behr, R., Sasaki, E., et al. (2015). Lineage-specific profiling delineates the emergence and progression of naive pluripotency in mammalian embryogenesis. *Developmental Cell*, *35*, 366–382.

Bou, G., Liu, S., Sun, M., Zhu, J., Xue, B., Guo, J., et al. (2017). CDX2 is essential for cell proliferation and polarity in porcine blastocysts. *Development*, *144*(7), 1296–1306. https://doi.org/10.1242/dev.141085.

Braude, P., Bolton, V., & Moore, S. (1988). Human gene expression first occurs between the four- and eight-cell stages of preimplantation development. *Nature*, *332*, 459–461.

Briggs, S. F., Dominguez, A. A., Chavez, S. L., & Reijo Pera, R. A. (2015). Single-cell XIST expression in human preimplantation embryos and newly reprogrammed female induced pluripotent stem cells. *Stem Cells*, *33*, 1771–1781.

Brockdorff, N., Ashworth, A., Kay, G. F., McCabe, V. M., Norris, D. P., Cooper, P. J., et al. (1992). The product of the mouse Xist gene is a 15 kb inactive X-specific transcript containing no conserved ORF and located in the nucleus. *Cell*, *71*, 515–526.

Brosens, J. J., Salker, M. S., Teklenburg, G., Nautiyal, J., Salter, S., Lucas, E. S., et al. (2014). Uterine selection of human embryos at implantation. *Scientific Reports*, *4*, 3894.

Brown, C. J., Hendrich, B. D., Rupert, J. L., Lafrenière, R. G., Xing, Y., Lawrence, J., et al. (1992). The human XIST gene: Analysis of a 17 kb inactive X-specific RNA that contains conserved repeats and is highly localized within the nucleus. *Cell*, *71*, 527–542.

Brown, S., Teo, A., Pauklin, S., Hannan, N., Cho, C. H., Lim, B., et al. (2011). Activin/Nodal signaling controls divergent transcriptional networks in human embryonic stem cells and in endoderm progenitors. *Stem Cells*, *29*, 1176–1185.

Budnik, B., Levy, E., & Slavov, N. (2017). Mass-spectrometry of single mammalian cells quantifies proteome heterogeneity during cell differentiation. *bioRxiv*. https://doi.org/10.1101/102681.

Burkart, A. D., Xiong, B., Baibakov, B., Jimenez-Movilla, M., & Dean, J. (2012). Ovastacin, a cortical granule protease, cleaves ZP2 in the zona pellucida to prevent polyspermy. *Journal of Cell Biology*, *197*, 37–44.

Calarco, P. G., & Epstein, C. J. (1973). Cell surface changes during preimplantation development in the mouse. *Developmental Biology*, *32*, 208–213.

Cao, Z., Carey, T. S., Ganguly, A., Wilson, C. A., Paul, S., & Knott, J. G. (2015). Transcription factor AP-2gamma induces early Cdx2 expression and represses HIPPO signaling to specify the trophectoderm lineage. *Development*, *142*, 1606–1615.

Carter, A. M., Enders, A. C., & Pijnenborg, R. (2015). The role of invasive trophoblast in implantation and placentation of primates. *Philosophical Transactions of the Royal Society of London. Series B, Biological Sciences*, *370*, 20140070.

Carver, J., Martin, K., Spyropoulou, I., Barlow, D. H., Sargent, I. L., & Mardon, H. J. (2003). An in-vitro model for stromal invasion during implantation of the human blastocyst. *Human Reproduction*, *18*, 283–290.

Cauffman, G., De Rycke, M., Sermon, K., Liebaers, I., & Van de Velde, H. (2009). Markers that define stemness in ESC are unable to identify the totipotent cells in human preimplantation embryos. *Human Reproduction*, *24*, 63–70.

Chambers, I., Silva, J., Colby, D., Nichols, J., Nijmeijer, B., Robertson, M., et al. (2007). Nanog safeguards pluripotency and mediates germline development. *Nature*, *450*, 1230–1234.

Chan, Y. S., Goke, J., Ng, J. H., Lu, X., Gonzales, K. A., Tan, C. P., et al. (2013). Induction of a human pluripotent state with distinct regulatory circuitry that resembles preimplantation epiblast. *Cell Stem Cell, 13*, 663–675.

Chazaud, C., Yamanaka, Y., Pawson, T., & Rossant, J. (2006). Early lineage segregation between epiblast and primitive endoderm in mouse blastocysts through the Grb2-MAPK pathway. *Developmental Cell, 10*, 615–624.

Chen, A. E., Egli, D., Niakan, K., Deng, J., Akutsu, H., Yamaki, M., et al. (2009). Optimal timing of inner cell mass isolation increases the efficiency of human embryonic stem cell derivation and allows generation of sibling cell lines. *Cell Stem Cell, 4*, 103–106.

Chen, G., Gulbranson, D. R., Hou, Z., Bolin, J. M., Ruotti, V., Probasco, M. D., et al. (2011). Chemically defined conditions for human iPSC derivation and culture. *Nature Methods, 8*, 424–429.

Cheng, A. M., Saxton, T. M., Sakai, R., Kulkarni, S., Mbamalu, G., Vogel, W., et al. (1998). Mammalian Grb2 regulates multiple steps in embryonic development and malignant transformation. *Cell, 95*, 793–803.

Cho, L. T., Wamaitha, S. E., Tsai, I. J., Artus, J., Sherwood, R. I., Pedersen, R. A., et al. (2012). Conversion from mouse embryonic to extra-embryonic endoderm stem cells reveals distinct differentiation capacities of pluripotent stem cell states. *Development, 139*, 2866–2877.

Chobotova, K., Spyropoulou, I., Carver, J., Manek, S., Heath, J. K., Gullick, W. J., et al. (2002). Heparin-binding epidermal growth factor and its receptor ErbB4 mediate implantation of the human blastocyst. *Mechanisms of Development, 119*, 137–144.

Choi, I., Carey, T. S., Wilson, C. A., & Knott, J. G. (2012). Transcription factor AP-2gamma is a core regulator of tight junction biogenesis and cavity formation during mouse early embryogenesis. *Development, 139*, 4623–4632.

Clift, D., & Schuh, M. (2013). Restarting life: Fertilization and the transition from meiosis to mitosis. *Nature Reviews Molecular Cell Biology, 14*, 549–562.

Cockburn, K., Biechele, S., Garner, J., & Rossant, J. (2013). The Hippo pathway member Nf2 is required for inner cell mass specification. *Current Biology, 23*, 1195–1201.

Cockburn, K., & Rossant, J. (2010). Making the blastocyst: Lessons from the mouse. *The Journal of Clinical Investigation, 120*, 995–1003.

Collier, A. J., Panula, S. P., Schell, J. P., Chovanec, P., Plaza Reyes, A., Petropoulos, S., et al. (2017). Comprehensive cell surface protein profiling identifies specific markers of human naive and primed pluripotent states. *Cell Stem Cell, 20*, 874–890.

Coucouvanis, E., & Martin, G. R. (1995). Signals for death and survival: A two-step mechanism for cavitation in the vertebrate embryo. *Cell, 83*, 279–287. https://doi.org/10.1016/0092-8674(95)90169-8.

Cowan, C., Klimanskaya, I., McMahon, J., Atienza, J., Witmeyer, J., Zucker, J. P., et al. (2004). Derivation of embryonic stem-cell lines from human blastocysts. *New England Journal of Medicine, 350*, 1353–1356.

Daniels, R., Zuccotti, M., Kinis, T., Serhal, P., & Monk, M. (1997). XIST expression in human oocytes and preimplantation embryos. *The American Journal of Human Genetics, 61*, 33–39.

Das, P., Ezashi, T., Schulz, L. C., Westfall, S. D., Livingston, K. A., & Roberts, R. M. (2007). Effects of fgf2 and oxygen in the bmp4-driven differentiation of trophoblast from human embryonic stem cells. *Stem Cell Research, 1*, 61–74.

De Iaco, A., Planet, E., Coluccio, A., Verp, S., Duc, J., & Trono, D. (2017). DUX-family transcription factors regulate zygotic genome activation in placental mammals. *Nature Genetics, 49*, 941–945.

De Paepe, C., Cauffman, G., Verloes, A., Sterckx, J., Devroey, P., Tournaye, H., et al. (2013). Human trophectoderm cells are not yet committed. *Human Reproduction, 28*, 740–749.

Deglincerti, A., Croft, G. F., Pietila, L. N., Zernicka-Goetz, M., Siggia, E. D., & Brivanlou, A. H. (2016). Self-organization of the in vitro attached human embryo. *Nature*, *533*, 251–254.

Delhanty, J. D., & Handyside, A. H. (1995). The origin of genetic defects in the human and their detection in the preimplantation embryo. *Human Reproduction Update*, *1*, 201–215.

Dobson, A. T., Raja, R., Abeyta, M. J., Taylor, T., Shen, S., Haqq, C., et al. (2004). The unique transcriptome through day 3 of human preimplantation development. *Human Molecular Genetics*, *13*, 1461–1470.

Drukker, M., Tang, C., Ardehali, R., Rinkevich, Y., Seita, J., Lee, A. S., et al. (2012). Isolation of primitive endoderm, mesoderm, vascular endothelial and trophoblast progenitors from human pluripotent stem cells. *Nature Biotechnology*, *30*, 531–542.

Ducibella, T., Albertini, D. F., Andersen, E., & Biggers, J. D. (1975). The preimplantation mammalian embryo: Characterization of intercellular junctions and their appearance during development. *Developmental Biology*, *45*, 231–250.

Ducibella, T., Ukena, T., Karnovsky, M., & Andersen, E. (1977). Changes in cell surface and cortical cytoplasmic organization during early embryogenesis in the preimplantation mouse embryo. *Journal of Cell Biology*, *74*, 153–167.

Duncan, F. E., Que, E. L., Zhang, N., Feinberg, E. C., O'Halloran, T. V., & Woodruff, T. K. (2016). The zinc spark is an inorganic signature of human egg activation. *Scientific Reports*, *6*, 24737.

Ebner, T., Maurer, M., Shebl, O., Moser, M., Mayer, R. B., Duba, H. C., et al. (2012). Planar embryos have poor prognosis in terms of blastocyst formation and implantation. *Reproductive Biomedicine Online*, *25*, 267–272.

Edwards, R. G., Purdy, J. M., Steptoe, P. C., & Walters, D. E. (1981). The growth of human preimplantation embryos in vitro. *American Journal of Obstetrics and Gynecology*, *141*, 408–416.

Enders, A. C., Schlafke, S., & Hendrickx, A. G. (1986). Differentiation of the embryonic disc, amnion and yolk sac in the rhesus monkey. *The American Journal of Anatomy*, *177*, 161–185.

Erlebacher, A., Price, K. A., & Glimcher, L. H. (2004). Maintenance of mouse trophoblast stem cell proliferation by TGF-β/activin. *Developmental Biology*, *275*, 158–169.

Etemad-Moghadam, B., Guo, S., & Kemphues, K. J. (1995). Asymmetrically distributed PAR-3 protein contributes to cell polarity and spindle alignment in early *C. elegans* embryos. *Cell*, *83*, 743–752.

Falco, G., Lee, S. L., Stanghellini, I., Bassey, U. C., Hamatani, T., & Ko, M. S. (2007). Zscan4: A novel gene expressed exclusively in late 2-cell embryos and embryonic stem cells. *Developmental Biology*, *307*, 539–550.

Feldman, B., Poueymirou, W., Papaioannou, V. E., DeChiara, T. M., & Goldfarb, M. (1995). Requirement of FGF-4 for postimplantation mouse development. *Science*, *267*, 246–249.

Fierro-Gonzalez, J. C., White, M. D., Silva, J. C., & Plachta, N. (2013). Cadherin-dependent filopodia control preimplantation embryo compaction. *Nature Cell Biology*, *15*, 1424–1433.

Flach, G., Johnson, M. H., Braude, P. R., Taylor, R. A., & Bolton, V. N. (1982). The transition from maternal to embryonic control in the 2-cell mouse embryo. *The EMBO Journal*, *1*, 681–686.

Fleming, T. P., McConnel, J., Johnson, M. H., & Stevenson, B. R. (1989). Development of tight junctions de novo in the mouse early embryo: Control of assembly of the tight junction-specific protein ZO-1. *Journal of Cell Biology*, *108*, 1407–1418.

Fogarty, N. M. E., McCarthy, A., Snijders, K. E., Powell, B. E., Kubikova, N., Blakeley, P., et al. (2017). Genome editing reveals a role for OCT4 in human embryogenesis. *Nature*, *550*, 67–73.

Fort, A., Hashimoto, K., Yamada, D., Salimullah, M., Keya, C. A., Saxena, A., et al. (2014). Deep transcriptome profiling of mammalian stem cells supports a regulatory role for retrotransposons in pluripotency maintenance. *Nature Genetics*, *46*, 558–566.

Frankenberg, S., Gerbe, F., Bessonnard, S., Belville, C., Pouchin, P., Bardot, O., et al. (2011). Primitive endoderm differentiates via a three-step mechanism involving Nanog and RTK signaling. *Developmental Cell*, *21*, 1005–1013.

French, D. B., Sabanegh, E. S., Jr., Goldfarb, J., & Desai, N. (2010). Does severe teratozoospermia affect blastocyst formation, live birth rate, and other clinical outcome parameters in ICSI cycles? *Fertility and Sterility*, *93*, 1097–1103.

Friedli, M., & Trono, D. (2015). The developmental control of transposable elements and the evolution of higher species. *Annual Review of Cell and Developmental Biology*, *31*, 429–451.

Frum, T., Halbisen, M. A., Wang, C., Amiri, H., Robson, P., & Ralston, A. (2013). Oct4 cell-autonomously promotes primitive endoderm development in the mouse blastocyst. *Developmental Cell*, *25*, 610–622.

Fujikura, J., Yamato, E., Yonemura, S., Hosoda, K., Masui, S., Nakao, K., et al. (2002). Differentiation of embryonic stem cells is induced by GATA factors. *Genes and Development*, *16*, 784–789.

Fukuda, M. N., Sato, T., Nakayama, J., Klier, G., Mikami, M., Aoki, D., et al. (1995). Trophinin and tastin, a novel cell adhesion molecule complex with potential involvement in embryo implantation. *Genes and Development*, *9*, 1199–1210.

Gafni, O., Weinberger, L., Mansour, A. A., Manor, Y. S., Chomsky, E., Ben-Yosef, D., et al. (2013). Derivation of novel human ground state naive pluripotent stem cells. *Nature*, *504*, 282–286.

Galan, A., Montaner, D., Poo, M. E., Valbuena, D., Ruiz, V., Aguilar, C., et al. (2010). Functional genomics of 5- to 8-cell stage human embryos by blastomere single-cell cDNA analysis. *PLoS One*, *5*, e13615.

Gardner, R. L. (1985). Clonal analysis of early mammalian development. *Philosophical Transactions of the Royal Society of London. Series B, Biological Sciences*, *312*, 163–178.

Gardner, D. K., Lane, M., Stevens, J., Schlenker, T., & Schoolcraft, W. B. (2000). Blastocyst score affects implantation and pregnancy outcome: Towards a single blastocyst transfer. *Fertility and Sterility*, *73*, 1155–1158.

Gardner, R. L., Papaioannou, V. E., & Barton, S. C. (1973). Origin of the ectoplacental cone and secondary giant cells in mouse blastocysts reconstituted from isolated trophoblast and inner cell mass. *Journal of Embryology and Experimental Morphology*, *30*, 561–572.

Gardner, R. L., & Rossant, J. (1979). Investigation of the fate of 4-5 day post-coitum mouse inner cell mass cells by blastocyst injection. *Journal of Embryology and Experimental Morphology*, *52*, 141–152.

Gellersen, B., & Brosens, J. J. (2014). Cyclic decidualization of the human endometrium in reproductive health and failure. *Endocrine Reviews*, *35*, 851–905.

Genbacev, O., Donne, M., Kapidzic, M., Gormley, M., Lamb, J., Gilmore, J., et al. (2011). Establishment of human trophoblast progenitor cell lines from the chorion. *Stem Cells*, *29*, 1427–1436.

Ghassemifar, M. R. (2003). Gene expression regulating epithelial intercellular junction biogenesis during human blastocyst development in vitro. *Molecular Human Reproduction*, *9*, 245–252.

Goke, J., Lu, X., Chan, Y. S., Ng, H. H., Ly, L. H., Sachs, F., et al. (2015). Dynamic transcription of distinct classes of endogenous retroviral elements marks specific populations of early human embryonic cells. *Cell Stem Cell*, *16*, 135–141.

Grabarek, J. B., Zyzynska, K., Saiz, N., Piliszek, A., Frankenberg, S., Nichols, J., et al. (2012). Differential plasticity of epiblast and primitive endoderm precursors within the ICM of the early mouse embryo. *Development*, *139*, 129–139.

Granier, C., Gurchenkov, V., Perea-Gomez, A., Camus, A., Ott, S., Papanayotou, C., et al. (2011). Nodal cis-regulatory elements reveal epiblast and primitive endoderm heterogeneity in the peri-implantation mouse embryo. *Developmental Biology*, *349*, 350–362.

Greber, B., Coulon, P., Zhang, M., Moritz, S., Frank, S., Muller-Molina, A. J., et al. (2011). FGF signalling inhibits neural induction in human embryonic stem cells. *EMBO Journal*, *30*, 4874–4884.

Greber, B., Lehrach, H., & Adjaye, J. (2007). Fibroblast growth factor 2 modulates transforming growth factor beta signaling in mouse embryonic fibroblasts and human ESCs (hESCs) to support hESC self-renewal. *Stem Cells*, *25*, 455–464.

Greber, B., Wu, G., Bernemann, C., Joo, J. Y., Han, D. W., Ko, K., et al. (2010). Conserved and divergent roles of FGF signaling in mouse epiblast stem cells and human embryonic stem cells. *Cell Stem Cell*, *6*, 215–226.

Grow, E. J., Flynn, R. A., Chavez, S. L., Bayless, N. L., Wossidlo, M., Wesche, D. J., et al. (2015). Intrinsic retroviral reactivation in human preimplantation embryos and pluripotent cells. *Nature*, *522*, 221–225.

Gualtieri, R., Santella, L., & Dale, B. (1992). Tight junctions and cavitation in the human pre-embryo. *Molecular Reproduction and Development*, *32*, 81–87.

Guo, G., Huss, M., Tong, G. Q., Wang, C., Li Sun, L., Clarke, N. D., et al. (2010). Resolution of cell fate decisions revealed by single-cell gene expression analysis from zygote to blastocyst. *Developmental Cell*, *18*, 675–685.

Guo, S., & Kemphues, K. J. (1995). par-1, a gene required for establishing polarity in *C. elegans* embryos, encodes a putative Ser/Thr kinase that is asymmetrically distributed. *Cell*, *81*, 611–620.

Guo, G., von Meyenn, F., Santos, F., Chen, Y., Reik, W., Bertone, P., et al. (2016). Naive pluripotent stem cells derived directly from isolated cells of the human inner cell mass. *Stem Cell Reports*, *6*, 437–446.

Guo, H., Zhu, P., Yan, L., Li, R., Hu, B., Lian, Y., et al. (2014). The DNA methylation landscape of human early embryos. *Nature*, *511*, 606–610.

Gupta, S. K., Bhandari, B., Shrestha, A., Biswal, B. K., Palaniappan, C., Malhotra, S. S., et al. (2012). Mammalian zona pellucida glycoproteins: Structure and function during fertilization. *Cell and Tissue Research*, *349*, 665–678.

Hamada, H., Okae, H., Toh, H., Chiba, H., Hiura, H., Shirane, K., et al. (2016). Allele-specific methylome and transcriptome analysis reveals widespread imprinting in the human placenta. *The American Journal of Human Genetics*, *99*, 1045–1058.

Hamatani, T., Carter, M. G., Sharov, A. A., & Ko, M. S. H. (2004). Dynamics of global gene expression changes during mouse preimplantation development. *Developmental Cell*, *6*, 117–131.

Hardy, K., Warner, A., Winston, R. M. L., & Becker, D. L. (1996). Expression of intercellular junctions during preimplantation development of the human embryo. *Molecular Human Reproduction*, *2*, 621–632.

Harrison, R. H., Kuo, H.-C., Scriven, P. N., Handyside, A. H., & Ogilvie, C. M. (2000). Lack of cell cycle checkpoints in human cleavage stage embryos revealed by a clonal pattern of chromosomal mosaicism analysed by sequential multicolour FISH. *Zygote*, *8*, 217–224.

Hassold, T., & Hunt, P. (2001). To err (meiotically) is human: The genesis of human aneuploidy. *Nature Reviews Genetics*, *2*, 280–291.

Heath, J. R., Ribas, A., & Mischel, P. S. (2015). Single-cell analysis tools for drug discovery and development. *Nature Reviews Drug Discovery*, *15*, 204–216.

Hemberger, M., Udayashankar, R., Tesar, P., Moore, H., & Burton, G. J. (2010). ELF5-enforced transcriptional networks define an epigenetically regulated trophoblast stem cell compartment in the human placenta. *Human Molecular Genetics*, *19*, 2456–2467.

Hendrickson, P. G., Dorais, J. A., Grow, E. J., Whiddon, J. L., Lim, J. W., Wike, C. L., et al. (2017). Conserved roles of mouse DUX and human DUX4 in activating cleavage-stage genes and MERVL/HERVL retrotransposons. *Nature Genetics*, *49*, 925–934.

Hertig, A. T., Rock, J., & Adams, E. C. (1956). A description of 34 human ova within the first 17 days of development. *The American Journal of Anatomy*, *98*, 435–493.

Hertig, A. T., Rock, J., Adams, E. C., & Mulligan, W. J. (1954). On the preimplantation stages of the human ovum-A description of 4 normal and 4 abnormal specimens ranging from the 2nd to the 5th day of development. *Contributions to Embryology*, *35*, 201–220.

Hirate, Y., Cockburn, K., Rossant, J., & Sasaki, H. (2012). Tead4 is constitutively nuclear, while nuclear vs. cytoplasmic Yap distribution is regulated in preimplantation mouse embryos. *Proceedings of the National Academy of Sciences of the United States of America*, *109*, E3389–E3390.

Hirate, Y., Hirahara, S., Inoue, K., Suzuki, A., Alarcon, V. B., Akimoto, K., et al. (2013). Polarity-dependent distribution of angiomotin localizes Hippo signaling in preimplantation embryos. *Current Biology*, *23*, 1181–1194.

Huang, K., Maruyama, T., & Fan, G. (2014). The naive state of human pluripotent stem cells: A synthesis of stem cell and preimplantation embryo transcriptome analyses. *Cell Stem Cell*, *15*, 410–415.

Hurov, J. B., Watkins, J. L., & Piwnica-Worms, H. (2004). Atypical PKC phosphorylates PAR-1 kinases to regulate localization and activity. *Current Biology*, *14*, 736–741.

Hyslop, L., Stojkovic, M., Armstrong, L., Walter, T., Stojkovic, P., Przyborski, S., et al. (2005). Downregulation of NANOG induces differentiation of human embryonic stem cells to extraembryonic lineages. *Stem Cells*, *23*, 1035–1043.

Hyun, I., Wilkerson, A., & Johnston, J. (2017). Revisit the 14-day rule. *Nature*, *533*, 169–171.

Inman, G. J., Nicolas, F. J., Callahan, J. F., Harling, J. D., Gaster, L. M., Reith, A. D., et al. (2002). SB-431542 is a potent and specific inhibitor of transforming growth factor-superfamily type I activin receptor-like kinase (ALK) receptors ALK4, ALK5, and ALK7. *Molecular Pharmacology*, *62*, 65–74.

Iwata, K., Yumoto, K., Sugishima, M., Mizoguchi, C., Kai, Y., Iba, Y., et al. (2014). Analysis of compaction initiation in human embryos by using time-lapse cinematography. *Journal of Assisted Reproduction and Genetics*, *31*, 421–426.

James, D., Levine, A. J., Besser, D., & Brivanlou, A. H. (2005). TGFβ/activin/nodal signaling is necessary for the maintenance of pluripotency in human embryonic stem cells. *Development*, *132*, 1273–1282.

Jinek, M., Chylinski, K., Fonfara, I., Hauer, M., Doudna, J. A., & Charpentier, E. (2012). A programmable dual-RNA-guided DNA endonuclease in adaptive bacterial immunity. *Science*, *337*, 816–821.

Joberty, G., Petersen, C., Gao, L., & Macara, I. G. (2000). The cell-polarity protein Par6 links Par3 and atypical protein kinase C to Cdc42. *Nature Cell Biology*, *2*, 531–539.

Johnson, G. A., Burghardt, R. C., Bazer, F. W., & Spencer, T. E. (2003). Osteopontin: Roles in implantation and placentation. *Biology of Reproduction*, *69*, 1458–1471.

Johnson, M. H., & Ziomek, C. A. (1981a). The foundation of two distinct cell lineages within the mouse morula. *Cell*, *24*, 71–80.

Johnson, M. H., & Ziomek, C. A. (1981b). Induction of polarity in mouse 8-cell blastomeres: Specificity, geometry, and stability. *Journal of Cell Biology*, *91*, 303–308.

Jouhilahti, E. M., Madissoon, E., Vesterlund, L., Tohonen, V., Krjutskov, K., Plaza Reyes, A., et al. (2016). The human PRD-like homeobox gene LEUTX has a central role in embryo genome activation. *Development*, *143*, 3459–3469.

Jukam, D., Shariati, S. A. M., & Skotheim, J. M. (2017). Zygotic genome activation in vertebrates. *Developmental Cell*, *42*, 316–332.

Kalantry, S., Purushothaman, S., Bowen, R. B., Starmer, J., & Magnuson, T. (2009). Evidence of Xist RNA-independent initiation of mouse imprinted X-chromosome inactivation. *Nature*, *460*, 647–651. https://doi.org/10.1038/nature08161.

Kim, A. M., Bernhardt, M. L., Kong, B. Y., Ahn, R. W., Vogt, S., Woodruff, T. K., et al. (2011). Zinc sparks are triggered by fertilization and facilitate cell cycle resumption in mammalian eggs. *ACS Chemical Biology*, *6*, 716–723.

Kim, A. M., Vogt, S., O'Halloran, T. V., & Woodruff, T. K. (2010). Zinc availability regulates exit from meiosis in maturing mammalian oocytes. *Nature Chemical Biology*, *6*, 674–681.

Ko, M. S. (2016). Zygotic genome activation revisited: Looking through the expression and function of Zscan4. *Current Topics in Developmental Biology*, *120*, 103–124.

Kobold, S., Guhr, A., Kurtz, A., & Loser, P. (2015). Human embryonic and induced pluripotent stem cell research trends: Complementation and diversification of the field. *Stem Cell Reports*, *4*, 914–925.

Kono, K., Tamashiro, D. A., & Alarcon, V. B. (2014). Inhibition of RHO-ROCK signaling enhances ICM and suppresses TE characteristics through activation of Hippo signaling in the mouse blastocyst. *Developmental Biology*, *394*, 142–155.

Korotkevich, E., Niwayama, R., Courtois, A., Friese, S., Berger, N., Buchholz, F., et al. (2017). The apical domain is required and sufficient for the first lineage segregation in the mouse embryo. *Developmental Cell*, *40*, 235-247 e237.

Kubaczka, C., Senner, C., Arauzo-Bravo, M. J., Sharma, N., Kuckenberg, P., Becker, A., et al. (2014). Derivation and maintenance of murine trophoblast stem cells under defined conditions. *Stem Cell Reports*, *2*, 232–242.

Kubaczka, C., Senner, C. E., Cierlitza, M., Araúzo-Bravo, M. J., Kuckenberg, P., Peitz, M., et al. (2015). Direct induction of trophoblast stem cells from murine fibroblasts. *Cell Stem Cell*, *17*, 557–568.

Kuckenberg, P., Buhl, S., Woynecki, T., van Fürden, B., Tolkunova, E., Seiffe, F., et al. (2010). The transcription factor TCFAP2C/AP-2γ cooperates with CDX2 to maintain trophectoderm formation. *Molecular and Cellular Biology*, *30*, 3310–3320.

Kuijk, E. W., van Tol, L. T., Van de Velde, H., Wubbolts, R., Welling, M., Geijsen, N., et al. (2012). The roles of FGF and MAP kinase signaling in the segregation of the epiblast and hypoblast cell lineages in bovine and human embryos. *Development*, *139*, 871–882.

Kunarso, G., Chia, N. Y., Jeyakani, J., Hwang, C., Lu, X., Chan, Y. S., et al. (2010). Transposable elements have rewired the core regulatory network of human embryonic stem cells. *Nature Genetics*, *42*, 631–634.

Kunath, T., Arnaud, D., Uy, G. D., Okamoto, I., Chureau, C., Yamanaka, Y., et al. (2005). Imprinted X-inactivation in extra-embryonic endoderm cell lines from mouse blastocysts. *Development*, *132*, 1649–1661.

Kunath, T., Yamanaka, Y., Detmar, J., MacPhee, D., Caniggia, I., Rossant, J., et al. (2014). Developmental differences in the expression of FGF receptors between human and mouse embryos. *Placenta*, *35*, 1079–1088.

Kurimoto, K., Yabuta, Y., Ohinata, Y., Ono, Y., Uno, K. D., Yamada, R. G., et al. (2006). An improved single-cell cDNA amplification method for efficient high-density oligonucleotide microarray analysis. *Nucleic Acids Research*, *34*, e42.

Lanner, F., & Rossant, J. (2010). The role of FGF/Erk signaling in pluripotent cells. *Development*, *137*, 3351–3360.

Latos, P. A., Sienerth, A. R., Murray, A., Senner, C. E., Muto, M., Ikawa, M., et al. (2015). Elf5-centered transcription factor hub controls trophoblast stem cell self-renewal and differentiation through stoichiometry-sensitive shifts in target gene networks. *Genes & Development*, *29*, 2435–2448.

Le Bin, G. C., Munoz-Descalzo, S., Kurowski, A., Leitch, H., Lou, X., Mansfield, W., et al. (2014). Oct4 is required for lineage priming in the developing inner cell mass of the mouse blastocyst. *Development*, *141*, 1001–1010.

Leach, R. E., Khalifa, R., Ramirez, N. D., Das, S. K., Wang, J., Dey, S. K., et al. (1999). Multiple roles for heparin-binding epidermal growth factor-like growth factor are suggested by its cell specific expression during the human endometrial cycle and early placentation. *Journal of Clinical Endocrinology and Metabolism*, *84*, 3355–3363.

Lee, K. Y., & DeMayo, F. J. (2004). Animal models of implantation. *Reproduction*, *128*, 679–695.

Lee, H. J., Hore, T. A., & Reik, W. (2014). Reprogramming the methylome: Erasing memory and creating diversity. *Cell Stem Cell*, *14*, 710–719.

Lefievre, L., Conner, S. J., Salpekar, A., Olufowobi, O., Ashton, P., Pavlovic, B., et al. (2004). Four zona pellucida glycoproteins are expressed in the human. *Human Reproduction*, *19*, 1580–1586.

Lessey, B. A., Damjanovich, L., Coutifaris, C., Castelbaum, A., Albelda, S. M., & Buck, C. A. (1992). Integrin adhesion molecules in the human endometrium. Correlation with the normal and abnormal menstrual cycle. *Journal of Clinical Investigation*, *90*, 188–195.

Levenstein, M. E., Ludwig, T. E., Xu, R. H., Llanas, R. A., VanDenHeuvel-Kramer, K., Manning, D., et al. (2006). Basic fibroblast growth factor support of human embryonic stem cell self-renewal. *Stem Cells*, *24*, 568–574.

Li, Y., Moretto-Zita, M., Soncin, F., Wakeland, A., Wolfe, L., Leon-Garcia, S., et al. (2013). BMP4-directed trophoblast differentiation of human embryonic stem cells is mediated through a DeltaNp63+ cytotrophoblast stem cell state. *Development*, *140*, 3965–3976.

Lin, D., Edwards, A. S., Fawcett, J. P., Mbamalu, G., Scott, J. D., & Pawson, T. (2000). A mammalian PAR-3-PAR-6 complex implicated in Cdc42/Rac1 and aPKC signalling and cell polarity. *Nature Cell Biology*, *2*, 540–547.

Loser, P., Schirm, J., Guhr, A., Wobus, A. M., & Kurtz, A. (2010). Human embryonic stem cell lines and their use in international research. *Stem Cells*, *28*, 240–246.

Lu, Z., Carter, A. C., & Chang, H. Y. (2017). Mechanistic insights in X-chromosome inactivation. *Philosophical Transactions of the Royal Society of London. Series B, Biological Sciences*, *372*, 20160356.

Luckett, W. P. (1975). The development of primordial and definitive amniotic cavities in early rhesus monkey and human embryos. *The American Journal of Anatomy*, *144*, 149–168.

Ludwig, T. E., Levenstein, M. E., Jones, J. M., Berggren, W. T., Mitchen, E. R., Frane, J. L., et al. (2006). Derivation of human embryonic stem cells in defined conditions. *Nature Biotechnology*, *24*, 185–187.

Lyon, M. F. (1961). Gene action in the X-chromosome of the mouse (Mus musculus L.). *Nature*, *190*, 372–373.

Madissoon, E., Jouhilahti, E. M., Vesterlund, L., Töhönen, V., Krjutškov, K., Petropoulous, S., et al. (2016). Characterization and target genes of nine human PRD-like homeobox domain genes expressed exclusively in early embryos. *Scientific Reports*, *6*, 28995. https://doi.org/10.1038/srep28995.

Magnuson, T., Demsey, A., & Stackpole, C. W. (1977). Characterization of intercellular junctions in the preimplantation mouse embryo by freeze-fracture and thin-section electron microscopy. *Developmental Biology*, *61*, 252–261.

Maitre, J. L., Turlier, H., Illukkumbura, R., Eismann, B., Niwayama, R., Nedelec, F., et al. (2016). Asymmetric division of contractile domains couples cell positioning and fate specification. *Nature*, *536*, 344–348.

Mak, W., Nesterova, T. B., de Napoles, M., Appanah, R., Yamanaka, S., Otte, A. P., et al. (2004). Reactivation of the paternal X chromosome in early mouse embryos. *Science*, *303*, 666–669.

Meseguer, M., Aplin, J. D., Caballero-Campo, P., O'Connor, J. E., Martin, J. C., Remohi, J., et al. (2001). Human endometrial mucin MUC1 is up-regulated by progesterone and down-regulated in vitro by the human blastocyst. *Biology of Reproduction*, *64*, 590–601.

Mitalipova, M., Calhoun, J., Shin, S., Wininger, D., Schulz, T., Noggle, S., et al. (2003). Human embryonic stem cell lines derived from discarded embryos. *Stem Cells*, *21*, 521–526.

Mitsui, K., Tokuzawa, Y., Itoh, N., Segawa, K., Murakami, M., Takahashi, K., et al. (2003). The homeoprotein nanog is required for maintenance of pluripotency in mouse epiblast and ES cells. *Cell*, *113*, 631–642.

Miyazaki, S., & Ito, M. (2006). Calcium signals for egg activation in mammals. *Journal of Pharmacological Sciences*, *100*, 545–552.

Moreira de Mello, J. C., de Araújo, É. S. S., Stabellini, R., Fraga, A. M., de Souza, J. E. S., Sumita, D. R., et al. (2010). Random X inactivation and extensive mosaicism in human placenta revealed by analysis of allele-specific gene expression along the X chromosome. *PLoS One*, *5*, e10947.

Moreira de Mello, J. C., Fernandes, G. R., Vibranovski, M. D., & Pereira, L. V. (2017). Early X chromosome inactivation during human preimplantation development revealed by single-cell RNA-sequencing. *Scientific Reports*, 7, 10794.

Nakamura, T., Okamoto, I., Sasaki, K., Yabuta, Y., Iwatani, C., Tsuchiya, H., et al. (2016). A developmental coordinate of pluripotency among mice, monkeys and humans. *Nature*, *537*, 57–62.

Namekawa, S. H., Payer, B., Huynh, K. D., Jaenisch, R., & Lee, J. T. (2010). Two-step imprinted X inactivation: repeat versus genic silencing in the mouse. *Molecular and Cellular Biology*, *30*(13), 3187–3205. https://doi.org/10.1128/MCB.00227-10.

Ng, R. K., Dean, W., Dawson, C., Lucifero, D., Madeja, Z., Reik, W., et al. (2008). Epigenetic restriction of embryonic cell lineage fate by methylation of Elf5. *Nature Cell Biology*, *10*, 1280–1290.

Niakan, K. K., & Eggan, K. (2013). Analysis of human embryos from zygote to blastocyst reveals distinct gene expression patterns relative to the mouse. *Developmental Biology*, *375*, 54–64.

Niakan, K. K., Han, J., Pedersen, R. A., Simon, C., & Pera, R. A. (2012). Human preimplantation embryo development. *Development*, *139*, 829–841.

Nichols, J., Branko, Z., Anastassiadis, K., Niwa, H., Kiewe-Nebenius, D., Chambers, I., et al. (1998). Formation of pluripotent stem cells in the mammalian embryo depends on the POU transcription factor Oct4. *Cell*, *95*, 379–391.

Nichols, J., Silva, J., Roode, M., & Smith, A. (2009). Suppression of Erk signalling promotes ground state pluripotency in the mouse embryo. *Development*, *136*, 3215–3222.

Nikas, G., Ao, A., Winston, R. M. L., & Handyside, A. H. (1996). Compaction and surface polarity in the human embryo in vitro. *Biology of Reproduction*, *55*, 32–37.

Nishioka, N., Inoue, K., Adachi, K., Kiyonari, H., Ota, M., Ralston, A., et al. (2009). The Hippo signaling pathway components Lats and Yap pattern Tead4 activity to distinguish mouse trophectoderm from inner cell mass. *Developmental Cell*, *16*, 398–410.

Nishioka, N., Yamamoto, S., Kiyonari, H., Sato, H., Sawada, A., Ota, M., et al. (2008). Tead4 is required for specification of trophectoderm in pre-implantation mouse embryos. *Mechanisms of Development*, *125*, 270–283.

Noli, L., Capalbo, A., Ogilvie, C., Khalaf, Y., & Ilic, D. (2015). Discordant growth of monozygotic twins starts at the blastocyst stage: A case study. *Stem Cell Reports*, *5*, 946–953.

Norwitz, E. R., Schust, D. J., & Fisher, S. J. (2001a). Implantation and the survival of early pregnancy. *New England Journal of Medicine, 345*, 1400–1408.
Norwitz, E. R., Schust, D. J., & Fisher, S. J. (2001b). Mechanisms of disease—Implantation and the survival of early pregnancy. *New England Journal of Medicine, 345*(19), 1400–1408. https://doi.org/10.1056/NEJMra000763.
Okabe, M. (2013). The cell biology of mammalian fertilization. *Development, 140*, 4471–4479.
Okae, H., Toh, H., Sato, T., Hiura, H., Takahashi, S., Shirane, K., et al. (2018). Derivation of human trophoblast stem cells. *Cell Stem Cell, 22*(1), 50.e6–63.e6. https://doi.org/10.1016/j.stem.2017.11.004.
Okamoto, I., Otte, A. P., Allis, C. D., Reinberg, D., & Heard, E. (2004). Epigenetic dynamics of imprinted X inactivation during early mouse development. *Science, 303*, 644–649.
Okamoto, I., Patrat, C., Thepot, D., Peynot, N., Fauque, P., Daniel, N., et al. (2011). Eutherian mammals use diverse strategies to initiate X-chromosome inactivation during development. *Nature, 472*, 370–374.
O'Rahilly, R., & Muller, F. (1987). *Developmental stages in human embryos*. Carnegie Institution of Washington.
Park, I. H., Lerou, P. H., Zhao, R., Huo, H., & Daley, G. Q. (2008). Generation of human-induced pluripotent stem cells. *Nature Protocols, 3*, 1180–1186.
Pastor, W. A., Chen, D., Liu, W., Kim, R., Sahakyan, A., Lukianchikov, A., et al. (2016). Naive human pluripotent cells feature a methylation landscape devoid of blastocyst or germline memory. *Cell Stem Cell, 18*, 323–329.
Pauken, C. M., & Capco, D. G. (1999). Regulation of cell adhesion during embryonic compaction of mammalian embryos: Roles for PKC and beta-catenin. *Molecular Reproduction and Development, 54*, 135–144.
Penny, G. D., Kay, G. F., Sheardown, S. A., Rastan, S., & Brockdorff, N. (1996). Requirement for Xist in X chromosome inactivation. *Nature, 379*, 131–137.
Pera, M. F. (2017). Human embryo research and the 14-day rule. *Development, 144*, 1923–1925.
Perera, A., Eisen, D., Wagner, M., Laube, S. K., Künzel, A. F., Koch, S., et al. (2015). TET3 is recruited by REST for context-specific hydroxymethylation and induction of gene expression. *Cell Reports, 11*, 283–294.
Petropoulos, S., Edsgard, D., Reinius, B., Deng, Q., Panula, S. P., Codeluppi, S., et al. (2016). Single-cell RNA-Seq reveals lineage and X chromosome dynamics in human preimplantation embryos. *Cell, 165*, 1012–1026.
Plusa, B., Piliszek, A., Frankenberg, S., Artus, J., & Hadjantonakis, A. K. (2008). Distinct sequential cell behaviours direct primitive endoderm formation in the mouse blastocyst. *Development, 135*, 3081–3091.
Que, E. L., Bleher, R., Duncan, F. E., Kong, B. Y., Gleber, S. C., Vogt, S., et al. (2015). Quantitative mapping of zinc fluxes in the mammalian egg reveals the origin of fertilization-induced zinc sparks. *Nature Chemistry, 7*, 130–139.
Que, E. L., Duncan, F. E., Bayer, A. R., Philips, S. J., Roth, E. W., Bleher, R., et al. (2017). Zinc sparks induce physiochemical changes in the egg zona pellucida that prevent polyspermy. *Integrative Biology: Quantitative Biosciences From Nano to Macro, 9*, 135–144.
Quesada, V., Sanchez, L. M., Alvarez, J., & Lopez-Otin, C. (2004). Identification and characterization of human and mouse ovastacin: A novel metalloproteinase similar to hatching enzymes from arthropods, birds, amphibians, and fish. *Journal of Biological Chemistry, 279*, 26627–26634.
Ralston, A., Cox, B. J., Nishioka, N., Sasaki, H., Chea, E., Rugg-Gunn, P., et al. (2010). Gata3 regulates trophoblast development downstream of Tead4 and in parallel to Cdx2. *Development, 137*, 395–403.

Ralston, A., & Rossant, J. (2008). Cdx2 acts downstream of cell polarization to cell-autonomously promote trophectoderm fate in the early mouse embryo. *Developmental Biology*, *313*, 614–629.
Ray, P. F., Winston, R. M., & Handyside, A. H. (1997). XIST expression from the maternal X chromosome in human male preimplantation embryos at the blastocyst stage. *Human Molecular Genetics*, *6*, 1323–1327.
Rayon, T., Menchero, S., Nieto, A., Xenopoulos, P., Crespo, M., Cockburn, K., et al. (2014). Notch and Hippo converge on Cdx2 to specify the trophectoderm lineage in the mouse blastocyst. *Developmental Cell*, *30*, 410–422.
Reeve, W. J. D., & Ziomek, C. A. (1981). Distribution of microvilli on dissociated blastomeres from mouse embryos: Evidence for surface polarization at compaction. *Journal of Embryology and Experimental Morphology*, *62*, 339–350.
Reubinoff, B., Pera, M. F., Fong, C.-Y., Trounson, A., & Bongso, A. (2000). Embryonic stem cell lines from human blastocysts: Somatic differentiation in vitro. *Nature Biotechnology*, *18*, 399–404.
Robertson, S. A., Sjoblom, C., Jasper, M. J., Norman, R. J., & Seamark, R. F. (2001). Granulocyte-macrophage colony-stimulating factor promotes glucose transport and blastomere viability in murine preimplantation embryos. *Biology of Reproduction*, *64*, 1206–1215.
Roode, M., Blair, K., Snell, P., Elder, K., Marchant, S., Smith, A., et al. (2012). Human hypoblast formation is not dependent on FGF signalling. *Developmental Biology*, *361*, 358–363.
Rossant, J. (2014). Mouse and human blastocyst-derived stem cells: Vive les differences. *Development*, *142*, 9.
Rugg-Gunn, P. J., Cox, B. J., Lanner, F., Sharma, P., Ignatchenko, V., McDonald, A. C., et al. (2012). Cell-surface proteomics identifies lineage-specific markers of embryo-derived stem cells. *Developmental Cell*, *22*, 887–901.
Russ, A. P., Wattler, S., Colledge, W. H., Aparicio, S. A., Carlton, M. B., Pearce, J. J., et al. (2000). Eomesodermin is required for mouse trophoblast development and mesoderm formation. *Nature*, *404*, 95–99.
Schrode, N., Saiz, N., Di Talia, S., & Hadjantonakis, A. K. (2014). GATA6 levels modulate primitive endoderm cell fate choice and timing in the mouse blastocyst. *Developmental Cell*, *29*, 454–467.
Seguin, C. A., Draper, J. S., Nagy, A., & Rossant, J. (2008). Establishment of endoderm progenitors by SOX transcription factor expression in human embryonic stem cells. *Cell Stem Cell*, *3*, 182–195.
Sengoku, K., Tamate, K., Horikawa, M., Takaoka, Y., Ishikawa, M., & Dukelow, W. R. (1995). Plasma membrane block to polyspermy in human oocytes and preimplantation embryos. *Journal of Reproduction and Fertility*, *105*, 85–90.
Sengoku, K., Tamate, K., Takaoka, Y., Horikawa, M., Katsutoshi, G., Okada, R., et al. (1999). Requirement of sperm-oocyte plasma membrane fusion for establishment of the plasma membrane block to polyspermy in human pronuclear oocytes. *Molecular Reproduction and Development*, *52*, 183–188.
Shahbazi, M. N., Jedrusik, A., Vuoristo, S., Recher, G., Hupalowska, A., Bolton, V., et al. (2016). Self-organization of the human embryo in the absence of maternal tissues. *Nature Cell Biology*, *18*, 700–708.
Shahbazi, M. N., Scialdone, A., Skorupska, N., Weberling, A., Recher, G., Zhu, M., et al. (2017). Pluripotent state transitions coordinate morphogenesis in mouse and human embryos. *Nature*, *552*(7684), 239–243. https://doi.org/10.1038/nature24675.
Shao, Y., Taniguchi, K., Townshend, R. F., Miki, T., Gumucio, D. L., Fu, J., et al. (2017). A pluripotent stem cell-based model for post-implantation human amniotic sac development. *Nature Communications*, *8*(1), 208. https://doi.org/10.1038/s41467-017-00236-w.

Shimosato, D., Shiki, M., & Niwa, H. (2007). Extra-embryonic endoderm cells derived from ES cells induced by GATA factors acquire the character of XEN cells. *BMC Developmental Biology*, *7*, 80.

Shirayoshi, Y., Okada, T. S., & Takeichi, M. (1983). The calcium-dependent cell-cell adhesion system regulates inner cell mass formation and cell surface polarization in early mouse development. *Cell*, *35*, 631–638.

Sjoblom, C., Wikland, M., & Robertson, S. A. (2002). Granulocyte-macrophage colony-stimulating factor (GM-CSF) acts independently of the beta common subunit of the GM-CSF receptor to prevent inner cell mass apoptosis in human embryos. *Biology of Reproduction*, *67*, 1817–1823.

Smith, Z. D., Chan, M. M., Humm, K. C., Karnik, R., Mekhoubad, S., Regev, A., et al. (2014). DNA methylation dynamics of the human preimplantation embryo. *Nature*, *511*, 611–615.

Spanos, S., Becker, D. L., Winston, R. M. L., & Hardy, K. (2000). Anti-apoptotic action of insulin-like growth factor-I during human preimplantation embryo development. *Biology of Reproduction*, *63*, 1413–1420.

Stephenson, R. O., Yamanaka, Y., & Rossant, J. (2010). Disorganized epithelial polarity and excess trophectoderm cell fate in preimplantation embryos lacking E-cadherin. *Development*, *137*, 3383–3391.

Steptoe, P. C., Edwards, R. G., & Purdy, J. M. (1971). Human blastocysts grown in culture. *Nature*, *229*, 132–133.

Strumpf, D., Mao, C.-A., Yamanaka, Y., Ralston, A., Chawengsaksophak, K., Beck, F., et al. (2005). Cdx2 is required for correct cell fate specification and differentiation of trophectoderm in the mouse blastocyst. *Development*, *132*, 2093–2102.

Suemori, H., Yasuchika, K., Hasegawa, K., Fujioka, T., Tsuneyoshi, N., & Nakatsuji, N. (2006). Efficient establishment of human embryonic stem cell lines and long-term maintenance with stable karyotype by enzymatic bulk passage. *Biochemical and Biophysical Research Communications*, *345*, 926–932.

Sugihara, K., Sugiyama, D., Byrne, J., Wolf, D. P., Lowitz, K. P., Kobayashi, Y., et al. (2007). Trophoblast cell activation by trophinin ligation is implicaed in human embryo implantation. *Proceedings of the National Academy of Sciences of the United States of America*, *104*, 3799–3804.

Suzuki, A., Hirata, M., Kamimura, K., Maniwa, R., Yamanaka, T., Mizuno, K., et al. (2004). aPKC acts upstream of PAR-1b in both the establishment and maintenance of mammalian epithelial polarity. *Current Biology*, *14*, 1425–1435.

Tahiliani, M., Koh, K. P., Shen, Y., Pastor, W. A., Bandukwala, H., Brudno, Y., et al. (2009). Conversion of 5-methylcytosine to 5-hydroxymethylcytosine in mammalian DNA by MLL partner TET1. *Science*, *324*, 930–935.

Takagi, N., & Sasaki, M. (1975). Preferential inactivation of the paternally derived X chromosome in the extraembryonic membranes of the mouse. *Nature*, *256*, 640–642.

Takahashi, K., & Yamanaka, S. (2006). Induction of pluripotent stem cells from mouse embryonic and adult fibroblast cultures by defined factors. *Cell*, *126*, 663–676.

Takashima, Y., Guo, G., Loos, R., Nichols, J., Ficz, G., Krueger, F., et al. (2014). Resetting transcription factor control circuitry toward ground-state pluripotency in human. *Cell*, *158*, 1254–1269.

Tanaka, S., Kunath, T., Hadjantonakis, A. K., Nagy, A., & Rossant, J. (1998). Promotion of trophoblast stem cell proliferation by FGF4. *Science*, *282*, 2072–2075.

Taniguchi, K., Shao, Y., Townshend, R. F., Tsai, Y.-H., DeLong, C. J., Lopez, S. A., et al. (2015). Lumen formation is an intrinsic property of isolated human pluripotent stem cells. *Stem Cell Reports*, *5*(6), 954–962. https://doi.org/10.1016/j.stemcr.2015.10.015.

Taylor, D. M., Ray, P., Ao, A., Winston, R. M., & Handyside, A. H. (1997). Paternal transcripts for glucose-6-phosphate dehydrogenase and adenosine deaminase are first detectable in the human preimplantation embryo at the three- to four-cell stage. *Molecular Reproduction and Development*, *48*, 442–448.

Teklenburg, G., & Macklon, N. S. (2009). Review: In vitro models for the study of early human embryo-endometrium interactions. *Reproductive Sciences*, *16*, 811–818.

Theunissen, T. W., Friedli, M., He, Y., Planet, E., O'Neill, R. C., Markoulaki, S., et al. (2016). Molecular criteria for defining the naive human pluripotent state. *Cell Stem Cell*, *19*, 502–515.

Theunissen, T. W., Powell, B. E., Wang, H., Mitalipova, M., Faddah, D. A., Reddy, J., et al. (2014). Systematic identification of culture conditions for induction and maintenance of naive human pluripotency. *Cell Stem Cell*, *15*, 471–487.

Thomson, J. A., Itskovitz-Eldor, J., Shapiro, S. S., Waknitz, M. A., Swiergiel, J. J., Marshall, V. S., et al. (1998). Embryonic stem cell lines derived from human blastocysts. *Science*, *282*, 1145–1147.

Thomson, J. A., Kalishman, J., Golos, T. G., Durning, M., Harris, C. P., Becker, R. A., et al. (1995). Isolation of a primate embryonic stem cell line. *Proceedings of the National Academy of Sciences of the United States of America*, *92*, 7844–7848.

Thomson, J. A., Kalishman, J., Golos, T. G., Durning, M., Harris, C. P., & Hearn, J. P. (1996). Pluripotent cell lines derived from common marmoset (Callithrix jacchus) blastocysts. *Biology of Reproduction*, *55*, 254–259.

Tohonen, V., Katayama, S., Vesterlund, L., Jouhilahti, E. M., Sheikhi, M., Madissoon, E., et al. (2015). Novel PRD-like homeodomain transcription factors and retrotransposon elements in early human development. *Nature Communications*, *6*, 8207.

Töhönen, V., Katayama, S., Vesterlund, L., Sheikhi, M., Antonsson, L., Filippini-Cattaneo, G., et al. (2017). Transcription activation of early human development suggests Dux4 as an embryonic regulator. *bioRxiv*. https://doi.org/10.1101/123208.

Tukiainen, T., Villani, A.-C., Yen, A., Rivas, M. A., Marshall, J. L., Satija, R., et al. (2017). Landscape of X chromosome inactivation across human tissues. *Nature*, *550*, 244–248.

Turner, D. A., Girgin, M., Alonso-Crisostomo, L., Trivedi, V., Baillie-Johnson, P., Glodowski, C. R., et al. (2017). Anteroposterior polarity and elongation in the absence of extra-embryonic tissues and of spatially localised signalling in gastruloids: Mammalianembryonic organoids. *Development*, *144*(21), 3894–3906. https://doi.org/10.1242/dev.150391.

Turner, D. A., Glodowski, C. R., Luz, A.-C., Baillie-Johnson, P., Hayward, P. C., Collignon, J., et al. (2016). Interactions between Nodal and Wnt signalling drive robust symmetry breaking and axial organisation in *Gastruloids* (embryonic organoids). *bioRxiv*. https://doi.org/10.1101/051722.

Uy, G. D., Downs, K. M., & Gardner, R. L. (2002). Inhibition of trophoblast stem cell potential in chorionic ectoderm coincides with occlusion of the ectoplacental cavity in the mouse. *Development*, *129*, 3913–3924.

Valamehr, B., Robinson, M., Abujarour, R., Rezner, B., Vranceanu, F., Le, T., et al. (2014). Platform for induction and maintenance of transgene-free hiPSCs resembling ground state pluripotent stem cells. *Stem Cell Reports*, *2*, 366–381.

Vallier, L., Alexander, M., & Pedersen, R. A. (2005). Activin/Nodal and FGF pathways cooperate to maintain pluripotency of human embryonic stem cells. *Journal of Cell Science*, *118*, 4495–4509.

Vallier, L., Touboul, T., Brown, S., Cho, C., Bilican, B., Alexander, M., et al. (2009). Signaling pathways controlling pluripotency and early cell fate decisions of human induced pluripotent stem cells. *Stem Cells*, *27*, 2655–2666.

Vallot, C., Patrat, C., Collier, A. J., Huret, C., Casanova, M., Liyakat Ali, T. M., et al. (2017). XACT noncoding RNA competes with XIST in the control of X chromosome activity during human early development. *Cell Stem Cell*, *20*, 102–111.

van den Berg, I. M., Laven, J. S. E., Stevens, M., Jonkers, I., Galjaard, R.-J., Gribnau, J., et al. (2009). X chromosome inactivation is initiated in human preimplantation embryos. *The American Journal of Human Genetics, 84*, 771–779.

Van der Jeught, M., O'Leary, T., Ghimire, S., Lierman, S., Duggal, G., Versieren, K., et al. (2013). The combination of inhibitors of FGF/MEK/Erk and GSK3b signaling increases the number of OCT3/4- and NANOG-positive cells in the human inner cell mass, but does not improve stem cell derivation. *Stem Cells and Development, 22*, 296–306.

Vanneste, E., Voet, T., Le Caignec, C., Ampe, M., Konings, P., Melotte, C., et al. (2009). Chromosome instability is common in human cleavage-stage embryos. *Nature Medicine, 15*, 577–583.

Vassena, R., Boue, S., Gonzalez-Roca, E., Aran, B., Auer, H., Veiga, A., et al. (2011). Waves of early transcriptional activation and pluripotency program initiation during human preimplantation development. *Development, 138*, 3699–3709.

Vestweber, D., Gossler, A., Boller, K., & Kemler, R. (1987). Expression and distribution of cell adhesion molecule uvomorulin in mouse preimplantation embryos. *Developmental Biology, 124*, 451–456.

Wakai, T., Vanderheyden, V., & Fissore, R. A. (2011). $Ca2+$ signaling during mammalian fertilization: Requirements, players, and adaptations. *Cold Spring Harbor Perspectives in Biology, 3*, 1–23.

Wamaitha, S. E., Del Valle, I., Cho, L. T., Wei, Y., Fogarty, N. M., Blakeley, P., et al. (2015). Gata6 potently initiates reprograming of pluripotent and differentiated cells to extraembryonic endoderm stem cells. Genes and. *Development, 29*, 1239–1255.

Wang, H., & Dey, S. K. (2006). Roadmap to embryo implantation: Clues from mouse models. *Nature Reviews Genetics, 7*, 185–199.

Wang, Z., Oron, E., Nelson, B., Razis, S., & Ivanova, N. (2012). Distinct lineage specification roles for NANOG, OCT4, and SOX2 in human embryonic stem cells. *Cell Stem Cell, 10*, 440–454.

Ware, C. B., Nelson, A. M., Mecham, B., Hesson, J., Zhou, W., Jonlin, E. C., et al. (2014). Derivation of naive human embryonic stem cells. *Proceedings of the National Academy of Sciences of the United States of America, 111*, 4484–4489.

Wassarman, P. M., Jovine, L., & Litscher, E. S. (2004). Mouse zona pellucida genes and glycoproteins. *Cytogenetic and Genome Research, 105*, 228–234.

Watanabe, T., Biggins, J. S., Tannan, N. B., & Srinivas, S. (2014). Limited predictive value of blastomere angle of division in trophectoderm and inner cell mass specification. *Development, 141*, 2279–2288.

Weimar, C. H., Post Uiterweer, E. D., Teklenburg, G., Heijnen, C. J., & Macklon, N. S. (2013). In-vitro model systems for the study of human embryo-endometrium interactions. *Reproductive Biomedicine Online, 27*, 461–476.

White, M. D., Bissiere, S., Alvarez, Y. D., & Plachta, N. (2016). Mouse embryo compaction. *Current Topics in Developmental Biology, 120*, 235–258.

White, M. D., & Plachta, N. (2015). How adhesion forms the early mammalian embryo. *Current Topics in Developmental Biology, 112*, 1–17.

Xie, D., Chen, C. C., Ptaszek, L. M., Xiao, S., Cao, X., Fang, F., et al. (2010). Rewirable gene regulatory networks in the preimplantation embryonic development of three mammalian species. *Genome Research, 20*, 804–815.

Xu, R.-H., Chen, X., Li, D. S., Li, R., Addicks, G. C., Glennon, C., et al. (2002). BMP4 initiates human embryonic stem cell differentiation to trophoblast. *Nature Biotechnology, 20*, 1261–1264.

Xu, C., Inokuma, M. S., Denham, J., Golds, K., Kundu, P., Gold, J. D., et al. (2001). Feeder-free growth of undifferentiated human embryonic stem cells. *Nature Biotechnology, 19*, 971–974.

Xue, Z., Huang, K., Cai, C., Cai, L., Jiang, C. Y., Feng, Y., et al. (2013). Genetic programs in human and mouse early embryos revealed by single-cell RNA sequencing. *Nature, 500*, 593–597.

Yagi, R., Kohn, M. J., Karavanova, I., Kaneko, K. J., Vullhorst, D., DePamphilis, M. L., et al. (2007). Transcription factor TEAD4 specifies the trophectoderm lineage at the beginning of mammalian development. *Development, 134*, 3827–3836.

Yamanaka, Y., Lanner, F., & Rossant, J. (2010). FGF signal-dependent segregation of primitive endoderm and epiblast in the mouse blastocyst. *Development, 137*, 715–724.

Yan, L., Yang, M., Guo, H., Yang, L., Wu, J., Li, R., et al. (2013). Single-cell RNA-Seq profiling of human preimplantation embryos and embryonic stem cells. *Nature Structural & Molecular Biology, 20*, 1131–1139.

Yang, F., Babak, T., Shendure, J., & Disteche, C. M. (2010). Global survey of escape from X inactivation by RNA-sequencing in mouse. *Genome Research, 20*, 614–622.

Yu, J., Vodyanik, M. A., Smuga-Otto, K., Antosiewicz-Bourget, J., Frane, J. L., Tian, S., et al. (2007). Induced pluripotent stem cell lines derived from human somatic cells. *Science, 318*, 1917–1920.

Zdravkovic, T., Nazor, K. L., Larocque, N., Gormley, M., Donne, M., Hunkapillar, N., et al. (2015). Human stem cells from single blastomeres reveal pathways of embryonic or trophoblast fate specification. *Development, 142*, 4010–4025.

Zhang, N., Duncan, F. E., Que, E. L., O'Halloran, T. V., & Woodruff, T. K. (2016). The fertilization-induced zinc spark is a novel biomarker of mouse embryo quality and early development. *Scientific Reports, 6*, 22772.

CHAPTER THIRTEEN

The Mitochondria and the Regulation of Cell Fitness During Early Mammalian Development

Ana Lima*,†,2, Jörg Burgstaller*,‡,2, Juan M. Sanchez-Nieto*,3, Tristan A. Rodríguez*,1

*British Heart Foundation Centre for Research Excellence, National Heart and Lung Institute, Imperial Centre for Translational and Experimental Medicine, Imperial College London, London, United Kingdom
†Cell Stress Group, MRC London Institute of Medical Sciences (LMS), London, United Kingdom
‡Biotechnology in Animal Production, Department for Agrobiotechnology, IFA Tulln, Tulln, Austria
[1]Corresponding author: e-mail address: tritstan.rodriguez@imperial.ac.uk

Contents

Abstract

From fertilization until the onset of gastrulation the early mammalian embryo undergoes a dramatic series of changes that converts a single fertilized cell into a remarkably complex organism. Much attention has been given to the molecular changes occurring during this process, but here we will review what is known about the changes affecting the mitochondria and how they impact on the energy metabolism and apoptotic response of the embryo. We will also focus on understanding what quality control mechanisms

[2] Equal contributing authors.
[3] Present address: Cell and Gene Therapy Catapult, 12th Floor Tower Wing, Guy's Hospital, Great Maze Pond, London SE1 9RT, United Kingdom.

Current Topics in Developmental Biology, Volume 128
ISSN 0070-2153
https://doi.org/10.1016/bs.ctdb.2017.10.012

ensure optimal mitochondrial activity in the embryo, and in this way provide an overview of the importance of the mitochondria in determining cell fitness during early mammalian development.

1. INTRODUCTION

The early stages of mammalian development comprise a fascinating series of changes that will convert, in a robust and highly reproducible manner, the newly fertilized zygote into an embryo that is patterned across its three axes: anterior–posterior, left–right, and dorso–ventral. These changes involve a substantial remodeling at all the cellular levels, from the transcriptional and epigenetic networks that control gene expression to the cytoskeletal organization that determines cell shape and the signaling pathways that govern cell interactions. The importance of these cellular aspects is comprehensively covered in other reviews in this issue, but here we would like to concentrate on one further set of changes that often go unnoticed but that are of fundamental importance in determining cell behavior, the changes affecting the mitochondria.

Mitochondria are cellular organelles that contain their own DNA and most likely originated from the endosymbiosis of an archaea and an α-proteobacteria and can be found in all eukaryotic organisms (reviewed in Baum & Baum, 2014). The evolutionary conservation of mitochondria is explained by their key function in determining cell fitness, as the many roles that they perform, from energy generation to the apoptotic response, ultimately determine the ability of the cell to thrive in its environment. Consequently, stringent quality controls also exist to ensure optimal mitochondrial performance. In this review we will give an overview of what is known about both the mitochondrial changes and the fitness selection pressures which act upon mitochondria during the early stages of mouse development. We will focus on the period between fertilization and early postimplantation stages of development (in mouse 6.5 days postcoitum, dpc), as this is the time frame when the first cell fate decisions and patterning events occur, and therefore when mitochondrial activity is most likely to help shape the cellular changes occurring during the onset of differentiation, from those that take place at the transcriptional and epigenetic level (Kojima, Tam, & Tam, 2014), to those involving the acceleration of the proliferation rate that occurs upon exit of pluripotency (Snow, 1977). Finally, although there is a wealth of literature emerging on the roles of mitochondria in stem cells, here we will concentrate on those studies that directly analyze the embryo.

2. MITOCHONDRIAL BIOGENESIS AND SHAPE IN THE EARLY EMBRYO

The biogenesis of the mitochondria in the embryo is a vastly coordinated process that begins even before fertilization. During oocyte maturation maternal mitochondria largely increase in numbers to compensate for the degradation of the paternal mitochondria that occurs shortly after fertilization (Jansen, 2000). The degradation of paternal mitochondria is mediated by ubiquitin labeling which targets these organelles for degradation within the proteasome or lysosomes (Sutovsky et al., 1999, 2000), although it has also been suggested to be a passive process resulting from the postfertilization elimination of sperm mitochondrial DNA (mtDNA) (Luo et al., 2013). This degradation is conserved as paternal mitochondria are undetectable by around the eight-cell stage in all mammalian embryos analyzed, including human, rhesus monkeys, cattle and mice (Cummins, Wakayama, & Yanagimachi, 1997). Interestingly, leakages of paternal mitochondria and mtDNA can occur in the rare eventuality of an interspecies cross—in this case the parental mtDNA can even be detected in the neonates (Kaneda et al., 1995) or transmitted to the next generation (Gyllensten, Wharton, Josefsson, & Wilson, 1991). This implies that some species-specific mechanism operates early in the embryogenesis to exclusively remove paternal mitochondria (reviewed by Sato & Sato, 2013), but so far the molecular basis of this phenomenon remains uncharacterized.

A second interesting feature of the early mouse embryo is the total number of mitochondria remains constant in the embryo from fertilization until blastocyst formation. An implication of this finding is that as long as the segregation of mitochondria is equal between blastomeres, each cleavage from the zygote to the blastocyst halves the number of mitochondria per cell (Van Blerkom, 2009). Consequently late preimplantation embryos from different mammalian species including mouse (Batten, Albertini, & Ducibella, 1987), pig (Spikings, Alderson, & John, 2007), cattle (Stojkovic et al., 2001), monkey (Squirrell, Schramm, Paprocki, Wokosin, & Bavister, 2003), and human (Wilding et al., 2001) contain reduced numbers of mitochondria per cell compared to the zygote. Given these observations it was proposed that the number and activity of the mitochondria in the oocyte correlate with the competence of this oocyte to generate an embryo after fertilization (reviewed in Van Blerkom, 2011). Although no precise calculation of mitochondrial mass has been performed for early postimplantation embryos, mtDNA copy number (Wai et al., 2010) (also see below for a more detailed discussion) and the expression of regulators of mitochondrial biogenesis (Ren et al., 2017) both increase at these stages when compared to preimplantation embryos, suggesting that mitochondrial mass increases upon implantation.

Importantly, the progression through preimplantation development can be predicted not only by the mitochondria numbers but also by their distribution within the cell. While homogeneously distributed mitochondria are found in normal cleaving embryos, the mitochondria from developmentally arrested embryos aggregate in perinuclear regions and in the pericortical regions of the cytoplasm, a pattern that is likely to be indicative of an underlying aberrant cytoskeletal and microtubular organization (Muggleton-Harris & Brown, 1988; Van Blerkom, 2009; Wilding et al., 2001). Another form of microzonation observed in mouse and human cleaving embryos is with respect to the membrane potential of the mitochondria ($\Delta\varphi m$). The $\Delta\varphi m$ is a readout of mitochondrial activity, and low mitochondrial activity in early embryos correlates with poor developmental potential (Wilding et al., 2001). In human and mouse cleaving embryos, mitochondria with low $\Delta\varphi m$ tend to aggregate in perinuclear regions, while highly polarized mitochondria localize in the periphery of the cytoplasm without neighbor cell contact (Van Blerkom, Davis, Mathwig, & Alexander, 2002). Furthermore, the disruption of this microzonation is detrimental to development, as a homogeneous distribution of mitochondria in the blastomeres (regardless of their polarization status) is also indicative of low development potential (Wilding et al., 2001). Further differences in $\Delta\varphi m$ are observed at the blastocyst stage, with the trophectoderm (TE) mitochondria being highly polarized and the mitochondria of the inner cell mass (ICM) showing a lower $\Delta\varphi m$ (Van Blerkom, 2009), but as will be discussed later this is due to the metabolic difference that exists between these cell types.

A final particularity of the mitochondria of the preimplantation embryo is their shape. Electron microscopy analysis of human and mouse early preimplantation mitochondria revealed that these are rounded and with a low density of cristae, which are folds in the inner membrane of the mitochondria that harbor proteins required for oxidative phosphorylation (OXPHOS) such as the ATP synthase and cytochromes. In humans as the blastocyst hatches the mitochondria elongate and fuse to form networks full of cristae, and in mouse this change takes place shortly after implantation (Shepard, Muffley, & Smith, 2000; Smith & Alcivar, 1993; Zhou et al., 2012; our unpublished observations). This process is thought to reflect the progressive maturation of the mitochondria occurring during early development.

3. MITOCHONDRIAL DNA AND SELECTION FOR OPTIMAL MITOCHONDRIAL PERFORMANCE

mtDNA is a multicopy genome that can be found within the mitochondria of most eukaryotic cells, and encodes for vital components of their bioenergetic machinery. Replication and transcription of mtDNA are tightly regulated, and synchronized with mitosis (Chatre & Ricchetti, 2013), but also "relaxed" (i.e., mtDNAs are chosen randomly for replication (Birky, 1994) and cells are able to vary the copy number to meet changes in OXPHOS energy requirements (Hood, 2001)). Recent discoveries showed that both replication and transcription of mtDNA are far more complex than previously thought (Agaronyan, Morozov, Anikin, & Temiakov, 2015; Reyes et al., 2013), and a detailed description is beyond the scope of this review (for detailed information, see Gustafsson, Falkenberg, & Larsson, 2016).

During embryogenesis mtDNA transcription starts at the same time as the zygotic genome activation (two-cell stage in mouse; four-/eight-cell stage in human; and morula stage in cattle (May-Panloup et al., 2005)). mtDNA copy number is thought to be regulated by the energetic demands of the cell, with high-performing tissues like muscle, heart, liver, and brain showing copy numbers of several thousand per cell (Kelly, Mahmud, McKenzie, Trounce, & St John, 2012; Miller, Rosenfeldt, Zhang, Linnane, & Nagley, 2003; Viscomi et al., 2009). Oocytes are the cells with the highest mtDNA copy number, with species-specific variation (mouse: 160,000 (Wai et al., 2010); human: 750,000 (Cotterill et al., 2013)). mtDNA copy number is important for fertility; oocytes retrieved from patients with ovarian insufficiency contain, on average, threefold fewer copies of mtDNA (Folmes, Ma, Mitalipov, & Terzic, 2016). While oocytes from mice are capable of developing to the blastocyst stage with as few as 4000 mtDNAs, the critical threshold for postimplantation development seems to be around 50,000 mtDNAs (Wai et al., 2010). As is the case for mitochondria, mtDNA is strictly maternally inherited as paternal mitochondria that reach the ooplasm during fertilization are digested (Al Rawi et al., 2011), with sperm mtDNA already starting to degrade within the sperm cell (Luo et al., 2013). This mechanism ensures that the embryo starts with a set of identical mtDNAs; however, as discussed above, very rarely this system can fail, mainly during interspecies fertilization (Gyllensten et al., 1991; Luo et al., 2013; Schwartz & Vissing, 2002; St John & Schatten, 2004).

After fertilization and until the early blastocyst stage although the total number of mtDNAs of the embryo stays constant, the number of mtDNA copies per cell is decreased with each embryonic cell division (Thundathil, Filion, & Smith, 2005). Nevertheless, the exact details of mtDNA replication and degradation during this time are species specific. In the mouse, a short period of mtDNA synthesis occurs shortly after fertilization and up to the two-cell stage (McConnell & Petrie, 2004), but as the mtDNA copy number of the embryo stays constant this synthesis must be accompanied by degradation (McConnell & Petrie, 2004). In pig and cattle mtDNA is diminished from the two-cell to the morula stage, with a marked increase at the blastocyst stage (May-Panloup et al., 2005; Spikings et al., 2007). In humans there seems to be a reduction between oocytes and the two-cell stage, with a sharp increase again at the blastocyst stage (Hashimoto et al., 2017). In cattle it was additionally found that embryos that are deprived of a considerable amount of their mtDNA at the one-cell stage (shortly after fertilization) were able to develop into blastocysts with normal mtDNA copy number, showing that the embryo is able to increase mtDNA replication under these specific conditions (Chiaratti et al., 2009; Ferreira et al., 2010). The reasons for preimplantation mtDNA synthesis and destruction are unclear, but destruction of paternal mtDNA (May-Panloup et al., 2005; McConnell & Petrie, 2004), selection of the "fittest" mtDNAs, and the bottleneck effect (see below) (Wai et al., 2010) have all been proposed as possible mechanisms.

At the blastocyst stage a large-scale increase in mtDNA replication and copy number occurs in TE cells as the mitochondria mature and increase OXPHOS activity. In contrast to this, cells from the ICM continue to dilute their mtDNA with each round of division until they finally reach the "mtDNA set point," i.e., the lowest level of mtDNA prior to differentiation (Kelly et al., 2012). The number of mtDNAs per cell is below 300 at this time, with estimates as low as around 30 in embryonic stem cells (ESCs) (Facucho-Oliveira, Alderson, Spikings, Egginton, & John, 2007). This set point is currently understood to be a mitochondrial checkpoint that helps to segregate an even amount of mtDNA between all cells once replication has started. Moreover in cells with lower mtDNA copy number, mutated mtDNAs can be selected against, as their relative effect on cell fitness is higher (Wai et al., 2010). Most interestingly, the mtDNA copy number at this point could also help to define the cells differentiation potential, as both in induced pluripotent stem cells (iPSCs) and cancer cells their differentiation status is correlated with their mtDNA copy number, with lower copy numbers indicating lower levels of differentiation. Importantly, differentiation stage and mtDNA copy number need to be exactly orchestrated, both by reaching the mtDNA set point and by subsequent mtDNA replication to increase mtDNA copy number, to allow the cell to switch between differentiation stages (e.g., in iPS cells; reviewed in John, 2016).

Later in development, after reaching the mitochondrial set point, the cell lineage dictates the mtDNA copy number: stem cells retain a low mtDNA copy number, seemingly to keep their stemness and high proliferation potential (Folmes et al., 2016). Cells that initiate lineage commitment generally increase their mtDNA copy number until they reach the fully differentiated state with a cell-specific mtDNA copy number, which mainly depends on the OXPHOS requirements of that specific cell type (Kelly et al., 2012). An important exception are mammalian erythrocytes, which lose both nuclear and mtDNA (Long, 2007).

Germ cells show a more complex development. Mouse primordial germ cells (PGCs) are laid aside shortly after implantation, around day 6.5 in embryonic development (Hayashi, de Sousa Lopes, & Surani, 2007) and initially maintain a low mtDNA copy number. In the male germ line the mtDNA copy number stays low, as each sperm cell contains only approximately 10–100 mtDNAs, and being that higher quality sperm cells have lower mtDNA copy numbers (Amaral, Ramalho-Santos, & St John, 2007; Wai et al., 2010). By the time of their specification, female PGCs contain between 100 and 1000 mtDNAs per cell, the exact number being debated (Cao et al., 2007, 2009; Cree et al., 2008; Johnston et al., 2015; Wai, Teoli, & Shoubridge, 2008). Female germ cells then increase their mtDNA copy number to around 5000 in the primary oocytes (Cao et al., 2007) and to 160,000 in mature oocytes (Wai et al., 2010).

Besides mtDNA copy number, an important part of mtDNA genetics is genomic integrity. Ideally, all copies of mtDNA within a cell are genetically identical (homoplasmy). However, the mutation rate of mtDNA is about 100 times higher than that of nuclear DNA (Khrapko et al., 1997). This is mainly caused by replication errors during cell division and constant turnover within cells, but also due to its location at the inner membrane where high concentrations of reactive oxygen species (ROS) are released during OXPHOS, damaging the DNA (Bratic & Larsson, 2013). While in somatic tissues damaged mtDNA may play an important part in aging (Bratic & Larsson, 2013), in the germ line such mutations can lead to (heritable) genetic diseases (Burgstaller, Johnston, & Poulton, 2015). As mtDNA rarely recombines (Fan, Lin, Potluri, Procaccio, & Wallace, 2012; Hagström, Freyer, Battersby, Stewart, & Larsson, 2014) and is strictly maternally inherited in mammals (Luo et al., 2013), at least two types of mechanisms protect the mtDNA from the accumulation of deleterious mutations (also known as Mueller's ratchet) (Wallace & Chalkia, 2013). These are the mtDNA bottleneck and mutation-specific mechanisms and both take place mainly at early stages of development.

The mitochondrial genetic bottleneck is one of the most important mechanisms ensuring mtDNA genetic homogeneity (Johnston et al., 2015). Due to the high mutation rate, mutated and unmutated mtDNA are frequently present in both somatic cells and the germ line (heteroplasmy) (Bratic & Larsson, 2013; Rebolledo-Jaramillo et al., 2014). Moreover, with strict maternal inheritance of mtDNA, and in the absence of recombination, deleterious mutations are likely to accumulate in the long run (Mueller's ratchet) (Wallace & Chalkia, 2013). The mitochondrial bottleneck counteracts this mechanism, as the low mtDNA copy number in PGCs represents only a small subset of the total mtDNA population of the entire embryo. This physical bottleneck, possibly intensified by replication of only a subset of mtDNAs within these cells, strongly increases heteroplasmy variance between PGCs and consequently the offspring (Johnston et al., 2015). Over generations, this random variance increase enhances the chance of a return to a homoplasmic (i.e., one type of mtDNA) state.

In addition to the nonspecifically acting bottleneck, mtDNA segregation is also influenced by several mutation-specific mechanisms. While the details of these mechanisms need to be elucidated, they function by either eliminating deleterious mutations, or else shifting cells toward homoplasmy irrespective of functionality. The following types, classified by mutation type (deleterious/nondeleterious) and the time they act during development, can be found in the literature: (i) purifying selection against deleterious mutations in tRNA and protein coding regions, acting in the germ line (Fan et al., 2008; Stewart et al., 2008) and during gestation (Freyer et al., 2012); (ii) biased segregation of different physiological mtDNA haplotypes acting in the germ line and during gestation (Burgstaller et al., 2014; Sharpley et al., 2012); and (iii) biased segregation of different physiological mtDNA haplotypes acting in the ICM, with selected cells contributing to the embryo proper (Lee et al., 2012). However, in spite of these mechanisms inheritable mtDNA-based diseases are reported with prevalence rates of 1 in

300 in humans, showing both the importance and limitations of these mechanisms (Burgstaller et al., 2015).

Mitochondrial performance is not only influenced by the integrity of its own genome but also by the nuclear genome. Encoding for important subunits of the respiratory chain, the mtDNA has to be compatible with the nuclear genes encoding the major part of the mitochondrion. As mtDNA is segregated with the female genome, the paternal half of the nuclear genome is "alien" to the mtDNA. It has therefore been proposed that the mtDNA has to be functionally flexible, with little functional difference between existing mtDNA haplotypes (Chinnery et al., 2014). Nevertheless, maternal transmission can cause accumulation of mutations that cause severe effects in males, but only mild effects in females (mother's curse) (Ballard & Pichaud, 2014). Differences between mtDNA haplotypes are also shown in adaption, e.g., to climate and altitude, along with incompatibilities between mtDNAs in heteroplasmic conditions and incompatibilities between nuclear and mtDNA genomes (reviewed in Ballard & Pichaud, 2014; Wallace & Chalkia, 2013). Moreover, mtDNA haplotypes were found to influence gene expression in ESCs (Kelly et al., 2013), and even modulate tumor formation and metastasis (Kenny & Germain, 2017).

4. MITOCHONDRIAL METABOLISM IN EARLY EMBRYOGENESIS

4.1 Oxidative Phosphorylation

Above we have described the biogenesis of mitochondria and the replication of its DNA in the early embryo and what potential mechanisms select for mitochondrial fitness. Next we would like to turn the table around and concentrate on how the mitochondria determine the fitness of the cell during early embryogenesis. Specifically, we will focus on what is known about two primary mitochondrial roles, the regulation of OXPHOS and the execution of apoptosis. There are other important roles that mitochondria play in the cell, such as in calcium homeostasis, but much less is known about this regulation in early development and for this reason these roles will not be covered here.

Probably the best-known mitochondrial role is the generation of ATP via OXPHOS. In this process ATP is produced from inorganic phosphate and ADP using the proton motor force (proton gradient) generated by the electron transport chain (ETC) across the inner membrane of the mitochondria. To generate this proton gradient the mitochondria require the consumption of oxygen as the final acceptor of electrons transported across the respiratory chain (Mitchell, 1961) and the oxidation of substrates generated by the Krebs cycle from metabolites coming from glycolysis, glutamate, or fatty acid oxidation (Björntorp, 1966).

The importance of OXPHOS in the early embryo is highlighted by the fact that the ATP produced in the late preimplantation embryo derives mostly from oxidative metabolism (Brinster, 1973). However, prior to blastocyst formation the case is very different. In cleavage stage embryos the mitochondria are still structurally and functionally immature and for this reason the oxygen consumption rate is limited in all mammalian species analyzed, including in human, rabbit, rat, mouse, porcine, and bovine embryos (Kurosawa et al., 2016; Leese, 2012). Due to this low oxygen consumption rate, cleavage embryos feed the substantial energy demands of the organelle remodeling that accompanies cleavage by an alternative pathway of ATP production—the adenosine salvage pathway—regulated by the activity of adenylate kinase and creatine kinase for AMP and ADP, respectively (Scantland et al., 2014). The low oxygen consumption rates of cleavage stage embryos have led to them being referred to as quiescent (Cho et al., 2006; Spikings et al., 2007; Squirrell et al., 2003; Stojkovic et al., 2001; Wilding et al., 2001), also known as the "quiet embryo" hypothesis (Leese, 2012). A practical consequence of the low oxygen consumption rate of early preimplantation embryos is that these embryos benefit from low oxygen culture conditions (Gardner, 2016; Leese, 2012). As will be discussed later, this is likely because at the early stages of development embryos are very sensitive to the oxidative stress that accompanies high oxygen consumption.

After cleavage and accompanying the maturation of the mitochondrial cristae, the rate of oxidative metabolism increases and by the blastocyst stage this consumption rate has not only doubled, but accounts for 70% of the total oxygen consumed by the embryo (Manes & Lai, 1995; Trimarchi, Liu, Porterfield, Smith, & Keefe, 2000). Interestingly, at this stage the metabolic activity of the ICM and TE differs. Cells from the ICM are less oxidative and therefore their ATP production is lower compared to the cells from the TE that produce about 80% of the ATP of the mouse blastocyst (Houghton, 2006). The high respiration and ATP generation at the blastocyst stage are most likely due to the energy required to power the Na^+/K^+ ATPase activity needed for cavitation and for the protein synthesis that sustains the net embryo growth that starts at this stage of development (Donnay & Leese, 1999; Houghton, Humpherson, Hawkhead, Hall, & Leese, 2003).

The substrates required to sustain OXPHOS in addition to oxygen also change during preimplantation development. From the zygote to the morula stage, pyruvate is the most important metabolite for Krebs cycle activity and pyruvate is required for the first cleavage division (Biggers, Whittingham, & Donahue, 1967; Whittingham, 1969; Whittingham & Biggers, 1967). After this division lactate can also sustain development (Brinster, 1965). At the blastocyst stage, although other substrates are also likely to be important, glucose consumption becomes predominant (Barbehenn, Wales, & Lowry, 1978; Houghton, Thompson, Kennedy, & Leese, 1996; Leese & Barton, 1984; Thompson, Partridge, Houghton, Cox, & Leese, 1996) and the coupling of glycolysis to oxidative phosphorylation ensures embryo viability (Lane & Gardner, 1996).

Upon implantation the oxidative capacity of the embryo changes again and reverts to cleavage stage levels (Houghton et al., 1996), as the embryo switches to a glycolytic metabolism (Clough & Whittingham, 1983; Shepard, Tanimura, & Park, 1997) (Fig. 1). This switch in metabolism is required to sustain the increase in proliferation rate that occurs between pre- and postimplantation, from a mean 12–14-h cell cycle time in the blastocyst (Barlow, Owen, & Graham, 1972; MacQueen & Johnson, 1983) to a 6-h cycle at 6.5 dpc (Snow, 1977). This is because glycolysis allows for a more rapid ATP generation than OXPHOS and feeds into pathways that produce the building blocks that sustain cell division, such as nucleotides, lipids, and amino acids. Regarding the molecular mechanisms controlling the switch to glycolysis, it has been suggested that the hypoxic environment that the embryo faces during implantation activates HIF-1alpha, which stimulates the expression of glycolytic genes and represses those involved in oxidative phosphorylation (Zhou et al., 2012). Finally, it is worth noting that the metabolic differences observed before and after implantation can also be captured in vitro by naive mouse and human ESCs, which resemble the preimplantation metabolism, and epiblast stem cells and primed human ESCs, which have a metabolism that is similar to the postimplantation embryo (Takashima et al., 2014; Zhou et al., 2012).

4.2 OXPHOS-Independent Roles of Mitochondrial Metabolites

Although the mitochondria are not mature organelles during early preimplantation development, the metabolic activity associated with OXPHOS is also important for fine-tuning embryogenesis independently of ATP generation at these stages. For example, Krebs cycle metabolites such

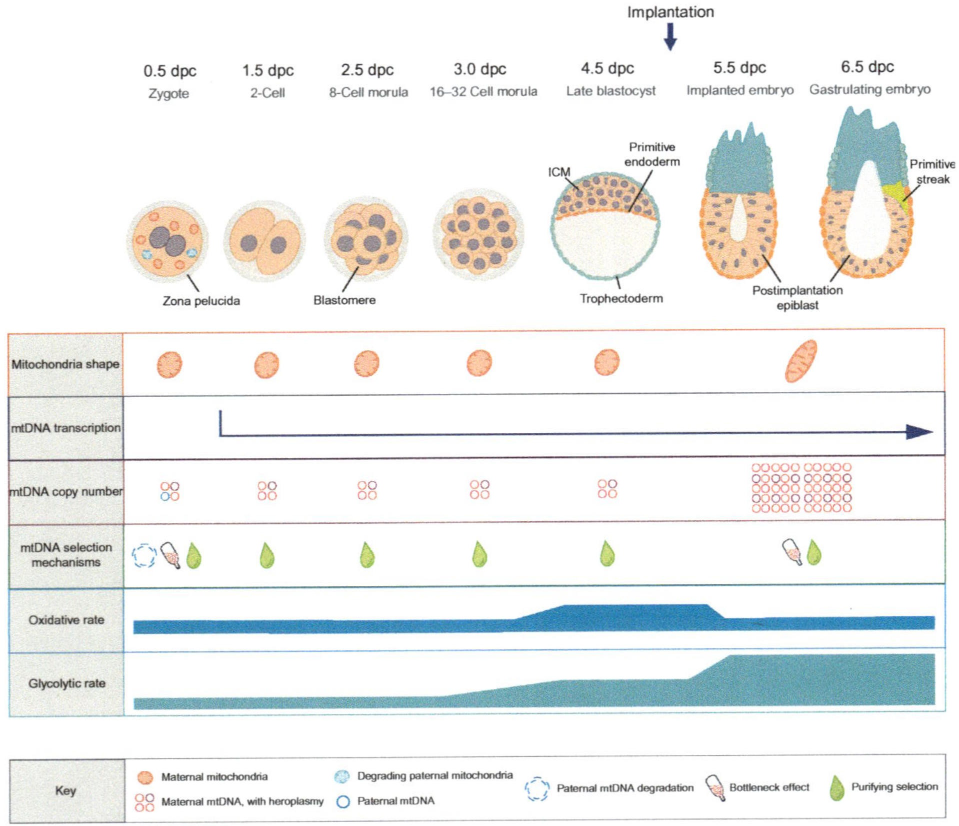

Fig. 1 See legend on opposite page.

as α-ketoglutarate and acetyl-CoA are required for the epigenetic changes (demethylation and acetylation, respectively) occurring during genomic activation in the cleavage stage mouse (Aoki, Worrad, & Schultz, 1997) and human embryo (Niakan & Eggan, 2013). Also pyruvate, O-glycosylation, and chaperones have been shown to transiently drive the nuclear localization of the metabolites and enzymes from the first half of the Krebs cycle (class I enzymes and metabolites) that regulate this zygotic genome activation (Nagaraj et al., 2017), providing a mechanism by which mitochondrial metabolites can regulate gene expression. Finally, the importance of α-ketoglutarate in regulating gene expression has also been shown in mESCs, where high intracellular levels of α-ketoglutarate is required for pluripotency by promoting DNA demethylation (Carey, Finley, Cross, Allis, & Thompson, 2015).

4.3 Impact of the Mitochondria on the Maintenance of the Reducing/Oxidizing (Redox) Balance

The increase in oxidative capacity that occurs during preimplantation development involves an increased activity of the ETC whereby the oxidation of NADH and $FADH_2$ culminates in the transfer of electrons to molecular oxygen and the generation of ATP. However, this process is not completely effective, as some electrons can "spill" and react with molecular oxygen generating ROS. It is for this reason that most of the ROS in the cell is produced in the mitochondria as by-products of the ETC, including the primary superoxide anion (O_2^-) along with its derivates hydrogen peroxide (H_2O_2) and hydroxyl radical (OH) (Balaban, Nemoto, & Finkel, 2005;

Fig. 1 Changes in the mitochondria and its mtDNA occurring during early mouse development. In the early embryo all mitochondria are maternally inherited due to the degradation of paternal mitochondria occurring after fertilization. From the zygote until the blastocyst the mitochondria are round with a low cristae density, but upon implantation they elongate and mature their cristae organization. mtDNA transcription is initiated at the two-cell stage with zygotic genome activation. As it occurs with mitochondrial numbers, the mtDNA copy number remains constant during preimplantation stages and then replicates upon implantation. Two selection mechanisms, the bottleneck effect and the purifying selection, protect from the accumulation of deleterious mtDNA mutations (Mueller's ratchet) during early embryogenesis. During early development the metabolic rate of the embryo also changes. The glycolytic rate is limited in cleaving embryos, increasing with the glucose uptake as the embryo progresses to the blastocyst stage. Similarly, the oxygen consumption in cleaving embryos is low but at the blastocyst stage the oxidative rate doubles, primarily due to the requirements of the trophectoderm. Upon implantation the oxidative rate returns to cleavage stage levels as the embryo switches to a glycolytic metabolism.

Takahashi, 2012). The formation of ROS has beneficial roles for signaling and cell homeostasis, but if excessive it poses a hazard by causing oxidative stress (a skewed redox potential toward oxidized potential (Balaban et al., 2005)). This hazard of ROS is very evident in the in vitro cultured early mammalian embryo, where oxidative stress is the cause of mitochondrial dysfunction, DNA damage, developmental arrest, and cell death (Qian et al., 2016; Takahashi, 2012; Wu et al., 2017). These observations imply that during early development the reducing/oxidizing balance must be finely tuned and therefore as the activity of OXPHOS increases so must the defense mechanisms against ROS.

Cells rely on enzymatic and nonenzymatic antioxidants to maintain a redox homeostasis. In the early embryo both enzymatic mitochondrial antioxidants, such as superoxide dismutases (SODs), and the nonenzymatic tripeptide glutathione (γ-glutamyl-cysteinyl-glycine; GSH-oxidized glutathione; GSSG-reduced glutathione), supported by glutathione peroxidase (Gpx), appear to be important for maintaining the redox state. The synthesis of GSH increases after zygotic genome activation (Calvin, Grosshans, & Blake, 1986; Nasr-Esfahani, Winston, & Johnson, 1992), and 90% of GSH is consumed by the blastocyst stage (Luberda, 2005). Perturbations of the GSH/GSSH balance cause apoptosis in mouse blastocysts (Ren et al., 2015) and correlate with loss of developmental potential in human embryos (Lee et al., 2010). The expression of SODs is also high in the early embryo (Donnay & Knoops, 2007; Guerin, El Mouatassim, & Menezo, 2001), suggesting that this antioxidant also plays a role. Together, these observations highlight the delicate redox balance that exists in the early embryo and indicate that a variety of mechanisms must be at play to allow the embryo to adapt to the changing energy requirements and oxygen environment that occurs during early development.

5. MITOCHONDRIAL CONTROL OF APOPTOTIC CELL DEATH DURING EARLY EMBRYOGENESIS

5.1 Regulation of Apoptotic Response

One of the most fascinating facets of the mitochondria is that it not only provides the energy for life, but it is also one of the most important mediators of the death of the cell. This importance is clearly illustrated by the essential role that the mitochondria play in activating caspase proteases through the intrinsic or mitochondrial pathway of apoptosis (reviewed in Tait & Green, 2013). There is not a wealth of knowledge regarding the regulation of mitochondrial apoptotic death during early mammalian development, primarily as much of the research in this field has concentrated on unraveling the mechanisms that govern pluripotency and differentiation. However, there is an acute need for apoptosis in development, not only to remove damaged cells but also to shape morphogenesis. For these reasons the observation made a number of years back that the apoptotic threshold dramatically changes during the early stages mammalian embryogenesis (Heyer, MacAuley, Behrendtsen, & Werb, 2000) has broad implications for establishing the importance of the mitochondria in shaping pre- and early postimplantation development.

The first conclusive evidence that the apoptotic response changes during early mammalian embryogenesis came from the observation that low doses of UV irradiation do not induce apoptosis in mouse preimplantation embryos, but lead to a strong apoptotic response in the postimplantation epiblast (Heyer et al., 2000). Subsequently, it was proposed that the postimplantation epiblast is primed for death, and therefore in a state that makes it hyper-responsive to apoptotic signals (Laurent & Blasi, 2015; Pernaute et al., 2014). In agreement with this hypothesis, a range of cellular defects, including chromosome fragmentation or chromosome missegregation, lead to apoptosis specifically during early postimplantation development (Brown & Baltimore, 2000; Dobles, Liberal, Scott, Benezra, & Sorger, 2000; Kalitsis, Earle, Fowler, & Choo, 2000), indicating that a checkpoint is acting at this stage to eliminate aberrant cells. The observation that hESCs are also hypersensitive to cell death signals and display a high "mitochondrial apoptotic priming" (Liu et al., 2013) suggests that the changes in apoptotic threshold may be conserved.

The mitochondrial apoptotic pathway is finely balanced at the mitochondria by the interaction of pro- and antiapoptotic factors of the BCL2 family. The binding of the antiapoptotic proteins (e.g., BCL-2, BCL-XL, or MCL-1) to their proapoptotic BCL2 family counterparts (e.g., BIM, BID, PUMA, NOXA, or BAD) prevents the induction of apoptosis. However, when this balance is broken, BIM or BID binds to the apoptotic effector molecules BAX and BAK which induce outer mitochondrial membrane permeabilization (MOMP) and caspase activation. There are at least two separate ways in which the balance of pro- and antiapoptotic factors is tilted to favor apoptosis in the postimplantation epiblast. The first is through microRNAs (miRNAs) of the miR-20, miR-92, and miR-302 families. These miRNAs target *Bim* and maintain it in a state that is poised for activation (Pernaute et al., 2014). This allows for the signals that induce apoptosis to rapidly induce death by stabilizing *Bim* RNA or protein. Furthermore, the expression of these miRNAs increases with the developmental progression of the epiblast, indicating that there is an increased requirement to dampen BIM activity in the postimplantation epiblast.

The second mechanism by which apoptosis is facilitated postimplantation is likely via the transcriptional regulation of BCL2 family factors. During mouse postimplantation development low doses of UV irradiation induce p53 activation at 5.5 and 6.5 dpc, but this irradiation only efficiently induces apoptosis at the later stage of development. One possible explanation for this difference is that 6.5 dpc embryos display lower transcriptional levels of expression of the antiapoptotic BCL-XL and higher expression of the proapoptotic BAK and BIM than their 5.5 dpc counterparts (Laurent & Blasi, 2015). Therefore during postimplantation development changes in the transcriptional and posttranscriptional regulation of BCL2 family members help sensitize the epiblast to apoptotic stimuli and in this way set the low apoptotic threshold that can be found in the primed pluripotent state (Fig. 2).

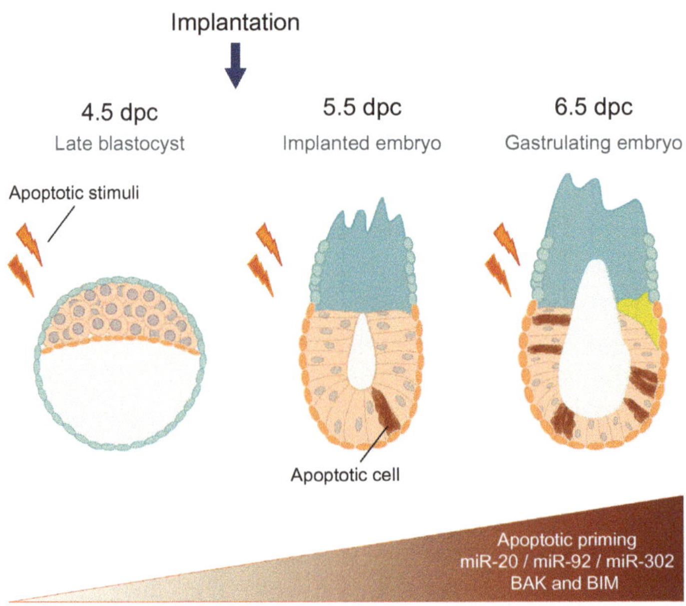

Fig. 2 Diagrammatic representation of the changes in the apoptotic response that occurs during early mammalian development. At preimplantation stages the apoptotic threshold is high, but after implantation epiblast cells become hypersensitive to death stimuli due to the increased expression of the proapoptotic proteins BIM and BAK. During this process the expression of miRNAs that target *Bim*, such as those belonging to the miR-20, miR-92, and miR-302 families, also increases and this prevents apoptosis but allow a rapid response to death stimuli.

5.2 Importance of Mitochondrial-Induced Apoptosis in the Early Embryo

The lowering of the apoptotic threshold occurring at early postimplantation stages has functional consequences, as a wave of apoptosis can be found to spread through the epiblast between 5.5 and 6.5 dpc (Manova et al., 1998; Spruce et al., 2010). But what is the purpose of this apoptosis and why do primed epiblast cells become hypersensitive to death? One obvious possibility relates to the fact that germ cell specification occurs shortly after cells transit to the primed epiblast state (Leitch, Tang, & Surani, 2013). It is likely that one mechanism to quickly eliminate abnormal cells is by lowering the mitochondrial death threshold, as this would ensure that even cells with moderate levels of damage are removed before they can contribute to the germline. Other possibilities have also been suggested for this apoptosis, such as driving the formation of the cavity that lies inside the cup-shaped epiblast epithelia at 6.5 dpc (Coucouvanis & Martin, 1995). However, this role seems unlikely as there is little correlation between apoptosis and the timing of cavitation during postimplantation development, and instead of death the apical constriction of polarized epiblast cells signaled by integrins appears to be required for this morphogenetic process (Bedzhov & Zernicka-Goetz, 2014). One final possibility is that this apoptosis helps explain the intriguing observation that the initiation of gastrulation is governed by the attainment of either a threshold cell number or tissue mass of the epiblast (recently reviewed by Kojima et al., 2014). It is possible that the culling of excess cells is part of the mechanism by which the embryo can coordinate growth and patterning to ensure that both occur with the appropriate developmental timing.

6. CONCLUDING REMARKS

What all the data above reveals are the fundamental roles that the mitochondria play in the regulation of cell fitness during early embryonic development, from controlling the energy output and redox state to determining the cell's apoptotic response. However, they also indicate not only how dynamically these roles change with developmental progression but also how coordinated these changes are. A prime example is the changes affecting OXPHOS and apoptosis. For example, at the blastocyst stage oxidative phosphorylation is high and the apoptotic response is muted.

In contrast to this, upon implantation OXPHOS activity is decreased and the embryo becomes hypersensitive to death stimuli. What these observations suggest is that either the apoptotic machinery is controlled by the oxidative state of the cell or/and that both apoptosis and OXPHOS activity are regulated in parallel by the same signal (Green, Galluzzi, & Kroemer, 2014). Alternatively, of course, it is possible that these two processes are not directly linked. Going forward into the future it will be important not only to unravel the mechanisms of this possible cross-regulation but also to uncover how this regulation is wired into the networks that govern differentiation. Answering if, in addition to environmental cues, cell identity determines mitochondrial activity, or if alternatively the mitochondria provide instructive signals for fate specification, and therefore coordinate environmental changes with patterning, will be fundamental to be able to understand how the early embryo develops.

The second implication of the reviewed studies is that not only do the mitochondria control cell fitness, but also that there are a variety of fitness quality control mechanisms that act directly on the mitochondria to optimize cell fitness. For example, the mtDNA bottleneck and purifying selection of mtDNAs both contribute at the mtDNA level to optimize mitochondrial activity. Similarly, mitophagy contributes to mitochondrial selection by removing those mitochondria that are damaged, although little is known about mitophagy during early development. All these mechanisms are cell intrinsic, but the synchronicity of developmental progression coupled with the remarkable plasticity that early embryonic cells show (see other reviews in this issue) suggests that there must also be cell-to-cell mechanisms that communicate relative mitochondrial fitness levels. This communication could be either cooperative, to synchronize growth, or competitive, to eliminate less metabolically active cells. Uncovering the nature of these interactions will shed essential light into how the embryo ensures that coordination of growth during development.

ACKNOWLEDGMENTS

We would like to thank Sarah Bowling, Katie Lawlor, and Tamzin Zawadzki for critical reading of the manuscript, Aida di Gregorio for help with the design of the figures, and all members of the Molecular Embryology lab (Heart Science) for helpful discussion. Tristan Rodriguez's lab is supported by the MRC project grant (MR/N009371/1) and by the British Heart Foundation Centre for research excellence. A.L. is supported by a British Heart Foundation PhD studentship.

REFERENCES

Agaronyan, K., Morozov, Y. I., Anikin, M., & Temiakov, D. (2015). Replication-transcription switch in human mitochondria. *Science*, *347*, 548–551.

Al Rawi, S., Louvet-Vallee, S., Djeddi, A., Sachse, M., Culetto, E., Hajjar, C., et al. (2011). Postfertilization autophagy of sperm organelles prevents paternal mitochondrial DNA transmission. *Science*, *334*, 1144–1147.

Amaral, A., Ramalho-Santos, J., & St John, J. C. (2007). The expression of polymerase gamma and mitochondrial transcription factor A and the regulation of mitochondrial DNA content in mature human sperm. *Human Reproduction*, *22*, 1585–1596.

Aoki, F., Worrad, D. M., & Schultz, R. M. (1997). Regulation of transcriptional activity during the first and second cell cycles in the preimplantation mouse embryo. *Developmental Biology*, *181*, 296–307.

Balaban, R. S., Nemoto, S., & Finkel, T. (2005). Mitochondria, oxidants, and aging. *Cell*, *120*, 483–495.

Ballard, J. W. O., & Pichaud, N. (2014). Mitochondrial DNA: More than an evolutionary bystander. *Functional Ecology*, *28*, 218–231.

Barbehenn, E. K., Wales, R. G., & Lowry, O. H. (1978). Measurement of metabolites in single preimplantation embryos; a new means to study metabolic control in early embryos. *Journal of Embryology and Experimental Morphology*, *43*, 29–46.

Barlow, P., Owen, D. A., & Graham, C. (1972). DNA synthesis in the preimplantation mouse embryo. *Journal of Embryology and Experimental Morphology*, *27*, 431–445.

Batten, B. E., Albertini, D. F., & Ducibella, T. (1987). Patterns of organelle distribution in mouse embryos during preimplantation development. *The American Journal of Anatomy*, *178*, 204–213.

Baum, D. A., & Baum, B. (2014). An inside-out origin for the eukaryotic cell. *BMC Biology*, *12*, 76.

Bedzhov, I., & Zernicka-Goetz, M. (2014). Self-organizing properties of mouse pluripotent cells initiate morphogenesis upon implantation. *Cell*, *156*, 1032–1044.

Biggers, J. D., Whittingham, D. G., & Donahue, R. P. (1967). The pattern of energy metabolism in the mouse oocyte and zygote. *Proceedings of the National Academy of Sciences of the United States of America*, *58*, 560–567.

Birky, C. (1994). Relaxed and stringent genomes: Why cytoplasmic genes don't obey Mendel's laws. *The Journal of Heredity*, *85*, 355–365.

Björntorp, P. (1966). The oxidation of fatty acids combined with albumin by isolated rat liver mitochondria. *Journal of Biological Chemistry*, *241*, 1537–1543.

Bratic, A., & Larsson, N. G. (2013). The role of mitochondria in aging. *The Journal of Clinical Investigation*, *123*, 951–957.

Brinster, R. L. (1965). Studies on the development of mouse embryos in vitro. II. The effect of energy source. *The Journal of Experimental Zoology*, *158*, 59–68.

Brinster, R. (1973). Nutrition and metabolism of the ovum, zygote and blastocyst. *Handbook of Physiology*, *2*, 165–170.

Brown, E. J., & Baltimore, D. (2000). ATR disruption leads to chromosomal fragmentation and early embryonic lethality. *Genes & Development*, *14*, 397–402.

Burgstaller, J. P., Johnston, I. G., Jones, N. S., Albrechtova, J., Kolbe, T., Vogl, C., et al. (2014). mtDNA segregation in heteroplasmic tissues is common in vivo and modulated by haplotype differences and developmental stage. *Cell Reports*, *7*, 2031–2041.

Burgstaller, J. P., Johnston, I. G., & Poulton, J. (2015). Mitochondrial DNA disease and developmental implications for reproductive strategies. *Molecular Human Reproduction*, *21*, 11–22.

Calvin, H. I., Grosshans, K., & Blake, E. J. (1986). Estimation and manipulation of glutathione levels in prepuberal mouse ovaries and ova: Relevance to sperm nucleus transformation in the fertilized egg. *Gamete Research*, *14*, 265–275.

Cao, L., Shitara, H., Horii, T., Nagao, Y., Imai, H., Abe, K., et al. (2007). The mitochondrial bottleneck occurs without reduction of mtDNA content in female mouse germ cells. *Nature Genetics*, *39*, 386–390.

Cao, L., Shitara, H., Sugimoto, M., Hayashi, J., Abe, K., & Yonekawa, H. (2009). New evidence confirms that the mitochondrial bottleneck is generated without reduction of mitochondrial DNA content in early primordial germ cells of mice. *PLoS Genetics*, *5*, e1000756.

Carey, B. W., Finley, L. W., Cross, J. R., Allis, C. D., & Thompson, C. B. (2015). Intracellular alpha-ketoglutarate maintains the pluripotency of embryonic stem cells. *Nature*, *518*, 413–416.

Chatre, L., & Ricchetti, M. (2013). Prevalent coordination of mitochondrial DNA transcription and initiation of replication with the cell cycle. *Nucleic Acids Research*, *41*, 3068–3078.

Chiaratti, M. R., Bressan, F. F., Ferreira, C. R., Caetano, A. R., Smith, L. C., Vercesi, A. E., et al. (2009). Embryo mitochondrial DNA depletion is reversed during early embryogenesis in cattle. *Biology of Reproduction*, *82*, 76–85.

Chinnery, P. F., Craven, L., Mitalipov, S., Stewart, J. B., Herbert, M., & Turnbull, D. M. (2014). The challenges of mitochondrial replacement. *PLoS Genetics*, *10*, e1004315.

Cho, Y., Kwon, S., Pak, Y., Seol, H., Choi, Y., Park, D., et al. (2006). Dynamic changes in mitochondrial biogenesis and antioxidant enzymes during the spontaneous differentiation of human embryonic stem cells. *Biochemical and Biophysical Research Communications*, *348*, 1472–1478.

Clough, J. R., & Whittingham, D. G. (1983). Metabolism of [14C]glucose by postimplantation mouse embryos in vitro. *Journal of Embryology and Experimental Morphology*, *74*, 133–142.

Cotterill, M., Harris, S. E., Fernandez, E. C., Lu, J., Huntriss, J. D., Campbell, B. K., et al. (2013). The activity and copy number of mitochondrial DNA in ovine oocytes throughout oogenesis in vivo and during oocyte maturation in vitro. *Molecular Human Reproduction*, *19*, 444–450.

Coucouvanis, E., & Martin, G. R. (1995). Signals for death and survival: A two-step mechanism for cavitation in the vertebrate embryo. *Cell*, *83*, 279–287.

Cree, L. M., Samuels, D. C., de Sousa Lopes, S. C., Rajasimha, H. K., Wonnapinij, P., Mann, J. R., et al. (2008). A reduction of mitochondrial DNA molecules during embryogenesis explains the rapid segregation of genotypes. *Nature Genetics*, *40*, 249–254.

Cummins, J. M., Wakayama, T., & Yanagimachi, R. (1997). Fate of microinjected sperm components in the mouse oocyte and embryo. *Zygote*, *5*, 301–308.

Dobles, M., Liberal, V., Scott, M. L., Benezra, R., & Sorger, P. K. (2000). Chromosome missegregation and apoptosis in mice lacking the mitotic checkpoint protein Mad2. *Cell*, *101*, 635–645.

Donnay, I., & Knoops, B. (2007). Peroxiredoxins in gametogenesis and embryo development. *Sub-Cellular Biochemistry*, *44*, 345–355.

Donnay, I., & Leese, H. J. (1999). Embryo metabolism during the expansion of the bovine blastocyst. *Molecular Reproduction and Development*, *53*, 171–178.

Facucho-Oliveira, J. M., Alderson, J., Spikings, E. C., Egginton, S., & John, J. C. S. (2007). Mitochondrial DNA replication during differentiation of murine embryonic stem cells. *Journal of Cell Science*, *120*, 4025–4034.

Fan, W., Lin, C. S., Potluri, P., Procaccio, V., & Wallace, D. C. (2012). mtDNA lineage analysis of mouse L-cell lines reveals the accumulation of multiple mtDNA mutants and intermolecular recombination. *Genes & Development*, *26*, 384–394.

Fan, W., Waymire, K. G., Narula, N., Li, P., Rocher, C., Coskun, P. E., et al. (2008). A mouse model of mitochondrial disease reveals germline selection against severe mtDNA mutations. *Science*, *319*, 958–962.

Ferreira, C. R., Burgstaller, J. P., Perecin, F., Garcia, J. M., Chiaratti, M. R., Meo, S. C., et al. (2010). Pronounced segregation of donor mitochondria introduced by bovine ooplasmic transfer to the female germ-line. *Biology of Reproduction, 82*, 563–571.

Folmes, C. D., Ma, H., Mitalipov, S., & Terzic, A. (2016). Mitochondria in pluripotent stem cells: Stemness regulators and disease targets. *Current Opinion in Genetics & Development, 38*, 1–7.

Freyer, C., Cree, L. M., Mourier, A., Stewart, J. B., Koolmeister, C., Milenkovic, D., et al. (2012). Variation in germline mtDNA heteroplasmy is determined prenatally but modified during subsequent transmission. *Nature Genetics, 44*, 1282–1285.

Gardner, D. K. (2016). The impact of physiological oxygen during culture, and vitrification for cryopreservation, on the outcome of extended culture in human IVF. *Reproductive Biomedicine Online, 32*, 137–141.

Green, D. R., Galluzzi, L., & Kroemer, G. (2014). Cell biology. Metabolic control of cell death. *Science, 345*, 1250256.

Guerin, P., El Mouatassim, S., & Menezo, Y. (2001). Oxidative stress and protection against reactive oxygen species in the pre-implantation embryo and its surroundings. *Human Reproduction Update, 7*, 175–189.

Gustafsson, C. M., Falkenberg, M., & Larsson, N.-G. (2016). Maintenance and expression of mammalian mitochondrial DNA. *Annual Review of Biochemistry, 85*, 133–160.

Gyllensten, U., Wharton, D., Josefsson, A., & Wilson, A. C. (1991). Paternal inheritance of mitochondrial DNA in mice. *Nature, 352*, 255–257.

Hagström, E., Freyer, C., Battersby, B. J., Stewart, J. B., & Larsson, N.-G. (2014). No recombination of mtDNA after heteroplasmy for 50 generations in the mouse maternal germline. *Nucleic Acids Research, 42*, 1111–1116.

Hashimoto, S., Morimoto, N., Yamanaka, M., Matsumoto, H., Yamochi, T., Goto, H., et al. (2017). Quantitative and qualitative changes of mitochondria in human preimplantation embryos. *Journal of Assisted Reproduction and Genetics, 34*, 573–580.

Hayashi, K., de Sousa Lopes, S. M. C., & Surani, M. A. (2007). Germ cell specification in mice. *Science, 316*, 394–396.

Heyer, B. S., MacAuley, A., Behrendtsen, O., & Werb, Z. (2000). Hypersensitivity to DNA damage leads to increased apoptosis during early mouse development. *Genes & Development, 14*, 2072–2084.

Hood, D. A. (2001). Invited review: Contractile activity-induced mitochondrial biogenesis in skeletal muscle. *Journal of Applied Physiology, 90*, 1137–1157.

Houghton, F. D. (2006). Energy metabolism of the inner cell mass and trophectoderm of the mouse blastocyst. *Differentiation, 74*, 11–18.

Houghton, F. D., Humpherson, P. G., Hawkhead, J. A., Hall, C. J., & Leese, H. J. (2003). Na+, K+, ATPase activity in the human and bovine preimplantation embryo. *Developmental Biology, 263*, 360–366.

Houghton, F. D., Thompson, J. G., Kennedy, C. J., & Leese, H. J. (1996). Oxygen consumption and energy metabolism of the early mouse embryo. *Molecular Reproduction and Development, 44*, 476–485.

Jansen, R. P. (2000). Germline passage of mitochondria: Quantitative considerations and possible embryological sequelae. *Human Reproduction, 15*(Suppl. 2), 112–128.

John, J. C. S. (2016). Mitochondrial DNA copy number and replication in reprogramming and differentiation. *Seminars in Cell and Developmental Biology, 52*, 93–101.

Johnston, I. G., Burgstaller, J. P., Havlicek, V., Kolbe, T., Rulicke, T., Brem, G., et al. (2015). Stochastic modelling, Bayesian inference, and new in vivo measurements elucidate the debated mtDNA bottleneck mechanism. *eLife, 4*, e07464.

Kalitsis, P., Earle, E., Fowler, K. J., & Choo, K. H. (2000). Bub3 gene disruption in mice reveals essential mitotic spindle checkpoint function during early embryogenesis. *Genes & Development, 14*, 2277–2282.

Kaneda, H., Hayashi, J., Takahama, S., Taya, C., Lindahl, K. F., & Yonekawa, H. (1995). Elimination of paternal mitochondrial DNA in intraspecific crosses during early mouse embryogenesis. *Proceedings of the National Academy of Sciences of the United States of America*, *92*, 4542–4546.

Kelly, R. D., Mahmud, A., McKenzie, M., Trounce, I. A., & St John, J. C. (2012). Mitochondrial DNA copy number is regulated in a tissue specific manner by DNA methylation of the nuclear-encoded DNA polymerase gamma A. *Nucleic Acids Research*, *40*, 10124–10138.

Kelly, R. D., Rodda, A. E., Dickinson, A., Mahmud, A., Nefzger, C. M., Lee, W., et al. (2013). Mitochondrial DNA haplotypes define gene expression patterns in pluripotent and differentiating embryonic stem cells. *Stem Cells*, *31*, 703–716.

Kenny, T. C., & Germain, D. (2017). mtDNA, metastasis, and the mitochondrial unfolded protein response (UPRmt). *Frontiers in Cell and Developmental Biology*, *5*, 37.

Khrapko, K., Coller, H. A., André, P. C., Li, X.-C., Hanekamp, J. S., & Thilly, W. G. (1997). Mitochondrial mutational spectra in human cells and tissues. *Proceedings of the National Academy of Sciences of the United States of America*, *94*, 13798–13803.

Kojima, Y., Tam, O. H., & Tam, P. P. (2014). Timing of developmental events in the early mouse embryo. *Seminars in Cell & Developmental Biology*, *34*, 65–75.

Kurosawa, H., Utsunomiya, H., Shiga, N., Takahashi, A., Ihara, M., Ishibashi, M., et al. (2016). Development of a new clinically applicable device for embryo evaluation which measures embryo oxygen consumption. *Human Reproduction*, *31*, 2321–2330.

Lane, M., & Gardner, D. K. (1996). Selection of viable mouse blastocysts prior to transfer using a metabolic criterion. *Human Reproduction*, *11*, 1975–1978.

Laurent, A., & Blasi, F. (2015). Differential DNA damage signalling and apoptotic threshold correlate with mouse epiblast-specific hypersensitivity to radiation. *Development*, *142*, 3675–3685.

Lee, M. S., Liu, C. H., Lee, T. H., Wu, H. M., Huang, C. C., Huang, L. S., et al. (2010). Association of creatin kinase B and peroxiredoxin 2 expression with age and embryo quality in cumulus cells. *Journal of Assisted Reproduction and Genetics*, *27*, 629–639.

Lee, H. S., Ma, H., Juanes, R. C., Tachibana, M., Sparman, M., Woodward, J., et al. (2012). Rapid mitochondrial DNA segregation in primate preimplantation embryos precedes somatic and germline bottleneck. *Cell Reports*, *1*, 506–515.

Leese, H. J. (2012). Metabolism of the preimplantation embryo: 40 years on. *Reproduction*, *143*, 417–427.

Leese, H. J., & Barton, A. M. (1984). Pyruvate and glucose uptake by mouse ova and preimplantation embryos. *Journal of Reproduction and Fertility*, *72*, 9–13.

Leitch, H. G., Tang, W. W., & Surani, M. A. (2013). Primordial germ-cell development and epigenetic reprogramming in mammals. *Current Topics in Developmental Biology*, *104*, 149–187.

Liu, J. C., Guan, X., Ryan, J. A., Rivera, A. G., Mock, C., Agarwal, V., et al. (2013). High mitochondrial priming sensitizes hESCs to DNA-damage-induced apoptosis. *Cell Stem Cell*, *13*, 483–491.

Long, C. A. (2007). In *Evolution of function and form in camelid erythrocytes. Actas de The 2007 WSEAS international conference on cellular & molecular biology, biophysics & bioengineering* (pp. 26–28).

Luberda, Z. (2005). The role of glutathione in mammalian gametes. *Reproductive Biology*, *5*, 5–17.

Luo, S. M., Ge, Z. J., Wang, Z. W., Jiang, Z. Z., Wang, Z. B., Ouyang, Y. C., et al. (2013). Unique insights into maternal mitochondrial inheritance in mice. *Proceedings of the National Academy of Sciences of the United States of America*, *110*, 13038–13043.

MacQueen, H. A., & Johnson, M. H. (1983). The fifth cell cycle of the mouse embryo is longer for smaller cells than for larger cells. *Journal of Embryology and Experimental Morphology*, 77, 297–308.

Manes, C., & Lai, N. C. (1995). Nonmitochondrial oxygen utilization by rabbit blastocysts and surface production of superoxide radicals. *Journal of Reproduction and Fertility, 104*, 69–75.

Manova, K., Tomihara-Newberger, C., Wang, S., Godelman, A., Kalantry, S., Witty-Blease, K., et al. (1998). Apoptosis in mouse embryos: Elevated levels in pregastrulae and in the distal anterior region of gastrulae of normal and mutant mice. *Developmental Dynamics, 213*, 293–308.

May-Panloup, P., Vignon, X., Chrétien, M.-F., Heyman, Y., Tamassia, M., Malthièry, Y., et al. (2005). Increase of mitochondrial DNA content and transcripts in early bovine embryogenesis associated with upregulation of mtTFA and NRF1 transcription factors. *Reproductive Biology and Endocrinology, 3*, 65.

McConnell, J. M., & Petrie, L. (2004). Mitochondrial DNA turnover occurs during preimplantation development and can be modulated by environmental factors. *Reproductive Biomedicine Online, 9*, 418–424.

Miller, F. J., Rosenfeldt, F. L., Zhang, C., Linnane, A. W., & Nagley, P. (2003). Precise determination of mitochondrial DNA copy number in human skeletal and cardiac muscle by a PCR-based assay: Lack of change of copy number with age. *Nucleic Acids Research, 31*, e61.

Mitchell, P. (1961). Coupling of phosphorylation to electron and hydrogen transfer by a chemi-osmotic type of mechanism. *Nature, 191*, 144–148.

Muggleton-Harris, A. L., & Brown, J. J. G. (1988). Cytoplasmic factors influence mitochondrial reorganization and resumption of cleavage during culture of early mouse embryos. *Human Reproduction, 3*, 1020–1028.

Nagaraj, R., Sharpley, M. S., Chi, F., Braas, D., Zhou, Y., Kim, R., et al. (2017). Nuclear localization of mitochondrial TCA cycle enzymes as a critical step in mammalian zygotic genome activation. *Cell, 168*, 210–223.

Nasr-Esfahani, M. H., Winston, N. J., & Johnson, M. H. (1992). Effects of glucose, glutamine, ethylenediaminetetraacetic acid and oxygen tension on the concentration of reactive oxygen species and on development of the mouse preimplantation embryo in vitro. *Journal of Reproduction and Fertility, 96*, 219–231.

Niakan, K. K., & Eggan, K. (2013). Analysis of human embryos from zygote to blastocyst reveals distinct gene expression patterns relative to the mouse. *Developmental Biology, 375*, 54–64.

Pernaute, B., Spruce, T., Smith, K. M., Sanchez-Nieto, J. M., Cobb, B. S., & Rodríguez, T. A. (2014). MicroRNAs control the apoptotic threshold in primed pluripotent stem cells through regulation of BIM. *Genes & Development, 28*, 1873–1878.

Qian, D., Li, Z., Zhang, Y., Huang, Y., Wu, Q., Ru, G., et al. (2016). Response of mouse zygotes treated with mild hydrogen peroxide as a model to reveal novel mechanisms of oxidative stress-induced injury in early embryos. *Oxidative Medicine and Cellular Longevity, 2016*, 11.

Rebolledo-Jaramillo, B., Su, M. S., Stoler, N., McElhoe, J. A., Dickins, B., Blankenberg, D., et al. (2014). Maternal age effect and severe germ-line bottleneck in the inheritance of human mitochondrial DNA. *Proceedings of the National Academy of Sciences of the United States of America, 111*, 15474–15479.

Ren, L., Wang, Z., An, L., Zhang, Z., Tan, K., Miao, K., et al. (2015). Dynamic comparisons of high-resolution expression profiles highlighting mitochondria-related genes between in vivo and in vitro fertilized early mouse embryos. *Human Reproduction, 30*, 2892–2911.

Ren, L., Zhang, C., Tao, L., Hao, J., Tan, K., Miao, K., et al. (2017). High-resolution profiles of gene expression and DNA methylation highlight mitochondrial modifications during early embryonic development. *The Journal of Reproduction and Development, 63*, 247–261.

Reyes, A., Kazak, L., Wood, S. R., Yasukawa, T., Jacobs, H. T., & Holt, I. J. (2013). Mitochondrial DNA replication proceeds via a 'bootlace' mechanism involving the incorporation of processed transcripts. *Nucleic Acids Research*, *41*, 5837–5850.

Sato, M., & Sato, K. (2013). Maternal inheritance of mitochondrial DNA by diverse mechanisms to eliminate paternal mitochondrial DNA. *Biochimica et Biophysica Acta*, *1833*, 1979–1984.

Scantland, S., Tessaro, I., Macabelli, C. H., Macaulay, A. D., Cagnone, G., Fournier, É., et al. (2014). The denosine salvage pathway as an alternative to mitochondrial production of ATP in maturing mammalian oocytes1. *Biology of Reproduction*, *91*, 75.

Schwartz, M., & Vissing, J. (2002). Paternal inheritance of mitochondrial DNA. *The New England Journal of Medicine*, *347*, 576–580.

Sharpley, M. S., Marciniak, C., Eckel-Mahan, K., McManus, M., Crimi, M., Waymire, K., et al. (2012). Heteroplasmy of mouse mtDNA is genetically unstable and results in altered behavior and cognition. *Cell*, *151*, 333–343.

Shepard, T. H., Muffley, L. A., & Smith, L. T. (2000). Mitochondrial ultrastructure in embryos after implantation. *Human Reproduction*, *15*(Suppl. 2), 218–228.

Shepard, T. H., Tanimura, T., & Park, H. W. (1997). Glucose absorption and utilization by rat embryos. *The International Journal of Developmental Biology*, *41*, 307–314.

Smith, L. C., & Alcivar, A. A. (1993). Cytoplasmic inheritance and its effects on development and performance. *Journal of Reproduction and Fertility. Supplement*, *48*, 31–43.

Snow, M. H. L. (1977). Gastrulation in the mouse: Growth and regionalization of the epiblast. *Journal of Embryology and Experimental Morphology*, *42*, 293–303.

Spikings, E. C., Alderson, J., & John, J. C. S. (2007). Regulated mitochondrial DNA replication during oocyte maturation is essential for successful porcine embryonic development. *Biology of Reproduction*, *76*, 327–335.

Spruce, T., Pernaute, B., Di-Gregorio, A., Cobb, B. S., Merkenschlager, M., Manzanares, M., et al. (2010). An early developmental role for miRNAs in the maintenance of extraembryonic stem cells in the mouse embryo. *Developmental Cell*, *19*, 207–219.

Squirrell, J. M., Schramm, R. D., Paprocki, A. M., Wokosin, D. L., & Bavister, B. D. (2003). Imaging mitochondrial organization in living primate oocytes and embryos using multiphoton microscopy. *Microscopy and Microanalysis*, *9*, 190–201.

St John, J. C., & Schatten, G. (2004). Paternal mitochondrial DNA transmission during nonhuman primate nuclear transfer. *Genetics*, *167*, 897–905.

Stewart, J. B., Freyer, C., Elson, J. L., Wredenberg, A., Cansu, Z., Trifunovic, A., et al. (2008). Strong purifying selection in transmission of mammalian mitochondrial DNA. *PLoS Biology*, *6*, e10.

Stojkovic, M., Machado, S. A., Stojkovic, P., Zakhartchenko, V., Hutzler, P., Gonçalves, P. B., et al. (2001). Mitochondrial distribution and adenosine triphosphate content of bovine oocytes before and after in vitro maturation: Correlation with morphological criteria and developmental capacity after in vitro fertilization and culture. *Biology of Reproduction*, *64*, 904–909.

Sutovsky, P., Moreno, R. D., Ramalho-Santos, J., Dominko, T., Simerly, C., & Schatten, G. (1999). Ubiquitin tag for sperm mitochondria. *Nature*, *402*, 371–372.

Sutovsky, P., Moreno, R. D., Ramalho-Santos, J. o., Dominko, T., Simerly, C., & Schatten, G. (2000). Ubiquitinated sperm mitochondria, selective proteolysis, and the regulation of mitochondrial inheritance in mammalian embryos. *Biology of Reproduction*, *63*, 582–590.

Tait, S. W., & Green, D. R. (2013). Mitochondrial regulation of cell death. *Cold Spring Harbor Perspectives in Biology*, *5*, 1–15.

Takahashi, M. (2012). Oxidative stress and redox regulation on *in vitro* development of mammalian embryos. *Journal of Reproduction and Development*, *58*, 1–9.

Takashima, Y., Guo, G., Loos, R., Nichols, J., Ficz, G., Krueger, F., et al. (2014). Resetting transcription factor control circuitry toward ground-state pluripotency in human. *Cell, 158*, 1254–1269.

Thompson, J. G., Partridge, R. J., Houghton, F. D., Cox, C. I., & Leese, H. J. (1996). Oxygen uptake and carbohydrate metabolism by in vitro derived bovine embryos. *Journal of Reproduction and Fertility, 106*, 299–306.

Thundathil, J., Filion, F., & Smith, L. C. (2005). Molecular control of mitochondrial function in preimplantation mouse embryos. *Molecular Reproduction and Development, 71*, 405–413.

Trimarchi, J. R., Liu, L., Porterfield, D. M., Smith, P. J., & Keefe, D. L. (2000). Oxidative phosphorylation-dependent and -independent oxygen consumption by individual preimplantation mouse embryos. *Biology of Reproduction, 62*, 1866–1874.

Van Blerkom, J. (2009). Mitochondria in early mammalian development. *Seminars in Cell & Developmental Biology, 20*, 354–364.

Van Blerkom, J. (2011). Mitochondrial function in the human oocyte and embryo and their role in developmental competence. *Mitochondrion, 11*, 797–813.

Van Blerkom, J., Davis, P., Mathwig, V., & Alexander, S. (2002). Domains of high-polarized and low-polarized mitochondria may occur in mouse and human oocytes and early embryos. *Human Reproduction, 17*, 393–406.

Viscomi, C., Spinazzola, A., Maggioni, M., Fernandez-Vizarra, E., Massa, V., Pagano, C., et al. (2009). Early-onset liver mtDNA depletion and late-onset proteinuric nephropathy in Mpv17 knockout mice. *Human Molecular Genetics, 18*, 12–26.

Wai, T., Ao, A., Zhang, X., Cyr, D., Dufort, D., & Shoubridge, E. A. (2010). The role of mitochondrial DNA copy number in mammalian fertility. *Biology of Reproduction, 83*, 52–62.

Wai, T., Teoli, D., & Shoubridge, E. A. (2008). The mitochondrial DNA genetic bottleneck results from replication of a subpopulation of genomes. *Nature Genetics, 40*, 1484–1488.

Wallace, D. C., & Chalkia, D. (2013). Mitochondrial DNA genetics and the heteroplasmy conundrum in evolution and disease. *Cold Spring Harbor Perspectives in Medicine, 3*, a021220.

Whittingham, D. G. (1969). The failure of lactate and phosphoenolpyruvate to support development of the mouse zygote in vitro. *Biology of Reproduction, 1*, 381–386.

Whittingham, D. G., & Biggers, J. D. (1967). Fallopian tube and early cleavage in the mouse. *Nature, 213*, 942–943.

Wilding, M., Dale, B., Marino, M., di Matteo, L., Alviggi, C., Pisaturo, M. L., et al. (2001). Mitochondrial aggregation patterns and activity in human oocytes and preimplantation embryos. *Human Reproduction, 16*, 909–917.

Wu, Y., Zhang, N., Li, Y. H., Zhao, L., Yang, M., Jin, Y., et al. (2017). Citrinin exposure affects oocyte maturation and embryo development by inducing oxidative stress-mediated apoptosis. *Oncotarget, 8*, 34525–34533.

Zhou, W., Choi, M., Margineantu, D., Margaretha, L., Hesson, J., Cavanaugh, C., et al. (2012). HIF1alpha induced switch from bivalent to exclusively glycolytic metabolism during ESC-to-EpiSC/hESC transition. *The EMBO Journal, 31*, 2103–2116.

CHAPTER FOURTEEN

The Head's Tale: Anterior-Posterior Axis Formation in the Mouse Embryo

Matthew J. Stower, Shankar Srinivas[1]
University of Oxford, Oxford, United Kingdom
[1]Corresponding author: e-mail address: shankar.srinivas@dpag.ox.ac.uk

Contents

Abstract

The establishment of the anterior–posterior (A–P) axis is a fundamental event during early development and marks the start of the process by which the basic body plan is laid down. This axial information determines where gastrulation, that generates and positions cells of the three-germ layers, occurs. A–P patterning requires coordinated interactions between multiple tissues, tight spatiotemporal control of signaling pathways, and the coordination of tissue growth with morphogenetic movements. In the mouse, a specialized population of cells, the anterior visceral endoderm (AVE) undergoes a migration event critical for correct A–P pattern. In this review, we summarize our understanding of the generation of anterior pattern, focusing on the role of the AVE. We will also outline some of the many questions that remain regarding the mechanism by which the first axial asymmetry is established, how the AVE is induced, and how it moves within the visceral endoderm epithelium.

Current Topics in Developmental Biology, Volume 128
ISSN 0070-2153
https://doi.org/10.1016/bs.ctdb.2017.11.003

1. INTRODUCTION: FORMATION OF THE EGG CYLINDER

At roughly three and a half days postfertilization, the mouse blastocyst consists of two subpopulations of cells: an outer shell of trophectoderm (TE) surrounding a hollow fluid-filled blastocoel cavity and an asymmetrically located aggregation of cells, termed the inner cell mass (ICM) (Gardner, Papaioannou, & Barton, 1973; Rossant & Tam, 2009). Around 1 day later, the cells of the ICM have differentiated into two cell types, the epiblast and the primitive endoderm, and the blastocyst implants into the maternal endometrial wall. The epiblast contains the pluripotent progenitors that will give rise to the three embryonic germ layers endoderm, mesoderm, and ectoderm that form the fetus (Tam & Loebel, 2007). It also forms the extraembryonic mesoderm, including cells that generate the allantois, the fetal portion of the placenta (Downs, 2009; Downs et al., 2009). The primitive endoderm will differentiate into two tissues, the parietal and visceral endoderm (VE) (Gardner & Rossant, 1979; Hermitte & Chazaud, 2014). The latter will ultimately generate visceral yolk sac as well as contribute cells to the definitive endoderm that forms the inner lining of the gut (Kwon, Viotti, & Hadjantonakis, 2008; Viotti, Foley, & Hadjantonakis, 2014). The trophectoderm meanwhile will give rise only to extraembryonic tissues, including the extraembryonic ectoderm (ExE), chorion, and ectoplacental cone (EPC) (Gardner, 1983; Gardner et al., 1973). While all these extraembryonic tissues have key trophic and homeostatic roles throughout development, studies have revealed that many also play essential roles in lineage segregation and patterning embryonic tissues during these early stages of development.

Implantation stimulates the polar trophectoderm to proliferate to form the EPC. The EPC acts as the point of attachment of the embryo to the maternal endometrial wall. Concomitant with the implantation of the embryo, maternal endometrial cells respond by proliferating and decidualizing so as to completely envelope the forming conceptus (Ramathal et al., 2010). During this phase of development from embryonic day (E) 4.5 to E5.0, the embryo also undergoes a change in both size and morphology, elongating from its point of attachment, the EPC (considered the proximal region) to form an elongated structure with a clear proximal–distal axis (Rossant, 2004). The polar trophectoderm also forms the ExE distal to the EPC. Around this time, the epiblast increases in cell number and cavitates to form a "cup-like" morphology with the margin of the cup apposed to the newly formed ExE (Tam & Loebel, 2007). As the embryo develops, due to the formation of the ExE and expansion of the epiblast, these tissues expand distally into the blastocoel cavity (Copp, 1979, 1981). The cells of the primitive endoderm that were initially only in contact with epiblast cells now expand in concert with the distalward growth of these tissues so that they envelope the entire outer surface of the egg cylinder as a continuous single-cell layered epithelium termed the VE (Fig. 1).

In this review, we refer to the region of the VE covering the epiblast at the distal half of the embryo as "Epi-VE" and the more proximal VE covering the ExE as "ExE-VE." Thus, from implantation at E4.5–E5.0 the morphology of the embryo changes dramatically to form a configuration often referred to as the "egg cylinder," a structure that while uniquely characteristic of rodents, still shares the same relative tissue topology to other mammals.

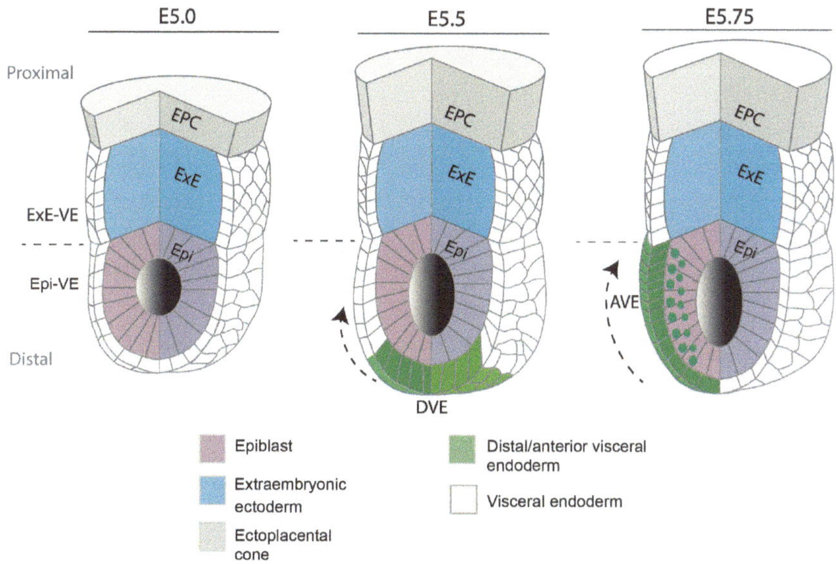

Fig. 1 Egg cylinder stage mouse embryos at preinduction, DVE-induction, and migration stages showing the major tissues. Embryos are shown in 3D, with an *en face* quarter section of each removed to reveal internal tissues. At E5.0 a single continuous layer of visceral endoderm covers the extraembryonic ectoderm (ExE), proximally and the epiblast distally. DVE cells (*green*) are induced at the distal tip of the egg cylinder at E5.5 and convert from squamous to columnar epithelium. DVE cells migrate unidirectionally from the distal tip to one side of the egg cylinder, forming the AVE and defining the anterior of the embryo (*dashed arrow*). The site of gastrulation (primitive streak) forms at E6.25 on the side of the epiblast opposite to and furthest from the AVE as a result of signals from the AVE that repress streak markers (*green dots* at E5.75). At this most proximal position the AVE cells retain columnar morphology.

The outer VE layer is a continuous single-cell monolayer with varying columnar, cuboidal, or squamous epithelial morphology depending on location and stage. By day E5.5 a morphological difference in the subset of Epi-VE at the distal tip of the egg cylinder becomes apparent as they change morphology from cuboidal to columnar (Kimura et al., 2000; Rivera-Perez, Mager, & Magnuson, 2003; Srinivas et al., 2004). These cells are known as the distal visceral endoderm (DVE) and have a key role in converting the proximal–distal asymmetry in the embryo to an anterior–posterior asymmetry (Thomas & Beddington, 1996). This is achieved by the directional proximal movement of the DVE cells from this distal position toward one side of the egg cylinder so that they come to reside at the boundary between the epiblast and ExE (Srinivas et al., 2004). At this position, these cells are now termed the anterior visceral endoderm (AVE) as anterior pattern is induced in the underlying epiblast on this side of the egg cylinder. AVE cells comprise a specialized signaling center that is morphologically distinguishable at E5.75 by columnar morphology and is positioned on one side of the epiblast (Thomas & Beddington, 1996; Thomas, Brown, & Beddington, 1998).

Between day E5.0 and day E6.25 the epiblast continues to proliferate and appears morphologically uniform until day E6.25 when the first sign in the epiblast of anterior–posterior asymmetry becomes evident as a thickening of the epiblast on the side opposite to the AVE (Beddington & Robertson, 1999). This thickening is the beginning of the formation of the "primitive streak," a dynamic cellular structure formed by cells in the proximal epiblast undergoing epithelial to mesenchymal transition, detaching from the pseudostratified epithelium of the epiblast and migrating away from their site of ingression to form the mesoderm and endoderm cells layers. The primitive streak is progressively built in a distalward manner from the "waistline" of the embryo—the margin of the epiblast in contact with the ExE—toward the tip of the epiblast (Tam & Loebel, 2007). This structure is the avenue for gastrulation, the fundamental process during which a diversity of cell types are generated in a controlled manner from the epiblast, while at the same time being positioned correctly to contribute to the basic body plan. Thus, by E6.5 the epiblast is starting to show fundamental asymmetries. On one side generally referred to as the posterior, the streak forms and gives rise to progenitors of the embryonic and extraembryonic mesoderm, and the definitive endoderm. On the opposite side, referred to as the anterior side, epiblast cells retain their pseudostratified epithelial morphology and will form predominantly neuroectodermal tissues. The establishment of the anterior–posterior axis in the epiblast and the site of the primitive streak is the subject of this review and requires an understanding of the cell behavior, cell states, and tissue interactions between both the embryonic cells of the epiblast, and those of extraembryonic tissues including the ExE and VE during development.

2. SIGNALING CENTERS THAT PATTERN THE EPIBLAST

Nearly all the cells that give rise to the fetus and adult are derived from the epiblast, which abuts the ExE proximally and is surrounded by a continuous epithelial monolayer of the VE. The epiblast is unpatterned at E5.0 and expresses several pluripotency factors including SOX2, NANOG, SALL4, and POU5F1 (Chew et al., 2005; Okumura-Nakanishi et al., 2005; Wu et al., 2006; Zhang et al., 2006). This pluripotent state is maintained by Nodal (Sun et al., 2014) and BMP signaling (Di-Gregorio et al., 2007). During the course of the next day, these epiblast cells are patterned as a result of reciprocal interactions with the ExE and VE cell population and come to express markers of the primitive streak such as *Cripto*, *Brachyury*, and *Nodal* in a localized asymmetric pattern in the presumptive posterior. Interestingly, single cell sequencing of the E5.5 epiblast does not reveal any distinct subgroups of epiblast cells that can be clustered based on primitive streak or other markers, suggesting that patterning at this stage is based on the asymmetric expression of relatively few, but potent, regulatory factors (Mohammed et al., 2017).

The main tissues influencing the epiblast are the ExE, the posterior visceral endoderm (PVE), and the AVE. We will briefly discuss the roles of the ExE and PVE before moving on to a more detailed discussion of the AVE, about which the most is known.

2.1 Extraembryonic Ectoderm

The ExE forms from the polar trophectoderm, and while this tissue has an extraembryonic fate, it also has a key role in patterning the adjacent epiblast and VE cell layers during early development. The role of this patterning was revealed through dissecting the early E5.5 embryo into its embryonic and abembryonic halves, which leads to the Epi-VE expressing markers of the DVE in an abnormally unrestricted manner (Rodriguez et al., 2005). This suggests that an inhibitory signal from the ExE normally prevents the differentiation of proximal VE cells toward this fate. BMP signaling from the ExE has been suggested to be the repressive signal that prevents DVE-like fate in the VE. Thus, in this model the DVE only forms at the distal tip of the egg cylinder, furthest from the repressive ExE signals (Rodriguez et al., 2005; Yamamoto et al., 2009).

In support of this idea, the ExE expresses BMPs such as BMP4 and BMP8b (Winnier et al., 1995; Ying & Zhao, 2001), and knockout experiments show that BMP4 plays an important role as mutants show a loss of expression of the mesodermal marker *Brachyury* and fail to gastrulate (Ben-Haim et al., 2006; Winnier et al., 1995). BMP in the ExE is maintained by NODAL signaling from the epiblast, and in a reciprocal manner BMP from the ExE is required for the correct development of the epiblast and formation of the primitive streak (Ben-Haim et al., 2006). NODAL is required for correct development of the epiblast as well as to induce the formation of the DVE (Brennan et al., 2001). The VE also expresses NODAL, and loss of its expression exclusively in this tissue, as shown through VE-specific knockout mice (Kumar et al., 2015), leads to a reduction of NODAL in the epiblast and subsequent developmental arrest (Kumar et al., 2015). Subsequently the DVE feeds back on NODAL in a negative manner through secreting inhibitors of the pathway including CER1 and LEFTY1 (Brennan et al., 2001). Thus, there is an initial reciprocal interaction between the VE and epiblast to maintain NODAL signaling in the epiblast, followed by repressive interaction when the epiblast is patterned.

Interestingly, the autoregulatory loop between the ExE and epiblast has been shown to require another interactive tissue, the ExE-VE (Uez et al., 2008), as knockouts of the *Sall4-a* isoform, a Spalt-like zinc finger transcription factor expressed exclusively in ExE-VE at E5.0, leads to loss of *Bmp4* expression in the underlying ExE. This loss of *Bmp4* then results in a loss of NODAL expression in the epiblast and a failure of DVE induction (Uez et al., 2008). Thus, a complex series of interactions exists between these tissues acting to both induce and maintain expression of *Bmp4*, and *Nodal*, and to induce the DVE at the distal tip that subsequently feeds back to pattern the epiblast by inhibiting Nodal and WNT.

2.2 Posterior Visceral Endoderm

While much progress has been made in understanding both the functional role and mechanism of movement of the AVE subset of cells in the VE, less attention has been focused on the opposite side, the PVE. This region of the VE overlays the forming posterior primitive streak, that is, the most proximally located cells in the epiblast. This region of the epiblast is of importance not only for the formation of the embryonic mesoderm but also the extraembryonic mesoderm that forms in the proximal abembryonic half of the embryo and gives rise to structures including the allantois, the embryonic contribution to the placenta (Downs, 2009).

The precise function of the PVE as a signaling center is currently unclear. However, it is known to express genes distinct from that of the rest of the VE, as well as show distinctive behavior. The PVE shows localized expression of *Wnt3* and *Mixl1*, both of which are also expressed in the posterior epiblast. *Wnt3* is required in the epiblast for the formation of the primitive streak (Liu et al., 1999), but the expression in the epiblast is preceded by expression in the PVE at E5.5 (Ferrer-Vaquer et al., 2010; Rivera-Perez & Magnuson, 2005). The question of whether WNT3 signaling from the PVE induces epiblast expression has been addressed by creating a tissue-specific *Wnt3* knockout using a VE-specific Transthyretin-Cre driver (Kwon & Hadjantonakis, 2009). Using this cross Yoon et al. (2015) showed that VE-specific knockouts of *Wnt3* still induce a primitive streak and appear normal until at least E9.5. However, these embryos appear developmentally retarded and had significantly smaller regions expressing the primitive streak marker *Brachyury* at E6.5 compared to litter mates (Yoon et al., 2015), suggesting that the PVE may act to reinforces the patterning of the posterior epiblast during development.

The behavior of cells throughout the Epi-VE has recently been imaged using time-lapse microscopy of a R26-H2B-EGFP line in which cell nuclei are marked by EGFP (Shioi et al., 2017). Unlike cells in other regions of the Epi-VE (AVE and lateral VE) that show stereotypic migratory movements, cells of the PVE were found to remain relatively stationary (Shioi et al., 2017). This combined with the expression data above suggests that the PVE may in some way be a functionally distinct group, though its precise function remains to be discovered.

3. THE ROLE OF THE AVE IN A–P AXIS FORMATION

While the ExE and PVE are thought to have instructive roles for the formation of posterior epiblast fates, the AVE cell population is thought to act by inhibiting posterior fates in the adjacent epiblast. Experiments that have revealed the role of the AVE include microsurgical dissection (Martinez Barbera et al., 2000; Thomas & Beddington, 1996), loss of function knockouts (Acampora et al., 1998; Kimura et al., 2000;

Perea-Gomez et al., 2002), and genetically ablation of the AVE cell population (Stuckey et al., 2011a). Such interventionist experiments lead to a loss of forebrain derivatives of the epiblast by E6.5 and ultimately to a nonviable embryo (Thomas & Beddington, 1996).

Due to the loss of rostral neural derivatives of the epiblast in ablation experiments, the AVE has been defined as a head organizer, acting to induce the expression of forebrain markers in the underlying ectodermal tissue. However, it is not clear that it induces forebrain markers directly, rather than acting to protect the tissue from caudalizing signals (Foley, Skromne, & Stern, 2000; Kimura et al., 2000; reviewed in Stern & Downs, 2012). Indeed, in mutants of some genes such as in *Frs2α* and *Elft5* (Gotoh et al., 2005) where posterior markers are not expressed or are aberrantly positioned, the proximal epiblast now expresses markers indicative of an anterior neural fate (*Sox1* and *Otx2*) (Gotoh et al., 2005), supporting the view that the epiblast is developmentally primed toward anterior ectodermal differentiation, so that in the absence of posteriorizing signals, it will differentiate into an anterior neural character.

The AVE as a signaling center is also functionally active earlier in development as it is essential for correctly positioning the primitive streak, by restricting the expression of key primitive streak markers including *Nodal*, *Brachyury*, and *Cripto* to the region of the epiblast on the opposite side (Ding et al., 1998; Norris et al., 2002; Perea-Gomez et al., 2004) (Fig. 1). This is achieved through secretion of inhibitors of the WNT (DKK1, CER-1), and NODAL (LEFTY, CER1) pathways (Belo et al., 1997; Yamamoto et al., 2004). Furthermore, the AVE may also act through extruding extracellular matrix components that act to inhibit epithelial–mesenchymal transition (Egea et al., 2008). Finally, evidence, for the importance of not only the signaling but also the asymmetric migration of the AVE relative to the epiblast, has been shown in knockout embryos in which AVE cells arrest at the distal tip of the egg cylinder, or fail to be induced (Fig. 2). This generally leads to the primitive streak being aberrantly localized to the proximal epiblast or in rare cases, can even lead to the formation of more than one, generally disorganized, primitive streaks. Thus, there is an abundance of evidence for an essential role for the AVE in the embryo in generating A–P asymmetry, in patterning the A–P axis and for normal development to proceed.

Adding an interesting twist to our understanding from these embryonic data has been the development of several in vitro models which appear to generate some aspects of axial polarity in the absence of VE or VE-like

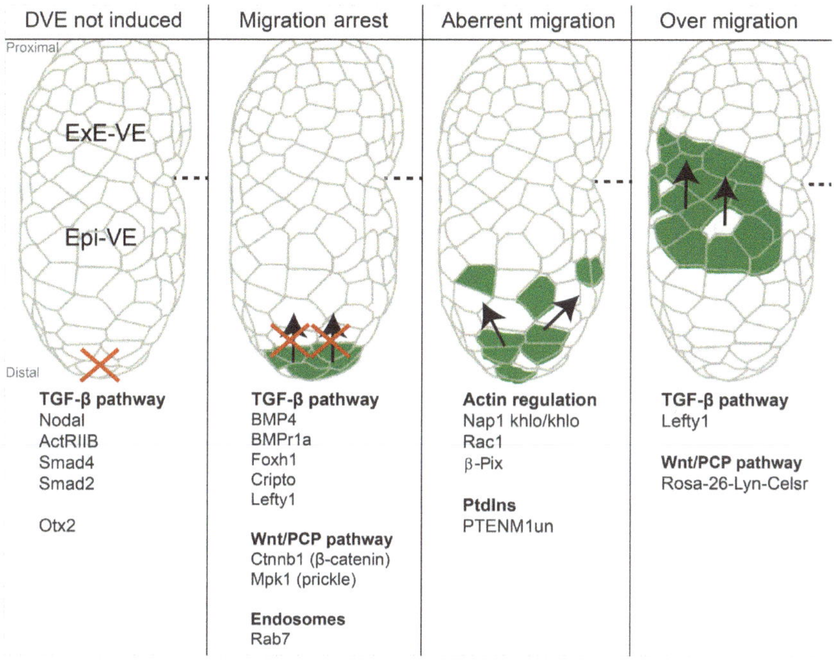

Fig. 2 Diagram of classes of AVE mutants. Embryos are categorized into four classes based on their knockout phenotype. Class I mutants do not induce expression of DVE markers. Class II mutants induce some, but in several cases not all, markers of the DVE and cells fail to migrate, being arrested at the distal tip. Class III mutants undergo migration of AVE cells, but do so inefficiently and do not make it to their normal proximal position. Class IV mutants undergo migration but continue beyond the normal boundary to their migration, into ExE-VE.

tissues. The first report of such a system was using P19 mouse embryonal carcinoma cells (Marikawa et al., 2009). When such cells were aggregated and cultured in hanging drops, the aggregates developed an elongated morphology and showed localized expression of several posterior mesoderm markers such as *Brachyury*, *Cdx2*, *Wnt5a*, and *Fgf8* (Marikawa et al., 2009).

Similar elongated structures have been achieved using embryonic stem cells (ESCs), by precisely determining the number of starting cells in the hanging drops. When van den Brink et al. (2014) cultured small clusters of ~300 mouse ES cells in suspension culture in N2B27 medium for 2 days followed by continuous treatment with mesodermal-inducing conditions, Activin A and a Wnt/β-catenin signaling agonist, the EBs self-organized into elongated 3D "gastruloid" aggregates (van den Brink et al., 2014). These gastruloid cell aggregates also lack a VE-like layer but still show an asymmetry in the expression of the mesodermal genes *Brachyury* to one end of the gastruloid, suggesting that the cells self-organize in this culture system to break their initial symmetry (van den Brink et al., 2014).

Self-organizing of ES cells on flat surfaces is also possible. Warmflash et al. (2014) have reported that human ESCs, when seeded on to a micropatterned discs, will develop into radially arranged concentric rings of trophectoderm, endoderm, mesoderm, and ectodermal cell populations from outer to inner, each expressing markers characteristic of each germ layer (Warmflash et al., 2014).

Recently it has been shown that coculture of mouse ESCs together in close apposition with extraembryonic trophoblast stem cells (TSCs) can self-organize in approximately 20% of cases into structures reminiscent of egg cylinder stage embryos (Harrison et al., 2017). These structures have been termed ETS-embryos as they are formed from a combination of ESC and TSC. The ESC component gives rise to an epiblast-like structure made of a pseudostratified epithelium that expresses OCT4 and has a central cavity similar to the proamniotic cavity (Harrison et al., 2017). Proximally apposed to this is an aggregate of ExE-like cells, expressing EOMES, and derived exclusively from the TSC cells. These ETS-embryos lack an outer VE-like layer as shown by the absence of GATA4 and yet are able to generate regionalized mesoderm, as shown by the expression of *Brachyury, Mixl1, and Hand1* in an asymmetric fashion on one side of the epiblast-like cells (Harrison et al., 2017).

How such self-organizing localized expression of mesoderm markers is achieved in the absence of the AVE is currently unclear. Furthermore, how are we to reconcile the localized expression of mesoderm markers in these in vitro systems lacking an AVE with the essential role of the AVE in embryos developing in utero? One explanation is that independent of the polarized signaling from the AVE, a mechanism exists within the epiblast to bias or limit streak formation to one side. It might be the activity of such a mechanism that is being manifest in the above in vitro systems. Experiments in the chick embryo have shown that removal of the hypoblast (the avian homologue of the AVE) leads to the formation of multiple primitive streaks, each forming from the margin of the epiblast (Bertocchini & Stern, 2002). Importantly, these ectopic streaks are often located far from one another, suggesting that there is an epiblast-based mechanism that inhibits two streaks from forming near each other (Bertocchini & Stern, 2002). The chick embryonic disk comprises approximately 30,000 cells (Stern, 2004) and is far larger than the mouse egg cylinder stage epiblast with fewer than 1000 epiblast cells prior to gastrulation (Snow, 1977). Perhaps a similar inhibitory mechanism acts within the mouse epiblast and ESC-derived in vitro models, but, due to their far smaller size, restricts the entire system to a single streak. This might also explain why the formation of gastruloids is sensitive to starting cell number and indeed, when made with a starting number of 800 cells or more, tend to show multiple elongations (van den Brink et al., 2014).

There is always variation in the size of embryos, even within a single litter. Therefore, another possibility is that the AVE as a mechanism of A–P patterning might have the advantage that the AVE size scales with the size of the embryo, so even if an epiblast is able to generate a streak autonomously, tying the process to the AVE ensures that streak size is tailored to each embryo. While ETS-embryos induce mesodermal expression on one side of the ES-derived cells, the spread of expressing cells appears much broader than that seen in equivalently staged embryo, perhaps due to the absence of a regulatory AVE cell population.

Another related possible role for the AVE might be to coordinate the growth of the embryo with its patterning. In embryos, the DVE is understood to form only when the egg cylinder has reached a specific size, thereby moving presumptive DVE cells in the VE via bulk growth sufficiently away from inhibitory signals from the ExE (Mesnard, Guzman-Ayala, & Constam, 2006; Rodriguez et al., 2005). After their formation, DVE cells migrate to the anterior over the course of the next 3–5 h (Srinivas et al., 2004). The timing of these series of events linked to the AVE may ensure that the formation of the primitive streak is coordinated with sufficient growth and maturity of the embryo prior to gastrulation to allow proper subsequent development.

Another possibility that must be considered is that the asymmetric expression of mesoderm markers in gastruloids and ETS-embryos is not a reflection of an underlying axial specification mechanism present in embryos in utero but an unrelated emergent phenomenon of the complex regulatory interactions between ES cells. There are very many ways to culture ES cells aggregates that do not result in gastruloids or ETS-embryos. A process of (artificial) selection has been at play in designing the specific culture conditions to channel ESC aggregates into forming gastruloids or ETS-embryos. It is not clear that these culture conditions reflect the in utero developmental experience of the implantation-stage embryo given that in vitro, the ES cell aggregates are literally immersed in several potent nondirectional signals. The behavior of ES cell aggregates is ultimately the result of a complex set of biochemical reactions. In the same way that one can get a particular biochemical end product through several different reaction pathways or with different starting chemicals by modulating reaction kinetics or reactants, it is conceivable that it is possible to engineer conditions under which a collection of ESC partially mimics (see below) the morphology or behavior of embryos in specific respects, but arrives at this point through a fundamentally different developmental pathway than that used by embryos developing in utero.

Related to this point is that neither gastruloids nor ETS-embryos appear to show localized expression of anterior markers, raising the possibility that rather than a broader A–P pattern being generated, they are systems where mesoderm cells differentiate across the bulk of the structure and then sort to one end due to differences in adhesiveness or their response to mechanical properties across the structure. Regardless, these recent approaches show the remarkable properties of ES cells to self-organize if provided with the correct starting conditions and provide powerful in vitro models for addressing questions about cell differentiation and movement. Understanding more about them will provide increased insight into how complex systems can self-organize.

4. THE FORMATION OF THE DVE

While the role of the AVE in patterning the A–P axis has been the source of much investigation, the mechanism by which it first forms has also attracted attention. At E5.5 a subset of VE cells undergoes differentiation into the DVE (later referred to as the AVE after their movement to the future anterior). DVE induced at the distal tip requires the interaction of NODAL and MAPK signaling pathways (Brennan et al., 2001; Clements et al., 2011; Ding et al., 1998; Norris et al., 2002) from the epiblast. Additional signals from the epiblast must also be required as epiblast-specific knockouts of *HCF-1*, a transcriptional coregulator involved in cell-cycle progression, show incomplete DVE patterning, and DVE cells remain at the distal tip despite the expression of *Nodal* in the epiblast (Minocha et al., 2016).

Upon induction of the DVE, cells become columnar and express characteristic markers including *Otx2, Lim1, Lefty1 (left–right determination factor 1), Cer1 (Cerberus-like 1),* and *Hex (haematopoietically expressed homeobox)* (Belo et al., 1997; Thomas et al., 1998; Yamamoto et al., 2004). Although

NODAL is expressed throughout the epiblast at this stage (Brennan et al., 2001), DVE differentiation is restricted to the distal tip by repressive signals from the ExE, as removal of the extraembryonic half of the embryo leads to the ectopic expression of DVE markers throughout the Epi-VE (Rodriguez et al., 2005).

It has been hypothesized that the growth of the egg cylinder takes the cells at the distal tip beyond the repressive influence of the ExE, since the DVE is only induced after the egg cylinder is roughly 180 μm long (Mesnard et al., 2006). However, more recently, Hiramatsu et al. have proposed that the growth of the embryo in combination with compressive forces imposed by the uterine tissue surrounding the embryo might be responsible for the onset of expression of DVE markers. They find that when E5.0 embryos are cultured in narrow, 90 μm diameter cavities, they extend along their proximal–distal axis and express *Cer1* at the distal tip (Hiramatsu et al., 2013). In contrast, when E5.0 embryos were cultured in wider, 180 μm diameter cavities, the embryos elongated to a lesser extent and did not induce DVE markers. The authors have proposed a model whereby mechanical stress is required for the induction of the DVE. This mechanical stimulus has been proposed to come from either the embryo's external environment—the maternal decidual tissue that closely covers the embryo postimplantation—or is generated internally by the proximal–distal growth of the egg cylinder focusing tension at the point of highest curvature, the distal tip of the embryo.

There are multiple examples in the literature of embryos being cultured in vitro without confinement from E4.5 or E5.0 until the induction, and in some case migration, of the AVE takes place (Bedzhov et al., 2015; Morris et al., 2012; Takaoka, Yamamoto, & Hamada, 2011; Torres-Padilla et al., 2007). One interpretation of these differences is that embryos do require external mechanical cues but that those cues come from the ability of embryos to attach to their substrate via their EPC. This might, it has been argued (Matsuo & Hiramatsu, 2017), then provide an anchor point from which autonomous mechanical pressure could be built up during the growth and elongation of the egg cylinder. However, the results of Bedzhov et al. (2015) argue again this interpretation as they culture embryos in hanging drops, showing that they can induce DVE markers and migrate in the absence of a substrate. In summary, there is strong evidence to support the conclusion that external mechanical stimuli are not required for the induction and development of the AVE under specific in vitro culture conditions. Whether there is a requirement for a self-generated internal mechanical force remains for further investigation.

Hiramatsu et al. have also suggested that the physical consequence of the mechanical stimuli is a breakdown of the basement membrane (BM) layer between the VE and epiblast in embryos cultured in narrow channels (Hiramatsu et al., 2013; Matsuo & Hiramatsu, 2017). Given that the site of this BM breakdown is at the distal tip of the embryo, the authors investigated whether this breach was associated with the movement of cells between the epiblast and VE layers. Using a Sox2-Cre driver line to mark epiblast cells, the authors report the presence of β-galactosidase reporter labeled cells in the VE at E5.5 and E6.5 (Hiramatsu et al., 2013). Time-lapse recordings using a Sox2-Venus reporter line can be interpreted as showing epiblast cells undergoing transmigration into the overlying VE cell layer at the distal-most region of the embryo (Hiramatsu et al., 2013), though the Venus-labeled structures are relatively small, raising the possibility that they are cell fragments rather than intact cells. At E6.0 they consistently observed labeled a small number of epiblast cells in the proximal-most, anterior VE (Hiramatsu et al., 2013). In support of these transmigratory events, a more recent study using a Rainbow reporter crossed to the Sox2-Cre driver similarly reported epiblast-derived cells in the VE—approximately two cells per embryo at all the stages examined (E5.25, E5.75, E6.5) (Bedzhov et al., 2015). This study also provides time-lapse evidence for the transmigration of epiblast cells into the VE cell layer; however, in contrast to Hiramatsu et al., they report that the position of transmigrating cells was random and not restricted to just the distal tip.

Most recently, a detailed time-lapse analysis of global VE cell behavior using H2B-GFP expressing embryos found no evidence for the transmigration of epiblast cells into the VE (Shioi et al., 2017). Both Hiramatsu et al. and Bedzhov et al. used the same Sox2-Cre driver line (Hayashi et al., 2002) to label epiblast cells in their lineage tracing experiments. It is possible that this driver has a low level of leaky expression in the VE lineage, resulting in what might appear to be the contribution of epiblast cells to the VE. However, the different findings in the time-lapse experiments are more difficult to explain and might require more detailed time-lapse imaging at high temporal resolution to resolve. Finally, even if epiblast cells do transmigrate into the VE, the question of whether there is a functional importance to these putative epiblast cells in the VE remains open.

5. WHAT CONTROLS THE DIRECTION OF AVE CELL MIGRATION?

The formation of the A–P axis has been considered a symmetry breaking event, as the initial proximal–distal asymmetry, visible at the time of formation of the DVE at the distal tip of the egg cylinder, is converted into an A–P asymmetry through the movement of these cells toward the future anterior side of the embryo. However, the question of when and how the first symmetry-breaking event in A–P axis formation occurs, that controls the direction of AVE cell migration, has been the subject of much investigation.

Although the embryo is described as a cylinder at postimplantation stages, it is not radially symmetrical and when considered in transverse section the embryo appears slightly squashed along one axis, so that it is oval in shape. Interestingly at E6.25 the primitive streak forms at one of the ends of the transverse long axis (Perea-Gomez et al., 2004). In fact this oval-like bilateral morphology can be initially observed prior to implantation in the blastocyst, raising the question of whether the movement of the AVE cells respects the morphological asymmetry observed in the preimplantation embryo? Studies investigating the direction of movement of the AVE at E5.5 suggest that something more complex occurs. The direction of migration of the AVE in fact appears to be perpendicular to that expected from the morphology, migrating along its short transverse axis (Perea-Gomez et al., 2004). This short axis will then become one end of the long transverse axis, as the entire embryo changes in shape around the time that the AVE migrates (Perea-Gomez et al., 2004). What drives this morphological change in the shape of the egg cylinder along its future A–P axis is currently unknown, though it is tempting to speculate that the epithelial visceral endoderm layer surrounding the epiblast may exert a mechanical influence on it during the initial migration and subsequent growth of the AVE cell population.

Another early asymmetry in the embryo that has been suggested as a potential indicator of the direction of AVE migration is in the positioning of the EPC (Gardner, Meredith, & Altman, 1992; Smith, 1985) at the most proximal region of the egg cylinder. The EPC is often askew, clearly displaced to one side of the P–D axis. However, there is a wide heterogeneity between embryos in the degree of this lopsidedness, being obviously pronounced in some embryos, but not in others, suggesting that this is not a reliable indicator and no causal link between EPC asymmetry and AVE migration direction has been proved.

In the E5.5 embryo the direction of migration does not appear to respect either the bilateral symmetry of the transverse axis nor the tilt of the EPC. However, at E5.5 there are molecular asymmetries in the expression domains of the AVE markers and NODAL antagonists *Lefty1* and *Cer1* which are tilted toward the prospective anterior prior to migration (Yamamoto et al., 2004). This asymmetry in expression is thought to cause an asymmetry in NODAL signaling that thereby providing a directional impetus for AVE migration. This model is supported by experiments showing that AVE cells migrate toward ectopically expressed NODAL antagonists (Yamamoto et al., 2004). Initially it was thought that this asymmetry in *Nodal* signaling caused proliferative imbalance in the VE cell layer, causing a passive movement of the AVE toward the future anterior (Yamamoto et al., 2004), although more recent studies found no such proliferative difference in AVE vs PVE (Shioi et al., 2017; Stuckey et al., 2011b).

This asymmetry in *Lefty1* and *Cer1* has been traced back to the forming primitive endoderm in the preimplantation blastocyst (Takaoka et al., 2011, 2006; Torres-Padilla et al., 2007). At this stage the ICM and polar TE are tilted with respect to the proximal–distal axis (Rossant, 2004; Smith, 1980), and the domain of expression of *Cer1* and *Lefty1* is tilted within the PrE (Takaoka et al., 2011, 2006). Furthermore, by lineage tracing experiments it has been shown that these Cer1/Lefty1 cell population at E4.5 are fated to give rise to the later asymmetrically located *Lefty1* and *Cerl1* expressing cells of the E5.5 egg cylinder, suggesting that the observed asymmetry a E5.5 has a preimplantation origin (Takaoka et al., 2011).

While there is evidence for the involvement of *Lefty1* and *Cer1* in controlling the direction of AVE migration, knockout studies suggest that there is more complexity to this process. Mouse knockouts of *Lefty1* (Meno et al., 1998) and *Cer1* (Belo et al., 2000; Simpson et al., 1999; Stanley et al., 2000) will still gastrulate, suggesting that the AVE migrates directionally even in their absence. While this could simply reflect some degrees of functional redundancy between these antagonists in single knockouts, 30% of double mutants still have correct streak positioning (Perea-Gomez et al., 2002), although the majority show an AVE arrest at the distal tip, suggesting that our understanding of the role of *Lefty1* and *Cer1* in controlling AVE cell migration is still not complete.

In additional to NODAL antagonists, the secreted WNT antagonist DICKKOPF 1 (DKK1) has also been suggested to control the direction of AVE movement (Kimura-Yoshida et al., 2005). DKK1 binds LRP5 receptors, thereby antagonizing canonical WNT signaling. It is expressed at E5.5 as a stripe at the most proximal Epi-VE highest at the future anterior side, and by E6.5 expressed in an arc of cells at the most proximal region of the AVE (Kimura-Yoshida et al., 2005). Kimura-Yoshida et al. (2005) showed that an exogenous source of DKK1 can act as a guidance cue to AVE cells. Furthermore, *Dkk1* knocked into the *Otx2* locus is sufficient to rescue the AVE arrest phenotype of *Otx2* mutants (Kimura-Yoshida et al., 2005). However, in *Dkk1* mutants, although some markers of the AVE are lost, *Otx2* is still correctly positioned and the primitive streak forms correctly (Kimura-Yoshida et al., 2005), suggesting the AVE is still able to migrate even in the absence of DKK1.

Though the AVE can still migrate in the absence of DKK1, the question remains open whether cells migrate to regions of low WNT. The pattern of WNT activity during AVE cell migration has been explored in live embryos using a TCF/Lef:H2B fluorescent reporter line (Ferrer-Vaquer et al., 2010). At the time of DVE induction, WNT activity appears exclusively in the VE layer, in both the ExE-VE and Epi-VE (including the distal tip cells), in a heterogeneous fashion (Ferrer-Vaquer et al., 2010). Postmigration cells that show WNT activity are restricted to the posterior VE, suggesting that the pattern of WNT activity changes between DVE induction and migration (Ferrer-Vaquer et al., 2010). From this study it is not clear whether there is a prepatterned gradient of WNT activity prior to AVE migration. Given the change in the pattern of WNT activity, it is likely that it has more than one role in the VE during these key phases of induction, migration, and patterning making it harder to separate out each of its potential roles.

Adding complexity to the picture, a detailed study of the localization, and crucially, colocalization of proteins of a panel of key markers, including HHEX, OTX2, CER1 and LEFTY1, and DKK1 between E4.5 and E6.5 has suggested an alternative sequence of events (Hoshino, Shioi, & Aizawa, 2015). In contrast to previous studies, suggesting that *Lefty1* and *Cer1* transcripts are asymmetrically localized in the tilted PrEn at E4.5 embryos and later at E5.5 (Takaoka et al., 2006), Hoshino et al. (2015) report a mosaic localization of LEFTY1 protein in the PrEn with no obvious asymmetry in localization (Hoshino et al., 2015). At E5.0, they found that both LEFTY1 and CER1 were undetectable, while at E5.5 they are located distally in the DVE, but show no asymmetry in their localization among DVE cells (Hoshino et al., 2015).

OTX2 was found to be expressed in DVE cells at the distal tip and throughout their migration between E5.5 and E6.5, largely colocalizing with HHEX (Hoshino et al., 2015). Hishino et al. also report a previously unknown asymmetry in OTX2 at E5.25, with a region of weakly expressing OTX2-positive Epi-VE cells extending proximally up the posterior side of the embryo (Hoshino et al., 2015). The expression of OTX2 was later downregulated from these more posterior Epi-VE cells by E5.75 (Hoshino et al., 2015).

In addition to OTX2 asymmetric localization, the work of Hoshino et al. also reports that DKK1 is also asymmetric with the location of early DKK1-positive cells being similar to that of OTX2 extending more proximally on the future posterior side of the Epi-VE at E5.25 (Hoshino et al., 2015). At E5.5 DKK1 appears localized to two regions in the Epi-VE, the most proximally located posterior Epi-VE and the most anterior OTX2-positive cells (Hoshino et al., 2015). Thus, in contrast to previous studies, Hoshino et al. suggest that the expression in the AVE occurs de novo only after egg cylinder formation after E5.25, which supports previous RNA in situ hybridization studies (Mesnard et al., 2006). One caveat to this work is the technical difficulty of using immunohistochemistry to identify secreted molecules such as LEFTY1, CER1, and DKK1 as it is unclear whether the signal observed is in cells that are secreting or responding to the signal. Nevertheless this study raises interesting questions of how the pattern of asymmetric expression of LEFTY1, CER1, OTX1, and DKK1 arise. If it is the case that the expression of LEFTY1 and CER1 is lost at E5.0, this does not necessarily rule out the possibility that subsequent asymmetries at E5.5 are still a consequence of their cell lineage from their initial expression in the PrEn at E4.75.

6. WHAT IS THE MECHANISM OF AVE CELL MIGRATION?

The migratory behavior of AVE cells has been studied through lineage labeling experiments (Thomas et al., 1998), genetic reporter lines, and time-lapse imaging (Rodriguez et al., 2005). These studies have shown that although the VE is maintained as a continuous single-cell monolayer with intact tight and adherent junctions (Migeotte et al., 2010; Trichas et al., 2011), migratory AVE cells are able to move through the plane of the epithelium via cell–cell rearrangement (Trichas et al., 2011). AVE cells show characteristics of active migration, sending out long, basally localized cellular projections in the direction of migration (Migeotte et al., 2010; Srinivas et al., 2004). Cells move proximally until they reach the epiblast-ExE boundary, whereupon they move apparently passively to lateral regions, as they no longer show cellular projections (Migeotte et al., 2010; Srinivas et al., 2004; Takaoka et al., 2011; Trichas et al., 2011). Movement in the VE is confined largely to the VE overlying the epiblast, whereas the VE overlying the ExE (ExE-VE) remains largely static and undergoes very few cellular rearrangements (Shioi et al., 2017; Trichas et al., 2011).

The mechanism by which AVE cells respect the boundary between the two regions of the VE is unclear. To date there are very few mutants that show an "overmigration" phenotype, where the AVE cells are able to invade the ExE-VE region of the epithelial sheet. Examples included a proportion of *Lefty1* homozygous null mutants and $ROSA26^{Lyn\text{-}Celsr1}$ embryos in which a membrane tethered form of Cesr1 (the homologue of Drosophila Flamingo) involved in the PCP pathway is overexpressed (Trichas et al., 2011), suggesting that Nodal and PCP pathways are involved in setting the boundary (Fig. 2).

Recent experiments tracking cells throughout the Epi-VE of embryos expressing H2B-GFP indicate that there is global movement throughout the majority of the E5.5 Epi-VE, with the DVE, AVE, and more lateral VE all showing a coordinated stereotypic movement (Shioi et al., 2017). Cells of the posterior Epi-VE are an exception to this, remaining relatively static throughout this global cell rearrangement (Shioi et al., 2017). It is not clear if this global movement is the cause or effect of the directional migration of AVE cells. Shioi and colleagues found that in *Otx2* mutant, in which the AVE fails to migrate, there is no global movement of any Epi-VE cells (Shioi et al., 2017). This finding however could be consistent with both possibilities—the *Otx2* mutant might primarily affect migration of AVE cells, the failure of which might result in a secondary defect in global VE movements if, for example, they are normally the result of VE cells accommodating AVE migration. Conversely, the mutant might directly affect global movement within the VE, which could as a result cause AVE migration to fail.

While much progress had been made in understanding the cell lineage and transcription factors involved in specifying AVE cell fate, much less is known about the factors that mechanistically drive their movement. Cell movements require regulation and dynamic control of the cytoskeleton to coordinate protrusion and contractility of cells, cell–cell adhesion, and cell substrate adhesion (Montell, 2008). F-actin and myosin are two crucial cytoskeletal elements required for cell movement, during AVE migration, there are clear regional differences in the localization of both actin and myosin IIA (Trichas et al., 2011). While F-actin is localized to subcortical rings within cells both before and during AVE migration in the Epi-VE, in the more static ExE-VE cells the localization of F-actin changes so that it is no longer confined to cortical rings but rather enriches the apical cortex, thereby forming an "apical shroud" of actin (Trichas et al., 2011), perhaps reflecting the different behavior of these two regions of the VE epithelium.

The question of what specifically controls F-actin localization and dynamics during AVE cell migration is an area of ongoing research. To date, four mutants have given us insights into the specific factors that control cytoskeletal dynamics during AVE migration (Bloomekatz et al., 2012; Migeotte et al., 2010; Omelchenko et al., 2014; Rakeman & Anderson, 2006). The first to be discovered in relation to the AVE was NAP1 (Rakeman & Anderson, 2006), a component of the WAVE complex, which acts through ARP2/3 to form a branched actin network at the leading edge of migrating cells (Takenawa & Suetsugu, 2007). This was discovered in a screen for embryonic lethality. *Nap1*^{khlo} mutants arrest at or prior to embryonic turning and by E8.5 half of the homozygous mutants show a striking axis duplication phenotype. This is thought to be due to aberrant AVE cell migration because at E5.75, half of NAP1 mutants fail to undergo AVE migration (Rakeman & Anderson, 2006) showing an arrest at the distal tip (Fig. 2). In a proportion of embryos, cells can migrate toward the anterior side, but some AVE cells are found aberrantly in a posterior position (Rakeman & Anderson, 2006). AVE cells in these mutants also appear rounded, lacking clear morphological front-back polarization (Rakeman & Anderson, 2006), suggesting the Nap1 is required for correct polarization of these cells during migration.

RAC1 is a member of the Rho-GTPases and functions through the WAVE complex, of which NAP1 is a member. In *Rac1* mutants AVE-specific markers including Hex and Cer1 are expressed, but these cells arrest at the distal tip of the embryo and cells lack cellular protrusions, suggesting that RAC1 is absolutely required to initiate cell migration (Migeotte et al., 2010). Rho-GTPases like RAC1 are regulated by Rho-GEFs (guanine nucleotide exchange factors) of which there are 82 known (Cherfils & Zeghouf, 2013; Rossman, Der, & Sondek, 2005). Rho-GEFs promote GDP–GTP exchange, GTPase, and are thought to provide spatial control over their activation (Cherfils & Zeghouf, 2013; Rossman et al., 2005). Omelchenko et al. (2014) used a migratory epithelial cell assay to identify Rho-GEFs through which they found β-Pix, a Cdc42/Rac Rho-GEF (Omelchenko et al., 2014). Homozygous mutants showed morphological defects from E6.5 and were embryonic lethal by E8.5. Live time-lapse imaging of *β-Pix* mutant embryos, in which AVE cells were labeled by the

Hex-GFP reporter, showed that mutants have defects in the directionality of their cellular projections, so that they no longer are protruded in the direction of migration (Omelchenko et al., 2014). Migratory cells also show a disorganized pattern of movements so that they no longer migrate collectively, as a group. Using the VE-specific Ttr-Cre-line (Kwon & Hadjantonakis, 2009) to ablate β-Pix specifically in the VE, Omelchenko et al. could show that β-Pix is required autonomously with in the VE cell layer for correct migration (Omelchenko et al., 2014).

While several factors have been identified that are associated with the leading edge of cells in the AVE, for example, RAC1 (Migeotte et al., 2010), β-Pix (Omelchenko et al., 2014), and NAP1 (Rakeman & Anderson, 2006), only one factor associated with the control cytoskeletal network at the rear of cells has been identified. PTEN (phosphatase and tensin homologue on chromosome 10) acts by effecting the distribution of phosphatidylinosides (PtdIns) in the plasma membrane (Nelson, 2009) with PtdIns(3,4,5)P_3 enriched at the front of migrating cells and PtdIns(3,4)P_2 localized to the rear. PTEN has been associated with the control of the direction of migration (Arimura & Kaibuchi, 2005; Funamoto et al., 2002; Iijima & Devreotes, 2002; Wang et al., 2002) through converting triphosphate PtdIns to their biphosphate form, creating an intracellular gradient of phophatidylinosides. Mouse *Pten*$^{-/-}$ mutants are embryonic lethal at E7.5–E8.5 and at E5.5 AVE cells show aberrant migration phenotype (Bloomekatz et al., 2012). AVE cells migrate in random directions, and do so as individuals and not as a collective, so that they fail to reach their proximal position and become more widely dispersed in the VE (Bloomekatz et al., 2012) (Fig. 2).

Through these studies the first effectors of cytoskeletal elements required for controlling AVE migration have been identified. How many more key components of these pathways are involved, and how Rho-GTPase, Rho-GEFs, WAVE complex members, and PTEN are correctly positioned to drive the movement of cells is still to be established. Finally, insights into how these processes are coordinated with dynamic cell–cell adhesion and cell substrate adhesion control is require for a more complete understanding of these morphogenetic events.

7. CONCLUSION

Research into anterior patterning and the AVE has burgeoned since the seminal labeling experiments by Rosa Beddington established the existence of this migratory cell population essential for anterior patterning.

Improvements in time-lapse imaging technology, fluorescent reporters, knockout mouse lines, and culture approaches have revealed a rich complexity of interactions between embryonic and extraembryonic tissues in specifying anterior pattern. In doing so, they have also shed light on key principles in biology relating to epithelial morphogenesis, tissue patterning, and cell fate determination. These exciting findings have also raised many questions relating to the origin of axial asymmetry, the mechanism of cell movement within epithelia, and the genetic control of axial patterning that will require increasingly multidisciplinary approaches to address in the future.

REFERENCES

Acampora, D., et al. (1998). Visceral endoderm-restricted translation of Otx1 mediates recovery of Otx2 requirements for specification of anterior neural plate and normal gastrulation. *Development*, *125*(24), 5091–5104.

Arimura, N., & Kaibuchi, K. (2005). Key regulators in neuronal polarity. *Neuron*, *48*(6), 881–884.

Beddington, R. S., & Robertson, E. J. (1999). Axis development and early asymmetry in mammals. *Cell*, *96*(2), 195–209.

Bedzhov, I., et al. (2015). Development of the anterior-posterior axis is a self-organizing process in the absence of maternal cues in the mouse embryo. *Cell Research*, *25*(12), 1368–1371.

Belo, J. A., et al. (1997). Cerberus-like is a secreted factor with neutralizing activity expressed in the anterior primitive endoderm of the mouse gastrula. *Mechanisms of Development*, *68*(1–2), 45–57.

Belo, J. A., et al. (2000). Cerberus-like is a secreted BMP and nodal antagonist not essential for mouse development. *Genesis*, *26*(4), 265–270.

Ben-Haim, N., et al. (2006). The nodal precursor acting via activin receptors induces mesoderm by maintaining a source of its convertases and BMP4. *Developmental Cell*, *11*(3), 313–323.

Bertocchini, F., & Stern, C. D. (2002). The hypoblast of the chick embryo positions the primitive streak by antagonizing nodal signaling. *Developmental Cell*, *3*(5), 735–744.

Bloomekatz, J., et al. (2012). Pten regulates collective cell migration during specification of the anterior-posterior axis of the mouse embryo. *Developmental Cell*, *364*(2), 192–201.

Brennan, J., et al. (2001). Nodal signalling in the epiblast patterns the early mouse embryo. *Nature*, *411*(6840), 965–969.

Cherfils, J., & Zeghouf, M. (2013). Regulation of small GTPases by GEFs, GAPs, and GDIs. *Physiological Reviews*, *93*(1), 269–309.

Chew, J. L., et al. (2005). Reciprocal transcriptional regulation of Pou5f1 and Sox2 via the Oct4/Sox2 complex in embryonic stem cells. *Molecular and Cellular Biology*, *25*(14), 6031–6046.

Clements, M., et al. (2011). Crosstalk between Nodal/activin and MAPK p38 signaling is essential for anterior-posterior axis specification. *Current Biology*, *21*(15), 1289–1295.

Copp, A. J. (1979). Interaction between inner cell mass and trophectoderm of the mouse blastocyst. II. The fate of the polar trophectoderm. *Journal of Embryology and Experimental Morphology*, *51*, 109–120.

Copp, A. J. (1981). The mechanism of mouse egg-cylinder morphogenesis in vitro. *Journal of Embryology and Experimental Morphology*, *61*, 277–287.

Di-Gregorio, A., et al. (2007). BMP signalling inhibits premature neural differentiation in the mouse embryo. *Development, 134*(18), 3359–3369.

Ding, J., et al. (1998). Cripto is required for correct orientation of the anterior-posterior axis in the mouse embryo. *Nature, 395*(6703), 702–707.

Downs, K. M. (2009). The enigmatic primitive streak: Prevailing notions and challenges concerning the body axis of mammals. *BioEssays, 31*(8), 892–902.

Downs, K. M., et al. (2009). The allantoic core domain: New insights into development of the murine allantois and its relation to the primitive streak. *Developmental Dynamics, 238*(3), 532–553.

Egea, J., et al. (2008). Genetic ablation of FLRT3 reveals a novel morphogenetic function for the anterior visceral endoderm in suppressing mesoderm differentiation. *Genes & Development, 22*(23), 3349–3362.

Ferrer-Vaquer, A., et al. (2010). A sensitive and bright single-cell resolution live imaging reporter of Wnt/beta-catenin signaling in the mouse. *BMC Developmental Biology, 10*, 121–139.

Foley, A. C., Skromne, I., & Stern, C. D. (2000). Reconciling different models of forebrain induction and patterning: A dual role for the hypoblast. *Development, 127*(17), 3839–3854.

Funamoto, S., et al. (2002). Spatial and temporal regulation of 3-phosphoinositides by PI 3-kinase and PTEN mediates chemotaxis. *Cell, 109*(5), 611–623.

Gardner, R. L. (1983). Origin and differentiation of extraembryonic tissues in the mouse. *International Review of Experimental Pathology, 24*, 63–133.

Gardner, R. L., Meredith, M. R., & Altman, D. G. (1992). Is the anterior-posterior axis of the fetus specified before implantation in the mouse? *The Journal of Experimental Zoology, 264*(4), 437–443.

Gardner, R. L., Papaioannou, V. E., & Barton, S. C. (1973). Origin of the ectoplacental cone and secondary giant cells in mouse blastocysts reconstituted from isolated trophoblast and inner cell mass. *Journal of Embryology and Experimental Morphology, 30*(3), 561–572.

Gardner, R. L., & Rossant, J. (1979). Investigation of the fate of 4-5 day post-coitum mouse inner cell mass cells by blastocyst injection. *Journal of Embryology and Experimental Morphology, 52*, 141–152.

Gotoh, N., et al. (2005). The docking protein FRS2alpha is an essential component of multiple fibroblast growth factor responses during early mouse development. *Molecular and Cellular Biochemistry, 25*(10), 4105–4116.

Harrison, S. E., et al. (2017). Assembly of embryonic and extraembryonic stem cells to mimic embryogenesis in vitro. *Science, 356*(6334), eaal1810.

Hayashi, S., et al. (2002). Efficient gene modulation in mouse epiblast using a Sox2Cre transgenic mouse strain. *Mechanisms of Development, 119*(Suppl. 1), S97–S101.

Hermitte, S., & Chazaud, C. (2014). Primitive endoderm differentiation: From specification to epithelium formation. *Philosophical Transactions of the Royal Society of London Series B, Biological Sciences, 369*(1657), 20130537.

Hiramatsu, R., et al. (2013). External mechanical cues trigger the establishment of the anterior-posterior axis in early mouse embryos. *Developmental Cell, 27*(2), 131–144.

Hoshino, H., Shioi, G., & Aizawa, S. (2015). AVE protein expression and visceral endoderm cell behavior during anterior-posterior axis formation in mouse embryos: Asymmetry in OTX2 and DKK1 expression. *Developmental Biology, 402*(2), 175–191.

Iijima, M., & Devreotes, P. (2002). Tumor suppressor PTEN mediates sensing of chemoattractant gradients. *Cell, 109*(5), 599–610.

Kimura, C., et al. (2000). Visceral endoderm mediates forebrain development by suppressing posteriorizing signals. *Developmental Biology, 225*(2), 304–321.

Kimura-Yoshida, C., et al. (2005). Canonical Wnt signaling and its antagonist regulate anterior-posterior axis polarization by guiding cell migration in mouse visceral endoderm. *Developmental Cell, 9*(5), 639–650.

Kumar, A., et al. (2015). Nodal signaling from the visceral endoderm is required to maintain Nodal gene expression in the epiblast and drive DVE/AVE migration. *Developmental Biology*, *400*(1), 1–9.
Kwon, G. S., & Hadjantonakis, A. K. (2009). Transthyretin mouse transgenes direct RFP expression or Cre-mediated recombination throughout the visceral endoderm. *Genesis*, *47*(7), 447–455.
Kwon, G. S., Viotti, M., & Hadjantonakis, A. K. (2008). The endoderm of the mouse embryo arises by dynamic widespread intercalation of embryonic and extraembryonic lineages. *Developmental Cell*, *15*(4), 509–520.
Liu, P., et al. (1999). Requirement for Wnt3 in vertebrate axis formation. *Nature Genetics*, *22*(4), 361–365.
Marikawa, Y., et al. (2009). Aggregated P19 mouse embryonal carcinoma cells as a simple in vitro model to study the molecular regulations of mesoderm formation and axial elongation morphogenesis. *Genesis*, *47*(2), 93–106.
Martinez Barbera, J. P., et al. (2000). The homeobox gene Hex is required in definitive endodermal tissues for normal forebrain, liver and thyroid formation. *Development*, *127*(11), 2433–2445.
Matsuo, I., & Hiramatsu, R. (2017). Mechanical perspectives on the anterior-posterior axis polarization of mouse implanted embryos. *Mechanisms of Development*, *144*(Pt. A), 62–70.
Meno, C., et al. (1998). lefty-1 is required for left-right determination as a regulator of lefty-2 and nodal. *Cell*, *94*(3), 287–297.
Mesnard, D., Guzman-Ayala, M., & Constam, D. B. (2006). Nodal specifies embryonic visceral endoderm and sustains pluripotent cells in the epiblast before overt axial patterning. *Development*, *133*(13), 2497–2505.
Migeotte, I., et al. (2010). Rac1-dependent collective cell migration is required for specification of the anterior-posterior body axis of the mouse. *PLoS Biology*, *8*(8) . e1000442.
Minocha, S., et al. (2016). Epiblast-specific loss of HCF-1 leads to failure in anterior-posterior axis specification. *Developmental Biology*, *418*(1), 75–88.
Mohammed, H., et al. (2017). Single-cell landscape of transcriptional heterogeneity and cell fate decisions during mouse early gastrulation. *Cell Reports*, *20*(5), 1215–1228.
Montell, D. J. (2008). Morphogenetic cell movements: Diversity from modular mechanical properties. *Science*, *322*(5907), 1502–1505.
Morris, S. A., et al. (2012). Dynamics of anterior-posterior axis formation in the developing mouse embryo. *Nature Communications*, *3*, 673.
Nelson, W. J. (2009). Remodeling epithelial cell organization: Transitions between front-rear and apical-basal polarity. *Cold Spring Harbor Perspectives in Biology*, *1*(1) . a000513.
Norris, D. P., et al. (2002). The Foxh1-dependent autoregulatory enhancer controls the level of Nodal signals in the mouse embryo. *Development*, *129*(14), 3455–3468.
Okumura-Nakanishi, S., et al. (2005). Oct-3/4 and Sox2 regulate Oct-3/4 gene in embryonic stem cells. *The Journal of Biological Chemistry*, *280*(7), 5307–5317.
Omelchenko, T., et al. (2014). beta-Pix directs collective migration of anterior visceral endoderm cells in the early mouse embryo. *Genes & Development*, *28*(24), 2764–2777.
Perea-Gomez, A., et al. (2002). Nodal antagonists in the anterior visceral endoderm prevent the formation of multiple primitive streaks. *Developmental Cell*, *3*(5), 745–756.
Perea-Gomez, A., et al. (2004). Initiation of gastrulation in the mouse embryo is preceded by an apparent shift in the orientation of the anterior-posterior axis. *Current Biology*, *14*(3), 197–207.
Rakeman, A. S., & Anderson, K. V. (2006). Axis specification and morphogenesis in the mouse embryo require Nap1, a regulator of WAVE-mediated actin branching. *Development*, *133*(16), 3075–3083.
Ramathal, C. Y., et al. (2010). Endometrial decidualization: Of mice and men. *Seminars in Reproductive Medicine*, *28*(1), 17–26.

Rivera-Perez, J. A., Mager, J., & Magnuson, T. (2003). Dynamic morphogenetic events characterize the mouse visceral endoderm. *Developmental Biology, 261*(2), 470–487.
Rivera-Perez, J. A., & Magnuson, T. (2005). Primitive streak formation in mice is preceded by localized activation of Brachyury and Wnt3. *Developmental Biology, 288*(2), 363–371.
Rodriguez, T. A., et al. (2005). Induction and migration of the anterior visceral endoderm is regulated by the extra-embryonic ectoderm. *Development, 132*(11), 2513–2520.
Rossant, J. (2004). Lineage development and polar asymmetries in the peri-implantation mouse blastocyst. *Seminars in Cell & Developmental Biology, 15*(5), 573–581.
Rossant, J., & Tam, P. P. (2009). Blastocyst lineage formation, early embryonic asymmetries and axis patterning in the mouse. *Development, 136*(5), 701–713.
Rossman, K. L., Der, C. J., & Sondek, J. (2005). GEF means go: Turning on RHO GTPases with guanine nucleotide-exchange factors. *Nature Reviews Molecular Cell Biology, 6*(2), 167–180.
Shioi, G., et al. (2017). Apical constriction in distal visceral endoderm cells initiates global, collective cell rearrangement in embryonic visceral endoderm to form anterior visceral endoderm. *Developmental Biology, 429*(1), 20–30.
Simpson, E. H., et al. (1999). The mouse Cer1 (Cerberus related or homologue) gene is not required for anterior pattern formation. *Developmental Biology, 213*(1), 202–206.
Smith, L. J. (1980). Embryonic axis orientation in the mouse and its correlation with blastocyst relationships to the uterus. Part 1. Relationships between 82 hours and 4 1/4 days. *Journal of Embryology and Experimental Morphology, 55*, 257–277.
Smith, L. J. (1985). Embryonic axis orientation in the mouse and its correlation with blastocyst relationships to the uterus. II. Relationships from 4 1/4 to 9 1/2 days. *Journal of Embryology and Experimental Morphology, 89*, 15–35.
Snow, M. H. L. (1977). Gastrulation in the mouse: Growth and regionalization of the epiblast. *Journal of Embryology and Experimental Morphology, 42*, 293–303.
Srinivas, S., et al. (2004). Active cell migration drives the unilateral movements of the anterior visceral endoderm. *Development, 131*(5), 1157–1164.
Stanley, E. G., et al. (2000). Targeted insertion of a lacZ reporter gene into the mouse Cer1 locus reveals complex and dynamic expression during embryogenesis. *Genesis, 26*(4), 259–264.
Stern, C. D. (2004). Gastrulation in the chick. In C. D. Stern (Ed.), *Gastrulation: From cells to embryo* (pp. 233–262). Cold Spring Harbor, NY: Cold Spring Harbor Laboratory Press.
Stern, C. D., & Downs, K. M. (2012). The hypoblast (visceral endoderm): An evo-devo perspective. *Development, 139*(6), 1059–1069.
Stuckey, D. W., et al. (2011a). Correct patterning of the primitive streak requires the anterior visceral endoderm. *PLoS One, 6*(3), e17620.
Stuckey, D. W., et al. (2011b). Coordination of cell proliferation and anterior-posterior axis establishment in the mouse embryo. *Development, 138*(8), 1521–1530.
Sun, L. T., et al. (2014). Nanog co-regulated by Nodal/Smad2 and Oct4 is required for pluripotency in developing mouse epiblast. *Developmental Biology, 392*(2), 182–192.
Takaoka, K., Yamamoto, M., & Hamada, H. (2011). Origin and role of distal visceral endoderm, a group of cells that determines anterior-posterior polarity of the mouse embryo. *Nature Cell Biology, 13*(7), 743–752.
Takaoka, K., et al. (2006). The mouse embryo autonomously acquires anterior-posterior polarity at implantation. *Developmental Cell, 10*(4), 451–459.
Takenawa, T., & Suetsugu, S. (2007). The WASP-WAVE protein network: Connecting the membrane to the cytoskeleton. *Nature Reviews Molecular Cell Biology, 8*(1), 37–48.
Tam, P. P., & Loebel, D. A. (2007). Gene function in mouse embryogenesis: Get set for gastrulation. *Nature Reviews. Genetics, 8*(5), 368–381.
Thomas, P., & Beddington, R. (1996). Anterior primitive endoderm may be responsible for patterning the anterior neural plate in the mouse embryo. *Current Biology, 6*(11), 1487–1496.

Thomas, P. Q., Brown, A., & Beddington, R. S. (1998). Hex: A homeobox gene revealing peri-implantation asymmetry in the mouse embryo and an early transient marker of endothelial cell precursors. *Development*, *125*(1), 85–94.

Torres-Padilla, M. E., et al. (2007). The anterior visceral endoderm of the mouse embryo is established from both preimplantation precursor cells and by de novo gene expression after implantation. *Developmental Biology*, *309*(1), 97–112.

Trichas, G., et al. (2011). Nodal dependent differential localisation of dishevelled-2 demarcates regions of differing cell behaviour in the visceral endoderm. *PLoS Biology*, *9*(2), e1001019.

Uez, N., et al. (2008). Sall4 isoforms act during proximal-distal and anterior-posterior axis formation in the mouse embryo. *Genesis*, *46*(9), 463–477.

van den Brink, S. C., et al. (2014). Symmetry breaking, germ layer specification and axial organisation in aggregates of mouse embryonic stem cells. *Development*, *141*(22), 4231–4242.

Viotti, M., Foley, A. C., & Hadjantonakis, A. K. (2014). Gutsy moves in mice: Cellular and molecular dynamics of endoderm morphogenesis. *Philosophical Transactions of the Royal Society of London. Series B, Biological Sciences*, *369*(1657), pii: 20130547.

Wang, F., et al. (2002). Lipid products of PI(3)Ks maintain persistent cell polarity and directed motility in neutrophils. *Nature Cell Biology*, *4*(7), 513–518.

Warmflash, A., et al. (2014). A method to recapitulate early embryonic spatial patterning in human embryonic stem cells. *Nature Methods*, *11*(8), 847–854.

Winnier, G., et al. (1995). Bone morphogenetic protein-4 is required for mesoderm formation and patterning in the mouse. *Genes & Development*, *9*(17), 2105–2116.

Wu, Q., et al. (2006). Sall4 interacts with Nanog and co-occupies Nanog genomic sites in embryonic stem cells. *The Journal of Biological Chemistry*, *281*(34), 24090–24094.

Yamamoto, M., et al. (2004). Nodal antagonists regulate formation of the anteroposterior axis of the mouse embryo. *Nature*, *428*(6981), 387–392.

Yamamoto, M., et al. (2009). Antagonism between Smad1 and Smad2 signaling determines the site of distal visceral endoderm formation in the mouse embryo. *The Journal of Cell Biology*, *184*(2), 323–334.

Ying, Y., & Zhao, G. Q. (2001). Cooperation of endoderm-derived BMP2 and extraembryonic ectoderm-derived BMP4 in primordial germ cell generation in the mouse. *Developmental Biology*, *232*(2), 484–492.

Yoon, Y., et al. (2015). Extra-embryonic Wnt3 regulates the establishment of the primitive streak in mice. *Developmental Biology*, *403*(1), 80–88.

Zhang, J., et al. (2006). Sall4 modulates embryonic stem cell pluripotency and early embryonic development by the transcriptional regulation of Pou5f1. *Nature Cell Biology*, *8*(10), 1114–1123.

FSC
www.fsc.org
MIX
Papier | Fördert gute Waldnutzung
FSC® C083411

Kolibri 360 GmbH
Ferdinand-Jühlke-Straße 7
99095 Erfurt, Deutschland
produktsicherheit@kolibri360.de